TRANSLATIONAL PAIN RESEARCH
FROM MOUSE TO MAN

FRONTIERS IN NEUROSCIENCE

Series Editors
Sidney A. Simon, Ph.D.
Miguel A.L. Nicolelis, M.D., Ph.D.

Published Titles

Apoptosis in Neurobiology
Yusuf A. Hannun, M.D., Professor of Biomedical Research and Chairman, Department
of Biochemistry and Molecular Biology, Medical University of South Carolina,
Charleston, South Carolina
Rose-Mary Boustany, M.D., tenured Associate Professor of Pediatrics and Neurobiology,
Duke University Medical Center, Durham, North Carolina

Neural Prostheses for Restoration of Sensory and Motor Function
John K. Chapin, Ph.D., Professor of Physiology and Pharmacology, State University
of New York Health Science Center, Brooklyn, New York
Karen A. Moxon, Ph.D., Assistant Professor, School of Biomedical Engineering, Science,
and Health Systems, Drexel University, Philadelphia, Pennsylvania

Computational Neuroscience: Realistic Modeling for Experimentalists
Eric DeSchutter, M.D., Ph.D., Professor, Department of Medicine, University of Antwerp,
Antwerp, Belgium

Methods in Pain Research
Lawrence Kruger, Ph.D., Professor of Neurobiology (Emeritus), UCLA School of Medicine
and Brain Research Institute, Los Angeles, California

Motor Neurobiology of the Spinal Cord
Timothy C. Cope, Ph.D., Professor of Physiology, Wright State University, Dayton, Ohio

Nicotinic Receptors in the Nervous System
Edward D. Levin, Ph.D., Associate Professor, Department of Psychiatry and Pharmacology
and Molecular Cancer Biology and Department of Psychiatry and Behavioral Sciences,
Duke University School of Medicine, Durham, North Carolina

Methods in Genomic Neuroscience
Helmin R. Chin, Ph.D., Genetics Research Branch, NIMH, NIH, Bethesda, Maryland
Steven O. Moldin, Ph.D., University of Southern California, Washington, D.C.

Methods in Chemosensory Research
Sidney A. Simon, Ph.D., Professor of Neurobiology, Biomedical Engineering,
and Anesthesiology, Duke University, Durham, North Carolina
Miguel A.L. Nicolelis, M.D., Ph.D., Professor of Neurobiology and Biomedical Engineering,
Duke University, Durham, North Carolina

The Somatosensory System: Deciphering the Brain's Own Body Image
Randall J. Nelson, Ph.D., Professor of Anatomy and Neurobiology,
University of Tennessee Health Sciences Center, Memphis, Tennessee

The Superior Colliculus: New Approaches for Studying Sensorimotor Integration
William C. Hall, Ph.D., Department of Neuroscience, Duke University,
   Durham, North Carolina
Adonis Moschovakis, Ph.D., Department of Basic Sciences, University of Crete,
   Heraklion, Greece

New Concepts in Cerebral Ischemia
Rick C. S. Lin, Ph.D., Professor of Anatomy, University of Mississippi Medical Center,
   Jackson, Mississippi

DNA Arrays: Technologies and Experimental Strategies
Elena Grigorenko, Ph.D., Technology Development Group, Millennium Pharmaceuticals,
   Cambridge, Massachusetts

Methods for Alcohol-Related Neuroscience Research
Yuan Liu, Ph.D., National Institute of Neurological Disorders and Stroke,
   National Institutes of Health, Bethesda, Maryland
David M. Lovinger, Ph.D., Laboratory of Integrative Neuroscience, NIAAA,
   Nashville, Tennessee

Primate Audition: Behavior and Neurobiology
Asif A. Ghazanfar, Ph.D., Princeton University, Princeton, New Jersey

Methods in Drug Abuse Research: Cellular and Circuit Level Analyses
Dr. Barry D. Waterhouse, Ph.D., MCP-Hahnemann University, Philadelphia, Pennsylvania

Functional and Neural Mechanisms of Interval Timing
Warren H. Meck, Ph.D., Professor of Psychology, Duke University, Durham, North Carolina

Biomedical Imaging in Experimental Neuroscience
Nick Van Bruggen, Ph.D., Department of Neuroscience Genentech, Inc.
Timothy P.L. Roberts, Ph.D., Associate Professor, University of Toronto, Canada

The Primate Visual System
John H. Kaas, Department of Psychology, Vanderbilt University
Christine Collins, Department of Psychology, Vanderbilt University, Nashville, Tennessee

Neurosteroid Effects in the Central Nervous System
Sheryl S. Smith, Ph.D., Department of Physiology, SUNY Health Science Center,
   Brooklyn, New York

Modern Neurosurgery: Clinical Translation of Neuroscience Advances
Dennis A. Turner, Department of Surgery, Division of Neurosurgery,
   Duke University Medical Center, Durham, North Carolina

Sleep: Circuits and Functions
Pierre-Hervé Luoou, Université Claude Bernard Lyon, France

Methods in Insect Sensory Neuroscience
Thomas A. Christensen, Arizona Research Laboratories, Division of Neurobiology,
   University of Arizona, Tuscon, Arizona

Motor Cortex in Voluntary Movements
Alexa Riehle, INCM-CNRS, Marseille, France
Eilon Vaadia, The Hebrew University, Jerusalem, Israel

Neural Plasticity in Adult Somatic Sensory-Motor Systems
Ford F. Ebner, Vanderbilt University, Nashville, Tennessee

Advances in Vagal Afferent Neurobiology
Bradley J. Undem, Johns Hopkins Asthma Center, Baltimore, Maryland
Daniel Weinreich, University of Maryland, Baltimore, Maryland

The Dynamic Synapse: Molecular Methods in Ionotropic Receptor Biology
Josef T. Kittler, University College, London, England
Stephen J. Moss, University College, London, England

Animal Models of Cognitive Impairment
Edward D. Levin, Duke University Medical Center, Durham, North Carolina
Jerry J. Buccafusco, Medical College of Georgia, Augusta, Georgia

The Role of the Nucleus of the Solitary Tract in Gustatory Processing
Robert M. Bradley, University of Michigan, Ann Arbor, Michigan

Brain Aging: Models, Methods, and Mechanisms
David R. Riddle, Wake Forest University, Winston-Salem, North Carolina

Neural Plasticity and Memory: From Genes to Brain Imaging
Frederico Bermudez-Rattoni, National University of Mexico, Mexico City, Mexico

Serotonin Receptors in Neurobiology
Amitabha Chattopadhyay, Center for Cellular and Molecular Biology, Hyderabad, India

Methods for Neural Ensemble Recordings, Second Edition
Miguel A.L. Nicolelis, M.D., Ph.D., Professor of Neurobiology and Biomedical Engineering,
  Duke University Medical Center, Durham, North Carolina

Biology of the NMDA Receptor
Antonius M. VanDongen, Duke University Medical Center, Durham, North Carolina

Methods of Behavioral Analysis in Neuroscience
Jerry J. Buccafusco, Ph.D., Alzheimer's Research Center, Professor of Pharmacology
  and Toxicology, Professor of Psychiatry and Health Behavior,
  Medical College of Georgia, Augusta, Georgia

In Vivo Optical Imaging of Brain Function, Second Edition
Ron Frostig, Ph.D., Professor, Department of Neurobiology,
  University of California, Irvine, California

Fat Detection: Taste, Texture, and Post Ingestive Effects
Jean-Pierre Montmayeur, Ph.D., Centre National de la Recherche Scientifique, Dijon, France
Johannes le Coutre, Ph.D., Nestlé Research Center, Lausanne, Switzerland

The Neurobiology of Olfaction
Anna Menini, Ph.D., Neurobiology Sector International School for Advanced
  Studies,(S.I.S.S.A.), Trieste, Italy

Neuroproteomics
Oscar Alzate, Ph.D., Department of Cell and Developmental Biology,
University of North Carolina, Chapel Hill, North Carolina

Translational Pain Research: From Mouse to Man
Lawrence Kruger, Ph.D., Department of Neurobiology, UCLA School of Medicine,
Los Angeles, California
Alan R. Light, Ph.D., Department of Anesthesiology, University of Utah,
Salt Lake City, Utah

TRANSLATIONAL PAIN RESEARCH
FROM MOUSE TO MAN

Edited by
Lawrence Kruger
UCLA School of Medicine
California

Alan R. Light
University of Utah
Utah

CRC Press
Taylor & Francis Group
Boca Raton London New York

CRC Press is an imprint of the
Taylor & Francis Group, an **informa** business

CRC Press
Taylor & Francis Group
6000 Broken Sound Parkway NW, Suite 300
Boca Raton, FL 33487-2742

First issued in paperback 2017

ISBN 13: 978-1-138-11604-7 (pbk)
ISBN 13: 978-1-4398-1209-9 (hbk)

Library of Congress Cataloging-in-Publication Data

Translational pain research : from mouse to man / editors, Lawrence Kruger, Alan R. Light.
    p. ; cm. -- (Frontiers in neuroscience)
    Includes bibliographical references and index.
    ISBN 978-1-4398-1209-9 (hardcover : alk. paper)
    1. Pain--Pathophysiology. 2. Pain Animal models. 3. Mice as laboratory animals. I. Kruger, Lawrence. II. Light, Alan R., 1950- III. Title. IV. Series: Frontiers in neuroscience (Boca Raton, Fla.)
    [DNLM: 1. Pain--physiopathology. 2. Research--methods. WL 704 T7225 2010]

RB127.T7315 2010
616'.0472--dc22
                                                               2009039338

Visit the Taylor & Francis Web site at
http://www.taylorandfrancis.com

and the CRC Press Web site at
http://www.crcpress.com

This volume is dedicated to the imagination, critical insights, and intellectual energy of Dr. Edward R. Perl, whose impact on the 20th century growth of pain research surpasses that of any individual in the history of this field and has continued into the modern era. His characterization of nociceptors and of specialized neuronal organization in the dorsal horn was instrumental in driving the interests and careers of both editors of this volume, and his scientific productivity and personal influence continues to provide inspirational leadership and critical guidance. By ignoring the naysayers who initially denied the existence of nociceptors, he propelled the focus into the contemporary scientific trends that have enabled a molecular approach to translational pain research.

Lawrence Kruger
Alan R. Light

Contents

Series Preface

Our goal in creating the Frontiers in Neuroscience Series is to present the insights of experts on emerging fields and theoretical concepts that are, or will be, in the vanguard of neuroscience. Books in the series cover genetics, ion channels, apoptosis, electrodes, neural ensemble recordings in behaving animals, and even robotics. The series also covers new and exciting multidisciplinary areas of brain research, such as computational neuroscience and neuroengineering, and describes breakthroughs in classical fields like behavioral neuroscience. We hope every neuroscientist will use these books in order to get acquainted with new ideas and frontiers in brain research. These books can be given to graduate students and postdoctoral fellows when they are looking for guidance to start a new line of research.

Each book is edited by an expert and consists of chapters written by the leaders in a particular field. Books are richly illustrated and contain comprehensive bibliographies. Chapters provide substantial background material relevant to the particular subject. We hope that as the volumes become available, the effort put in by us, the publisher, the book editors, and individual authors will contribute to the further development of brain research. The extent to which we achieve this goal will be determined by the utility of these books.

Sidney A. Simon, Ph.D.
Miguel A.L. Nicolelis, M.D., Ph.D.
Series Editors

Preface

The progress and remarkable expansion of pain research in the past decade prompted Taylor & Francis executive Barbara Norwitz to inquire about a new edition of their now decade-old *Methods in Pain Research*. The query was serendipitously timed to coincide with a highly successful session at the 2007 Spring Brain Conference in Sedona, Arizona, at which Jeff Kennedy had organized a session presenting the progress and ideas driving a burgeoning expansion of research programs in the pharmaceutical industry, focusing on translational aspects of pain research. Despite the general dominance of scientific meetings by researchers in academia, the impact of Big Pharma revealed a trend presaging a new era in which the practical consequences of basic research at the level of molecular biology was becoming relevant to clinical aspects of pain therapy and control. The industry refers to such studies as *preclinical*. While methodology continues to advance, and some spectacular new twists have emerged in the past decade, the gradual shift to "translation" of basic science knowledge to its applications relevant to the human condition appears to have become an important trend in pain research.

Translation in modern biomedical parlance seems approximately equivalent to *applied*—and not necessarily from animals to humans, because the reverse can occur. The earliest attempts at uncovering cerebral localization by producing focal cerebral lesions in dogs were published by the French military surgeon Francois Pourfour du Petit in 1710, based on post-mortem observations of head wounds and related neurological deficits in soldiers—something akin to a "retrograde translation." Surprisingly, the earliest prominent example of translation that we have uncovered does not employ the word but expresses the concept—the discovery by Edward Jenner that cowpox could provide a vaccine to protect humans from the serious rampant scourge of smallpox in the late 18th century.

A congratulatory letter to Jenner from Thomas Jefferson (from the University of Virginia Archive) reveals the following unexpected judgment:

A TRIBUTE OF GRATITUDE

To Dr. Edward Jenner from Thomas Jefferson,
Monticello, May 14, 1806

SIR, I have received a copy of the evidence at large respecting the discovery of the vaccine inoculation which you have been pleased to send me, and for which I return you my thanks. Having been among the early converts, in this part of the globe, to its efficiency, I took an early part in recommending it to my countrymen. I avail myself of this occasion of rendering you a portion of the tribute of gratitude due to you from the whole human family. Medicine has never before produced any single improvement of such utility. Harvey's discovery of the circulation of the blood was a beautiful addition to our knowledge of the animal economy, but on a review of the practice of medicine before and since that epoch, I do not see any great amelioration that has been derived from that discovery. You have erased from the calendar of human afflictions one of its greatest. Yours is the comfortable reflection that mankind can never forget that you have lived. Future nations will know by history only that the loathsome small-pox has existed and by you has been extirpated.

Accept my fervent wishes for your health and happiness and assurances of the greatest respect and consideration.

The devastating demise of a huge proportion of the Native American population, totally lacking in immunity to smallpox, as well as the protection of many of the English settlers, enabled Jefferson to easily recognize the profound importance of "translating" observations from cows to human welfare. But contrasting that with the seemingly less important understanding of the circulation of blood derived from William Harvey's observations might surprise modern scientists who recognize the profound importance of the inductive empiricism embodied in the scientific method developed by Harvey—the foundation of modern medicine. Two centuries later the long-term value of basic science is evident in its enormous governmental financial support. The practical payoff of translation is a relatively recent phenomenon, turning threat to promise.

The study of pain is of enormous human importance for the obvious reason that in the course of a lifetime, few individuals manage to evade disruption of their lives by consequential pain experiences. Pain is the most common clinical complaint for which people seek medical care. While it can be described as a sensation, it is *sensu stricto*, a response to a variety of stimuli, including some that are ordinarily innocuous and often unrelated to any detectable stimulus or visible sign of an inflammatory process. Pain behaviors are widely known in such diverse inflammatory diseases as rheumatoid arthritis, inflammatory bowel disease, psoriasis, and liver disease, but they are also in central nervous system (CNS) diseases such as multiple sclerosis. This book begins with a valuable and original systems approach to the organization of a "systems" approach and then moves on to the details of new information on nociceptor sensitization from the Koerber lab, visceral afferents in disease (Davis lab), and the innovation of a rodent model of bone cancer pain from the Mantyh laboratory providing an explicit pain model, easily equated with similar disease associated with considerable human suffering and leading to remarkable recent progress in treatment of cancer pain derived from animal experimentation.

We have only recently begun to identify insight into the signaling cytokines and the involvement of CNS microglia and their role in recruitment of monocytes in pain syndromes, and we are fortunate in presenting contributions from the laboratories of De Leo and of Shubayev, sources from which key knowledge has been derived by employing animal models. Similarly, the elusive, devastating fibromyalgia syndromes studied in the Light laboratories reveal the validity of exploiting animal models for translation to pain responsible for high-incidence human suffering, and a review by McCord and Kaufman provides original insights and practice in this field.

Several chapters in this book address new technical advances in areas that were dormant or nonexistent in the previous *Methods in Pain Research* volume in this series. These are largely based upon advances in molecular biology found in most of the contributions, but also include exceedingly important advances in tract tracing from the Basbaum lab, new detailed descriptions of visceral pain innervation (Davis lab), an account from the Spigelman lab of the significance of endocannabinoids, as well as cannabis as a therapeutic agent, the current state of advances in gene therapy for pain and the promise of siRNA gene knockdown (Nishimura lab), and some unexpected very recent discoveries about the role of P2X receptors and ATP from Khakh and his colleagues that presage important strides in human therapy. With the discovery of the capsaicin receptor and a specific ion channel related to cardiac pain, it has been anticipated that the tools of molecular biology will lead to broadly available new therapeutics and better understanding of relevant genetic implications, but this has not yet been fully realized despite realistic high hopes.

We are now at a crossroad, with methodologies derived from animal experiments that can impinge upon and perhaps profoundly alter human suffering. Somewhat surprisingly, we have also reached an intersection where methods of studying pain in humans are providing guidance for study of experimental animals where such reciprocity can generate unexpected new insights, such as a new understanding of the "thalamic pain" derived from recent brain-imaging studies from the Llinas laboratory. Critical commentaries from the Apkarian and Greenspan laboratories as well as others active in brain imaging should provide valuable guidance in how to interpret what the imaging methods can and cannot tell us about the localization of brain mechanisms underlying pain.

It should be noted that the emphasis of this volume is directed toward what we consider the noble enterprise of addressing the relief of human suffering in a field of scientific endeavor that must be carefully and sensibly scrutinized to avoid cruelty and needless suffering in experimental subjects. We here must reiterate the admonition introducing the prior *Methods* volume—"it is ... the responsibility of pain researchers to define their aims and to weigh carefully the ethical, legal, and political consequences of pain research with an enthusiasm that matches their desire to understand and control pain for its obvious human benefit." Myriad experiments of enormous importance in gaining insight into molecular mechanisms relevant to pain are based on mouse models that do not exist in nature but are *manufactured* (i.e., literally man-made) by laboratory scientists for the sole explicit purpose of understanding the biological mechanisms essential to our existence. It should be noted that pain research is regulated by federal and state statutes, as well as by serious, dedicated local committees at each research institution. Experimental protocols are designed to narrowly restrict the duration and magnitude of noxious stimuli or to avoid conditions of ascertainable misery. It is the special responsibility of pain researchers to reach out to critics of their endeavors, especially the anti-vivisectionists and animal rights advocates, to persuade them to comprehend that analyzing pain and its deleterious consequences in all animals, but especially the increasingly populous human species, is an enterprise worthy of their support and tolerant understanding.

There are other societal issues of perhaps greater concern because historically the suffering of pain has been largely controlled by opiates; potent drugs that are addictive, controlled illegal substances that constitute the lifeblood of a criminal underworld, the magnitude of which exceeds that of most large industries, and whose impact undermines national interests globally. The economic consequences of unregulated trade in illicit drugs heavily impacts attempts to reorganize the fiscal structure and availability of health care in almost every country. Public policy world-wide concerning pain therapy sometimes succumbs to assertion substituting for reasoned ideas. This year, a leader in clinical testing of pain medications admitted to being variously motivated to certify experimental medications and admitted to numerous publications based on faked data requiring withdrawal from anesthesiology and pain journals while some of these substances remain on the market. We are particularly pleased to present several thoughtful articles from Big Pharma in this volume. The pharmaceutical industry has proven remarkably sensitive to the difficulty and high cost of bringing new medications to the licensing stage in a world that is struggling with the concept of controlled substances, including several crucial for pain control. Despite the exigencies of mergers and acquisitions and the seemingly erratic rules imposed by controlling federal and state agencies, leadership in setting standards for thoughtful, caring use of large animals in pain research and most importantly, clinical human trials of medications derived from animal experimentation, have proceeded with sensitive restraint and concern for human welfare. We are particularly indebted to Jeff Kennedy for leadership in organizing the three insightful papers in this volume outlining the key role of pharmaceutical companies in rendering this subject truly susceptible to translation. We have omitted the various surgical and nerve block procedures in common use today, as well as nontraditional therapies, for practical reasons but also because these modalities have proven less amenable to the translational theme.

Other modalities of pain control are also subject to regulation and perhaps social structure changes susceptible to policy modification; for example, the role of cannabinoids or exotic employment of hypnotic states. Although this volume does not pretend to properly address such issues of social policy in medicinal treatment, it attempts to recognize and promulgate the importance of the legitimate, worldwide pharmaceutical industry. Unfortunately, pain is one end of a hedonic spectrum whose opposite end is exquisite pleasure—and often susceptible to crippling addiction. Thus, it is the aim of translational pain research to alleviate human suffering from pain without substituting the dangerous, deleterious consequences of drug addiction. This volume attempts to bring attention to the present status of recent strides in bringing laboratory science (much of it at the molecular level) to our understanding of pain phenomena in humans, with the ultimate goal of reducing the suffering that often accompanies pain and its indirect consequences.

The Editors

Lawrence Kruger has been active in various aspects of pain-related research for over 50 years. He continues his long career at the UCLA Geffen School of Medicine and Brain Research Institute as Distinguished Professor of Neurobiology, Emeritus in Recalled status and has held a joint appointment in the Department of Anesthesiology for several decades, serving as an advisor to the department's pain clinic.

Professor Kruger earned his Ph.D. in physiology from Yale University and completed postdoctoral training at the Johns Hopkins Medical School, the Institut Marey of the College de France, and Oxford University. He was among the founding members of the International Association for the Study of Pain (IASP) and the Society for Neuroscience, has served on numerous journal editorial boards and as founding editor of *Somatosensory and Motor Research,* and contributed reviews and several books dealing with pain, including *Methods in Pain Research* (2001) in the present CRC Press series. His pain research, supported by successive National Institute of Health Jacob Javits Neuroscience Investigator Awards, has ranged from early studies of thalamic and trigeminal nociceptive-specific neuron discharge properties to anatomical studies of pain pathways encompassing the tracing of peripheral innervation patterns labeled by peptides and other molecular markers of pain endings, and a 3-D electron microscopic reconstruction of physiologically identified nociceptor endings. Since closing his laboratory, he has devoted his efforts largely to service and writing on diverse areas of neuroscience, most recently on historical issues relevant to current trends in neuroscience.

Alan R. Light is professor of Anesthesiology and Neurobiology and Anatomy at the University of Utah. He earned a B.A. from Hamilton College in 1972, and a Ph.D. in physiology from State University of New York, Upstate Medical University, in 1976. He did his postdoctoral work under Dr. Edward Perl, then worked through the ranks to professor at the University of North Carolina at Chapel Hill. Dr. Light moved to the University of Utah School of Medicine in 2003, where he is a member of the Pain Research Center and Program in Neuroscience. He has published over 90 peer-reviewed research articles focused on the peripheral and spinal cord mechanisms of pain processing and its descending control. He has contributed to a number of advanced reviews and textbooks on pain, authored the monograph *The Initial Processing of Pain and Its Descending Control*, published in the Pain and Headache series by Karger, and received an NIH Javits Award for his research on descending control of pain. He is course director of molecular and cellular neuroscience at University of Utah, where his current research focuses on the mechanisms of the sensations of muscle pain and fatigue, and the plasticity they undergo following inflammation and injury and in functional disorders and syndromes such as chronic fatigue and fibromyalgia.

Contributors

Juan Miguel Jimenez Andrade
Department of Pharmacology and the
   Arizona Cancer Center
University of Arizona
Tucson, Arizona

A. Vania Apkarian
Departments of Physiology, Anesthesia,
   Surgery, and Neuroscience Institute
   at Northwestern University
Feinberg School of Medicine
Chicago, Illinois

Allan I. Basbaum
Departments of Anatomy and
   Physiology and W.M. Keck
   Foundation Center for Integrative
   Neuroscience
University of California, San Francisco
San Francisco, California

João M. Bráz
Departments of Anatomy and
   Physiology and W.M. Keck
   Foundation Center for Integrative
   Neuroscience
University of California, San Francisco
San Francisco, California

Nicholas I. Carruthers
Johnson & Johnson Pharmaceutical
   Research and Development, LLC
San Diego, California

Sandra R. Chaplan
Johnson & Johnson Pharmaceutical
   Research and Development, LLC
San Diego, California

C. Richard Chapman
Department of Anesthesiology
School of Medicine
University of Utah
Salt Lake City, Utah

Julie A. Christianson
Department of Medicine
Division of Gastroenterology,
   Hepatology and Nutrition
Pittsburgh Center for Pain Research
Pittsburgh School of Medicine
Pittsburgh, Pennsylvania

Brian M. Davis
Department of Medicine
Division of Gastroenterology,
   Hepatology and Nutrition
Pittsburgh Center for Pain Research
Department of Neurobiology
Pittsburgh School of Medicine
Pittsburgh, Pennsylvania

Joyce A. De Leo
Department of Pharmacology and
   Toxicology
Dartmouth College
Department of Anesthesiology
Neuroscience Center at Dartmouth
Dartmouth Medical School
Lebanon, New Hampshire

William A. Eckert III
Johnson & Johnson Pharmaceutical
   Research and Development, LLC
San Diego, California

J.D. Greenspan
Department of Neural and Pain
  Sciences
University of Maryland Dental School
Program in Neuroscience
University of Maryland
Baltimore, Maryland

Darrell A. Henze
Department of Pain Research
Merck Research Laboratories
West Point, Pennsylvania

Ryan J. Horvath
Department of Pharmacology and
  Toxicology
Dartmouth College
Neuroscience Center at Dartmouth
Dartmouth Medical School
Lebanon, New Hampshire

Kazuhide Inoue
Department of Molecular and System
  Pharmacology
Graduate School of Pharmaceutical
  Sciences
Kyushu University
Higashi, Fukuoka, Japan

Michael P. Jankowski
Department of Neurobiology
University of Pittsburgh
Pittsburgh, Pennsylvania

Kinshi Kato
Department of Orthopedic Surgery
Fukushima Medical University
School of Medicine
Fukushima City, Fukushima, Japan

Marc P. Kaufman
Penn State Heart and Vascular Institute
Pennsylvania State University College
  of Medicine
Hershey, Pennsylvania

Jeffrey D. Kennedy
Neuroscience Discovery Research
Wyeth Research
Princeton, New Jersey

Baljit S. Khakh
Department of Physiology
David Geffen School of Medicine
University of California, Los Angeles
Los Angeles, California

H. Richard Koerber
Department of Neurobiology
University of Pittsburgh
Pittsburgh, Pennsylvania

S.C. LaGraize
Department of Neural and Pain
  Sciences
University of Maryland Dental School
Program in Neuroscience
University of Maryland
Baltimore, Maryland

F.A. Lenz
Department of Neurosurgery
Johns Hopkins School of Medicine
Baltimore, Maryland

Alan R. Light
Department of Anesthesiology and
  Department of Neurobiology and
  Anatomy
University of Utah
Salt Lake City, Utah

Kathleen C. Light
Department of Anesthesiology
University of Utah
Salt Lake City, Utah

Audrey Lin
The Weintraub Center for
  Reconstructive Biotechnology
School of Dentistry
University of California, Los Angeles
Los Angeles, California

Rodolfo R. Llinás
Department of Physiology and
    Neuroscience
New York University School of
    Medicine
New York City, New York

Patrick Mantyh
Department of Pharmacology and the
    Arizona Cancer Center
University of Arizona
Tucson, Arizona
Research Service
VA Medical Center
Minneapolis, Minnesota

Jennifer L. McCord
Penn State Heart and Vascular Institute
Pennsylvania State University College
    of Medicine
Hershey, Pennsylvania

Robert R. Myers
Departments of Anesthesiology and
    Pathology
University of California, San Diego
La Jolla, California

Ichiro Nishimura
Division of Advanced Prosthodontics,
    Biomaterials and Hospital Dentistry
Division of Oral Biology and Medicine
The Weintraub Center for
    Reconstructive Biotechnology
School of Dentistry
University of California, Los Angeles
Los Angeles, California

Edgar Alfonso Romero-Sandoval
Department of Anesthesiology
Neuroscience Center at Dartmouth
Dartmouth Medical School
Lebanon, New Hampshire

Supanigar Ruangsri
The Weintraub Center for
    Reconstructive Biotechnology
School of Dentistry
University of California, Los Angeles
Los Angeles, California

Veronica I. Shubayev
University of California, San Diego
Department of Anesthesiology
San Diego VA Healthcare Center
La Jolla, California

Igor Spigelman
Division of Oral Biology and Medicine
School of Dentistry
Brain Research Institute
Dental Research Institute
University of California, Los Angeles
Los Angeles, California

Devang Thakor
U.S. Department of Veteran Affairs
Harvard Medical School
Brigham and Women's Hospital
Boston, Massachusetts

Estelle Toulme
Department of Physiology
David Geffen School of Medicine
University of California, Los Angeles
Los Angeles, California

Makoto Tsuda
Department of Molecular and System
    Pharmacology
Graduate School of Pharmaceutical
    Sciences
Kyushu University
Higashi, Fukuoka, Japan

Mark O. Urban
Department of Pain Research
Merck Research Laboratories
West Point, Pennsylvania

D.S. Veldhuijzen
Division of Perioperative Care and
  Emergency Medicine
University Medical Center Utrecht
Utrecht, Netherlands

Charles J. Vierck
Department of Neuroscience and
  McKnight Brain Institute
University of Florida College of
  Medicine
Gainesville, Florida

Kerry D. Walton
Department of Physiology and
  Neuroscience
New York University School of
  Medicine
New York City, New York

Garth T. Whiteside
Neuroscience Discovery Research
Wyeth Research
Princeton, New Jersey

1 Painful Multi-Symptom Disorders

A Systems Perspective

C. Richard Chapman

CONTENTS

Chronic multi-symptom disorders are persisting conditions characterized by distressing or disabling symptoms in multiple organ systems and for which no physiological or anatomical cause is evident. Pain is a feature of most such disorders. Examples of such syndromes include irritable bowel, chronic fatigue, fibromyalgia, multiple chemical sensitivity, interstitial cystitis, temporomandibular joint disorder, pelvic pain, and many other chronic pain conditions. The common denominator linking these disorders is a pattern. Each has a constellation of multiple symptoms that obvious pathophysiology cannot explain; emotionally distressing events exacerbate symptoms and strong resistance to conventional medical intervention. Multi-symptom disorders are neither surrogate manifestations of psychological problems nor symptom exaggerations, and physiological markers exist in many cases. For example, irritable bowel syndrome patients and interstitial cystitis patients both demonstrate abnormalities of the epithelium. Patients with multi-symptom disorders appear to suffer more distress than patients with similar symptoms due to identifiable organic disease. Such disorders compromise performance at work, prevent or limit recreation and travel, alter interpersonal relationships, and in general degrade quality of life. Some patients are partially or fully disabled by their condition.

Because each of the disorders involves symptoms in multiple organs and disturbed circadian rhythms, substantial comorbidity exists (Warren et al. 2009). Fibromyalgia patients suffer from muscle pain, but they typically complain as well of fatigue, bowel pain and dysfunction, sleep disturbance, headaches, and cognitive difficulties. Chronic fatigue syndrome and irritable bowel syndrome patients have many of the same conditions but identify fatigue or bowel dysfunction as the most salient problem. In addition to such comorbidities, multisymptom disorders may co-exist with well-defined disease states. For example, a person could have both metabolic syndrome and irritable bowel syndrome, and the two conditions could interact to produce even more complex symptom constellations.

Historically, physicians labeled multi-symptom syndromes as functional disorders and viewed them within a dualistic mind–body framework. Subsequently functional brain imaging and other advances have made it clear that brain activity and physiological processes are interdependent aspects of a single system. Biological systems are adaptive and self-regulating, but they are subject to dysregulation. Multi-symptom disorders manifest both organ system and chronobiological dysregulation.

This chapter provides a systems theoretical framework for health that bridges brain and body. Its purposes are to (1) introduce systems theory as an explanatory framework; (2) account for multi-symptom disorders within this framework, emphasizing stress mechanisms; and (3) introduce and discuss the concept of dys-regulation and its potential role in the genesis and perpetuation of multi-symptom disorders. At the core of this approach is the Complex Adaptive Systems aspect of Complexity Theory.

1.1 COMPLEX SYSTEMS FRAMEWORK

1.1.1 COMPLEXITY SCIENCE

A *system* is a group of independent but interrelated elements comprising a unified whole. A collection of elements comprises a complex system if open and dynamic connections and interactions exist between its components and contribute to the behavior of the collective. An open system is one that exists far from energetic equi-librium; that is, it takes in energy and expels waste. Formally, a complex system is any open system that involves a number of elements arranged in a structure and that requires many scales for adequate measurement. Such systems go through processes of change that defy description by a single rule or by reduction to a single level of explanation.

Complexity is a scientific approach for studying how the interacting parts of a system produce collective behaviors more complex than the sum of the contrib-uting parts, and how the system responds to, and interacts with, its environment. Complexity theory provides a framework and language for describing and modeling such processes.

Complexity as a science studies the behavior of complex systems as a class using mathematical tools such as differential equations, graph theory, neural networks, time series analyses, and genetic algorithms as well as descriptive, predictive, and simulation modeling. Complexity theorists differ from conventional scientists in that they address unpredictable, nondeterministic processes within systems that do not decompose into simpler elements. This allows them to engage natural phenomena in natural settings more readily than their conventional science counterparts. Broadly, the complexity approach employs multi-scale descriptors to characterize dynamic systems and their phenomena. Applications include the study of economies, ecol-ogies, weather, traffic flow, social organizations, and cultures, in addition to such physiological processes as gene and immune networks.

When living organisms engaged in adaptation and survival are the systems of interest, then complexity analysis falls under the heading of complex adaptive sys-tems (CAS) (Gell-Mann 1994; Kelso 1998; Kaneko 2006). Such systems are pur-poseful, pro-creative and pro-active in relationship to their environments rather than simply reactive. An insect hive exemplifies a CAS, as does the immune system. When elements of a system are of interest, for example worker ants within an ant colony or antigen-presenting cells within the immune system, then modelers may designate the CAS as individual based or *agent based*. CASs manifest ever-chang-ing, self-organizing behavior in response to a variable environment, and they move toward, but never sustain, equilibrium. In a classic paper, Prigogne and Stengers (1984) termed this behavior "order through fluctuations."

1.1.2 FEATURES OF COMPLEX SYSTEMS

Any complex system, including a CAS, has several fundamental distinguishing features. A CAS has additional properties because it continuously adapts to an environment. We first introduce these features here and subsequently explore their utility for describing and investigating the physiological and psychological impact of nociception.

1.1.3 LACK OF CENTRAL CONTROL

Complex systems differ from simple systems in that they lack central control. A control hierarchy with a leader at the top simply does not exist. Rather, the power spreads over a decentralized structure and multiple agents combine to generate the actual system behavior. A building heating system is a noncomplex, closed system in which a single component, a thermostat, controls system behavior. In this case, the whole can never be more complex than the sum of its parts. When control emerges from the collective in a way that exceeds the sum of the contribution of the individual agents, as it does in an insect swarm, then true complexity exists and the collective behaves in a manner more complex than the individual agent within it could ever achieve. In a complex system, control is an *emergent property*. That is, control appears spontaneously and is unpredictable solely on the basis of information about the individual components.

1.1.4 EMERGENCE

Complexity researchers regard emergent phenomena as normal properties of dynamic, self-organizing systems. In principle, *emergence* is the process of deriving some new and coherent structures, patterns, and properties in a complex system. For example, an insect colony exhibits purposeful and intelligent adaptive behavior that makes possible foraging for food, defense, and reproduction. This property, which we may loosely term intelligence, is unpredictable from what we know about the individual insect and appears spontaneously. Emergence is readily apparent in the behaviors of an insect swarm, a flock of migrating birds, a human crowd, and indeed in human culture. As a general principle, complexity theorists hold that emergent phenomena occur due to patterning of interactions (nonlinear and distributed) between the elements of the system over time. One might describe acute-phase tissue inflammation as an emergent property of the immune system. Complex behaviors emerge as a result of often nonlinear, spatiotemporal interactions among a large number of component systems at different levels of system organization.

1.1.5 STATES AND STATE TRANSITIONS

Complex systems of all types are dynamic and function in states; that is, relatively stable modes of operation. Complexity theorists refer to a collection of system properties as a *state* and the set of all possible states of a system is its *state space*.

Basically, the total number of properties transmitted by a system, and detected by an observer, defines the complexity of that system. For the sake of illustration, consider familiar objects rather than complex systems. For a coin toss, there are only two states, namely heads and tails, but for a computer screen with a resolution of 800 × 600 pixels and 256 colors, the number of states is 256 to the power of 480,000. Of course, in a CAS, some states are much more likely to occur than others, and experience shapes the probability of transitions to certain states.

Complex systems sometimes undergo abrupt and unpredictable shifts in states. *State transitions*, often called *phase transitions*, are everywhere in nature. These are abrupt, nonlinear changes in a system. Water can change from solid to liquid and then to vapor with increasing temperature. The human brain can shift from waking consciousness to slow-wave sleep, and from that to paradoxical sleep, as a function of circadian rhythm. During combat, a soldier may become totally insensitive to injury, a state transition that fosters survival.

1.1.6 ATTRACTORS

Although complex systems are dynamic and self-organizing, when perturbed they go into disorder and then settle back into relatively stable states with relatively simple behavioral patterns. The transition from disorder to order reduces complexity and defines the new state space. A common metaphor describes a ball falling onto a three-dimensional landscape surface with peaks and depressions. The ball will roll away from the peaks and eventually settle into a depression. The depression, or basin, represents a subset of a system's state space that the system can enter but not leave, unless boundary conditions or perturbations bring about reorganization. This is an *attractor.*

Systems are inherently dynamic, and so the interaction of the ball with the landscape may change over time, as the system's environment varies. In this respect, a state space has a trajectory over time and may change as its environment changes. For example, one might characterize an ion channel as a two-state, or on–off, system and the probability of the *on* state will vary across time as a function of change in the system's environment. Naturally, a CAS has a history, and the experience of prior states may influence the probability of occurrence of future states.

1.1.7 NESTING

A complex system always has the feature of *nesting*; that is, subsystems nest within it, and it nests within a higher-level complex system. Each system level can have its own state transitions, and these transitions occur within a higher-level system. Therefore, the first challenge we face in engaging the idea of multi-symptom disorder is deciding upon a *level of inquiry.* That is, we must single out one level of a hierarchically organized complex system as the *System of Interest* and define the levels above it as its environment. We could choose the sensory end organ, the dorsal horn of the spinal cord, the brain, the family, or the American culture. Because multi-symptom disorders happen to individual people, we normally select the individual as our System of Interest. Social systems such as the family comprise our system's environment, and various psychological and physiological subsystems nest within the individual.

Figure 1.1 broadly illustrates the principle of hierarchal system nesting, within which the investigator defines the System of Interest, depicted as the Wider System of Interest in Figure 1.1. For our purposes, the Wider System of Interest is the individual, or person. The Environment immediately surrounding the person is his or her social network: family, friends, work environment, and perhaps involved health care professionals. Figure 1.1 designates this as the Wider Environment; that is, the surrounding social community, its economy, its culture, and all of the influences, opportunities, support, and hassles that this can exert upon the individual and his/ her social network. Many stressors reside in the Wider Environment. The Narrow System of Interest, nested within the Wider System of Interest, refers to the physical and psychological health of the person. It is this to which health care providers normally attend. Of course, the Narrow System of Interest contains multiple physiological subsystems that are the concern of medicine and the targets of medical diagnosis and evaluation.

Causal influences are bidirectional and extend across system levels. For example, the interactions of the person, or Wider System of Interest, with the family, or Environment, if negative, can create stress at the psychological and physical level with negative effects on health. Conversely, improvements in health, or Narrow System of Interest, can positively influence the interactions of the person with his or her environment. The concept of dysregulation, which we discuss below, applies at all levels of the system. Dysregulation within the Wider Environment, or society, can evoke consequent dysregulation within the Environment, or family, and this subsequently can dysregulate the person and compromise health itself.

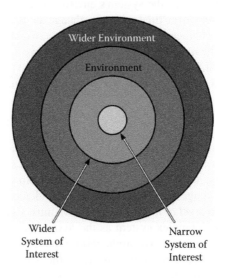

Wider Environment

Environment

Wider
System of
Interest

Narrow
System of
Interest

FIGURE 1.1   Hierarchal system nesting.

1.2 FEATURES OF COMPLEX ADAPTIVE SYSTEMS

In addition to the characteristics of all complex systems, CASs express the following features.

1.2.1 ADAPTATION AND AGENCY

Adaptation is the continual adjustment of an agent to its changing environment. An agent is a living entity, a self-organizing system, and an individual entity operating purposefully within its environment in the service of adaptation. The concept of *agent* equates with the individual when the focus of study is on the interaction of the individual with the environment, especially the social environment. Grimm and colleagues advocate the concept of agent-based complex system, sometimes termed individual-based complex system, which directly identifies the individual in the world as an agent (Grimm et al. 2005).

The complexity investigator imputes agency to aspects of nested subsystems whenever an element exhibits some degree of autonomy. For example, the migratory cells of the immune system serve as agents for the detection of toxins, invading microorganisms and tumor development, all of which are invisible to the nervous and endocrine systems. Moreover, dendritic cells serve as professional antigen-presenting agents. They appear in peripheral organs such as skin where they encounter and capture antigens. They then migrate to the T cell areas of lymphoid tissues and present the processed antigens in order to elicit antigen-specific T cell responses.

Whatever the level of inquiry, agents are semi-autonomous units that evolve over time and help to maximize adaptation. They scan their environment and develop *schemata*, which are perceptual and/or motor patterns comprising rules for interpretation and action. Therefore, multiple agent-based subsystems exist in principle, nested within and working in service of the CAS of interest.

1.2.2 EQUILIBRIUM AND HOMEOSTASIS

Complex systems are open, dissipative and operate far from equilibrium, but they tend to move toward equilibrium after disturbance and disorder. Physiologically, the bottom line for equilibrium is *homeostasis*. Although many writers equate homeostasis with adaptive adjustment, McEwen points out that homeostasis strictly applies to a limited set of systems concerned with maintaining the essentials of the internal milieu (McEwen 2000). The maintenance of homeostasis is the control of internal processes truly necessary for life such as thermoregulation, blood gases, acid base, fluid levels, metabolite levels, and blood pressure. McEwen's distinction is critical because homeostasis has no adaptive features.

Three interdependent systems control the process of homeostasis: neural, immune, and endocrine (Goetzl and Sreedharan 1992). From the CAS perspective, specific processes must exist to protect and preserve homeostasis. Generic threats to homeostasis include environmental extremes, excessive physical exertion, depletion of essential resources, abnormal feedback processes, aging, and disease. Perturbations from the environment can threaten homeostatic regulation at any time.

1.2.3 ALLOSTASIS AND STRESS

Allostasis is an adaptive process in the service of homeostasis; it dynamically adapts multiple internal systems to changes in the environment and coordinates their responses (McEwen 2000; Korte et al. 2005). Changes in the external or internal environment trigger physiological coping mechanisms. These mechanisms insure that the processes sustaining homeostasis stay within normal range. The allostatic process, which involves substantial autonomic activity, depends upon the coordinating effects of agent messenger substances that also serve as mediators and determinants of neural regulatory processes, particularly hormones, neurotransmitters, peptides, endocannabinoids, and cytokines. I describe this more fully below.

Stress is the resource-intensive process of mounting adaptive coping responses to challenges that occur in the external or internal environment. A *stressor* is any event that elicits a *stress response*. It may be a physical or social event, an invading microorganism, or a signal of tissue trauma. Selye (1936) first described this response as a syndrome produced by "diverse nocuous agents." He eventually characterized the stress response as having three stages: alarm reaction, resistance, and if the stressor does not relent, exhaustion. The normal stress responses of everyday life consist of the alarm reaction, resistance, and recovery. Stressors have as their primary features intensity, duration, and frequency. The impact of a stressor is the magnitude of the response it elicits. This impact involves cognitive mediation because it is a function of both the predictability and the controllability of the stressor.

Allostasis is the essence of the stress response because it mobilizes internal resources to meet the challenge that a stressor represents. Stressors may be multimodal and complex or unimodal and simple. When a stressor persists for a long period of time, or when repeated stressors occur in rapid succession, allostasis may burn resources faster than the body can replenish them. The cost to the body of allostatic adjustment, whether in response to extreme acute challenges or to lesser challenges over an extended period of time, is called *allostatic load*.

1.2.4 FEEDBACK

In open systems, self-regulation and self-organization depend upon feedback, which determines stability. That is, information about the output of a system passes back to the input and thereby dynamically controls the level of the output. Figure 1.2 illustrates two fundamental feedback principles: negative and positive feedback. These are essential constructs in all areas of the biological, behavioral, and social sciences as well as in engineering and complexity science (Jones 1973; Thomas and D'Ari 1990; Northrop 2000; Flood and Carson 1993).

Negative feedback generally involves a circuit and a controller with a set point, and it works toward establishing equilibrium. Figure 1.3 illustrates an adaptive negative feedback system that depends upon systemic circulation. Negative feedback regulation occurs throughout physiology and is a fundamental principle of endocrinology. Negative feedback acts to insure system stability and to maintain homeostasis. The difference between normal set point and current condition gauges allostatic load. Negative feedback continually moves a system away from imbalance and disorder

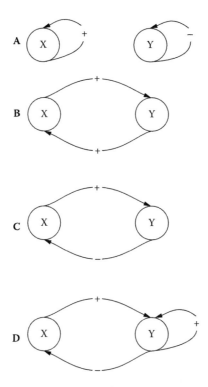

FIGURE 1.2 Negative and positive feedback.

toward balance and order. In principle, biological systems are always nested, and the set point for a negative feedback loop is generally under the control of a larger system within which it is embedded. Disturbance of a set point compromises negative feedback and is a potential cause of dysregulation.

Positive feedback loops also occur such that, when a variable changes, the system responds by changing that variable even more in the same direction, generating escalation and rapid acceleration (Ferrell 2002). This is a process that abandons stability for instability. From an adaptation point of view, positive feedback loop capability is

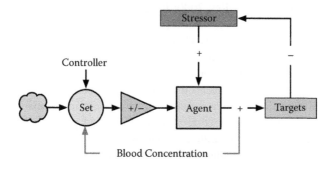

FIGURE 1.3 Negative feedback with a controller and set point.

essential for meeting acute threat with defensive arousal or reproductive opportunity with sexual arousal. Positive feedback loops make state change possible.

A simple example of positive feedback is autocrine signaling. A cell may produce a substance, such as an activated microglial secreting a pro-inflammatory cytokine. The presence of the secreted cytokine in the cell's environment stimulates the cell to produce still more of the cytokine. Such autocrine signaling, through a combination of strong nonlinearity and positive feedback, promotes cellular instability and allows transient inputs to shift the cellular system between two steady states, that is, bistability (Shvartsman et al. 2002). Brandman and colleagues pointed out that positive feedback allows systems to convert graded inputs to decisive all-or-none outputs (Brandman et al. 2005).

In this way positive feedback can move a CAS toward an adaptive state transition that by definition has an all-or-none quality (Brandman et al. 2005). Setting up stable transition is essential for rapid adaptation. In the case of the dorsal horn, this could be a biphasic state transition, as described below. Inhibitory systems may also have positive feedback components that can wind up inhibitory processes and eventually shut down an excitatory state. Such processes may play a role in sleep, fatigue, and other conditions of hypo-arousal or impaired cognition. In the CAS framework, positive feedback loops and bistable states are the products of evolution and are essential for adaptation and survival.

Positive feedback loops do not normally operate independently within a CAS. Because every system is embedded within a larger system, a positive feedback loop is typically under the control of an overarching negative feedback system that limits overshoot and can eventually terminate the positive feedback loop. Positive feedback can persist or terminate through state shift transition or in response to an overarching system that acts on the basic mechanism from which the positive feedback system arises or by initiating an opponent process. For example, to prevent overshoot in a positive feedback excitatory process, the superordinate system may initiate a competing inhibitory process. Overarching systems typically control the on–off state of a positive feedback loop.

Feedback loops appear to exist reciprocally across nervous, endocrine, and immune subsystems and thereby contribute to overall system regulation. For example, such processes clearly play key roles in the interdependence of endocrine and immune systems (Besedovsky and del Rey 2000; Rivest 2001). Glucocorticoid products of the hypothalamo-pituitary-adrenocortical (HPA) axis modulate the basal operations of cytokine-producing immune cells. Cytokines, in turn, influence the activity of the HPA axis. Thus, the products of one system provide messenger substances that serve a feedback function for another system.

Feedback loops are essential agents in system regulation. Negative feedback tends to sustain stability in an adaptive system despite changes in the external or internal environment, thereby minimizing allostatic load and protecting homeostasis. Positive feedback increases possibilities for change in system behavior and provides pathways to establish new set points for its negative feedback processes. More importantly, positive feedback is a mechanism for inducing rapid, adaptive state transitions that are necessary for emergency reactions in a threatening environment.

Negative and positive feedback can go awry within the nervous, endocrine, and immune systems. The result is a disease process. Negative feedback may fail when an endogenous messenger substance providing the feedback disappears, occurs in excess, or becomes confounded by exogenous products such as medications or substances of abuse that resemble the messenger substance in chemical structure. In some cases, negative feedback fails when an extraneous influence alters the set point. For example, chronic opioid pharmacotherapy in a male pain patient confuses the hypothalamo-pituitary-gonadal axis and results in hypogonadism (Daniell 2002; Bliesener et al. 2005; Daniell, Lentz, and Mazer 2006).

Positive feedback processes can also malfunction. When a positive feedback loop does not fulfill its natural purpose, it can generate an extreme shift in adaptive state. In some cases, this violates homeostasis and results in death. Although positive sensory input does not directly cause death, it can contribute to life-threatening conditions such as cardiovascular shock. Persistent, stressor-related positive feedback probably contributes to migraine headache, allodynia, severe idiopathic abdominal pain, noncardiac chest pain, and a variety of multi-symptom disorders.

1.2.5 AGENT CONNECTIVITY

By definition, connections and interactions exist among the components, or agents, of a CAS, and these linkages define self-organization and behavior. The *connectivity* of a system is the nature and extent of such connections and interactions. It is from these connections that patterns form and feedback occurs. The relationships between the components, or agents, within a system are generally more important than the agents themselves.

Following a stressful event, connectivity insures an extensive, systemic response. It encompasses all forms of physiological information exchange: neural, blood-borne, extracellular, and immune. Neurotransmitters, peptides, hormones, endocannabinoids, and cytokines are among the tools of connectivity. Neural, endocrine, and immune systems are able to mount a concerted, fully coordinated response because these messenger substances constantly exchange information and provide feedback. Connectivity makes possible many negative and positive feedback functions.

1.3 THE NERVOUS-ENDOCRINE-IMMUNE SUPERSYSTEM

Life depends on homeostasis, and the goal of adaptation on the part of a CAS is homeostatic maintenance. The three major body systems responsible for maintaining vertebrate homeostasis are the nervous, endocrine, and immune systems. Conventional, reductionistic science studies these systems as independent entities with each undertaking a unique function. Together with Robert Tuckett and Chan Woo Song, I postulated that the neuro-endocrine-immune ensemble operates as an overarching system, a supersystem, within which each individual system functions as a subsystem (Chapman, Tuckett, and Song 2008). A corollary is that the supersystem nests with a larger system that we characterize as the whole person, or individual. Figure 1.4 depicts the supersystem, emphasizing connectivity. It depicts a dynamic

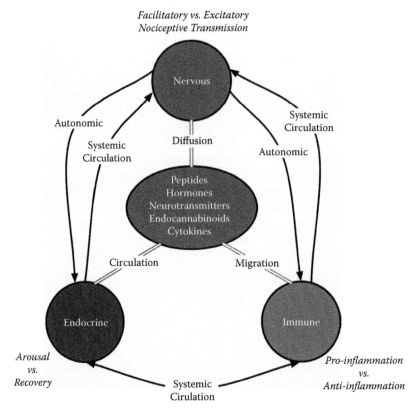

FIGURE 1.4 Supersystem and connectivity. (Reprinted with permission from Chapman, C. R., R. P. Tuckett, and C. W. Song. 2008. Pain and stress in a systems perspective: reciprocal neural, endocrine, and immune interactions. *J Pain* 9 (2):122–145. Copyright © 2008 American Pain Society, Elsevier, Inc.)

process of constant message interchange within the autonomic nervous system and through systemic circulation.

An extensive literature reveals the interconnectedness of the neuro-endocrine-immune subsystems (Chapman, Tuckett, and Song 2008). This literature addresses the pairwise connections of the three systems rather than a three-way interdependence; but taken as an aggregate, it is readily clear that each of these systems is dynamically listening to, signaling to, and coordinating with the others. As the figure indicates, this connectivity employs neurotransmitters, hormones, peptides, endocannabinoids, and cytokines.

Connectivity serves many purposes within the supersystem. One is chronobiological coordination. Infracadian, circadian, and ultracadian rhythms employ messenger substances that move across multiple subsystems. Sleep and appetite regulation are familiar examples. In order to regulate such rhythms, the supersystem must continuously move messages from one subsystem to another. Positive and negative feedback loops also commonly involve multiple subsystems.

I propose that the supersystem is the primary mechanism of allostasis; that is, any threatening or challenging event engages the supersystem, and its primary purpose is to meet the challenge in order to protect homeostasis. The supersystem responds to biological challenges such as microbial invasion, tissue trauma, or exposure to cold. It also responds to social threats such as anger on the part of another person, threat of ostracism or social criticism (e.g., public speaking situations), and economic hardship. Humans, because of our frontal lobes, are uniquely equipped to anticipate physical and social threats, and so it is possible to generate threat that activates the supersystem purely through anticipation, belief, or imagination. Uexküll (1973, 1928) defined the "Umwelt" as the subjective universe of the individual, emphasizing the unique nature of each individual's attribution of meaning to a given situation. Individuals identify and experience stressors according to their own histories, beliefs, and values; and so a social or physical event that is very disturbing to one person may be trivial to another. The higher-order cognitive processes of the brain greatly influence the response of the supersystem.

1.3.1 Stress and Hormesis

Stress is a normal aspect of human life, and some degree of daily stress is necessary for health. Mild stress, like that induced with exercise, caloric restriction, or alcohol, is clearly beneficial. However, a severe stressor, repetitive or relentless stressors, or multiple simultaneous stressors can overwhelm allostatic resources and cause deleterious consequences.

The concept of *hormesis* captures the nonlinear relationship of stress severity to benefit versus harm. In toxicology it refers to a dose response pattern whereby a substance that in a high dose inhibits or is toxic to a biological process stimulates or protects that same process in a smaller dose. The analogy of drug dose to the magnitude of a stressor is straightforward. Hormesis refers to the nonlinear benefit versus harm effect resulting from varying degrees of exposure to a stressor.

Figure 1.5 illustrates this concept. A stressor disrupts supersystem homeodynamics. This results in modest over-compensation and restoration of homeodynamics, or recovery. This represents successful adaptation.

At the apex of the curve, where the magnitude of the stressor is substantial, exposure to the stressor begins to harm rather than benefit the individual. It is clear from athletics that repeated exposure to a beneficial stressor increases resistance to subsequent stressors and increases overall fitness. This is as true for adaptive immunity as it is for the runner's endurance. The peak of the curve in Figure 1.5 shifts to the right with repeated exposure at modest levels. It shifts to the left for individuals who manage to avoid most stressors including exercise, and the result is that their supersystems become hypersensitive.

What happens when the magnitude of the stressor exceeds the peak of the hormetic curve? The answer lies in the nature of the stress response. The stress literature tends to group all reactions to a stressor under the single heading of stress response, as the response consisted solely of arousal. However, deKloet and Derijk (2004) characterize the stress response as having two modes of operation, or states. The first state is immediate arousal in response to the stressor in order to enable adaptive behaviors, and the second state is a slower process that promotes recovery, behavioral adaptation, and return to normalcy. They describe these phases as the fast and slow responding modes. Repeated stressors, multiple stressors, or a severe stressor

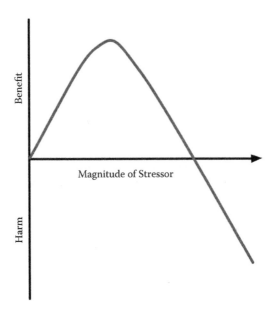

FIGURE 1.5   Hormesis.

may interfere with the transition from fast to slow mode or the completion of the slow mode. This can generate dysregulation in an organ system or chronobiological processes, as I describe below.

1.3.2   SUPERSYSTEM AROUSAL MECHANISMS

1.3.2.1   Nervous System

The major brain mechanisms of the stress response are the locus coeruleus (LC) noradrenergic system, the HPA axis based in the hypothalamic periventricular nucleus (PVN) (Tsigos and Chrousos 2002), and the sympatho-adreno-medullary (SAM) axis (Padgett and Glaser 2003). The peripheral effectors of these mechanisms are the autonomic nervous system, the SAM circulating hormones, principally the catecholamines epinephrine (E) and norepinephrine (NE) together with the sympathetic co-transmitter neuropeptide Y (NPY) (Zukowska et al. 2003), all of which originate in the chromaffin cells of the adrenal medulla. The stress response also involves hypothalamically induced release of peptides derived from pro-opiomelanocortin (POMC) at the anterior pituitary. The POMC-related family of anterior pituitary hormones includes adrenocorticotropin hormone (ACTH), β-lipotropin, β-melanocyte stimulating hormone, and β-endorphin.

Corticotropin-releasing hormone (CRH), produced at the hypothalamic PVN, initiates the stress response. CRH initiates and coordinates the stress response at many levels (Elenkov 2004), including the LC (Rassnick, Sved, and Rabin 1994). It is the key excitatory central neurotransmitter and regulator in the endocrine response to injury. Two receptors respond to CRH and CRH-related peptides, CRH-1 and CRH-2. These distribute widely in limbic brain (Leonard 2005). CRH-1 (deKloet

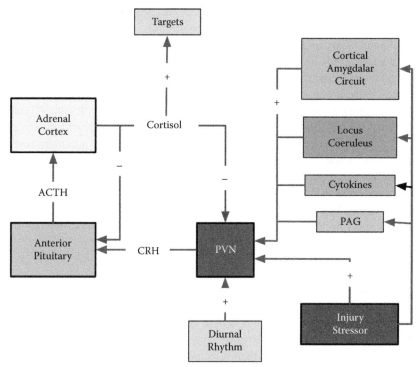

FIGURE 1.6 HPA axis response to a stressor. (Reprinted with permission from Chapman, C. R., R. P. Tuckett, and C. W. Song. 2008. Pain and stress in a systems perspective: reciprocal neural, endocrine, and immune interactions. *J Pain* 9 (2):122–145. Copyright © 2008 American Pain Society, Elsevier, Inc.)

and Derijk 2004) is the key mechanism of the *defensive arousal* response. Figure 1.6 illustrates the HPA axis response to a stressor such as tissue injury.

1.3.2.2 Central Noradrenergic Mechanisms

Most stressors, and tissue injury in particular, inevitably and reliably activate the LC noradrenergic neurons; and LC excitation appears to be a consistent response to nociception (Svensson 1987; Stone 1975). The LC heightens vigilance, attention, and fear, as well as facilitating general defensive reactions mediated through the sympathetic nervous system. Basically, any stimulus that threatens the biological, psychological, or psychosocial integrity of the individual increases the firing rate of the LC, and this in turn increases the release and turnover of NE in the brain areas having noradrenergic innervation. The LC exerts a powerful influence on cognitive processes such as attention and task performance (Berridge and Waterhouse 2003; Aston-Jones and Cohen 2005). In addition to directly receiving noxious signals during spinoreticular transmission, the LC also responds to CRH (Rassnick, Sved, and Rabin 1994). LC neurons increase firing rates in response to CRH, and this increases NE levels throughout the CNS (Jedema and Grace 2004).

How do psychological stressors work? Clearly, endogenous thought processes, unpleasant memories, and anticipations of negative situations can serve as stressors that are every bit as influential as environmental events. Basically, psychosocial stressors evoke cognitive responses such as appraisal, memory, expectation, and the attribution of meaning. These endogenous processes heavily involve the prefrontal and frontal cortices of the brain, and these cortices exert control over aspects of the hypothalamus including the hypothalamic PVN. The PVN initiates the HPA stress response and controls it through negative feedback mechanisms. The PVN triggers further stress response in the SAM axis by recruiting catecholaminergic cells in the rostral ventrolateral medulla. This structure is a cardiovascular regulatory area involved, together with the solitary nucleus, in the control of blood pressure. The rostral ventrolateral medulla activates the solitary nucleus and, together with it, provides tonic excitatory drive to sympathetic vasoconstrictor nerves that maintain resting blood pressure levels. Through this mechanism, a normal stress response involves a complex pattern of autonomic arousal that includes increased blood pressure followed by a period of recovery when blood pressure and other aspects of arousal return to normal. Briefly, cortical activity associated with cognitive-emotional processes can generate stress responses through central and autonomic mechanisms. These responses, in turn, influence endocrine and immune activity.

1.3.2.3   Endocrine Mechanisms

The adrenal medulla, an endocrine organ, is a functional expression of the sympathetic nervous system that broadcasts excitatory messages by secreting substances into the blood stream. Acetylcholine (Ach) released from pre-ganglionic sympathetic nerves during the stress response triggers secretion of E, NE, and NPY into the systemic circulation. E and NE exert their effects by binding to adrenergic receptors on the surface of target cells, and they induce a general systemic arousal that mobilizes fight-or-flight behaviors. These catecholamines increase heart rate and breathing, tighten muscles, constrict blood vessels in parts of the body, and initiate vasodilation in other parts such as muscle, brain, lung, and heart. They increase blood supply to organs involved in fighting or fleeing, but decrease flow in other areas.

1.3.2.4   Immune Mechanisms

Just as the nervous system is the primary agent for detecting and defending against threat arising in the external environment, the immune system is the primary agent of defense for the internal environment. Kohl (2006) described it as "a network of complex danger sensors and transmitters." This interactive network of lymphoid organs, cells, humoral factors, and cytokines works interdependently with the nervous and endocrine systems to protect homeostasis. Parkin and Cohen (2001) provide a detailed overview of the immune system.

Imagine the stressor as a tissue injury event. The immune system detects an injury event in at least three ways: (1) through blood-borne immune messengers originating at the wound; (2) through nociceptor-induced sympathetic activation and subsequent stimulation of immune tissues; and (3) through SAM and HPA endocrine signaling. Immune messaging begins with the acute phase reaction at the wound (Gruys et al. 2005). Local macrophages, neutrophils, and granulocytes produce and release into intracellular space and circulation the pro-inflammatory cytokines Il-1, Il-6, IL-8, and TNF-α. This alerts and activates other immune tissues and cells that have a complex systemic impact. The acute phase reaction to injury is the immune counterpart to nociception in the nervous system, as it encompasses transduction, transmission, and effector responses.

The immune and nervous systems cooperate at the wound focus. Tissue injury releases the immunostimulatory neuropeptides SP and NKA. These activate T cells and cause them to increase production of the pro-inflammatory cytokine IFN-α (Lambrecht 2001). In addition, another pro-inflammatory cytokine, IL1-β, stimulates the release of SP from primary afferent neurons (Inoue et al. 1999). The neurogenic inflammatory response helps initiate the immune defense response and at the same time is in part a product of that response (Eskandari, Webster, and Sternberg 2003). Immune-nervous system interaction is feedback dependent.

Although many cell types produce cytokines in response to an immune stimulus, classical description holds that their principal origin is leukocytes. Cytokines powerfully affect many tissues, but they are also major signaling compounds that recruit many cell types in response to injury. They bind specifically to cell surface receptors to achieve their effects, and exogenous antagonists can block their effects. Cytokines act upon (1) the cells that secrete them, autocrine mode; (2) nearby cells, paracrine mode; and (3) distant cells, endocrine mode. Chemokines are chemotactic cytokines that attract specific types of immune cells, mainly leukocytes, to an area of injury. Broadly, cytokines group into four families based on their receptor types: (1) hematopoietins, including IL-1 to IL-7 and the granulocyte macrophage colony stimulating factor (GM-CSF) group; (2) interferons including INFα and INFα; (3) tumor necrosis factors, including TNF-α; and (4) chemokines, including IL-8. For a basic review, see Elenkov and colleagues (2005) and Gosain and Garnelli (2005). Cytokines can act synergistically or antagonistically in many dimensions.

Sympathetic nervous system activity following a stressor can directly modulate many aspects of immune activity and provide feedback. This can occur because all lymphoid organs have sympathetic nervous system innervation (Elenkov et al. 2000) and because many immune cells express adrenoceptors (Vizi and Elenkov 2002; Kin and Sanders 2006). This is another potential link between higher central nervous system structures involved in cognitive-emotional processes and general health.

Inflammation assists the immune system in defense against the microbial invasion that normally accompanies any breach of the skin. If microorganisms reach the bloodstream, sepsis occurs. The inflammatory process creates a barrier against the invading microorganisms, activates various cells including macrophages and lymphocytes that find and destroy invaders, and sensitizes the wound, thereby minimizing the risk of further injury. Redness, pain, heat, and swelling are its cardinal signs. Inflammation reduces function and increases pain by sensitizing nociceptors. Tracey (2002) described the "inflammatory reflex" as an Ach-mediated process by which the nervous system recognizes the presence of, and exerts influence upon, peripheral inflammation. Through vagal and glossopharyngeal bidirectional nerves, the nervous system modulates circulating cytokine levels. The key point is that certain nervous structures sense the activities of the immune system.

1.3.3 SUPERSYSTEM RECOVERY MECHANISMS

The *recovery* phase of the supersystem stress response commences before the arousal, or alarm, phase ends to protect against arousal and inflammatory overshoot. The arousal state is catabolic, and if the allostatic response is too strong or goes on too long, it can deplete neurotransmitters and/or dysregulate system functions. The purposes of the recovery response are first to regulate the intensity of the alarm reaction, and second, when it is safe to stop arousal, to terminate allostasis, minimize the costs of allostatic load, and bring the body back to normalcy.

CRH synthesis and release occur in response to a stressor and also in response to levels of circulating cortisol (CORT) and normal diurnal rhythm. The neurons of the median eminence secrete CRH into the hypophyseal portal circulation, and this carries it to the anterior pituitary where it binds to CRH receptors on corticotropes. This generates POMC synthesis and release of ACTH (Miller and O'Callaghan 2002) into systemic circulation. Circulating ACTH stimulates production of CORT at the adrenal cortex with release into the systemic circulation. Circulating CORT, in turn, provides a negative feedback signal to the PVN and the anterior pituitary. (See Figure 1.6.)

The endocrine mechanism for recovery induces CRH-2 receptor expression. This receptor responds to the CRH family of peptides (deKloet 2004) including the urocortins. The anterior pituitary initiates production of adrenocortical glucocorticoids (GCs), including CORT, that bind to glucocorticoid receptors (GRs). The primary agent and classical marker for stress recovery in humans is CORT. It normally functions in concert with the catecholamines and CRH. GR activation promotes energy storage and termination of inflammation to prepare for future emergency. Although the recovery process is inherently protective, prolonged CORT can cause substantial damage (deKloet and Derijk 2004; deKloet 2004; Elenkov 2004).

Immune mechanisms demonstrate a similar process in response to a stressor. There is first an inflammatory reaction and then an extended anti-inflammatory reaction that prevents overshoot and damage (Elenkov et al. 2005; Menger and Vollmar 2004). Pro- and anti-inflammatory influences are essentially opponent processes, and ultimately with healing the system should return to a balance.

Cytokines classify into two categories. Soon after formation, helper T cells differentiate into two types in response to existing cytokines and then secrete their own cytokines with one of two profiles: Th1, pro-inflammatory; and Th2, anti-inflammatory. Most cytokines classify readily as either Th1 or Th2 according to the influence they exert. For example, IL-4 stimulates Th2 activity and suppresses Th1 activity, so it is anti-inflammatory. IL-12, on the other hand, promotes pro-inflammatory activity and is therefore Th1. Pro-inflammatory cytokines include IL1-β, IL-2, IL-6, IL-8, IL-12, IFN-α, and TNF-α. Anti-inflammatory cytokines include IL-4, IL-10, insulin-like growth factor 1 (IGF-10), and IL-13. Some investigators characterize an individual's immune response profile using a *Th1/Th2 ratio*.

The immune recovery process requires that the Th1 predominant state wind down and return to a Th1/Th2 ratio equal to 1.0. A persisting imbalance in the Th1 direction fosters harmful inflammation. Imbalance in the Th2 direction incurs risk of adventitious infection and tumor development.

1.4 STRESS RESPONSE, SICKNESS, AND DEPRESSION

Fever and sickness with pain is an immune systemic response to a stressor (Elenkov et al. 2005; Watkins and Maier 1999, 2005; Steinman 2004; Wieseler-Frank, Maier, and Watkins 2005). This *sickness response* is cytokine-mediated and depends on the CNS. Macrophages and other cells release pro-inflammatory cytokines including IL1-β, IL-6, IL-8, IL-12, IFN-α and TNF-α in response to an injury, microbial invasion, or another stressor. These substances act on the vagus and glossopharyngeal nerves, hypothalamus, and elsewhere to trigger a cascade of unpleasant, activity-limiting symptoms (Wieseler-Frank, Maier, and Watkins 2005a; Romeo et al. 2001).

The sickness response, a systemwide change in mode of operation triggered by cytokines, is a vivid and dysphoric subjective experience characterized by fever, malaise, fatigue, difficulty concentrating, excessive sleep, decreased appetite and libido, stimulation of the HPA axis, and hyperalgesia. The sickness-related hyperalgesia may reflect the contributions of spinal cord microglia and astrocytes (Wieseler-Frank, Maier, and Watkins 2005). Functionally, this state is adaptive; it minimizes risk by limiting normal behavior and social interactions and forcing recuperation.

Depression may be another complex immune response. Mounting evidence supports the hypothesis that cytokines are causal mechanisms of depression, even though specifics are still at issue (Reiche, Morimoto, and Nunes 2005). Pro-inflammatory cytokines instigate the behavioral, neuroendocrine, and neurochemical features of depressive disorders (Anisman and Merali 2003). The therapeutic use of pro-inflammatory cytokines INFα and IL-2 for cancer treatment produces depression (Cahalan and Gutman 2006), or more specifically, hyperactivity and dysregulation in the HPA axis, which are common features of severe depression. The sickness response and depression overlap in that many of the behavioral and sensory manifestations of sickness are also manifestations of a depressive disorder.

1.4.1 DYSREGULATION

Dysregulation is prolonged dysfunction in the ability of a system to recover its normal relationship to other systems and its usual level of operation following perturbation. This concept applies to any level of system focus, whether it is the hypothalamo-pituitary-gonadal axis or the response of an individual to a demanding social environment. An extensive literature addresses the relationships of trauma and prolonged stress with dysregulation of the HPA axis, the central noradrenergic system, and the SAM axis (Neumeister, Daher, and Charney 2005; deKloet 2004). The supersystem model proposes that multi-symptom disorders emerge and become chronic and disabling conditions as a result of regulatory problems developing over time within the supersystem. An important concept is that dysfunction arising in one subsystem is likely to lead to dysfunction in the others because they operate interdependently within the supersystem. Prolonged dysregulation can cause irreversible organ pathology that in turn can generate somatic distress. Dysregulation may manifest in at least four ways in multi-symptom disorders. These manifestations are not mutually exclusive.

1.4.2 BIORHYTHM DISTURBANCE

First, in a temporal frame of reference, dysregulation refers to deviation from or loss of normal biological rhythms: ultracadian, circadian, and infracadian. Humans secrete hormones, alter body temperature, eat, sleep, and work according to circadian rhythms; and social activity patterns reflect these rhythms. Rhythm is a fundamental feature of homeostasis, as temperature regulation demonstrates. Subsystems also operate according to rhythms. Hormones pulse at certain times, and the sinoatrial node gives the heart a rhythm. Dysregulation of temporal processes may play a role in peripheral neuropathy (Siau and Bennett 2006). The concept of cross-system rhythm is still poorly defined, but some substances participating in connectivity appear to coordinate biological rhythms at multiple system levels. The hormone melatonin is one example (Bella and Gualano 2006). Among its many effects is control of POMC gene expression (Rasmussen et al. 2003). The relationship of temporal rhythm dysregulation in multi-symptom syndrome is largely unexplored, apart from the documentation of sleep disturbances. Inquiry into multi-rhythm dysregulation at multiple system levels in multi-symptom syndrome patients could prove informative.

1.4.3 FEEDBACK DYSFUNCTION

Messenger substances play multiple roles, including feedback messaging. Subsystems like the HPA axis depend upon negative feedback to terminate recovery from stress processes. Subsystems also limit lower-level positive feedback loops that make possible emergency responses, thus protecting against overshoot. Positive feedback processes are not self-limiting by definition, and without such control they continue until either a state shift occurs or the system self-destructs. Allodynia is a familiar example of positive feedback in chronic pain, as is panic attack in emotional regulation. Within the immune system, positive and negative feedback play a central role in T cell discrimination of self from non-self ligands (Stefanova et al. 2003; Mueller 2003). This process, too, is subject to dysregulation with negative health consequences manifesting as auto-immune disorders.

The literature identifies many examples of disturbed feedback-dependent regulatory processes in stressed patients. For example, patients may develop HPA axis dysregulation (Herman et al. 2003; Dinan et al. 2006), autonomic dysregulation (Kodounis et al. 2005), peptide dysregulation (Staines 2006), Th1/Th2 cytokine dysregulation (Elenkov et al. 2005; Viveros-Paredes et al. 2006), endogenous opioid dysregulation (Ribeiro et al. 2005), and dysregulation of the relationship between pain and blood pressure (Bruehl, Burns, and McCubbin 1998). Basically, dysregulation occurs when a subsystem regulated by negative feedback breaks down in one way or another, for example, through depletion of a key neurotransmitter or peptide.

Feedback mechanisms may also falter under the opposite condition of resource excess. The medical introduction of substances that resemble biological messengers may interfere with normal allostasis and produce iatrogenic disorder. Opioid medications provide a strong example, as they resemble beta endorphin and other endogenous opioids. The hypothalamo-pituitary-gonadal axis responds to such products as though they were endogenous signals and the result is often hypogonadism (Daniell, Lentz, and Mazer 2006).

1.5   DISTURBED INTERSUBSYSTEM COORDINATION

Nervous, endocrine, and immune subsystems are interdependent and coordinate their response to a stressor. The connectivity essential for cross-subsystem coordination may falter or break down. Examples include the reciprocal relationship of cytokines with HPA axis regulation (Viveros-Paredes et al. 2006; Calcagni and Elenkov 2006; Rivest 2001; Dunn, Wang, and Ando 1999), the relationship of cytokine regulation to autonomic regulation (Czura and Tracey 2005), and the relationship of cytokine regulation to the LC response (Borsody and Weiss 2002). This is the mechanism for how dysregulation in one subsystem will tend to disrupt another, leading eventually to supersystem dysfunction.

1.5.1   INCOMPLETE STRESS RESPONSE RECOVERY

Dysregulation could occur if a system alters its set point in response to a stressor and then fails to readjust to the normal level after the stress has passed. This corresponds with McEwen's metaphor of failure to hear the all-clear signal (McEwen 2002). This explanatory model nicely describes the hypervigilance and hyperreactivity of post-traumatic stress disorder (Bedi and Arora 2007). Some multi-symptom syndrome patients have trauma histories.

Set points are often straightforward to define. For example, Vogeser and colleagues (Vogeser et al. 2003) studied major surgery as a stressor and chose the cortisol:cortisone ratio as a marker of HPA axis activity and as a stress-sensitive indicator of the overall set-point shift in the breakdown of cortisone to produce CORT, namely 11b-hydroxysteroid dehydrogenase activity. Surgery caused a shift in this set point that later returned to pre-surgical levels. Cardiac variability, MR/GR ratio, and Th1/Th2 ratio represent other potential system set-point indicators that may exhibit pathological shifts in chronic pain. The auditory startle response, which indicates excessive autonomic response activation to startling stimuli, may be a marker of past trauma (Siegelaar et al. 2006). Traumatic life events can permanently alter the set point of an individual's feedback-dependent HPA axis (deKloet et al. 2005; Bremner et al. 2003).

1.6   INDEXES OF DYSREGULATION

Psychological concepts of trait and state are useful for describing how dysregulation manifests. A *trait* is a relatively enduring predisposition to respond in certain ways when perturbed. It gauges the adaptive capability of an individual challenged by a stressor. A *state* is a transitory condition of the system, typically following perturbation. During chronic pain, dysregulation is likely to alter traits, and this alteration may manifest as abnormal state responses to perturbation. For example, a person with normal trait anxiety may undergo a traumatic event and afterward become highly anxious in response to small problems and produce abnormal startle responses. This is high state anxiety. By analogy, the trait–state distinction applies to neural, endocrine, and immune subsystems. Below are some examples of ways to quantify subsystem dysregulation. One can either quantify traits directly or infer them from challenge-induced changes in states.

1.6.1 AUTONOMIC DYSREGULATION

Cardiac variability, sometimes called vagal tone, gauges sympathetic/parasympathetic balance in the autonomic nervous system. It indexes behavioral, cognitive, and emotional function (Beauchaine 2001). Basically, cardiac variability reflects the balance of sympathetic and parasympathetic influence in autonomic function as evident in cardiac activity. The vagus nerve is bidirectional. Vagal afferent fibers from the heart project to the solitary nucleus. Efferent fibers from the brain stem terminate on the sinoatrial node, the cardiac pacemaker. Sympathetic activation accelerates heart rate, and parasympathetic activation decelerates heart rate.

Estimation of cardiac variability derives from respiratory sinus arrhythmia; that is, changes in heart rate during the respiratory cycle. During exhalation, vagal efferent activity modulates this rate and causes deceleration. Inhalation increases heart rate. Statistical indexes of instantaneous heart rate variability based on the R-to-R wave interval estimate of cardiac variability. Such estimates are stress sensitive, and some investigators postulate that early trauma may permanently diminish cardiac variability, leaving the individual less resilient to future stressors (Bracha 2004; Porges 1992). High cardiac variability, or vagal tone, may be an indirect marker of an individual's ability to respond effectively to a stressor and recover efficiently from it.

1.6.2 SENSORY DYSREGULATION

Tracey and Mantyh (2007) postulate that chronic pain patients may have dysfunction in either the facilitatory system or the inhibitory system for noxious signal modulation. One can assess these processes by looking at wind-up and diffuse noxious inhibitory control (DNIC). Wind-up, or temporal summation, occurs when a subject undergoes a series of identical noxious stimuli. Tracking the pain rating across trials reveals increased pain or sensitization. This process may be abnormal in some chronic pain populations (Staud et al. 2001). When the activation of one noxious stimulus causes a diminished response to a second noxious stimulus, DNIC exists. In the laboratory, one measures the response to a phasic stimulus at baseline, applies a tonic stimulus such as the cold pressor test, and then measures the response to the phasic stimulus again. The response to the phasic stimulus should diminish following the tonic stimulus. This is independent of segment (hence, diffuse); and not being naloxone reversible in most reports, it is probably independent of the HPA axis. DNIC is a laboratory predictor of clinical pain and quality of life (Edwards et al. 2003). Whether wind-up and DNIC are true opposing processes is uncertain but worth exploration.

1.6.3   ENDOCRINE DYSREGULATION

Potential trait measures exist for the HPA axis. DeKloet and Derijk postulated that mineralocorticoid receptor (MR)- and glucocorticoid receptor (GR)-mediated stress responses counterbalance. MR responses contribute to immediate arousal and coping, whereas GR responses attenuate emergency reactions and assist recovery from stress (deKloet and Derijk 2004). Normally, an individual possesses a characteristic MR/GR balance that is largely genetically determined.

Some approaches to diagnosing dysregulation involve challenging the HPA axis and looking for abnormal state responses to the challenges. The dexamethasone suppression test gauges HPA axis response in this way (Raison and Miller 2003). Dexamethasone is an exogenous steroid that provides negative feedback to the pituitary to suppress the secretion of ACTH. It does not cross the blood-brain barrier. Excessive CORT response to dexamethasone occurs in up to half of all severely depressed patients, indicating axis dysregulation. Alternatively, the CRH challenge involves the infusion of CRH and measurement of subsequent ACTH and cortisol responses (Dinan et al. 2006). It, too, can gauge HPA axis dysregulation.

Detection of biorhythm dysregulation necessitates examination of diurnal or other chronological variation in hormones. This typically requires multiple samples within a single day and examination of the resulting profile against a normal profile. CORT, for example, normally peaks shortly after arising, and then blood levels decline and are very low late in the day and evening. Any other pattern indicates dysregulation. In contrast, opponent process dysregulation indicators derive from a ratio of opposing processes like the Th1/Th2 ratio. For this, there are many possibilities.

In looking at the negative impact of sleep deprivation, Copinschi (2005) examined both types of dysregulation. Sleep-deprived subjects had increased cortisol levels in the late afternoon and evening. Examination of two brain-gut axis hormones related to appetite, ghrelin, and leptin also revealed dysregulation. Ghrelin increases appetite, but leptin decreases it. With sleep deprivation, the ghrelin-to-leptin ratio shifted in the direction of higher ghrelin and lower leptin; this correlated strongly with increased hunger.

1.6.4   IMMUNE DYSREGULATION

For the immune subsystem, Th1/Th2 balance has become a focus of attention in cytokine research (Kidd 2003; Elenkov 2004). The general view holds that stress is immunosuppressive, but as we have seen, stress may be beneficial at lower levels but detrimental at higher levels. Moreover, the initial arousal/inflammatory response to stress is different from the subsequent recovery response. Much uncertainty still exists, but it is becoming clear that glucocorticoids and catecholamines support inflammation locally in certain conditions; that is, they promote Th1 cytokine production. And yet, systemically these substances potentiate Th2 production while inhibiting Th1 production, thereby exerting an anti-inflammatory effect (Calcagni and Elenkov 2006). Because cytokine activity depends heavily upon stress hormones, such localized targeting of pro-inflammatory processes could be advantageous in promoting increased blood flow and cell trafficking to injured tissue. Th1/Th2 balance varies with the stress response. Regardless of whether that response is hyperactive or hypoactive, it may alter the course of immune-related disease. The Th1/Th2 ratio is skewed in several common diseases (Elenkov 2004) and is a useful parameter from a psychosomatic perspective. For example, Glaser and colleagues (2001) examined Th1/Th2 balance in chronically stressed caregivers of demented patients and found a shift in the Th2 direction, suggesting vulnerability to infection.

1.7   STRESS, DYSFUNCTION, AND CHRONIC DISORDERS

I asserted above that the individual patient is a system, and every system exists within a larger, encapsulating system that influences it. The psychosocial system surrounding the individual patient is a potential source of stressors that demands allostatic response above and beyond that elicited by injury. The biopsychosocial interactions of the individual with his/her environment and various psychosocial factors generate allostatic load. In the presence of psychosocial stressors, biological acute stress responses can fail to resolve properly, leading to chronic disorders. This can happen in three ways.

1.7.1   Failed Arousal-to-Recovery Transition

Clinicians managing chronic multi-symptom disorders sometimes see pain patients who report surviving a horrific accident or event that left them traumatized. A single trauma of sufficient magnitude can produce a stress response that does not resolve properly. McEwen (2002) described other allostatic load scenarios that might lead to system malfunction: (1) unremitting or chronic stressors, (2) inability to adjust to a stressor of modest duration and demand, and (3) not hearing the "all clear" in which the stress response persists after the stressor has disappeared. These concepts, collectively, describe an arousal or fast-response phase that fails to give way to a recovery or slow-response phase. When the arousal-to-recovery process does not progress to completion, the patient is likely to suffer some form of system dysregulation with corresponding organ dysfunction and unremitting, disordered chronobiology.

1.7.2   Dysfunctional Recovery

In some cases recovery may occur but fail to return the supersystem to an approximate balance. For example, the recovery process in the HPA axis can go wrong in two ways. CORT insufficiency and CORT excess are both damaging (deKloet 2004). Too little CORT means prolonged anabolism. Moreover, positive feedback arousal processes can go unchecked, and conversion to a proper recovery state may not occur. Conversely, too much CORT over time has negative catabolic consequences. Hypercortisolism is a well-known marker of severe depression. In both cases, loss of normal diurnal variation in CORT pulsing indicates dysregulation. Thus, a dysfunctional endocrine recovery process is a mechanism for chronic endocrine dysregulation with a related constellation of chronic symptoms.

The boundaries of endocrine dysregulation extend to the immune subsystem. GCs profoundly affect cytokine responses. Evidence indicates that GCs inhibit Th1 cytokine production while at the same time promoting Th2 cytokine production (Elenkov 2004). This is another form of protection against overshoot of positive-feedback-driven arousal responses (Elenkov and Chrousos 2002).

As noted above, many writers (Kidd 2003) characterize the immune system as operating in either Th1 dominant (pro-inflammatory) or Th2 dominant (anti-inflammatory). These modes roughly parallel the stress-response arousal and recovery phases. This is more than a parallel concept. Evidence indicates that pro-inflammatory cytokines activate the HPA axis (Besedovsky and del Rey 2000) and thereby elicit MR and GC responses, whereas inhibitory peptides support Th2 processes (Ganea, Rodriguez, and Delgado 2003; Delgado et al. 2003). In this way, chronic endocrine dysregulation may contribute to immune dysregulation and incur the related symptom burden.

1.7.3 DYSFUNCTIONAL SUBSYSTEM INTERFACE

The interface between subsystems can become chronically dysfunctional, impairing intersystem coordination. For example, Calcagni and Elenkov, in reviewing both endocrine and immune system response patterns during stress, raised the possibility of dysregulation in the neuroendocrine–immune interface (Calcagni and Elenkov 2006). Weber identified the same potential source of disease (Weber 2003). By extension, potential chronic dysfunction could occur in the nervous–immune interface or the nervous–endocrine interface as causal mechanisms for chronic multi-symptom syndromes.

REFERENCES

Anisman, H., and Z. Merali. 2003. Cytokines, stress and depressive illness: brain-immune interactions. *Ann Med* 35 (1):2–11.

Aston-Jones, G., and J. D. Cohen. 2005. Adaptive gain and the role of the locus coeruleus-norepinephrine system in optimal performance. *J Comp Neurol* 493 (1):99–110.

Beauchaine, T. 2001. Vagal tone, development, and Gray's motivational theory: toward an integrated model of autonomic nervous system functioning in psychopathology. *Dev Psychopathol* 13 (2):183–214.

Bedi, U. S., and R. Arora. 2007. Cardiovascular manifestations of posttraumatic stress disorder. *J Natl Med Assoc* 99 (6):642–49.

Bella, L. D., and L. Gualano. 2006. Key aspects of melatonin physiology: thirty years of research. *Neuro Endocrinol Lett* 27 (4).

Berridge, C. W., and B. D. Waterhouse. 2003. The locus coeruleus-noradrenergic system: modulation of behavioral state and state-dependent cognitive processes. *Brain Res Brain Res Rev* 42 (1):33–84.

Besedovsky, H. O., and A. del Rey. 2000. The cytokine-HPA axis feed-back circuit. *Z Rheumatol* 59 Suppl 2:II/26–30.

Bliesener, N., S. Albrecht, A. Schwager, K. Weckbecker, D. Lichtermann, and D. Klingmuller. 2005. Plasma testosterone and sexual function in men receiving buprenorphine maintenance for opioid dependence. *J Clin Endocrinol Metab* 90 (1):203–6.

Borsody, M. K., and J. M. Weiss. 2002. Alteration of locus coeruleus neuronal activity by interleukin-1 and the involvement of endogenous corticotropin-releasing hormone. *Neuroimmunomodulation* 10 (2):101–21.

Bracha, H. S. 2004. Can premorbid episodes of diminished vagal tone be detected via histological markers in patients with PTSD? *Int J Psychophysiol* 51 (2):127–33.

Brandman, O., J. E. Ferrell, Jr., R. Li, and T. Meyer. 2005. Interlinked fast and slow positive feedback loops drive reliable cell decisions. *Science* 310 (5747):496–98.

Bremner, J. D., M. Vythilingam, G. Anderson, E. Vermetten, T. McGlashan, G. Heninger, A. Rasmusson, S. M. Southwick, and D. S. Charney. 2003. Assessment of the hypothalamic-pituitary-adrenal axis over a 24-hour diurnal period and in response to neuroendocrine challenges in women with and without childhood sexual abuse and posttraumatic stress disorder. *Biol Psychiatry* 54 (7):710–718.

Bruehl, S., J. W. Burns, and J. A. McCubbin. 1998. Altered cardiovascular/pain regulatory relationships in chronic pain. *Int J Behav Med* 5 (1):63–75.

Cahalan, M. D., and G. A. Gutman. 2006. The sense of place in the immune system. *Nat Immunol* 7 (4):329–32.

Calcagni, E., and I. Elenkov. 2006. Stress system activity, innate and T helper cytokines, and susceptibility to immune-related diseases. *Ann N Y Acad Sci* 1069:62–76.

Chapman, C. R., R. P. Tuckett, and C. W. Song. 2008. Pain and stress in a systems perspective: reciprocal neural, endocrine, and immune interactions. *J Pain* 9 (2):122–45.

Copinschi, G. 2005. Metabolic and endocrine effects of sleep deprivation. *Essent Psychopharmacol* 6 (6):341–47.

Czura, C. J., and K. J. Tracey. 2005. Autonomic neural regulation of immunity. *J Intern Med* 257 (2):156–66.

Daniell, H. W. 2002. Hypogonadism in men consuming sustained-action oral opioids. *J Pain* 3 (5):377–84.

Daniell, H. W., R. Lentz, and N. A. Mazer. 2006. Open-label pilot study of testosterone patch therapy in men with opioid-induced androgen deficiency. *J Pain* 7 (3):200–210.

deKloet, C. S., E. Vermetten, E. Geuze, A. Kavelaars, C. J. Heijnen, and H. G. Westenberg. 2005. Assessment of HPA-axis function in posttraumatic stress disorder: pharmacological and non-pharmacological challenge tests, a review. *J Psychiatr Res* 40 (6):550–67.

deKloet, E. R. 2004. Hormones and the stressed brain. *Ann N Y Acad Sci* 1018:1–15.

deKloet, E.R., and R. Derijk. 2004. Signaling pathways in brain involved in predisposition and pathogenesis of stress-related disease: genetic and kinetic factors affecting the MR/GR balance. *Ann N Y Acad Sci* 1032:14–34.

Delgado, M., C. Abad, C. Martinez, M. G. Juarranz, J. Leceta, D. Ganea, and R. P. Gomariz. 2003. PACAP in immunity and inflammation. *Ann N Y Acad Sci* 992:141–57.

Dinan, T. G., E. M. Quigley, S. M. Ahmed, P. Scully, S. O'Brien, L. O'Mahony, S. O'Mahony, F. Shanahan, and P. W. Keeling. 2006. Hypothalamic-pituitary-gut axis dysregulation in irritable bowel syndrome: plasma cytokines as a potential biomarker? *Gastroenterology* 130 (2):304–11.

Dunn, A. J., J. Wang, and T. Ando. 1999. Effects of cytokines on cerebral neurotransmission. Comparison with the effects of stress. *Adv Exp Med Biol* 461:117–27.

Edwards, R. R., T. J. Ness, D. A. Weigent, and R. B. Fillingim. 2003. Individual differences in diffuse noxious inhibitory controls (DNIC): association with clinical variables. *Pain* 106 (3):427–37.

Elenkov, I. J. 2004. Glucocorticoids and the Th1/Th2 balance. *Ann N Y Acad Sci* 1024:138–46.

Elenkov, I. J., and G. P. Chrousos. 2002. Stress hormones, proinflammatory and antiinflammatory cytokines, and autoimmunity. *Ann N Y Acad Sci* 966:290–303.

Elenkov, I. J., D. G. Iezzoni, A. Daly, A. G. Harris, and G. P. Chrousos. 2005. Cytokine dysregulation, inflammation and well-being. *Neuroimmunomodulation* 12 (5):255–69.

Elenkov, I. J., R. L. Wilder, G. P. Chrousos, and E. S. Vizi. 2000. The sympathetic nerve—an integrative interface between two supersystems: the brain and the immune system. *Pharmacol Rev* 52 (4):595–638.

Eskandari, F., J. I. Webster, and E. M. Sternberg. 2003. Neural immune pathways and their connection to inflammatory diseases. *Arthritis Res Ther* 5 (6):251–65.

Ferrell, J. E., Jr. 2002. Self-perpetuating states in signal transduction: positive feedback, double-negative feedback and bistability. *Curr Opin Cell Biol* 14 (2):140–48.

Flood, R. L., and E. R. Carson. 1993. *Dealing with complexity: an introduction to the theory and application of systems science.* Second ed. New York: Plenum Press.

Ganea, D., R. Rodriguez, and M. Delgado. 2003. Vasoactive intestinal peptide and pituitary adenylate cyclase-activating polypeptide: players in innate and adaptive immunity. *Cell Mol Biol (Noisy-le-grand)* 49 (2):127–42.

Gell-Mann, M. 1994. *The quark and the jaguar: adventures in the simple and the complex.* London: Little, Brown and Company.

Glaser, R., R. C. MacCallum, B. F. Laskowski, W. B. Malarkey, J. F. Sheridan, and J. K. Kiecolt-Glaser. 2001. Evidence for a shift in the Th-1 to Th-2 cytokine response associated with chronic stress and aging. *J Gerontol A Biol Sci Med Sci* 56 (8):M477–82.

Goetzl, E. J., and S. P. Sreedharan. 1992. Mediators of communication and adaptation in the neuroendocrine and immune systems. *Faseb J* 6 (9):2646–52.

Gosain, A., and R. L. Garnelli. 2005. A primer in cytokines. *J Burn Care Rehabil* 26 (1):7–12.

Grimm, V., E. Revilla, U. Berger, F. Jeltsch, W. M. Mooij, S. F. Railsback, H. H. Thulke, J. Weiner, T. Wiegand, and D. L. DeAngelis. 2005. Pattern-oriented modeling of agent-based complex systems: lessons from ecology. *Science* 310 (5750):987–91.

Gruys, E., M.J.M. Toussaint, T. A. Niewold, and S. J. Koopmans. 2005. Acute phase reaction and acute phase proteins. *J Zhejiang Univ SCI* 6B (11):1045–56.

Herman, J. P., H. Figueiredo, N. K. Mueller, Y. Ulrich-Lai, M. M. Ostrander, D. C. Choi, and W. E. Cullinan. 2003. Central mechanisms of stress integration: hierarchical circuitry controlling hypothalamo-pituitary-adrenocortical responsiveness. *Front Neuroendocrinol* 24 (3):151–80.

Inoue, A., K. Ikoma, N. Morioka, K. Kumagai, T. Hashimoto, I. Hide, and Y. Nakata. 1999. Interleukin-1 beta induces substance P release from primary afferent neurons through the cyclooxygenase-2 system. *J Neurochem* 73 (5):2206–13.

Jedema, H. P., and A. A. Grace. 2004. Corticotropin-releasing hormone directly activates noradrenergic neurons of the locus ceruleus recorded in vitro. *J Neurosci* 24 (43):9703–13.

Jones, R. W. 1973. *Principles of biological regulation: an introduction to feedback systems.* New York: Academic Press.

Kaneko, K. 2006. *Life: an introduction to complex systems biology, understanding complex systems.* New York: Springer.

Kelso, J. A. S. 1998. *Dynamic patterns: the self-organization of brain and behavior (complex adaptive systems).* Cambridge: MIT Press.

Kidd, P. 2003. Th1/Th2 balance: the hypothesis, its limitations, and implications for health and disease. *Altern Med Rev* 8 (3):223–46.

Kin, N. W., and V. M. Sanders. 2006. It takes nerve to tell T and B cells what to do. *J Leukoc Biol* 79 (6):1093–1104.

Kodounis, A., E. Stamboulis, T. S. Constantinidis, and A. Liolios. 2005. Measurement of autonomic dysregulation in multiple sclerosis. *Acta Neurol Scand* 112 (6):403–8.

Kohl, J. 2006. The role of complement in danger sensing and transmission. *Immunol Res* 34 (2):157–76.

Korte, S. M., J. M. Koolhaas, J. C. Wingfield, and B. S. McEwen. 2005. The Darwinian concept of stress: benefits of allostasis and costs of allostatic load and the trade-offs in health and disease. *Neurosci Biobehav Rev* 29 (1):3–38.

Lambrecht, B. N. 2001. Immunologists getting nervous: neuropeptides, dendritic cells and T cell activation. *Respir Res* 2 (3):133–8.

Leonard, B. E. 2005. The HPA and immune axes in stress: the involvement of the serotonergic system. *Eur Psychiatry* 20 Suppl 3:S302–6.

McEwen, B. S. 2000. Allostasis and allostatic load: implications for neuropsychopharmacology. *Neuropsychopharmacology* 22 (2):108–24.

———. 2002. *The end of stress as we know it*. Washington, D.C.: Joseph Henry Press.

Menger, M. D., and B. Vollmar. 2004. Surgical trauma: hyperinflammation versus immunosuppression? *Langenbecks Arch Surg* 389 (6):475–84.

Miller, D. B., and J. P. O'Callaghan. 2002. Neuroendocrine aspects of the response to stress. *Metabolism* 51 (6 Suppl 1):5–10.

Mueller, D. L. 2003. Tuning the immune system: competing positive and negative feedback loops. *Nat Immunol* 4 (3):210–11.

Neumeister, A., R. J. Daher, and D. S. Charney. 2005. Anxiety disorders: noradrenergic neurotransmission. *Handb Exp Pharmacol* 2005 (169):205–23.

Northrop, R. B. 2000. Endogenous and exogenous regulation and control of physiological systems. In *Biomedical engineering*, edited by M. Neuman. Boca Raton: Chapman & Hall/CRC.

Padgett, D. A., and R. Glaser. 2003. How stress influences the immune response. *Trends Immunol* 24 (8):444–48.

Parkin, J., and B. Cohen. 2001. An overview of the immune system. *Lancet* 357 (9270):1777–89.

Porges, S. W. 1992. Vagal tone: a physiologic marker of stress vulnerability. *Pediatrics* 90 (3 Pt 2):498–504.

Prigogne, I., and E. Stengers. 1984. *Order out of chaos*. New York: Bantam Books.

Raison, C. L., and A. H. Miller. 2003. When not enough is too much: the role of insufficient glucocorticoid signaling in the pathophysiology of stress-related disorders. *Am J Psychiatry* 160 (9):1554–65.

Rasmussen, D. D., B. T. Marck, B. M. Boldt, S. M. Yellon, and A. M. Matsumoto. 2003. Suppression of hypothalamic pro-opiomelanocortin (POMC) gene expression by daily melatonin supplementation in aging rats. *J Pineal Res* 34 (2):127–33.

Rassnick, S., A. F. Sved, and B. S. Rabin. 1994. Locus coeruleus stimulation by corticotropin-releasing hormone suppresses in vitro cellular immune responses. *J Neurosci* 14 (10):6033–40.

Reiche, E. M., H. K. Morimoto, and S. M. Nunes. 2005. Stress and depression-induced immune dysfunction: implications for the development and progression of cancer. *Int Rev Psychiatry* 17 (6):515–27.

Ribeiro, S. C., S. E. Kennedy, Y. R. Smith, C. S. Stohler, and J. K. Zubieta. 2005. Interface of physical and emotional stress regulation through the endogenous opioid system and mu-opioid receptors. *Prog Neuropsychopharmacol Biol Psychiatry* 29 (8):1264–80.

Rivest, S. 2001. How circulating cytokines trigger the neural circuits that control the hypothalamic-pituitary-adrenal axis. *Psychoneuroendocrinology* 26 (8):761–88.

Romeo, H. E., D. L. Tio, S. U. Rahman, F. Chiappelli, and A. N. Taylor. 2001. The glossopharyngeal nerve as a novel pathway in immune-to-brain communication: relevance to neuroimmune surveillance of the oral cavity. *J Neuroimmunol* 115 (1–2):91–100.

Selye, H. 1936. A syndrome produced by diverse nocuous agents. *Nature (London)* 138:32.

Shvartsman, S. Y., M. P. Hagan, A. Yacoub, P. Dent, H. S. Wiley, and D. A. Lauffenburger. 2002. Autocrine loops with positive feedback enable context-dependent cell signaling. *Am J Physiol Cell Physiol* 282 (3):C545–59.

Siau, C., and G. J. Bennett. 2006. Dysregulation of cellular calcium homeostasis in chemotherapy-evoked painful peripheral neuropathy. *Anesth Analg* 102 (5):1485–90.

Siegelaar, S. E., M. Olff, L. J. Bour, D. Veelo, A. H. Zwinderman, G. van Bruggen, G. J. de Vries, S. Raabe, C. Cupido, J. H. Koelman, and M. A. Tijssen. 2006. The auditory startle response in post-traumatic stress disorder. *Exp Brain Res* 174 (1):1–6.

Staines, D. R. 2006. Postulated vasoactive neuropeptide autoimmunity in fatigue-related conditions: a brief review and hypothesis. *Clin Dev Immunol* 13 (1):25–39.

Staud, R., C. J. Vierck, R. L. Cannon, A. P. Mauderli, and D. D. Price. 2001. Abnormal sensitization and temporal summation of second pain (wind-up) in patients with fibromyalgia syndrome. *Pain* 91 (1–2):165–75.

Stefanova, I., J. R. Dorfman, M. Tsukamoto, and R. N. Germain. 2003. On the role of self-recognition in T cell responses to foreign antigen. *Immunol Rev* 191:97–106.

Steinman, L. 2004. Elaborate interactions between the immune and nervous systems. *Nat Immunol* 5 (6):575–81.

Stone, E. A. 1975. Stress and catecholamines. In *Catecholamines and behavior*, edited by A. J. Friedhoff. New York: Plenum Press.

Svensson, T. H. 1987. Peripheral, autonomic regulation of locus coeruleus noradrenergic neurons in brain: putative implications for psychiatry and psychopharmacology. *Psychopharmacology* 92:1–7.

Thomas, R., and R. D'Ari. 1990. *Biological feedback*. Boca Raton: CRC Press, Inc.

Tracey, I., and P. W. Mantyh. 2007. The cerebral signature for pain perception and its modulation. *Neuron* 55 (3):377–91.

Tracey, K. J. 2002. The inflammatory reflex. *Nature* 420 (6917):853–59.

Tsigos, C., and G. P. Chrousos. 2002. Hypothalamic-pituitary-adrenal axis, neuroendocrine factors and stress. *J Psychosom Res* 53 (4):865–71.

Uexküll von, J. 1973, 1928. *Theoretische biologic*. Frankfurt: Suhrkamp.

Viveros-Paredes, J. M., A. M. Puebla-Perez, O. Gutierrez-Coronado, L. Sandoval-Ramirez, and M. M. Villasenor-Garcia. 2006. Dysregulation of the Th1/Th2 cytokine profile is associated with immunosuppression induced by hypothalamic-pituitary-adrenal axis activation in mice. *Int Immunopharmacol* 6 (5):774–81.

Vizi, E. S., and I. J. Elenkov. 2002. Nonsynaptic noradrenaline release in neuro-immune responses. *Acta Biol Hung* 53 (1–2):229–44.

Vogeser, M., J. Groetzner, C. Kupper, and J. Briegel. 2003. The serum cortisol:cortisone ratio in the postoperative acute-phase response. *Horm Res* 59 (6):293–96.

Warren, J. W., F. M. Howard, R. K. Cross, J. L. Good, M. M. Weissman, U. Wesselmann, P. Langenberg, P. Greenberg, and D. J. Clauw. 2009. Antecedent nonbladder syndromes in case-control study of interstitial cystitis/painful bladder syndrome. *Urology* 73 (1):52–57.

Watkins, L. R., and S. F. Maier. 1999. Implications of immune-to-brain communication for sickness and pain. *Proc Natl Acad Sci U S A* 96 (14):7710–13.

Watkins L. R. and S. F. Maier. 2005. Immune regulation of central nervous system functions: from sickness responses to pathological pain. *J Intern Med* 257 (2):139–55.

Weber, K. T. 2003. A neuroendocrine-immune interface. The immunostimulatory state of aldosteronism. *Herz* 28 (8):692–701.

Wieseler-Frank, J., S. F. Maier, and L. R. Watkins. 2005a. Central proinflammatory cytokines and pain enhancement. *Neurosignals* 14 (4):166–74.

Wieseler-Frank, J., S. F. Maier, and L. R. Watkins. 2005b. Immune-to-brain communication dynamically modulates pain: physiological and pathological consequences. *Brain Behav Immun* 19 (2):104–11.

Zukowska, Z., J. Pons, E. W. Lee, and L. Li. 2003. Neuropeptide Y: a new mediator linking sympathetic nerves, blood vessels and immune system? *Can J Physiol Pharmacol* 81 (2):89–94.

2 Neurotrophic Factors and Nociceptor Sensitization

Michael P. Jankowski and H. Richard Koerber

CONTENTS

Chronic pain affects the lives of millions of people, and its treatment remains one of the most challenging problems faced by clinicians who can offer their patients few if any effective means of relief devoid of serious side effects. Chronic pain commonly arises following injury to the peripheral nervous system, and this is termed *neuropathic pain*. A major problem in developing effective treatments for chronic neuropathic pain lies in the translation of basic science research using animal models to the clinic. The consensus arising from both clinical and preclinical data is that peripheral neuropathic pain reflects aberrant activity in subsets of primary afferent neurons. However, there is little agreement on which subpopulations are responsible for each and/or all aspects of neuropathic pain, which afferents are necessary for the initiation of neuropathic pain, and whether the same afferents continue to play the same role over time. The unraveling of neuropathic pain requires the development of new approaches that allow investigators to selectively identify and modulate activity in specific subsets of sensory neurons believed to be involved in this process. Here we will discuss recent developments in our understanding of possible molecular mechanisms involved in modulation of primary sensory neuron function and new experimental methods for investigating unique subsets. In addition, we will discuss similarities in sensory neurons across species and the parallel changes in function observed in animal models and human pain disorders.

Primary afferent neurons have their cell bodies housed in the dorsal root ganglia (DRG) or trigeminal ganglia (TG) and convey somatosensory information from the periphery to the central nervous system (CNS). Specific subpopulations of afferents convey various modalities of sensory information such as mechanical, thermal, and chemical sensation. In general, afferents can be divided into cell body diameter and axon caliber, which correlates with myelin thickness (or absence of myelination) and conduction velocity. The large diameter and fastest conducting Aα- and Aβ-fibers, which immunocytochemically label for neurofilaments (e.g., NF150 and NF200), mostly respond to relatively innocuous stimuli, although there are some Aβ fibers that respond to noxious stimulation (rev. in Todd and Koerber). Examples of these fibers include rapidly and slowly adapting low-threshold mechanoreceptors. Medium diameter, mid-range conducting Aδ-fibers, some of which are immunoreactive for neurofilaments, can respond to both innocuous and noxious stimulation. Aδ-fibers respond to high- or low threshold mechanical stimulation and in some cases respond to thermal stimuli. Small-diameter, neurofilament-negative, slowly conducting C-fibers detect both noxious and innocuous stimuli and comprise additional subpopulations, with the greatest complexity within each category found in fibers that encode noxious stimulus intensities, generally referred to as nociceptors. The majority of nociceptive afferents respond to multiple types of stimuli (i.e., polymodal nociceptors), while others only respond to a single stimulus modality (Figure 2.1). In addition to subdivision by modality, nociceptors can be further differentiated based on target of innervation (cutaneous, muscle, viscera, etc.) and histological and biochemical properties. For example, some unmyelinated C-fiber nociceptors contain peptides, such as Substance P (SP) and calcitonin gene-related peptide (CGRP), while others are non-peptidergic but bind isolectin B4 (IB4; Bennett et al. 1996; Molliver et al. 1997). Myelinated nociceptors can also be divided into groups based on presence or lack of peptides. However, depending on the species, specific labeling patterns can vary. For example, in rats, there is a large overlap between the IB4 population and those that label for heat-transducing channel transient receptor vanilloid type 1 (TRPV1); however, in mouse, there is relatively little overlap between these markers.

Recent studies have made progress on the relationship between function and neurochemical phenotype (Figure 2.1). Using an *ex vivo* skin-nerve-DRG-spinal cord preparation, single DRG neurons have been recorded intracellularly and peripheral response characteristics determined. After characterization of a sensory neuron's peripheral response properties, the DRG cell body was filled with neurobiotin and subsequently processed immunocytochemically to determine its neurochemical identity. The results of these studies have provided new information on the relationship between functional and neurochemical phenotype. In uninjured (naïve) mice, polymodal C-fibers (CPMs) are mostly IB4+ and occasionally CGRP+, but lack TRPV1, which is known to overlap extensively with the peptidergic population of DRG neurons (Funakoshi et al. 2006; Price and Flores 2007). Mechanically insensitive C-fibers that respond to heating of the skin (CH) are consistently IB4 negative and TRPV1 positive (Woodbury et al. 2004; Lawson et al. 2008). A-fiber nociceptors do not fit into either of these categories exclusively, but most were positive for putative mechano-sensing channel, acid sensing ion channel 3 (ASIC3; McIlwrath et al. 2007).

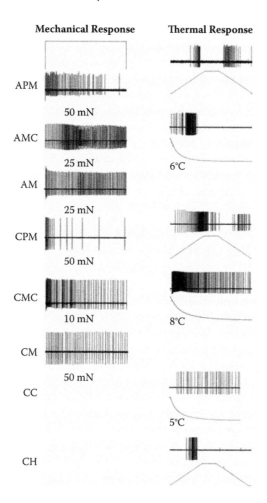

FIGURE 2.1 Response properties of cutaneous A- and C-fibers in mouse. Both A- and C-fibers in mouse can be polymodal or responsive to a single stimulus modality. A-fibers are mostly mechanically sensitive (AM) with some of these being responsive to both mechanical and thermal stimulation (APM and AMC) of the skin. C-fibers are more diverse in that they can be polymodal (CPM; mechanical and thermal responsiveness), mechanically sensitive and thermally insensitive (CM), mechanically sensitive and responsive to cooling of the skin (CMC), mechanically insensitive but cooling sensitive (CC), or mechanically insensitive but heat sensitive (CH). Traces represent the response of a particular cell to a square wave mechanical stimulation at a determined force or ramped thermal stimulation. Absence of a trace for some cells (CC and CH) denotes a non-response to that particular modality. APM: A-polymodal; AMC: A-mechano-cold/cooling; AM: A-mechanical; CPM: C-polymodal; CMC: C-mechano-cold/cooling; CM: C-mechanical; CC: C-cold/cooling; CH: C-heat.

During development, all sensory neurons are initially dependent on neurotrophic factor signaling from the periphery for survival. After migration from the neural crest, almost all neurons express the tyrosine kinase receptors trkB and trkC and are dependent on their ligands, neurotrophic factors brain-derived neurotrophic factor

(BDNF) and/or neurotrophin-3 (NT-3). In subsequent days, DRG neurons undergo extensive proliferation and some become trkA positive, the receptor for another neurotrophic factor, nerve growth factor (NGF). Following this period of proliferation is a phase of rapid cell death, and only cells that have access to sufficient amounts of a particular neurotrophic factor survive into adulthood.

Neurotrophic factor signaling from the periphery to DRG neurons is also thought to play a role in the maintenance of phenotype (Diamond et al. 1992), but under conditions of inflammation or nerve injury, neurotrophic factors play a role in sensory neuron sensitization and cause pain. The discussion below details the changes in neurotrophic factor expression and how they modify primary afferent response properties leading to conditions of acute and/or chronic pain.

2.1 NEUROTROPHIC FACTORS CAN SENSITIZE SPECIFIC POPULATIONS OF SENSORY NEURONS

2.1.1 Nerve Growth Factor (NGF)

NGF is the most commonly studied growth factor in relation to nociceptor sensitization and serves to promote the survival of DRG neurons during development that express its receptor, trkA (Averill et al. 1995; Huang et al. 2001; Patapoutian and Reichardt 2001). These neurons are generally part of the small and medium diameter DRG population, but some larger cells also express trkA (Wright and Snider 1995; Patapoutian and Reichardt 2001). In addition to its role in development and neuronal survival, it promotes sprouting and regulates innervation density of NGF-responsive neurons in peripheral targets in early post-natal and adult life. For example, it has been shown that ligation of a peripheral nerve induces NGF expression in its target area and these elevated levels are associated with sprouting of adjacent, non-injured afferents into the denervated region (Pertens et al. 1999). Other studies analyzing constitutive overexpression of NGF in the skin (NGF-OEs) report enhanced innervation of the epidermis by both sensory and sympathetic neurons (Albers et al. 1994; Davis et al. 1994, 1996; Goodness et al. 1997).

Although NGF appears to be necessary and beneficial for development and maintenance of the peripheral sensory neuron system (Diamond et al. 1992), it has also been shown to participate in the development of thermal and mechanical hyperalgesia (i.e., increased pain in response to normally painful stimuli; Malin et al. 2006; Pertens et al. 1999; Andreev et al. 1995; Lewin et al. 1993) and pain in disorders such as bone cancer and interstitial cystitis (Lowe et al. 1997; Sevcik et al. 2005). Rats chronically treated with NGF are hypersensitive to both mechanical and radiant heat stimulation (Lewin et al. 1993; Andreev et al. 1995; Pertens et al. 1999) in a dose-dependent fashion, and injection of NGF directly into the paw of mice induces a decrease in the paw withdrawal latency to radiant heat (Malin et al. 2006). This NGF sensitization is partially dependent on sympathetic neurons, as sympathectomy partly reduces the effect of NGF in causing hyperalgesia (Andreev et al. 1995). NGF also acts indirectly by activating mast cells and neutrophils, which in turn release additional inflammatory mediators causing hypersensitivity (Lewin et al. 1994; Andreev et al. 1995; Amann et al. 1996; Woolf et al. 1996; Bennett et al. 1998;

Bennett 2001). Regardless, it is clear that NGF levels in the target tissue participate in sensitization of nociceptors. For example, NGF-OEs display increases in afferent responses to thermal and mechanical stimulation in a skin–nerve preparation. Stucky and Lewin (1999) found that large diameter Aβ non-nociceptive afferents (typically trkA negative) were unaffected by NGF overexpression, but thermal responsiveness was significantly increased in nociceptive afferents as a result of enhanced cutaneous NGF levels.

NGF-sensitive, trkA positive neurons co-label with a variety of other molecules thought to be involved in pain processing. trkA overlaps with neurons containing peptides CGRP and SP (Averill et al. 1995; Molliver and Snider 1997), known mediators of pain behaviors (Koltzenburg et al. 1999; Reeh and Kress 2001; Li et al. 2008) shown to induce hyperalgesia (Oku et al. 1987; Nakamura-Craig and Gill 1991; McMahon, 1996; Sann and Pierau 1998). This population also co-labels with TRPV1, crucial for the development of heat hyperalgesia (Caterina et al. 2000).

NGF-induced hyperalgesia may also be mediated by sodium channel, Nav1.8. In mice lacking this channel, NGF does not induce heat hyperalgesia (Kerr et al. 2001), although Nav1.8 knockout mice display indistinguishable thermal thresholds under normal conditions compared to wildtypes (WTs). Since many NGF-responsive neurons contain TRPV1, this channel is suspected of a role in NGF-mediated hypersensitivity (Caterina et al. 1997; Tominaga et al. 1998; Michael and Priestley 1999). Cultured DRG neurons treated with NGF display enhanced inward current in response to application of the TRPV1 agonist capsaicin (Shu and Mendell 1999; Caterina et al. 2000; Zhu et al. 2004). NGF can increase TRPV1 expression (Donnerer et al. 2005; Xue et al. 2007) and promote TRPV1 insertion into the plasma membrane (Zhang et al. 2005). Furthermore, anti-NGF antibodies injected into the hindpaw after peripheral inflammation decrease levels of TRPV1 in DRGs and reduce inflammation-induced hyperalgesia (Ji et al. 2002; Cheng and Ji 2008).

Given a clear role for NGF in sensory neuron sensitization and hyperalgesia, anti-NGF treatments may constitute an effective means of treating pain in humans (Anand et al. 1997; Lowe et al. 1997; Saldanha et al. 1999; Sena et al. 2006; Jimenez-Andrade et al. 2007). These hypotheses, however, have not been extensively studied (Abdiche et al. 2008) or verified. Perhaps NGF may only affect a small proportion of nociceptors in the DRG, and other molecules and neurotrophic factors most certainly are involved in hyperalgesia and overall sensory neuron sensitization.

2.1.2 NEUROTROPHIN-3 (NT-3)

NT-3 has long been studied in association with survival and sprouting of neurons that express its receptor, trkC. trkC is mainly localized to large diameter DRG neurons although also found in some small and medium cells (Wright and Snider 1995; Molliver et al. 2005b). NT-3 plays a role in sympathetic neurons survival (Elshamy and Ernfors 1996), and increased levels of NT-3 may foster sympathetic neuron sprouting into the DRG (Zhou et al. 1999; Deng et al. 2000). Intrathecal NT-3 elicits selective regeneration of injured axons of NF200 positive DRG neurons through the dorsal root entry zone (DREZ) of the spinal cord (Ramer et al. 2000). This selective regeneration of NF200 neurons also was correlated with selective functional recovery of large myelinated A-fibers, suggesting that NT-3, like NGF, may play a dual role in regulating processes in both sympathetic and sensory neurons.

NT-3 may also play an anti-nociceptive role in neuropathic pain processing (White 2000; Park et al. 2003; Wilson-Gerwing et al. 2005; Wilson-Gerwing and Verge 2006). Delivery of NT-3 to nerve-injured rats in a rat model of neuropathic pain caused a decreased paw withdrawal threshold after thermal, but not mechanical, stimulation (Wilson-Gerwing et al. 2005). This correlates with other studies showing that under conditions of neuropathic pain, NT-3 reduces injury-induced expression of putative pain-related channels localized to small diameter nociceptors such as Nav1.8, Nav1.9 (Wilson-Gerwing et al. 2008) and TRPV1 (Wilson-Gerwing et al. 2005), although the mechanism of this action is unclear since the majority of small diameter nociceptors lack trkC. However, it has also been shown that NT-3 has low-affinity trkA binding (Kullander and Ebendal 1994; Ryden and Ibanez 1996); thus it is possible that under conditions of inflammation and nerve injury, NT-3 may have a greater affect on these cells compared to naïve conditions.

Although NT-3 may be anti-nociceptive, NT-3 may also play a pro-nociceptive role in conditions such as spondyloarthritis synovitus (Rihl et al. 2005) and colitis (Flamig et al. 2001). In support of this, another recent study has shown that cutaneous overexpression of NT-3 enhances the mechanical response properties of larger diameter non-nociceptive and nociceptive A-fibers (McIlwrath et al. 2007). Slowly adapting type 1 (SA1) and Aδ-nociceptors display significantly higher firing rates to mechanical stimulation in an *ex vivo* skin-nerve-DRG-spinal cord preparation compared to WTs (Woodbury et al. 2004; McIlwrath et al. 2007). These A-fibers are immunopositive for trkC and ASIC3, and the overexpression of NT-3 is correlated with increased ASIC1 and ASIC3 mRNA expression in DRGs. Although ASIC3 is co-expressed in both trkA and trkC neurons and overlaps with some peptidergic sensory neurons in the DRGs (Molliver et al. 2005b), these new results suggest an additional role for NT-3 in pain processing and sensory neuron sensitization that may not be anti-nociceptive and involves the myelinated DRG population.

2.1.3 GLIAL CELL LINE-DERIVED NEUROTROPHIC FACTOR

Another family of growth factors that may play a role in sensory neuron sensitivity is the glial cell line–derived neurotrophic factor (GDNF) family which includes GDNF, neurturin, and artemin. GDNF family ligands signal through a different tyrosine kinase receptor (ret), but neurotrophic factor specificity is conferred by a co-receptor. GDNF binds mainly to GDNF Family Receptor α1 (GFRα1; Treanor et al. 1996; Baloh et al. 1997; Klein et al. 1997; Enokido et al. 1998; Milbrandt et al. 1998; Trupp et al. 1998) and to a degree to GFRα2 (Sanicola et al. 1997) to promote the survival of this neurotrophic factor responsive population of sensory neurons. Neurturin has the highest affinity for GFRα2, although it binds GFRα1 to a lesser extent (Jing et al. 1997). GFRα3 is highly selective for the GDNF family member artemin, which does not bind to either GFRα1 or GFRα2 (Baloh et al. 1998; Orozco et al. 2001).

One of the main roles of GDNF family neurotrophic factors is in neuronal survival and maturation in early postnatal life. During the first week of postnatal development

(P0–7), a subset of small sensory neurons downregulate trkA (Molliver and Snider 1997) and become dependent on GDNF for survival (Molliver et al. 1997). During this period, ret-positive cells begin to co-label extensively with IB4, which is not co-localized with trkA. For some sensory neurons, this is maintained throughout life. In rodents, the trkA and ret-positive sensory neurons develop into separate populations (Molliver et al. 1997). For example, the NGF- and GDNF-responsive populations of sensory neurons also terminate in slightly different regions of the skin and spinal cord (Coimbra et al. 1974; Silverman and Kruger 1990; Zylka et al. 2003, 2005; Lindfors et al. 2006; Liu et al. 2008). In addition, constitutive overexpression of GDNF (GDNF-OE) in the skin causes selective survival of the IB4 population of sensory neurons and an increase in the innervation density of the epidermis and lamina II of the dorsal horn (Zwick et al. 2002), a result suggesting possible fundamental differences between the functional properties of the GDNF-responsive population of sensory neurons from other neurotrophic factor responsive populations.

There are many studies that support this possibility. For example, IB4-positive neurons appear to have smaller noxious heat-activated currents compared to IB4-negative neurons (Stucky and Lewin 1999). Furthermore, *in vitro* treatment of primary DRG neurons with GDNF does not cause an increase in the percentage of neurons that respond to heat in the same manner as NGF (Stucky and Lewin 1999). It has thus been hypothesized that these two populations of nociceptors are involved in different aspects of pain (Snider and McMahon 1998).

Acute exogenous GDNF can lead to the prevention of injury-induced pain behaviors (Boucher et al. 2000); however, other studies suggest a possible pro-nociceptive role for GDNF. Studies using GDNF-OEs in which constitutive GDNF is expressed in the skin did not alter heat or mechanical withdrawal responses (Zwick et al. 2002). In addition, GDNF-OEs also had higher levels of mRNA in the DRGs for kappa opioid receptor 1 (KOR1), metabotropic glutamate receptor R1 (mGluR1), and TRPV2; and a decrease in delta opioid receptor 1 (DOR1) compared to WTs (Zwick et al. 2003). After peripheral inflammation, GDNF-OEs express less TRPV1 and NMDA receptor subunit NR1 but higher levels of μ opioid receptor (MOR) and κ opioid receptor (KOR) in the DRGs compared to WTs (Molliver et al. 2005a).

These data make understanding the role of GDNF on nociceptor sensitization difficult, but a recent study analyzing single DRG neurons has provided some insight (Albers et al. 2006). Mice with enhanced levels of GDNF in the skin show that mechanically sensitive C-fibers have significantly lower mechanical thresholds compared to WTs. These mice also reveal a significant increase in the percentage of C-fibers that responded to heat (Albers et al. 2006). Neurochemical identification of individually characterized sensory neurons showed that only non-peptidergic, IB4 positive C-polymodal fibers were affected by the excess cutaneous GDNF. Other fiber types were unaffected, such as the high-threshold, mechanically sensitive, IB4- and CGRP+ afferents and low-threshold mechanically sensitive afferents that were neither IB4 nor CGRP positive (Albers et al. 2006). The mechanism of this sensitization in this population of CPM fibers is unclear, although GDNF-OEs did have increased levels of putative mechanosensitive ion channels, ASIC2a and ASIC2b, predominantly localized to the IB4 positive population.

These data caution that behavioral analysis studying the function of a single growth factor may fail to display its true purpose since single fiber analysis revealed a functional effect of chronic GDNF exposure. However, it should be noted that acute injection of GDNF into the hindpaw has been found to induce heat hyperalgesia in mice (Malin et al. 2006). Thus while GDNF may have a role in a sensory neuron sensitization, its role in pain processing is still unclear.

2.1.4 ARTEMIN

The most recently characterized neurotrophic factor related to sensory neuron sensitization is artemin, another member of the GDNF family. Unlike GDNF and neurturin, which are slightly more promiscuous in binding GFRα receptors in the DRGs (Baloh et al. 1997; Klein et al. 1997; Enokido et al. 1998; Milbrandt et al. 1998; Trupp et al. 1998), artemin is highly selective in binding its receptor GFRα3 (Baloh et al. 1998; Orozco et al. 2001). GFRα3 co-localizes extensively with the TRPV1+, peptidergic population of sensory neurons, and rarely co-localizes with IB4 (Orozco et al. 2001; Elitt et al. 2006; Lawson et al. 2008).

Constitutive overexpression of artemin in the skin did not cause a selective survival of sensory neurons as seen in other transgenic mice overexpressing NGF (Albers et al. 1994) or GDNF (Zwick et al. 2002). Developmentally, overexpression of artemin was not found to increase the percentage of GFRα3+ cells in the DRG, but it did cause hypertrophy of the GFRα3+ population. Overproduction of artemin did result in an increase in the innervation density of GFRα3+ fibers in the skin, but did not alter the peptidergic or IB4 staining patterns in the spinal cord dorsal horn (Elitt et al. 2006). This result suggests that artemin may selectively alter the response properties of certain sensory neurons in the DRGs, but may not be involved in neuronal survival and growth to the same degree as NGF, NT-3, GDNF, and other neurotrophic factors.

As with GDNF, artemin signaling may have both pro- and antinociceptive effects. In the spinal nerve ligation (SNL) model of nerve injury in which one spinal nerve that contributes axons to the sciatic nerve is ligated and the other spinal nerves remain intact, Gardell and colleagues (2003) found that systemic delivery of artemin reduces the injury-induced decrease in paw withdrawal threshold to both mechanical and thermal stimulation. They also found that artemin reduced the capsaicin-induced release of CGRP and dynorphin, and it prevented the decrease in IB4 binding, and substance P (SP) and P2X3 immunoreactivity in DRGs. This group suggested, based on these findings, that artemin may be therapeutically useful for patients with nerve injury-induced chronic pain. However, in a study of pancreatic cancer, Ceyhan and colleagues (2007) found that artemin was enhanced in human cancer cells and hypertrophic afferents of the pancreas and actually promoted cancer cell invasion in pancreatic ductal adenocarcinoma, but these changes in artemin did not correlate with pain. Additional future studies involving delivery of artemin in humans would facilitate the determination of the benefits of artemin in relieving chronic pain.

Although these studies may suggest that artemin could be antinociceptive, other studies also show that enhanced cutaneous artemin could also sensitize nociceptors. Polymodal C-fibers (CPM) in artemin overexpressors (ART-OEs) possess lower heat thresholds and higher firing rates per degree, and dissociated DRG neurons from these mice reveal significantly larger responses to capsaicin application, which correlated with the observed increase in TRPV1 expression in the DRGs in these animals (Elitt et al. 2006). These results conflict with data reported from naïve mice (Woodbury et al. 2004; Lawson et al. 2008) in which the vast majority of cutaneous CPM neurons were found to be IB4+ and TRPV1-. Recent data however have shown that several CPMs in ART-OEs contained TRPV1 (Albers K.M. and Koerber H.R. unpublished), suggesting that enhanced levels of artemin may induce a neurochemical switch in some neurons. One hypothesis is that the GFRα3+/ TRPV1+ CH neurons gain mechanical sensitivity in the presence of excess artemin. In any case, artemin is likely to have a significant role in nociceptive processing under normal conditions as well as after peripheral injury.

2.2 PERIPHERAL INJURY CAUSES SENSITIZATION OF NOCICEPTORS

The typical features of nociceptor sensitization include increased prevalence of cells with ongoing activity, an increase in the level of ongoing activity, changes in stimulus–response properties, and changes in the expression of transmitters and receptors in the sensory ganglia (Koltzenburg et al. 1999). These changes in afferent properties depend on a number of factors including the type and site of injury and the time after injury. Recently, it has also been determined that there are significant changes in various molecules at the site of tissue injury and in the targets of injured afferents that may lead to the changes observed in the sensory neurons. These include, but are not limited to, changes in inflammatory mediators, signaling molecules, CGRP, bradykinin, prostaglandins, serotonin, histamine, H+, ATP, and neurotrophic factors (Reeh and Kress 2001; Koltzenberg et al. 1999; Malin et al. 2006; Li et al. 2008; Jankowski et al. 2009a). Combinations of these changes probably underlie various aspects of thermal and mechanical pain, including ongoing pain, hyperalgesia, and the perception of pain in response to normally innocuous stimuli (allodynia).

These features seem to be common among many types of injury in animal models of persistent pain including plantar incision (Pogatzki et al. 2002; Woo et al. 2004; Banik and Brennan 2004, 2009; Mujenda et al. 2007), SNL (Wu et al. 2001; Shim et al. 2005), chronic constriction injury (Aley and Levine 2002; Milligan et al. 2004; Wilson-Gerwing et al. 2008; Zhang et al. 2008), and axotomy and regeneration (Michalski et al. 2008; Jankowski et al. 2009a). In these neuropathic pain models two distinct patterns of change emerge. One set of changes occurs in the injured afferents, which includes the emergence of spontaneous activity and alterations in the pattern of transmitter expression (see above) that appears to reflect, at least in part, the loss of access to neurotrophic factors supplied by the target of innervation. The second set of changes occurs in the "uninjured" neighboring afferents. These fibers are exposed to aforementioned inflammatory mediators released during the process of Wallerian degeneration of the injured axons and alterations in neurotrophic factors (among other molecules) present in the denervated or partially denervated target tissue and/or the injury site that serve to promote regeneration of the injured fibers.

The changes in the uninjured afferents are similar to those described in the presence of inflammation (Koltzenburg et al. 1999; Djouhri et al. 2006). There is also evidence of time-dependent changes in both injured and uninjured afferents, suggesting that the relative contribution of each to various aspects of pain may change with time following injury.

Although there are several molecular mechanisms that could be involved in nociceptor sensitization after injury, enhanced neurotrophic factor signaling from the periphery may be vital regardless of whether an afferent is injured or uninjured. After all of these injuries, neurotrophic factor levels are changed in the DRGs, injury site, and/or peripheral innervation field (e.g., Malin et al. 2006; Jankowski et al. 2009a; above). For example, NGF has been shown by many groups to be upregulated in the peripheral target tissue after inflammation and nerve injury (Weskamp and Otten 1987; Woolf et al. 1994; Braun et al. 1998; Malin et al. 2006; Nicol and Vasko 2007; Paterson et al. 2008; Jankowski et al. 2009a), which could play a role in the sensitization of some of the peptidergic, trkA+ neurons. Nerve lesion has also been shown to induce NT-3 upregulation in the nerve at the site of afferent injury (Lee et al. 2001; Campana et al. 2006; Hoke et al. 2006) and in the target regions (Jankowski et al. 2009a). This increase could play a role in the sensitization of trkC+/ ASIC3+ myelinated nociceptors (McIlwrath et al. 2007). GDNF has been shown to be upregulated in the cutaneous tissue after axotomy (Jankowski et al. 2009a) and after peripheral inflammation (Malin et al. 2006), and mechanically sensitive neurons in GDNF-OEs are sensitized to mechanical stimulation (Albers et al. 2006) and acute injection of GDNF into the skin of mice induces some pain behaviors (Malin et al. 2006). These actions may rely on sensitization of the IB4+ subset of sensory neurons (Albers et al. 2006). Artemin has also been found to be upregulated in the skin after inflammation (Malin et al. 2006) and nerve injury (Jankowski et al. 2009a) and may alter the GFRα3+/ TRPV1+ CH neurons by causing some of these cells to gain mechanical sensitivity (Elitt et al. 2006; Albers K.M. and Koerber H.R. unpublished). This phenotypic switch suggests that artemin may play a crucial role in nociceptor sensitization and development of pain. From these results, it is clear that altered levels of several neurotrophic factors can cause sensory neuron sensitization after nerve injury, which may lead to conditions of chronic pain.

Although it is apparent that altered nociceptor properties are critical for both the initiation and maintenance of pain, the relative contribution of specific subpopulations is unknown. Recent data from a mouse model of nociceptor sensitization after injury reveals that two specific populations of C-fibers are affected differently by nerve injury. GDNF-sensitive, IB4+/ TRPV1- CPM neurons (Albers et al. 2006) exhibit lower heat thresholds after axotomy and regeneration. On the other hand, artemin-sensitive GFRα3+/ TRPV1+/ IB4- CH fibers (Elitt et al. 2006; Lawson et al. 2008) were not sensitized but increased in number compared to the naïve condition, apparently recruited from a silent population. Interestingly, some CPM neurons also stained positive for TRPV1 after injury. Immunolabeling also showed that there was an increase in the overlap between IB4 and TRPV1 after regeneration. However, this was found to be due to an increase in the number of cells binding IB4, not an increase in the number of TRPV1-positive cells (Jankowski et al. 2009a). Taken together these results suggest that in addition to CH fiber recruitment, some of the artemin-responsive CH fibers may be gaining mechanical sensitivity after injury similar to what has been shown in the ART-OEs (Elitt et al. 2006).

2.3   CLINICAL RELEVANCE

Clinical data on the role of atypical afferent activity in pain states comes from the use of regional nerve blocks (Hsieh et al. 1995), surgical interventions and analyses of excised nerves (Hilz 2002; Brisby 2006), and microneurography (Orstavik et al. 2006). For example, microneurographic recordings suggest that a specific subpopulation of C-fiber nociceptors, referred to as mechanically insensitive afferents (MIAs) in healthy control subjects, play a major role in the maintenance of chronic neuropathic pain. Fibers that have the biophysical properties of MIAs in patients with erythromyalgia acquire mechanical responsiveness (Orstavik et al. 2003); and in patients with diabetic neuropathy, mechanically insensitive, heat-sensitive afferents increase in prevalence compared to normal control subjects (Orstavik et al. 2006). The MIAs found in humans (Schmidt et al. 1995) have several characteristics in common with the TRPV1+ CH neurons characterized in mouse (Lawson et al. 2008; Jankowski et al. 2009a) and non-human primates (Baumann et al. 1991). In addition to the lack of mechanical sensitivity, these fibers are among the slowest-conducting cutaneous fibers in the DRGs. Although murine CH fibers are the only cutaneous fibers to contain TRPV1 (Lawson et al. 2008), MIA fibers found in humans and CH fibers in non-human primates are the only cells to respond to TRPV1 agonist capsaicin in a way that correlates with pain assessments following capsaicin injection (Baumann et al. 1991; Schmelz et al. 2000; Ringkamp et al. 2001; Jankowski et al. 2009a). In addition, following nerve injury and regeneration in mice, some CH fibers apparently are recruited from a silent population of DRG neurons, and some gain mechanical sensitivity (Jankowski et al. 2009a) similar to that reported for MIA fibers in microneurographic recordings from chronic pain patients (Orstavik et al. 2003, 2006). These results suggest that the CH/MIA fibers may play the largest role in the development of chronic neuropathic pain.

2.4   FUTURE TREATMENTS FOR NEUROTROPHIC FACTOR-INDUCED NOCICEPTOR SENSITIZATION

These data suggest that different populations of sensory neurons may actually be involved in different aspects of tactile and thermal detection. The changes in the artemin/NGF-responsive, peptidergic, TRPV1+/ GFRα3+/ IB4-, mechanically insensitive, heat-sensitive afferents (CHs) may play a dominant role in thermal hyperalgesia; and the GDNF responsive, IB4+/TRPV1- polymodal nociceptors (CPMs) may be more important in mechanical and/or other aspects of nociceptive processing. A recent publication supports this hypothesis in which they have shown that ablation of the TRPV1+ (CH) population of sensory neurons only resulted in changes in thermal, not mechanical, sensitivity. Conversely, elimination of the IB4+ (CPMs) cells only caused a deficit in mechanical sensitivity (Cavanaugh et al. 2009). Further support for this hypothesis comes from studies using TRPV1 knockout mice that have normal heat sensitivity but do not develop heat hyperalgesia after inflammation (Caterina et al. 2000; Woodbury et al. 2004).

It is unclear why the GDNF-sensitive IB4+ CPM neurons (Albers et al. 2006) are sensitized to heat after injury and inflammation (Jankowski et al. 2008, 2009a) if they may not be involved in heat hyperalgesia. Injection of GDNF was found to induce heat hyperalgesia in mice, which should affect these neurons specifically (Malin et al. 2006). One possible explanation is that these fibers engage central networks that are more involved in processing mechanical, rather than thermal, information. Nevertheless, these data do suggest that modality-specific aspects of pain syndrome development may be determined by which subpopulation of nociceptors is most affected.

The difficulty in developing effective treatments for chronic pain syndromes has come in translating work performed in the laboratory to the clinic. The inability to selectively silence specific subpopulations of afferents has seemingly proven to be a major barrier to the development of effective therapeutic interventions devoid of side effects. Outside of conventional pharmacological treatments, which are not adequate at treating all pain conditions, recent developments have begun to test the validity of utilizing genetically engineered viruses to deliver various molecules involved in pain processing to sensory neurons in order to treat pain states. There are obvious limitations and potentially dangerous side effects from the use of these methodologies since viruses require host cell invasion and, in certain cases, manipulation of the cell's genome.

One way to develop appropriate treatments for chronic pain using animal models derives from recent studies using experimental manipulation of the endogenous RNA interference (RNAi) pathway. RNAi targets the transcriptional output of the cell rather than the genome itself. Short (19–21) base-pair RNA duplexes that have perfect complementarity to a particular mRNA induce degradation of the mRNA so that it cannot be translated into protein. Since there are a variety of changes in gene expression in DRG neurons under conditions of inflammation and nerve injury, targeted RNA degradation may be a better way to selectively alter the function of specific subsets of sensory neurons. A new study using injection of modified small interfering RNAs (siRNAs) into specific nerves *in vivo* has proven to be effective in inducing functional changes in identified nerves (Jankowski et al. 2008, 2009b). Another recent study using this technology has also shown that targeted inhibition of the inflammation-induced increase in purinergic receptor expression (P2Y1) had the effect of completely blocking the inflammation-induced decrease in CPM thermal threshold (Jankowski et al. 2008). Only IB4+ CPM neurons are thought to contain P2Y1 (Dussor et al. 2008), and P2Y1-negative sensory neurons did not appear to be affected by injection of P2Y1 targeting siRNAs. Moreover, the injection of modified siRNAs did not cause the levels of P2Y1 mRNA to drop below baseline and did not induce a hyposensitive state in CPM neurons. Thus this new strategy for specifically targeting particular molecules in identified afferents without inducing sub-baseline neuronal activity after inflammation and/or nerve injury may provide a new way to develop suitable therapeutic agents for neuropathic pain conditions. In addition, clinical use of RNAi may even offer a safe and effective means of treating conditions of chronic pain in the future.

REFERENCES

Abdiche YN, Malashock DS, Pons J (2008) Probing the binding mechanism and affinity of tanezumab, a recombinant humanized anti-NGF monoclonal antibody, using a repertoire of biosensors. *Protein Sci* 17:1326–1335.

Albers KM, Wright DE, Davis BM (1994) Overexpression of nerve growth factor in epidermis of transgenic mice causes hypertrophy of the peripheral nervous system. *J Neurosci* 14:1422–1432.

Albers KM, Woodbury CJ, Ritter AM, Davis BM, Koerber HR (2006) Glial cell-line-derived neurotrophic factor expression in skin alters the mechanical sensitivity of cutaneous nociceptors. *J Neurosci* 26:2981–2990.

Aley KO, Levine JD (2002) Different peripheral mechanisms mediate enhanced nociception in metabolic/toxic and traumatic painful peripheral neuropathies in the rat. *Neuroscience* 111:389–397.

Amann R, Schuligoi R, Lanz I, Peskar BA (1996) Effect of a 5-lipoxygenase inhibitor on nerve growth factor-induced thermal hyperalgesia in the rat. *Eur J Pharmacol* 306:89–91.

Anand P, Terenghi G, Birch R, Wellmer A, Cedarbaum JM, Lindsay RM, Williams-Chestnut RE, Sinicropi DV (1997) Endogenous NGF and CNTF levels in human peripheral nerve injury. *Neuroreport* 8:1935–1938.

Andreev N, Dimitrieva N, Koltzenburg M, McMahon SB (1995) Peripheral administration of nerve growth factor in the adult rat produces a thermal hyperalgesia that requires the presence of sympathetic post-ganglionic neurones. *Pain* 63:109–115.

Averill S, McMahon SB, Clary DO, Reichardt LF, Priestley JV (1995) Immunocytochemical localization of trkA receptors in chemically identified subgroups of adult rat sensory neurons. *Eur J Neurosci* 7:1484–1494.

Baloh RH, Gorodinsky A, Golden JP, Tansey MG, Keck CL, Popescu NC, Johnson EM, Jr., Milbrandt J (1998) GFRalpha3 is an orphan member of the GDNF/neurturin/persephin receptor family. *Proc Natl Acad Sci USA* 95:5801–5806.

Baloh RH, Tansey MG, Golden JP, Creedon DJ, Heuckeroth RO, Keck CL, Zimonjic DB, Popescu NC, Johnson EM, Jr., Milbrandt J (1997) TrnR2, a novel receptor that mediates neurturin and GDNF signaling through Ret. *Neuron* 18:793–802.

Banik RK, Brennan TJ (2004) Spontaneous discharge and increased heat sensitivity of rat C-fiber nociceptors are present in vitro after plantar incision. *Pain* 112: 204–213.

Banik RK, Brennan TJ (2009) Trpv1 mediates spontaneous firing and heat sensitization of cutaneous primary afferents after plantar incision. *Pain* 141: 41–51.

Baumann TK, Simone DA, Shain CN, LaMotte RH (1991) Neurogenic hyperalgesia: the search for the primary cutaneous afferent fibers that contribute to capsaicin-induced pain and hyperalgesia. *J Neurophysiol* 66:212–227.

Bennett DL (2001) Neurotrophic factors: important regulators of nociceptive function. *Neuroscientist* 7:13–17.

Bennett DL, Averill S, Clary DO, Priestley JV, McMahon SB (1996) Postnatal changes in the expression of the trkA high-affinity NGF receptor in primary sensory neurons. *Eur J Neurosci* 8: 2204–2208.

Bennett DL MG, Ramachandran N, Munson JB, Averill S, Yan Q, McMahon SB, Priestly JV. (1998) A distinct subgroup of small DRG cells express GDNFF receptor components and GDNF is protective for these neurons after nerve injury. *J Neurosci* 18:3059–3072.

Boucher TJ, Okuse K, Bennett DL, Munson JB, Wood JN, McMahon SB (2000) Potent analgesic effects of GDNF in neuropathic pain states. *Science* 290:124–127.

Braun A, Appel E, Baruch R, Herz U, Botchkarev V, Paus R, Brodie C, Renz H (1998) Role of nerve growth factor in a mouse model of allergic airway inflammation and asthma. *Eur J Immunol* 28: 3240–3251.

Brisby H (2006) Pathology and possible mechanisms of nervous system response to disc degeneration. *J Bone Joint Surg Am* 88 Suppl 2:68–71.

Campana WM, Li X, Dragojlovic N, Janes J, Gaultier A, Gonias SL (2006) The low-density lipoprotein receptor-related protein is a pro-survival receptor in Schwann cells: possible implications in peripheral nerve injury. *J Neurosci* 26:11197–11207.

Caterina MJ, Schumacher MA, Tominaga M, Rosen TA, Levine JD, Julius D (1997) The capsaicin receptor: a heat-activated ion channel in the pain pathway. *Nature* 389:816–824.

Caterina MJ, Leffler A, Malmberg AB, Martin WJ, Trafton J, Petersen-Zeitz KR, Koltzenburg M, Basbaum AI, Julius D (2000) Impaired nociception and pain sensation in mice lacking the capsaicin receptor. *Science* 288:306–313.

Cavanaugh DJ, Lee H, Lo L, Shields SD, Zylka MJ, Basbaum AI, Anderson DJ (2009) Distinct subsets of unmyelinated primary sensory fibers mediate behavioral responses to noxious thermal and mechanical stimuli. *PNAS*. In press.

Ceyhan GO, Bergmann F, Kadihasanoglu M, Erkan M, Park W, Hinz U, Giese T, et al. (2007) The neurotrophic factor artemin influences the extent of neural damage and growth in chronic pancreatitis. *Gut* 56:534–544.

Cheng JK, Ji RR (2008) Intracellular signaling in primary sensory neurons and persistent pain. *Neurochem Res* 33:1970–1978.

Coimbra A, Sodre-Borges BP, Magalhaes MM (1974) The substantia gelatinosa Rolandi of the rat. Fine structure, cytochemistry (acid phosphatase) and changes after dorsal root section. *J Neurocytol* 3: 199–217.

Davis BM, Albers KM, Seroogy KB, Katz DM (1994) Overexpression of nerve growth factor in transgenic mice induces novel sympathetic projections to primary sensory neurons. *J Comp Neurol* 349:464–474.

Davis BM, Wang HS, Albers KM, Carlson SL, Goodness TP, McKinnon D (1996) Effects of NGF overexpression on anatomical and physiological properties of sympathetic postganglionic neurons. *Brain Res* 724:47–54.

Deng YS, Zhong JH, Zhou XF (2000) Effects of endogenous neurotrophins on sympathetic sprouting in the dorsal root ganglia and allodynia following spinal nerve injury. *Exp Neurol* 164:344–350.

Diamond J, Holmes M, Coughlin M (1992) Endogenous NGF and nerve impulses regulate the collateral sprouting of sensory axons in the skin of the adult rat. *J Neurosci* 12:1454–1466.

Djouhri L, Koutsikou S, Fang X, McMullan S, Lawson SN (2006) Spontaneous pain, both neuropathic and inflammatory, is related to frequency of spontaneous firing in intact C-fiber nociceptors. *J Neurosci* 26:1281–1292.

Donnerer J, Liebmann I, Schicho R (2005) Differential regulation of 3-beta-hydroxysteroid dehydrogenase and vanilloid receptor TRPV1 mRNA in sensory neurons by capsaicin and NGF. *Pharmacology* 73:97–101.

Dussor G, Koerber HR, Oaklander AL, Rice FL, Molliver DC (2008) Nucleotide signaling and cutaneous mechanisms of pain transduction. *Brain Res Rev.*

Elitt CM, McIlwrath SL, Lawson JJ, Malin SA, Molliver DC, Cornuet PK, Koerber HR, Davis BM, Albers KM (2006) Artemin overexpression in skin enhances expression of TRPV1 and TRPA1 in cutaneous sensory neurons and leads to behavioral sensitivity to heat and cold. *J Neurosci* 26:8578–8587.

Elshamy WM, Ernfors P (1996) A local action of neurotrophin-3 prevents the death of proliferating sensory neuron precursor cells. *Neuron* 16: 963–972.

Enokido Y, de Sauvage F, Hongo JA, Ninkina N, Rosenthal A, Buchman VL, Davies AM (1998) GFR alpha-4 and the tyrosine kinase Ret form a functional receptor complex for persephin. *Curr Biol* 8:1019–1022.

Flamig G, Engele J, Geerling I, Pezeshki G, Adler G, Reinshagen M (2001) Neurotrophin and GDNF expression increases in rat adrenal glands during experimental colitis. *Neuro Endocrinol Lett* 22:461–466.

Funakoshi K, Nakano M, Atobe Y, Goris RC, Kadota T, Yazama F (2006) Differential development of TRPV1-expressing sensory nerves in peripheral organs. *Cell Tissue Res* 323:27–41.

Gardell LR, Wang R, Ehrenfels C, Ossipov MH, Rossomando AJ, Miller S, Buckley C, et al. (2003) Multiple actions of systemic artemin in experimental neuropathy. *Nat Med* 9:1383–1389.

Goodness TP, Albers KM, Davis FE, Davis BM (1997) Overexpression of nerve growth factor in skin increases sensory neuron size and modulates Trk receptor expression. *Eur J Neurosci* 9:1574–1585.

Hilz MJ (2002) Assessment and evaluation of hereditary sensory and autonomic neuropathies with autonomic and neurophysiological examinations. *Clin Auton Res* 12 Suppl 1:I33–I43.

Hoke A, Redett R, Hameed H, Jari R, Zhou C, Li ZB, Griffin JW, Brushart TM (2006) Schwann cells express motor and sensory phenotypes that regulate axon regeneration. *J Neurosci* 26:9646–9655.

Hsieh JC, Belfrage M, Stone-Elander S, Hansson P, Ingvar M (1995) Central representation of chronic ongoing neuropathic pain studied by positron emission tomography. *Pain* 63:225–236.

Huang EJ WG, Farinas I, Backus C, Zang K, Wong SL, Reichardt LF (2001) Expression of trk receptors in the developing mouse trigeminal ganglion: in vivo evidence for NT-3 activation of trkA and trkB in addition to trkC. *Development* 126:2191–2203.

Jankowski MP, Lawson JJ, McIlwrath SL, Rau KK, Anderson CE, Albers KM, Koerber HR (2009a) Sensitization of cutaneous nociceptors after nerve transection and regeneration: possible role for target derived neurotrophic factor signaling. *J Neurosci* 29: 1636–1647.

Jankowski MP, McIlwrath SL, Cornuet PK, Jing X, Salerno KM, Koerber HR, Albers KM (2009b) Sox11 transcription factor regulates peripheral nerve regeneration in the adult. *Brain Res* 1256: 43–54.

Jankowski MP, Rau KK, Soneji DJ, Anderson CE, Molliver DC, Koerber HR. (2008) Purinergic receptor P2Y1 regulates polymodal C-fiber sensitivity during peripheral inflammation. *Soc Neurosci Abs* 38.

Ji RR, Samad TA, Jin SX, Schmoll R, Woolf CJ (2002) p38 MAPK activation by NGF in primary sensory neurons after inflammation increases TRPV1 levels and maintains heat hyperalgesia. *Neuron* 36:57–68.

Jimenez-Andrade JM, Martin CD, Koewler NJ, Freeman KT, Sullivan LJ, Halvorson KG, Barthold CM, et al. (2007) Nerve growth factor sequestering therapy attenuates non-malignant skeletal pain following fracture. *Pain* 133:183–196.

Jing S, Yu Y, Fang M, Hu Z, Holst PL, Boone T, Delaney J, Schultz H, Zhou R, Fox GM (1997) GFRalpha-2 and GFRalpha-3 are two new receptors for ligands of the GDNF family. *J Biol Chem* 272:33111–33117.

Kerr BJ, Souslova V, McMahon SB, Wood JN (2001) A role for the TTX-resistant sodium channel Nav 1.8 in NGF-induced hyperalgesia, but not neuropathic pain. *Neuroreport* 12:3077–3080.

Klein RD, Sherman D, Ho WH, Stone D, Bennett GL, Moffat B, Vandlen R, et al. (1997) A GPI-linked protein that interacts with Ret to form a candidate neurturin receptor. *Nature* 387:717–721.

Koltzenburg M, Bennett DL, Shelton DL, McMahon SB (1999) Neutralization of endogenous NGF prevents the sensitization of nociceptors supplying inflamed skin. *Eur J Neurosci* 11:1698–1704.

Kullander K, Ebendal T (1994) Neurotrophin-3 acquires NGF-like activity after exchange to five NGF amino acid residues: molecular analysis of the sites in NGF mediating the specific interaction with the NGF high affinity receptor. *J Neurosci* Res 39:195–210.

Lawson JJ, McIlwrath SL, Woodbury CJ, Davis BM, Koerber HR (2008) TRPV1 unlike TRPV2 is restricted to a subset of mechanically insensitive cutaneous nociceptors responding to heat. *J Pain* 9:298–308.

Lee P, Zhuo H, Helke CJ (2001) Axotomy alters neurotrophin and neurotrophin receptor mRNAs in the vagus nerve and nodose ganglion of the rat. *Brain Res Mol Brain Res* 87:31–41.

Lewin GR, Ritter AM, Mendell LM (1993) Nerve growth factor-induced hyperalgesia in the neonatal and adult rat. *J Neurosci* 13:2136–2148.

Lewin GR, Rueff A, Mendell LM (1994) Peripheral and central mechanisms of NGF-induced hyperalgesia. *Eur J Neurosci* 6:1903–1912.

Li D, Ren Y, Xu X, Zou X, Fang L, Lin Q (2008) Sensitization of primary afferent nociceptors induced by intradermal capsaicin involves the peripheral release of calcitonin gene-related Peptide driven by dorsal root reflexes. *J Pain* 9:1155–1168.

Lindfors PH, Voikar V, Rossi J, Airaksinen MS (2006) Deficient nonpeptidergic epidermis innervation and reduced inflammatory pain in glial cell line-derived neurotrophic factor family receptor alpha2 knock-out mice. *J Neurosci* 26: 1953–1960.

Liu Y, Yang FC, Okuda T, Dong X, Zylka MJ, Chen CL, Anderson DJ, Kuner R, Ma Q (2008) Mechanisms of compartmentalized expression of Mrg class G-protein coupled sensory receptors. *J Neurosci* 28: 125–132.

Lowe EM, Anand P, Terenghi G, Williams-Chestnut RE, Sinicropi DV, Osborne JL (1997) Increased nerve growth factor levels in the urinary bladder of women with idiopathic sensory urgency and interstitial cystitis. *Br J Urol* 79:572–577.

Malin SA, Molliver DC, Koerber HR, Cornuet P, Frye R, Albers KM, Davis BM (2006) Glial cell line-derived neurotrophic factor family members sensitize nociceptors in vitro and produce thermal hyperalgesia in vivo. *J Neurosci* 26:8588–8599.

McIlwrath SL, Lawson JJ, Anderson CE, Albers KM, Koerber HR (2007) Overexpression of neurotrophin-3 enhances the mechanical response properties of slowly adapting type 1 afferents and myelinated nociceptors. *Eur J Neurosci* 26:1801–1812.

McMahon SB (1996) NGF as a mediator of inflammatory pain. *Philos Trans R Soc Lond B Biol Sci* 351:431–440.

Michael GJ, Priestley JV (1999) Differential expression of the mRNA for the vanilloid receptor subtype 1 in cells of the adult rat dorsal root and nodose ganglia and its downregulation by axotomy. *J Neurosci* 19:1844–1854.

Michalski B, Bain JR, Fahnestock M (2008) Long-term changes in neurotrophic factor expression in distal nerve stump following denervation and reinnervation with motor or sensory nerve. *J Neurochem* 105: 1244–1252.

Milbrandt J, de Sauvage FJ, Fahrner TJ, Baloh RH, Leitner ML, Tansey MG, Lampe PA, et al. (1998) Persephin, a novel neurotrophic factor related to GDNF and neurturin. *Neuron* 20:245–253.

Milligan ED, Zapata V, Chacur M, Schoeniger D, Biedenkapp J, O'Connor KA, Verge GM, et al. (2004) Evidence that exogenous and endogenous fractalkine can induce spinal nociceptive facilitation in rats. *Eur J Neurosci* 20:2294–2302.

Molliver DC, Immke DC, Fierro L, Pare M, Rice FL, McCleskey EW (2005b) ASIC3, an acid-sensing ion channel, is expressed in metaboreceptive sensory neurons. *Mol Pain* 1:35.

Molliver DC, Lindsay J, Albers KM, Davis BM (2005a) Overexpression of NGF or GDNF alters transcriptional plasticity evoked by inflammation. *Pain* 113:277–284.

Molliver DC, Snider WD (1997) Nerve growth factor receptor TrkA is down-regulated during postnatal development by a subset of dorsal root ganglion neurons. *J Comp Neurol* 381:428–438.

Molliver DC, Wright DE, Leitner ML, Parsadanian AS, Doster K, Wen D, Yan Q, Snider WD (1997) IB4-binding DRG neurons switch from NGF to GDNF dependence in early post-natal life. *Neuron* 19:849–861.

Mujenda FH, Duarte AM, Reilly EK, Strichartz GR (2007) Cutaneous endothelin-A receptors elevate post-incisional pain. *Pain* 133: 161–173.

Nagy JI, Hunt SP (1982) Fluoride-resistant acid phosphatase-containing neurones in dorsal root ganglia are separate from those containing substance P or somatostatin. *Neuroscience* 7: 89–97.

Nakamura-Craig M, Gill BK (1991) Effect of neurokinin A, substance P and calcitonin gene related peptide in peripheral hyperalgesia in the rat paw. *Neurosci Lett* 124:49–51.

Nicol GD, Vasko MR (2007) Unraveling the story of NGF-mediated sensitization of nociceptive sensory neurons: ON or OFF the Trks? *Mol Interv* 7: 26–41.

Oku R, Satoh M, Fujii N, Otaka A, Yajima H, Takagi H (1987) Calcitonin gene-related peptide promotes mechanical nociception by potentiating release of substance P from the spinal dorsal horn in rats. *Brain Res* 403:350–354.

Orozco OE, Walus L, Sah DW, Pepinsky RB, Sanicola M (2001) GFRalpha3 is expressed predominantly in nociceptive sensory neurons. *Eur J Neurosci* 13:2177–2182.

Orstavik K, Namer B, Schmidt R, Schmelz M, Hilliges M, Weidner C, Carr RW, Handwerker H, Jorum E, Torebjork HE (2006) Abnormal function of C-fibers in patients with diabetic neuropathy. *J Neurosci* 26:11287–11294.

Orstavik K, Weidner C, Schmidt R, Schmelz M, Hilliges M, Jorum E, Handwerker H, Torebjork E (2003) Pathological C-fibres in patients with a chronic painful condition. *Brain* 126:567–578.

Park SY, Choi JY, Kim RU, Lee YS, Cho HJ, Kim DS (2003) Downregulation of voltage-gated potassium channel alpha gene expression by axotomy and neurotrophins in rat dorsal root ganglia. *Mol Cells* 16:256–259.

Patapoutian A, Reichardt LF (2001) Trk receptors: mediators of neurotrophin action. *Curr Opin Neurobiol* 11:272–280.

Paterson S, Schmelz M, McGlone F, Turner G, Rukwied R (2008) Facilitated neurotrophin release in sensitized human skin. *Eur J Pain.*

Pertens E, Urschel-Gysbers BA, Holmes M, Pal R, Foerster A, Kril Y, Diamond J (1999) Intraspinal and behavioral consequences of nerve growth factor-induced nociceptive sprouting and nerve growth factor-induced hyperalgesia compared in adult rats. *J Comp Neurol* 410:73–89.

Pogatzki EM, Gebhart GF, Brennan TJ (2002) Characterization of Adelta- and C-fibers innervating the plantar rat hindpaw one day after an incision. *J Neurophysiolol* 87: 721–731.

Price TJ, Flores CM (2007) Critical evaluation of the colocalization between calcitonin gene-related peptide, substance P, transient receptor potential vanilloid subfamily type 1 immunoreactivities, and isolectin B4 binding in primary afferent neurons of the rat and mouse. *J Pain* 8:263–272.

Ramer MS, Priestley JV, McMahon SB (2000) Functional regeneration of sensory axons into the adult spinal cord. *Nature* 403:312–316.

Reeh PW, Kress M (2001) Molecular physiology of proton transduction in nociceptors. *Curr Opin Pharmacol* 1:45–51.

Rihl M, Kruithof E, Barthel C, De Keyser F, Veys EM, Zeidler H, Yu DT, Kuipers JG, Baeten D (2005) Involvement of neurotrophins and their receptors in spondyloarthritis synovitis: relation to inflammation and response to treatment. *Ann Rheum Dis* 64:1542–1549.

Ringkamp M, Peng YB, Wu G, Hartke TV, Campbell JN, Meyer RA (2001) Capsaicin responses in heat-sensitive and heat-insensitive A-fiber nociceptors. *J Neurosci* 21: 4460–4468.

Ryden M, Ibanez CF (1996) Binding of neurotrophin-3 to p75LNGFR, TrkA, and TrkB medi-
    ated by a single functional epitope distinct from that recognized by trkC. *J Biol Chem*
    271:5623–5627.
Saldanha G, Hongo J, Plant G, Acheson J, Levy I, Anand P (1999) Decreased CGRP, but pre-
    served Trk A immunoreactivity in nerve fibres in inflamed human superficial temporal
    arteries. *J Neurol Neurosurg Psychiatry* 66:390–392.
Sanicola M, Hession C, Worley D, Carmillo P, Ehrenfels C, Walus L, Robinson S, et al. (1997)
    Glial cell line-derived neurotrophic factor-dependent RET activation can be mediated by
    two different cell-surface accessory proteins. *Proc Natl Acad Sci U S A* 94:6238–6243.
Sann H, Pierau FK (1998) Efferent functions of C-fiber nociceptors. *J Rheumatol* 57 Suppl
    2:8–13.
Schmelz M, Schmid R, Handwerker HO, Torebjork HE (2000) Encoding of burning pain from
    capsaicin-treated human skin in two categories of unmyelinated nerve fibres. *Brain* 123
    Pt3: 560–571.
Schmidt R, Schmelz M, Forster C, Ringkamp M, Torebjork E, Handwerker H (1995) Novel
    classes of responsive and unresponsive C nociceptors in human skin. *J Neurosci*
    15:333–341.
Sena CB, Salgado CG, Tavares CM, Da Cruz CA, Xavier MB, Do Nascimento JL (2006)
    Cyclosporine A treatment of leprosy patients with chronic neuritis is associated
    with pain control and reduction in antibodies against nerve growth factor. *Lepr Rev*
    77:121–129.
Sevcik MA, Ghilardi JR, Peters CM, Lindsay TH, Halvorson KG, Jonas BM, Kubota K, et al.
    (2005) Anti-NGF therapy profoundly reduces bone cancer pain and the accompanying
    increase in markers of peripheral and central sensitization. *Pain* 115:128–141.
Shim B, Kim DW, Kim BH, Nam TS, Leem JW, Chung JM (2005) Mechanical and heat
    sensitization of cutaneous nociceptors in rats with experimental peripheral neuropathy.
    Neuroscience 132: 193–201.
Shu X, Mendell LM (1999) Nerve growth factor acutely sensitizes the response of adult rat
    sensory neurons to capsaicin. *Neurosci Lett* 274:159–162.
Silverman JD, Kruger L (1990) Selective neuronal glycoconjugate expression in sensory
    and autonomic ganglia: relation of lectin reactivity to peptide and enzyme markers.
    J Neurocytol 19: 789–801.
Snider WD, McMahon SB (1998) Tackling pain at the source: new ideas about nociceptors.
    Neuron 20: 629:32.
Stucky CL, Lewin GR (1999) Isolectin B(4)-positive and -negative nociceptors are function-
    ally distinct. *J Neurosci* 19:6497–6505.
Todd AJ, Koerber HR (2005) Neuroanatomical substrates of spinal nociception. Wall and
    Melzack's *Textbook of Pain*: 73–90. Churchill Livingstone.
Tominaga M, Caterina MJ, Malmberg AB, Rosen TA, Gilbert H, Skinner K, Raumann BE,
    Basbaum AI, Julius D (1998) The cloned capsaicin receptor integrates multiple pain-
    producing stimuli. *Neuron* 21:531–543.
Treanor JJ, Goodman L, de Sauvage F, Stone DM, Poulsen KT, Beck CD, Gray C, et al. (1996)
    Characterization of a multicomponent receptor for GDNF. *Nature* 382: 80–83.
Trupp M, Raynoschek C, Belluardo N, Ibanez CF (1998) Multiple GPI-anchored receptors
    control GDNF-dependent and independent activation of the c-Ret receptor tyrosine
    kinase. *Mol Cell Neurosci* 11:47–63.
Weskamp G, Otten U (1987) An enzyme-linked immunoassay for nerve growth factor (NGF):
    a tool for studying regulatory mechanisms involved in NGF production in brain and in
    peripheral tissues. *J Neurochem* 48: 1779–1786.
White DM (2000) Neurotrophin-3 antisense oligonucleotide attenuates nerve injury-induced
    A-beta-fibre sprouting. *Brain Res* 885:79–86.

Wilson-Gerwing TD, Dmyterko MV, Zochodne DW, Johnston JM, Verge VM (2005) Neurotrophin-3 suppresses thermal hyperalgesia associated with neuropathic pain and attenuates transient receptor potential vanilloid receptor-1 expression in adult sensory neurons. *J Neurosci* 25:758–767.

Wilson-Gerwing TD, Stucky CL, McComb GW, Verge VM (2008) Neurotrophin-3 significantly reduces sodium channel expression linked to neuropathic pain states. *Exp Neurol* 213:303–314.

Wilson-Gerwing TD, Verge VM (2006) Neurotrophin-3 attenuates galanin expression in the chronic constriction injury model of neuropathic pain. *Neuroscience* 141:2075–2085.

Woo YC, Park SS, Subieta AR, Brennan TJ (2004) Changes in tissue pH and temperature after incision indicate acidosis may contribute to postoperative pain. *Anesthesiology* 101:468–475.

Woodbury CJ, Zwick M, Wang S, Lawson JJ, Caterina MJ, Koltzenburg M, Albers KM, Koerber HR, Davis BM (2004) Nociceptors lacking TRPV1 and TRPV2 have normal heat responses. *J Neurosci* 24:6410–6415.

Woolf CJ, Ma QP, Allchorne A, Poole S (1996) Peripheral cell types contributing to the hyperalgesic action of nerve growth factor in inflammation. *J Neurosci* 16:2716–2723.

Woolf CJ, Safieh-Garabedian B, Ma QP, Crilly P, Winter J (1994) Nerve growth factor contributes to the generation of inflammatory sensory hypersensitivity. *Neuroscience* 62:327–331.

Wright DE, Snider WD (1995) Neurotrophin receptor mRNA expression defines distinct populations of neurons in rat dorsal root ganglia. *J Comp Neurol* 351:329–338.

Wu G, Ringkamp M, Murinson BB, Campbell JN, Griffin JW, Meyer RA (2001) Early onset of spontaneous activity in uninjured C-fiber nociceptors after injury to neighboring nerve fibers. *J Neurosci* 21: RC140.

Xue Q, Jong B, Chen T, Schumacher MA (2007) Transcription of rat TRPV1 utilizes a dual promoter system that is positively regulated by nerve growth factor. *J Neurochem.*

Zhang A, Xu C, Liang S, Gao Y, Li G, Wei J, Wan F, Liu S, Lin J (2008) Role of sodium ferulate in the nociceptive sensory facilitation of neuropathic pain injury mediated by P2X(3) receptor. *Neurochem Int* 53:278–282.

Zhang X, Huang J McNaughton PA (2005) NGF rapidly increases membrane expression of TRPV1 heat-gated ion channels. *EMBO* J 24: 4211–4223.

Zhou XF, Deng YS, Chie E, Xue Q, Zhong JH, McLachlan EM, Rush RA, Xian CJ (1999) Satellite-cell-derived nerve growth factor and neurotrophin-3 are involved in noradrenergic sprouting in the dorsal root ganglia following peripheral nerve injury in the rat. *Eur J Neurosci* 11:1711–1722.

Zhu W, Galoyan SM, Petruska JC, Oxford GS, Mendell LM (2004) A developmental switch in acute sensitization of small dorsal root ganglion (DRG) neurons to capsaicin or noxious heating by NGF. *J Neurophysiol* 92:3148–3152.

Zwick M, Davis BM, Woodbury CJ, Burkett JN, Koerber HR, Simpson JF, Albers KM (2002) Glial cell line-derived neurotrophic factor is a survival factor for isolectin B4-positive, but not vanilloid receptor 1-positive, neurons in the mouse. *J Neurosci* 22:4057–4065.

Zwick M, Molliver DC, Lindsay J, Fairbanks CA, Sengoku T, Albers KM, Davis BM (2003) Transgenic mice possessing increased numbers of nociceptors do not exhibit increased behavioral sensitivity in models of inflammatory and neuropathic pain. *Pain* 106:491–500.

Zylka MJ, Dong X, Southwell AL, Anderson DJ (2003) Atypical expansion in mice of the sensory neuron-specific Mrg G protein-coupled receptor family. *PNAS* 100: 10043–10048.

Zylka MJ, Rice FL, Anderson DJ (2005) Topographically distinct epidermal nociceptive circuits revealed by axonal tracers targeted to Mrgprd. *Neuron* 45: 17–25.

3 The Role of Visceral Afferents in Disease

Julie A. Christianson and Brian M. Davis

CONTENTS

3.1 INTRODUCTION

Visceral pain is the number one reason for patient visits in the United States. In many cases, visceral pain is not associated with obvious pathology. For example, irritable bowel syndrome (IBS), which can occur following inflammation (Gwee et al. 1996; Collins et al. 1999; Bercik et al. 2005), is a diagnosis of exclusion because its hallmarks include abdominal pain accompanied by diarrhea or constipation in the absence of any obvious pathophysiology. It has been proposed that one of the contributing factors to these persistent pain states is chronic hypersensitivity of visceral sensory neurons (Wood 2002; Cenac et al. 2007).

Afferents innervating somatic tissue, such as skin, muscle, or bone, can be categorized based on their response properties to stimulation. Large, myelinated afferents generally mediate information related to proprioception and light touch or vibration, whereas small, thinly myelinated or unmyelinated afferents, commonly termed nociceptors, detect noxious or potentially damaging stimuli, including thermal, high-threshold mechanical and chemical stimuli. This is in contrast to the sensory innervation of the viscera, which is mostly made up of small, thinly myelinated or unmyelinated afferents that display low mechanical thresholds, enabling them to code normal physiological stimuli (i.e., non-noxious), as well as an ability to code stimuli in the noxious range (Sengupta and Gebhart 1994a, 1994b; Wood 2002; Cenac et al. 2007). Thus, if one uses a functional definition for nociceptors (i.e., the ability to code noxious stimuli), most visceral afferents would be classified as nociceptors. To further separate themselves from somatic afferents, which receive sensory innervation only from neurons located in the dorsal root ganglia (DRG), visceral structures from the esophagus to the transverse colon are innervated not only by DRG located in the cervical, thoracic, and upper lumbar regions, but also by sensory neurons arising from the superior and inferior vagal ganglia (jugular and nodose ganglia, respectively; Figure 3.1) (Ricco et al. 1996; Undem et al. 2004; Yu et al. 2005; Zhong et al. 2008). Visceral structures located distal to the transverse colon, particularly the distal colon, rectum and bladder are also innervated by two populations of afferents; however, these are both of spinal origin arising from two different levels of the spinal cord (thoracolumbar and lumbosacral; Figure 3.1) (de Groat 1987; Keast and de Groat 1992; Wang et al. 1998; Traub et al. 1999; Christianson et al. 2006a, 2007). Sensory neurons arising from these two spinal locations appear to convey different aspects of the complex sensation that humans identify as visceral pain. The functional difference between these populations is not as obvious as that between vagal and spinal afferents, but evidence suggests that they may differentially respond to injury and disease (Traub 2000; Traub and Murphy 2002; Lin and Al-Chaer 2003).

In this chapter, we will discuss recent findings regarding the anatomy and physiology of visceral afferents and how these discoveries may lead to new treatments for visceral pain. In addition, we will discuss exciting new studies that suggest hyperactive visceral nociceptors might not only mediate persistent visceral pain, but that they may actually drive the initial visceral disease processes.

3.2 ANATOMICAL CHARACTERIZATION OF VISCERAL AFFERENTS

3.2.1 DISTRIBUTION AND NEUROCHEMICAL PHENOTYPE OF UPPER GI AND RESPIRATORY SENSORY AFFERENTS

Sensory innervation of upper abdominal viscera (e.g., stomach, pancreas, small intestine) arises via the vagus and spinal nerves with cell bodies in the nodose (NG) and dorsal root ganglia (DRG), respectively (Figure 3.1). Retrograde labeling has commonly been used to evaluate the distribution and neurochemical phenotype of visceral afferents. In a recent study, we used differently conjugated forms of the β-subunit of cholera toxin (CTB) to retrogradely label vagal and spinal afferents from the head and tail of the mouse pancreas (Fasanella et al. 2008). CTB-positive neurons were evenly distributed between the NG and DRG, with the latter containing pancreatic afferents from T5 to T13 with a peak between T9 and T12 (Figure 3.1). Injections into the pancreatic head produced the greatest number of retrogradely labeled afferents, coinciding with a previous study showing that the highest density of sensory innervation was observed in the head region for both myelinated and CGRP-positive afferents, with significantly diminished innervation in the pancreatic tail (Lindsay et al. 2006). A similar study by Won et al. (Won et al. 1998) also

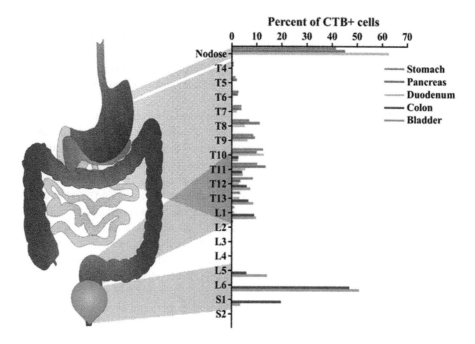

FIGURE 3.1 (see color insert following page 166) Vertebral distribution of visceral afferents innervating different organs. Sensory neurons innervating visceral structures in the mouse were retrogradely labeled using Alexa fluor-conjugated cholera toxin β (CTB) and the distribution of afferents innervating the stomach (Zhong et al. 2008), pancreas (Fasanella et al. 2008), duodenum (Zhong et al. 2008), colon (Christianson et al. 2006a, 2007), or bladder (Christianson et al. 2007) in nodose or dorsal root ganglia (DRG) in studies from our laboratories are expressed as a percentage of total CTB-positive afferents in the above graph..

reported a larger number of DRG neurons retrogradely labeled from the head of the rat pancreas versus the tail. Moreover, vagal afferents innervating the head of the pancreas were observed in the left NG, whereas the majority of the vagal pancreatic tail afferents were observed in the right NG (Sharkey and Williams 1983; Sharkey et al. 1984; Carobi 1987; Sternini et al. 1992; Fasanella et al. 2008). Pancreatic spinal afferents have been shown to have an opposite pattern of innervation, with the right DRG neurons innervating the head of the pancreas and the left DRG innervating the tail (Won et al. 1998). However, in our studies in mouse, we saw no preference among spinal pancreatic afferents (Fasanella et al. 2008).

Similar studies have been conducted to determine the distribution of gastric and small bowel sensory afferents (Sharkey et al. 1984; Green and Dockray 1987, 1988; Zhong et al. 2008). In our study of the mouse, approximately 40% of CTB-positive afferents innervating the fundus, corpus, and/or pylorus of the stomach were observed in the NG (Zhong et al. 2008). The remaining 60% originated from T4 to L2 DRG, with a peak in T10 or T11 (Figure 3.1). Similar distributions were observed in earlier publications of rat gastric afferents (Sharkey et al. 1984; Green and Dockray 1987, 1988). Vagal gastric afferents were more prevalent in the left NG in the rat with a propensity to locate toward the peripheral pole (Sharkey et al. 1984), an observation not reported in the mouse. Duodenal afferents in the mouse were much less prevalent than gastric afferents; however, they were also equally distributed between NG and DRG (Figure 3.1) (Zhong et al. 2008). Jejunal afferents in the mouse, retrogradely labeled with CTB, were observed in T9-T13 DRG; however, NG afferents were not evaluated in this study (Tan et al. 2008).

Sensory innervation of the respiratory and upper gastrointestinal tract has most thoroughly been examined using guinea pig, which unlike rodents, have distinguishable nodose and jugular (aka, superior vagal) ganglia. These two structures are embryologically distinct with nodose arising from placodal tissue and jugular arising from neural crest, similar to DRG. Retrogradely labeled esophageal afferents were observed in both nodose and jugular ganglia and to a much lesser extent in T1-T4 DRG (Yu et al. 2005; Kwong et al. 2008). Vagal afferents innervating the lungs are equally distributed between nodose and jugular ganglia (Undem et al. 2004) with a much smaller contribution of sensory afferents arising from T1-T4 DRG (Kwong et al. 2008). Similar distributions have also been observed when examining guinea pig esophageal (Yu et al. 2005; Kwong et al. 2008) and tracheal afferents (Ricco et al. 1996).

The neurochemical phenotype of these visceral afferents appears to be determined more by the ganglionic source of the afferent, rather than its peripheral target. In general, DRG afferents innervating the pancreas (Sharkey and Williams 1983; Sharkey et al. 1984; Won et al. 1998; Fasanella et al. 2008), stomach (Sharkey et al. 1984; Green and Dockray 1987, 1988; Zhong et al. 2008), duodenum (Zhong et al. 2008), and jejunum (Tan et al. 2008) are largely peptidergic, expressing calcitonin gene-related peptide (CGRP), Substance P, and/or transient receptor potential vanilloid 1 (TRPV1; Table 3.1). In contrast, very few NG afferents innervating the same viscera are peptidergic (Sharkey et al. 1984; Green and Dockray 1987, 1988; Fasanella et al. 2008; Zhong et al. 2008). A specific example of embryonic sources, rather than the innervation targets, driving the expression of nociception-related molecules is the expression patterns of specific P2X receptors among guinea pig lung afferents (Kwong et al. 2008). Neural crest-derived afferents (jugular/DRG) express the homomeric P2X3 receptor complex, which elicits a very short, sometimes nonexistent, inward current when exposed to ATP or a,β-methyl ATP. In contrast, the placodally derived nodose lung afferents express the heteromeric P2X2/P2X3 receptor complex, which elicits a long-lasting, action potential-driving inward current when exposed to the same agonist. The same expression patterns were observed in tracheal and esophageal afferents, further supporting the segregation between nodose and jugular/DRG afferents.

The anatomical complexity of innervation of these visceral structures almost certainly contributes to the difficulty experienced by clinicians when attempting to treat visceral pain. Each population of sensory neurons (i.e., nodose, jugular, spinal) has different central connections, neurochemistry, and physiological properties. Therapies that might positively affect one population of afferents could be ineffective or even potentiate signals within a separate population of afferents. As such, some investigators propose that pain might be due to an abnormal combination of signals arising from the periphery as opposed to an increased excitability within any one population of afferents (F. Rice, pers. comm.).

TABLE 3.1

The Percentage of Retrogradely-Identified Neurons in the Mouse or Guinea Pig That Show Immunohistochemical Reactivity to the Listed Molecules

	TRPV1 (Cap-R)	CGRP/SubP	GFRα3	N52	IB4
Stomach[1]	N: 51	N: 10	–	N: 23	N: 41
	D: 77	D: 83	–	D: 21	D: 12
Duodenum[1]	N: 27	N: 12	–	N: 19	N: 39
	D: 59	D: 82	–	D: 39	D: 3
Jejunum[2]	D: 82	D: 65	–	–	D: 10
Pancreas[3]	N: 35.2	N: 14.9	N: 1.0	–	N: 10.6
	D: 74.9	D: 64.5	D: 67.5	–	D: 2.5
Colon[4]	D (TL): 81.4	D (TL): 77.8	D (TL): 100#	D (TL): 19.0	D (TL): 4.3
	D (LS): 65.0	D (LS): 56.5	D (LS): 60.5#	D (LS): 19.1	D (LS): 6.4
Trachea[5,6]	V: 77*	N: 1.5	–	N: 98.5	–
		J: 50	–	J: 49	–
Lung[6]	V: 80*	N: 36.6	–	N: 36	N: 68.6
		J: 71.1	–	J: 21	J: 74.2
Esophagus[7]	N: 0 (A-fibers),	N: 8.4	–	N: 56	–
	94 (C-fibers)	J: 68	–	J: 8	–
	J: 100				

Note: Transient receptor potential vanilloid 1 (TRPV1) or capsaicin-responsiveness (Cap-R) indicative of TRPV1 activation; calcitonin gene-related peptide (CGRP) or substance P (SubP); the artemin coreceptor GFRα3; a marker of large, myelinated afferents, N52; isolectin B4 (IB4). Examined ganglia include nodose (N), jugular (J), combined vagal afferents (V), and dorsal root ganglia (DRG) including thoracolumbar (TL; T12–L1) and lumbosacral (LS; L5–S1) segments.

* B.J. Undem, personal communication.

\# J.A. Christianson, unpublished observation.

[1] Zhong et al. 2008.

[2] Tan et al. 2008.

[3] Fasanella et al. 2008.

[4] Christianson et al. 2006a.

[5] Ricco et al. 1996.

[6] Undem et al. 2004.

[7] Yu et al. 2005.

3.2.2 DISTRIBUTION AND NEUROCHEMICAL PHENOTYPE OF PELVIC VISCERAL SENSORY AFFERENTS

Sensory innervation of the pelvic viscera arises via the hypogastric/lumbar colonic nerves and pelvic nerve with central terminals in the thoracolumbar (TL) and lumbosacral (LS) spinal cord, respectively (Al-Chaer and Traub 2002) (Figure 3.1). Some discrepancies exist within the literature as to the relative contribution of these two nerves toward the sensory innervation of the pelvic viscera. In a study comparing the effectiveness of CTB, wheat germ agglutinin (WGA), and isolectin B4 (IB4)) to retrogradely label bladder afferents in the rat, the majority of all retrogradely labeled DRG neurons were observed in the L6 and S1 ganglia with a smaller percentage in T13-L2 (Wang et al. 1998). Our studies using CTB to retrogradely label bladder afferents showed a similar distribution (greater LS contribution than TL) in both mouse and rat (Figure 3.1) (Christianson et al. 2007). Fast blue similarly retrogradely labeled a large population of bladder-specific lumbosacral ganglia in the rat; however, they did not look at the thoracolumbar ganglia for comparison (Bennett et al. 1996). Our studies of CTB-labeled colon afferents in mouse revealed a larger LS contribution than TL (Figure 3.1) (Christianson et al. 2006a, 2007), as did a separate study by Tan et al. (2008). However, we obtained contradictory results in the rat, with one study showing a larger population in the TL ganglia (Christianson et al. 2006a) and a second study showing a larger population in the LS ganglia (Christianson et al. 2007). This discrepancy may be explained by a more caudal injection site for the latter study, as a loose rostrocaudal orientation of sensory innervation was observable in both the mouse and rat colon (Figure 3.2) (Christianson et al. 2006a). Additional studies using various retrograde tracers to evaluate sensory innervation of the colon have reported similar disparate results. Fast blue injected into the descending colon has been shown to label predominantly the TL afferents in the mouse (Robinson et al. 2004) and rat (Ness and Gebhart 1988); however, the latter study was not meant to provide a quantitative measure. Similarly, a study using FluoroGold to evaluate colon afferents in the rat showed a higher percentage of labeled afferents in the TL ganglia than in the LS ganglia (Traub et al. 1999).

Our study revealed that the majority of the colon afferents in the mouse were immunopositive for TRPV1 (67%) or CGRP (63%) (Table 3.1) (Christianson et al. 2006a). The high percentage of neurons expressing each marker makes it likely that many neurons expressed both CGRP and TRPV1. Indeed, the study by Tan et al. (2008) revealed a 75%–100% overlap between Substance P- and TRPV1-positive colon afferents. In our study, a significantly greater percentage of TL afferents expressed these markers than LS afferents (Christianson et al. 2006a). This finding compliments a study by Brierley et al. (2004), which reported that approximately 50% of hypogastric afferents in the mouse innervate the mesentery surrounding the colon, whereas no mesenteric afferents were observed in the pelvic nerve. The majority of mesenteric afferents has been observed near or on blood vessels (Morrison 1973; Brunsden et al. 2002) and are also known to express CGRP and TRPV1 (Silverman and Kruger 1988; Stander et al. 2004). This notion is supported by studies examining the distribution and neurochemical phenotype of colon afferents in the mouse. The study by Tan et al. (2008) reported a similar distribution (LS>TL) of colon afferents, as well as a similar percentage of TRPV1-positive colon afferents as our study. Following the same pattern is the study by Robinson et al. (2004), which reported a greater population of TL colon afferents, as compared with LS afferents, with a correspondingly higher percentage of both TRPV1- (82%) and CGRP-positive (81%) mouse colon afferents than observed in our study. These three studies combined suggest that the peripheral target of the afferent may, at least partially, determine receptor expression and neurochemical phenotype.

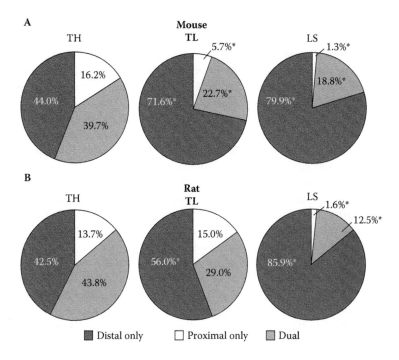

FIGURE 3.2 Innervation of colon by thoracolumbar and lumbosacral afferents and the percent of neurons that project to both proximal and distal colon. In both the mouse and rat, the percentage of distally projecting colonic afferents (black segment) was significantly higher within the most caudal ganglia (lumbosacral, LS) than within the most rostral ganglia (TH; T10–T12). Likewise, the percentage of proximally (white segment) and dually (gray segment) projecting colonic afferents was significantly higher in the most rostral ganglia (TH) than in the most caudal ganglia (LS; L5–S1). In the mouse, the composition of the LS ganglia is significantly different from that of both the TH and thoracolumbar (TL; T13–L1) ganglia. In the rat, the LS ganglionic composition is only significantly different from the TH ganglia. These results indicate that the rostral ganglia send a larger proportion of afferents to the rostral portion of the descending colon, whereas the caudal ganglia preferentially innervate the more caudal regions of the descending colon, suggesting a viscerotopic organization of colonic innervation. Data represented as means. * $P < 0.05$ vs. TH. $^+$ $P < 0.05$ vs. mouse (Christianson et al. 2006a).

In our study, most colon afferents in the rat were TRPV1- (83%) or CGRP-positive (87%), significantly higher than what we observed in the mouse (Christianson et al. 2006a). This is a slightly higher percentage of CGRP-positive afferents than reported in a previous study by Traub et al. (1999) using FluoroGold to identify colon afferents in the rat. However, neither study reported a difference in expression between TL and LS colon afferents. A higher percentage of IB4-positive colon afferents was observed in the rat, compared with mouse, which may underlie the higher percentage of TRPV1-positive afferents considering that IB4/TRPV1-positive afferents are commonly found in rat, but are rarely observed in mouse (Guo et al. 1999; Zwick et al. 2002; Lawson et al. 2004). Both species had a relatively low percentage of colon afferents that expressed neurofilament heavy-chain (N52-positive). This was

not surprising considering that N52 immunoreactivity generally indicates myelination and colon afferents have been reported to be mostly unmyelinated C or A∂
fibers (Sengupta and Gebhart 1994a). It is likely that the N52-positive colon afferents observed in our study were thinly myelinated A∂-fiber nociceptors, considering
that approximately 20% of mouse and 60% of rat N52-positive colon afferents also
expressed TRPV1 (Christianson et al. 2006a).

Similar to the innervation pattern of more proximal GI organs, those of the lower
GI tract, including those located in the pelvis, have complex innervation patterns
arising from ganglia located at multiple vertebral levels. This diffuse innervation
pattern presents a unique challenge to clinicians attempting to use nerve blocks to
treat chronic pain arising from these structures. For more proximal structures, such
as the stomach or pancreas, injections of analgesics or neurolytics into the celiac
ganglion can affect the majority of spinal afferents, as well as the vagal afferents that
project through this structure. No similar peripheral nexus exists for lower abdominal or pelvic structures, although hypogastric blocks can be effective (Kapural et
al. 2006). An alternative approach is disruption of the dorsal column visceral pain
pathway (Willis et al. 1999; Willis and Westlund 2001; Palecek 2004). This pathway was originally identified in patients being treated for intractable cancer pain
(Gildenberg and Hirshberg 1984). The goal of this study was to sever commissural
axons contributing to the classical pain pathway of the spinothalamic tract (STT).
One patient in particular had superior relief from pain arising from colon cancer. A
collaboration between the clinicians involved in the original study and the laboratory
of Dr. William Willis found that the success of the lesion was due not to disruption
of the STT but to a previously unidentified pathway traveling in the medial margin of
the dorsal columns, synapsing in the nuclei gracilus and cuneatus (Willis et al. 1999).
Transection of this pathway has now been shown to be effective for blocking visceral pain arising from both pelvic and abdominal structures (Kim and Kwon 2000;
Houghton et al. 2001). Because this surgery involves irreversible lesions of the spinal
cord, it is used only in terminal patients or patients with intractable, severe pain. As
discussed below, an alternative approach would be to identify distinguishing features of visceral afferents that would allow these fibers to be selectively targeted.
This type of approach is the focus of both academic and pharmaceutical industry
investigations.

3.2.3   DICHOTOMIZING VISCERAL SENSORY AFFERENTS

Single afferents that innervate two different visceral structures have been identified by both dual retrograde labeling studies and/or electrophysiological recordings
(Bahns et al. 1987; de Groat 1987; Keast and de Groat 1992; Malykhina et al. 2004,
2006; Christianson et al. 2007; Zhong et al. 2008). In earlier studies using retrograde tracers, de Groat found that 3%–6% of afferents innervating the colon and
urogenital organs in both the Wistar rat and the cat were dually labeled (de Groat
1987; Keast and de Groat 1992). Physiological existence of dichotomizing afferents
was also observed by Bahns et al. (1987) who found single units that were excited
by stimulation of two separate pelvic organs. Similar findings were also noted by
Berkley et al. (1990), and more recently, by Malykhina et al. (2004, 2006) who have

physiologically characterized dually labeled DRG neurons projecting to both the bladder and colon. Sengupta and Gebhart (1994b) reported no evidence of individual sacral dorsal root afferents responding to distension of both the colon and bladder; however, it is possible that afferents innervating more than one organ are either not responsive to mechanical stretch (e.g., they are mucosal or serosal afferents) (Brierley et al. 2004) or that they are "silent" nociceptors, which become mechanically sensitive following inflammation or other insult.

In our studies, we have observed dichotomous afferents innervating viscera in both the upper abdomen and pelvis. Injection of differently conjugated forms of CTB into the mouse stomach and duodenum resulted in double-labeled neurons in both NG and DRG (Zhong et al. 2008). The extent of double labeling was greatest following injection into consecutive areas of the stomach and decreased as the injection sites grew further apart. The pattern of dichotomous afferents followed that of single-organ afferents, in that there were more observed in DRG than NG with a similar distribution among spinal ganglia. In a similar study, differently conjugated forms of CTB were injected into the colon and bladder of both mouse and rat (Christianson et al. 2007). We observed that 17% and 21% of the total CTB-positive DRG neurons were dually labeled in the rat and mouse, respectively, suggesting that a single afferent was capable of supplying both the bladder and bowel. Dichotomizing afferents in the mouse were distributed similarly to those that innervated only the colon or bladder, with a significantly higher prevalence among LS afferents than TL. However, dichotomizing afferents in the rat were most prevalent in the TL ganglia with a smaller proportion observed in the LS ganglia. This observation provides a possible anatomical substrate for cross-organ sensitization, considering previous studies have shown that pelvic nerve afferents in the rat respond to acute, mechanical stimuli in the pelvic viscera, whereas hypogastric afferents do not respond to acute stimuli but rather mediate inflammatory pain and become mechanically sensitive following injury or inflammation (Berkley et al. 1993a, 1993b; Traub 2000; Mitsui et al. 2001; Sandner-Kiesling et al. 2002; Lin and Al-Chaer 2003).

Dually projecting visceral afferents have been proposed to contribute to a number of human chronic visceral pain syndromes. For example, chronic pelvic pain (CPP) encompasses a number of potentially debilitating syndromes and accounts for 10% of all gynecological visits (Reiter 1990). Common syndromes associated with CPP are irritable bowel syndrome (IBS), interstitial cystitis (IC), vulvadynia, and endometriosis. Although the highest prevalence of CPP is seen among women of reproductive age, men with CPP often present with symptoms of IBS, IC, or chronic prostatitis (CP). Comorbidity of CPP-associated syndromes is remarkably high with 40%–60% of IBS patients also presenting with IC (Prior et al. 1989) and 30% and 26% of IC patients also presenting with IBS (Alagiri et al. 1997) and vulvadynia (Fitzpatrick et al. 1993), respectively. The symptoms of IC and CP are often overlapping, making it difficult to properly diagnosis the correct syndrome (Eisenberg and Moldwin 2003). Cross-sensitization between pelvic organs has been demonstrated following acute colonic inflammation in rats. Instillation of trinitrobenzene sulfonic acid (TNBS) into the distal colon produces a robust inflammatory response used to model inflammatory bowel disease. Rats treated with TNBS show an increased sensitivity to both colonic distension, as well as bladder distension (Pezzone et al.

2005). Further investigation by a second group revealed an increased excitability in DRG neurons that innervate both the colon and bladder, identified by retrograde tracing (Malykhina et al. 2006). Most importantly, there was no evidence of bladder inflammation, indicating that information relayed from an inflamed or damaged peripheral target can affect the interpretation of signals from an unaffected organ or perhaps change the efferent functions of peripheral endings in those unaffected organs, thereby initiating a "noxious state."

3.3   ROLE OF SPECIFIC CHANNELS IN VISCERAL SENSATION AND PERSISTENT VISCERAL PAIN

Anatomically, visceral nociceptors have not been reported to have any obvious specialization of their endings that allow them to differentially respond to the various stimuli that have been shown to excite these afferents. However, recent studies have identified a number of channels that may confer modality-specific sensitivity of visceral afferents. These include purinergic channels including P2X and P2Y families (Cockayne et al. 2000; Burnstock 2001; Yiangou et al. 2001; Dang et al. 2005; Xu et al. 2008), the family of acid sensitive ion (ASIC) channels (Jones et al. 2005; Sugiura et al. 2005; Hughes et al. 2007; Jones et al. 2007; Bielefeldt and Davis 2008), protease activated receptor 2 (PAR2) (Kawabata et al. 2006; Cenac et al. 2007; Laukkarinen et al. 2008), and members of the transient receptor potential (TRP) family (see below). This review will focus on the TRP channels, as these have recently been implicated as central players in a number of visceral diseases and may be excellent targets for future therapies.

As noted above, visceral nociceptors are comprised of a heterogeneous population of afferents that, for some organs, have been classified based on which layer of tissue they innervate. For example, in the colon, there are separate classes of afferents that appear to monitor changes in the mucosa, muscle layers, serosa, and mesenteries. Some afferents appear to innervate more than one tissue layer (e.g., muscular–mucosal afferents described by Lynn and Blackshaw, 1999). Similar afferent classes have also been described in the bladder (Xu and Gebhart 2008) and esophagus (Page and Blackshaw 1998). One theme that does appear to be common among all visceral afferents is the presence of low- and high-threshold afferents (Sengupta and Gebhart 1994a; Rong et al. 2002; Daly et al. 2007; Malin et al. 2009); that is, one population of afferents that can code non-noxious, as well as noxious stimuli, and a second population that is not engaged unless more intense and potentially damaging stimuli are encountered. It has been proposed that these relatively insensitive fibers are normally silent and only become active following injury or in disease. For example, studies by Gebhart and colleagues have reported that polymodal mechanosensitive afferents in the rat colon could be distinguished based on their response to colorectal distension (induced by inflating a small balloon in the colon). Low-threshold (LT) afferents had extrapolated thresholds that were 10 times smaller than high-threshold (HT) afferents and exhibited twice the firing frequency to noxious distension (Sengupta and Gebhart 1994a). Similar observations have been made in studies of mouse jejunal afferents (Rong et al. 2002), rat bladder (Sengupta and

Gebhart 1994b), rat duodenal (Qin et al. 2007), and rat stomach (Ozaki and Gebhart 2001) afferents. Assuming that noxious distension pressures are rarely encountered in the course of normal function, these fibers may not be involved in signaling normal mechanosensory information to the CNS. However, there are data from experimental models that show that both LT and HT afferents exhibit higher firing rates and that HT afferents exhibit lower thresholds in response to inflammatory soup (a combination of histamine, bradykinin, prostaglandin E2, and serotonin) (Rong et al. 2002; Xu and Gebhart 2008). This suggestion that sensitized HT afferents exhibit decreased thresholds may explain the allodynia experienced during certain human disease states where stimuli that are normally non-noxious induce considerable pain (Chan et al. 2003) (see below).

The molecular basis underlying the difference in mechanical thresholds between LT and HT visceral afferents has yet to be definitively proven. However, a recent study from our lab (Malin et al. 2009) found at least two potential correlates from the TRP family: TRPV1 and TRPA1 (Figure 3.3). As noted above, the majority of visceral afferents, regardless of target organ, are peptidergic and express TRPV1 (Table 3.1). All of the mechanosensitive colon neurons in our study that responded

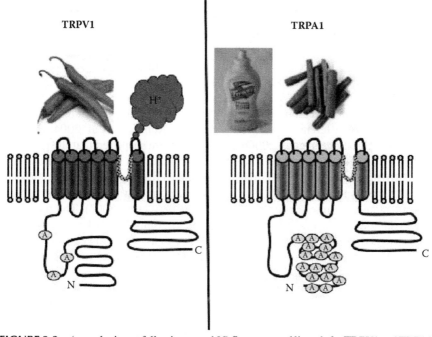

FIGURE 3.3 (see color insert following page 166) Structure and ligands for TRPV1 and TRPA1. Both TRPA1 and TRPV1 are non-selective cation channels and have six transmembrane domains with the pore-forming region contained in a loop between the fifth and sixth transmembrane domain. TRPV1 respond to exogenous compounds like the active ingredient in chili peppers (capsaicin) and endogenous molecules like protons. TRPA1 binds a number of exogenous pungent compounds like isothiocyanate (mustard oil) and cinnamonaldehyde (found in cinnamon). It also binds a large number of derivatives of membrane lipids (see text). (Reprinted with permission from Christianson JA, Traub RJ, Davis BM (2006a) Differences in spinal distribution and neurochemical phenotype of colonic afferents in mouse and rat. *J Comp Neurol* 494:246–259.)

to the TRPA1 agonist mustard oil were also TRPV1-positive indicating a high level of coexpression for these two receptors in visceral afferents (Malin et al. 2009). Both TRPV1 and TRPA1 are non-selective cation channels that respond to multiple stimuli. TRPV1 conducts inward current in response to protons, noxious heat, exogenous vanilloid compounds like capsaicin, lipids (e.g., anademide and N-arachidonoyl-dopamine) (Van Der Stelt and Di Marzo 2004; Suh and Oh 2005; Starowicz et al. 2007), and more recently the second messenger signal molecule diacylglycerol DAG (Woo et al. 2008). TRPA1 responds to noxious cold, exogenous compounds including mustard oil and cinnamaldehyde, and to endogenous products derived from plasma membrane including 4-hydroxynonenal and to inflammatory mediators such as H_2O_2, and 15-deoxy-delta(12,14)-prostaglandin J(2) (15d-PGJ(2)) (Trevisani et al. 2007) Andersson et al. 2008; Cruz-Orengo et al. 2008; Sawada et al. 2008). Using a novel *ex vivo* preparation that allowed us to record intracellularly from intact colon afferents while distending the colon in a manner similar to studies of *in vivo* colorectal distension, we were able to physiologically and neurochemically characterize mechanosensitive colon afferents (Malin et al. 2009). Similar to previous reports, we found that the colon was innervated by both LT and HT colon afferents; HT afferents had average thresholds four times greater than LT fibers and fired at significantly lower frequencies. Surprisingly, 87% of all HT afferents were TRPV1-positive, whereas 87% of LT were TRPV1-negative. Functional expression of TRPV1 was demonstrated by the ability of capsaicin to potentiate the response of TRPV1-immunopositive neurons to distension. Furthermore, the vast majority of TRPV1-responsive afferents were also potentiated by the TRPA1 agonist mustard oil.

Although our initial emphasis on TRPV1 was based on reports showing that the majority of colon afferents were TRPV1-positive (Robinson et al. 2004; Brierley et al. 2005; Christianson et al. 2006b, 2006a), we were surprised that the presence or absence of TRPV1 would be so closely correlated with mechanical sensitivity considering that neuronal expression of this channel has not typically been associated with mechanosensation. TRPA1, on the other hand, was originally proposed as a potential mechanical sensor in hair cells (Corey et al. 2004). The theory of TRPV1 only being involved in regulation of thermosensation is largely rooted in studies involving inflammation and damage of somatic tissue. Data from studies of visceral afferents have often suggested that TRPV1 might affect mechanosensation. For example, studies by Birder and colleagues found altered bladder micturation in TRPV1-/- mice (Birder et al. 2002; Birder 2006) and more recently, Jones et al. (2005) found that both TRPV1 and ASIC3 null mice had decreased visceromotor and single-fiber responses to colorectal distension. ASIC3 has been proposed to be a mechanical sensor (Page et al. 2005; Sugiura et al. 2005; Hughes et al. 2007; Bielefeldt and Davis 2008); however, the mechanical phenotype of ASIC3 null mice, with respect to colon sensitivity, was milder than that observed in TRPV1 null mice. Neither the ASIC3 or TRPV1 null mice displayed a total loss of mechanical sensitivity within visceral afferents; however, the mechanical sensitivity of certain populations of somatic sensory afferents was either increased (rapidly adapting mechanoreceptors) or decreased (slowly conducting myelinated mechanoreceptors) (Price et al. 2001). Taken together, these results suggest that

TRPV1 is not directly transducing mechanical stimuli. Rather, its detection of and subsequent excitation by specific chemical ligands might be regulating the overall sensitivity of sensory neurons to a variety of stimuli, including mechanosensory activation.

The role of TRPV1 in mediating responses to mechanical stimulation is best illustrated in the presence of injury or inflammation. A number of recent studies have shown that mice lacking TRPV1 do not develop hyperalgesia associated with visceral insult. Jones et al. (2005) reported that TRPV1 null mice, unlike controls, do not display visceral hypersensitivity following an installation of inflammatory mediators into the colon. Infusion of lipopolysaccarides (LPS) into the rat or mouse bladder is a common model for human cystitis and produces significant hypersensitivity when performed in wildtype mice. When this is done in TRPV1-/-mice, distension of the bladder does not produce an increased expression of the early immediate gene, fos, in the dorsal horn of the spinal cord, indicating that LPS-induced inflammation is not painful in the absence of functional TRPV1 (Charrua et al. 2007).

Another TRP channel, TRPV4, has been shown to affect visceral mechanosensation. Like TRPV1, TRPV4 is expressed in the majority of colon afferents and is coexpressed with CGRP (Brierley et al. 2008; Cenac et al. 2008) and therefore most likely with TRPV1. TRPV4 is also expressed in vagal afferents, including those innervating the stomach (Zhang et al. 2004). Deletion of TRPV4 selectively decreases the mechanical sensitivity of serosal and mesenteric colon afferents, but not that of muscular or mucosal colon afferents or vagal esophageal afferents (Brierley et al. 2008). This suggests that the ability of TRPV4 to regulate mechanosensation is determined by the context in which it is expressed. Interestingly, two different laboratories have shown that TRPV4 is required for PAR2-induced hypersensitivity (Cenac et al. 2008; Sipe et al. 2008). Increased release of activators of PAR2 has been observed in biopsies of patients with IBS (Cenac et al. 2007). Supernatants of these biopsies produced significant visceral hypersensitivity when instilled into the colons of control mice, but not in PAR2 null mice (Cenac et al. 2007). In mice lacking TRPV4 or in which TRPV4 has been downregulated using siRNA, dramatically attenuated responses to PAR2 activation are observed (Brierley et al. 2008; Cenac et al. 2008). This observation is reminiscent of the inability of TRPV1-/- mice to develop inflammatory hyperalgesia, suggesting the presence of redundant mechanisms for induction of visceral hypersensitivity.

3.4 ROLE OF VISCERAL AFFERENTS IN DISEASE

3.4.1 Effect of Neonatal Injury on Adult Pain Processing

Exposure to noxious stimuli during early neonatal periods can have long-term effects on nociceptive processing (Fitzgerald 1995; Anand 2001; Fitzgerald and Beggs 2001; Lidow 2002). Several studies have shown that children exposed to excessive noxious stimulation in the neonatal intensive care unit (NICU) can later develop decreased behavioral responses and increased physiological responses to painful events (Anand 1998; Whitfield and Grunau 2000). The effect of neonatal organ insult on adult nociceptive processing has not been directly addressed in humans; however, a possible clinical correlation exists within IC patients as 10%–30% report having experienced childhood bladder problems, including infection (Jones and Nyberg 1997). The effect of early bladder inflammation has recently been investigated by treating young rats with zymosan, a cell wall component of yeast that produces a robust inflammatory response (Randich et al. 2006; DeBerry et al. 2007). Intravesical instillation of zymosan was performed on postnatal days (PN) 14–16 or 28–30, to determine the consequences of bladder inflammation during neonatal or adolescent periods, respectively. When the rats were reinflamed with zymosan as adults and tested for abdominal electromyographic (EMG) responses to urinary bladder distension (UBD), the rats that received neonatal zymosan treatments exhibited a much higher mean EMG response than controls or rats treated with zymosan as adolescents. The rats that received neonatal zymosan treatments also exhibited a higher arterial blood pressure response to UBD (Randich et al. 2006). Additional studies have shown that the endogenous opioid-inhibitory system may be disrupted in these neonatally inflamed rats, as treatment with naloxone did not enhance EMG responses to UBD, unlike its effect in acute bladder inflammation in naïve rats (DeBerry et al. 2007).

Similar phenomena have also been demonstrated following neonatal colon irritation. Neonatal rats that receive noxious mechanical or chemical irritation of the colon during the second week of life display more severe abdominal contractions in response to colorectal distension (CRD) during adulthood (Al-Chaer et al. 2000; Winston et al. 2007). The study by Al-Chaer et al. (2000) reported increased firing rates of dorsal horn neurons in response to either CRD or cutaneous stimulation. A subsequent study revealed that the number of TL spinal afferents responding to CRD was increased and LS spinal afferents exhibit increased background activity and decreased activation thresholds (Lin and Al-Chaer 2003). Studies from the Pasricha lab have investigated the roles of specific receptors in mediating neonatally induced changes in viscerosensitivity, particularly TRPV1 (Winston et al. 2007) and P2X3 (Xu et al. 2008). The percentage of colon-specific afferents expressing these two receptors was increased, and treatment with antagonists for either receptor, either prior to the neonatal insult or in adults, was shown to decrease visceral hypersensitivity. It is important to note that the neonatal injuries sustained in these experiments occurred after the critical period that has been established for somatic afferents to induce long-term changes in nociceptive processing (Ruda et al. 2000; Fitzgerald and Beggs 2001; Lidow 2002). However, when injury occurred after PN21, no changes in visceral sensitivity were observed in the adult rats, indicating that visceral and somatic afferents may have different periods of vulnerability to alterations in nociceptive processing.

These animal studies have had a real impact on the treatment of pain in infants and children, especially those that require extended stays in NICUs. Until as recently as a few decades ago, it was widely believed that young infants could not perceive or process nociceptive input; therefore pain in neonatal patients was largely left untreated. Recent studies have proven this theory to be untrue, as responses to noxious stimuli have been recorded in cortical areas of neonates as young as 25 weeks gestational age (Bartocci et al. 2006; Slater et al. 2006). This finding is especially important considering that a preterm neonate receives, on average, 14 painful interventions a day during the first two weeks in the NICU (Simons et al. 2003). Assessment of neonatal pain relies on measuring behavioral indicators, such as facial expression, body movement and vocalization, and physiological indicators, such as heart and respiratory rate, blood pressure, and oxygen desaturation. Neonates can be treated pharmacologically with local anesthetics for acute procedures, such as a heel lance, or with systemic analgesics for chronic pain conditions or following surgery. However, the pharmacokinetics for many analgesics is unknown for neonates and infants. In addition, there is still concern that prolonged use could lead to complications arising from lack of sensory input (Walker 2008). These concerns have lead to the use of non-pharmacological interventions, such as "non-nutritive suckling," "swaddling and positioning," and "facilitated tucking." These therapies have been shown to significantly attenuate pain scores; however, the effectiveness of these procedures differs depending on whether neonates are evaluated for behavioral or physiological indicators (reviewed in Cignacco et al. 2007). One of the most effective non-pharmacological treatments for alleviating neonatal procedural pain is nutritive suckling with sucrose that is thought to be mediated by endogenous opioid pathways activated by sweet taste (Stevens et al. 2005). Regardless the treatment type, the long-term effects of treating neonatal pain have yet to be determined, underscoring the need for further examination in this important area of pain research.

3.4.2 Changes in Visceral Afferents Can Contribute to the Development of Visceral Disease

As noted in the Introduction, recent studies have implicated changes in visceral afferents as not only responsible for ongoing visceral pain but also in a role of initiating or contributing to disease progression. Some of the best examples come from studies of pancreatic diseases including diabetes, pancreatic cancer, and pancreatitis. In all of these cases the evidence points to a central role for TRPV1-expressing pancreatic afferents. For example, TRPV1 mRNA levels were significantly increased in pancreata from patients suffering from either pancreatic cancer or pancreatitis, compared to healthy control tissue (Hartel et al. 2006). In cancer patients, the level of TRPV1 mRNA was positively correlated with their reported pain scores; however, the same observation was not made among pancreatitis patients. Underlying this difference might be the expression of TRPV1 on the cancer cells themselves, as observed histologically. Treatment of pancreatic cancer cell lines with resiniferotoxin (RTX), which binds TRPV1 and ablates sensory neurons, reduced cancer cell growth and induced apoptosis above what was observed following treatment with chemotherapeutic agents (Hartel et al. 2006). It has long been known that pancreatic tumors are accompanied by hypertrophy of pancreatic nerve bundles, which is likely due to the reported increased production of neurotrophic factors (including nerve growth factor [NGF] and artemin) that occurs in cancerous pancreata (Zhu et al. 1999, 2001; Okada et al. 2004; Ito et al. 2005; Ceyhan et al. 2006; Ma et al. 2008). Receptors for these growth factors are expressed on the same neurons that express TRPV1 and TRPA1 (Orozco et al. 2001; Elitt et al. 2006; Malin et al. 2006; Malin et al. 2009), and brief exposure of NGF or artemin can potentiate TRPV1 and

TRPA1 function in dissociated neurons (Malin et al. 2006). The hypertrophied nerve bundles in combination with potentiated TRPV1 and TRPA1 likely contribute to the intense pain that accompanies most pancreatic cancers. In addition, the hypertrophied nerve bundles have been shown to form a conduit on which tumor cells travel to metastasize to adjacent organs (Ceyhan et al. 2006). This observation has led to the proposal that sprouting and hypertrophy of neuronal processes is the third leg of a metastatic triad, also including proliferation of blood and lymphatic vessels, which serves to promote survival and spread of tumor cells (Ceyhan et al. 2008b, 2008a; Schneider et al. 2008).

Evidence that visceral afferents contribute to the development of disease via their efferent function (e.g., via peripheral release of peptides) comes from studies of pancreatitis and diabetes. Animal models of acute pancreatitis have demonstrated that afferents expressing TRPV1 are essential for the onset and maintenance of pancreatitis and associated pain. TRPV1 mRNA and protein expression were both increased in pancreatic DRG neurons in a rat model of chronic pancreatitis, and treatment with TRPV1 antagonist has been shown to reduce both visceral and referred somatic pain behaviors (Xu et al. 2007). Ablation of the TRPV1-positive afferent fibers by either neonatal capsaicin treatment (Nathan et al. 2002) or via resiniferatoxin (Noble et al. 2006; Romac et al. 2008) significantly reduced substance P release and pancreatic inflammation and/or prevented the development of pancreatitis. It is important to note that the expression of TRPV1 is not required for afferent-induced pancreatic inflammation, as mice lacking functional TRPV1 developed pancreatitis similar to control mice (Xu et al. 2007). The efferent functions of the afferent, as well as the expression of other receptors, is sufficient for the initiation and maintenance of pancreatitis. PAR2, previously mentioned for its role in IBS, is activated endogenously by release of trypsin or tryptase, both of which are increased in pancreatitis (Namkung et al. 2008). PAR2 activation has been shown to sensitize TRPV1 in vitro (Hoogerwerf et al. 2001) and has been implicated in pancreatitis (Hoogerwerf et al. 2004; Kawabata et al. 2006; Ishikura et al. 2007; Laukkarinen et al. 2008; Namkung et al. 2008).

Another intriguing observation implicating efferent function of primary afferents in visceral disease comes from studies of a mouse model of type 1 diabetes (Razavi et al. 2006). In diabetes-prone non-obese diabetic (NOD) mice, islet inflammation develops spontaneously as the animal ages, leading to insulin resistance, degeneration of the pancreas, and early mortality. Ablation of TRPV1-positive fibers with neonatal capsaicin blocks the development of islet inflammation and abnormal glucose regulation. However, the exact role of these afferents in disease progression is complicated. The *nod* locus that is responsible for development of diabetes contains TRPV1, and TRPV1 function in these mice is reduced both with respect to capsaicin-induced currents and amount of mRNA produced. Moreover, infusion of substance P, which would normally be released by the TRPV1-expressing afferents, also prevents progression of diabetes. Thus, the diabetic phenotype in these mice can be reversed by either ablating the defective fibers or replacing the neuroactive peptide (SP) that these fibers would normally release. These data indicate that these fibers are intimately involved in pancreatic health. The authors of this report suggest that the apparently conflicting results can be explained by the need for a balance between inflammatory β-cells and primary afferents. This balance is disrupted in the NOD mice by the TRPV1 hypomorph and can be restored by either ablation of these fibers or by restoration of SP levels in the pancreas (Razavi et al. 2006). Clearly, more work needs to be done, but these results are exciting inasmuch as they open up a completely new way to treat a complicated and debilitating disease.

3.5 SUMMARY

Visceral afferents have received significantly less attention than their somatic counterparts. This is in large part because they have been more difficult to study due to their small numbers and inaccessibility. However, development of new anatomical, physiological, and molecular techniques has made it possible to examine these neurons at the same level of detail as other sensory neurons. The changes that occur in visceral afferents not only play central roles in the development of chronic pain states, but in some cases they may initiate or maintain disease states, as well. Phenotypic characterization of visceral neurons is already well underway, and initial studies have identified unique neurochemical properties, which might make them attractive targets for new pharmacological approaches. Potential treatment options may pharmacologically silence specific visceral afferents or, in extreme cases, ablate pathological neurons while leaving the sensory neurons that are required for maintaining homeostasis untouched.

REFERENCES

Al-Chaer ED, Kawasaki M, Pasricha PJ (2000) A new model of chronic visceral hypersensitivity in adult rats induced by colon irritation during postnatal development. *Gastroenterology* 119:1276–1285.

Al-Chaer ED, Traub RJ (2002) Biological basis of visceral pain: recent developments. *Pain* 96:221–225.

Alagiri M, Chottiner S, Ratner V, Slade D, Hanno PM (1997) Interstitial cystitis: unexplained associations with other chronic disease and pain syndromes. *Urology* 49:52–57.

Anand KJ (1998) Clinical importance of pain and stress in preterm neonates. *Biol Neonate* 73:1–9.

Anand KJ (2001) Consensus statement for the prevention and management of pain in the newborn. *Arch Pediatr Adolesc Med* 155:173–180.

Andersson DA, Gentry C, Moss S, Bevan S (2008) Transient receptor potential A1 is a sensory receptor for multiple products of oxidative stress. *J Neurosci* 28:2485–2494.

Bahns E, Halsband U, Janig W (1987) Responses of sacral visceral afferents from the lower urinary tract, colon and anus to mechanical stimulation. *Pflugers Arch* 410:296–303.

Bartocci M, Bergqvist LL, Lagercrantz H, Anand KJ (2006) Pain activates cortical areas in the preterm newborn brain. *Pain* 122:109–117.

Bennett DL, Dmietrieva N, Priestley JV, Clary D, McMahon SB (1996) trkA, CGRP and IB4 expression in retrogradely labelled cutaneous and visceral primary sensory neurones in the rat. *Neurosci Lett* 206:33–36.

Bercik P, Verdu EF, Collins SM (2005) Is irritable bowel syndrome a low-grade inflammatory bowel disease? *Gastroenterol Clin North Am* 34:235–245, vi–vii.

Berkley KJ, Hotta H, Robbins A, Sato Y (1990) Functional properties of afferent fibers supplying reproductive and other pelvic organs in pelvic nerve of female rat. *J Neurophysiol* 63:256–272.

Berkley KJ, Hubscher CH, Wall PD (1993a) Neuronal responses to stimulation of the cervix, uterus, colon, and skin in the rat spinal cord. *J Neurophysiol* 69:545–556.

Berkley KJ, Robbins A, Sato Y (1993b) Functional differences between afferent fibers in the hypogastric and pelvic nerves innervating female reproductive organs in the rat. *J Neurophysiol* 69:533–544.

Bielefeldt K, Davis BM (2008) Differential effects of ASIC3 and TRPV1 deletion on gastroesophageal sensation in mice. *Am J Physiol Gastrointest Liver Physiol* 294:G130–138.

Birder LA (2006) Urinary bladder urothelium: molecular sensors of chemical/thermal/mechanical stimuli. *Vascul Pharmacol* 45:221–226.

Birder LA, Nakamura Y, Kiss S, Nealen ML, Barrick S, Kanai AJ, Wang E, et al. (2002) Altered urinary bladder function in mice lacking the vanilloid receptor TRPV1. *Nat Neurosci* 5:856–860.

Brierley SM, Jones RC, 3rd, Gebhart GF, Blackshaw LA (2004) Splanchnic and pelvic mechanosensory afferents signal different qualities of colonic stimuli in mice. *Gastroenterology* 127:166–178.

Brierley SM, Jones RC, 3rd, Xu L, Gebhart GF, Blackshaw LA (2005) Activation of splanchnic and pelvic colonic afferents by bradykinin in mice. *Neurogastroenterol Motil* 17:854–862.

Brierley SM, Page AJ, Hughes PA, Adam B, Liebregts T, Cooper NJ, Holtmann G, Liedtke W, Blackshaw LA (2008) Selective role for TRPV4 ion channels in visceral sensory pathways. *Gastroenterology*.

Brunsden AM, Jacob S, Bardhan KD, Grundy D (2002) Mesenteric afferent nerves are sensitive to vascular perfusion in a novel preparation of rat ileum in vitro. *Am J Physiol Gastrointest Liver Physiol* 283:G656–665.

Burnstock G (2001) Purine-mediated signalling in pain and visceral perception. *Trends Pharmacol Sci* 22:182–188.

Carobi C (1987) Capsaicin-sensitive vagal afferent neurons innervating the rat pancreas. *Neurosci Lett* 77:5–9.

Cenac N, Altier C, Chapman K, Liedtke W, Zamponi G, Vergnolle N (2008) Transient receptor potential vanilloid-4 has a major role in visceral hypersensitivity symptoms. *Gastroenterology* 135:937–946, 946 e931–932.

Cenac N, Andrews CN, Holzhausen M, Chapman K, Cottrell G, Andrade-Gordon P, Steinhoff M, et al. (2007) Role for protease activity in visceral pain in irritable bowel syndrome. *J Clin Invest* 117:636–647.

Ceyhan GO, Demir IE, Altintas B, Rauch U, Thiel G, Muller MW, Giese NA, Friess H, Schafer KH (2008b) Neural invasion in pancreatic cancer: a mutual tropism between neurons and cancer cells. *Biochem Biophys Res Commun* 374:442–447.

Ceyhan GO, Giese NA, Erkan M, Kerscher AG, Wente MN, Giese T, Buchler MW, Friess H (2006) The neurotrophic factor artemin promotes pancreatic cancer invasion. *Ann Surg* 244:274–281.

Ceyhan GO, Michalski CW, Demir IE, Muller MW, Friess H (2008a) Pancreatic pain. *Best Pract Res Clin Gastroenterol* 22:31–44.

Chan CL, Facer P, Davis JB, Smith GD, Egerton J, Bountra C, Williams NS, Anand P (2003) Sensory fibres expressing capsaicin receptor TRPV1 in patients with rectal hypersensitivity and faecal urgency. *Lancet* 361:385–391.

Charrua A, Cruz CD, Cruz F, Avelino A (2007) Transient receptor potential vanilloid subfamily 1 is essential for the generation of noxious bladder input and bladder overactivity in cystitis. *J Urol* 177:1537–1541.

Christianson JA, Liang R, Ustinova EE, Davis BM, Fraser MO, Pezzone MA (2007) Convergence of bladder and colon sensory innervation occurs at the primary afferent level. *Pain* 128:235–243.

Christianson JA, McIlwrath SL, Koerber HR, Davis BM (2006b) Transient receptor potential vanilloid 1-immunopositive neurons in the mouse are more prevalent within colon afferents compared to skin and muscle afferents. *Neuroscience* 140:247–257.

Christianson JA, Traub RJ, Davis BM (2006a) Differences in spinal distribution and neurochemical phenotype of colonic afferents in mouse and rat. *J Comp Neurol* 494:246–259.

Cignacco E, Hamers JP, Stoffel L, van Lingen RA, Gessler P, McDougall J, Nelle M (2007) The efficacy of non-pharmacological interventions in the management of procedural pain in preterm and term neonates. A systematic literature review. *Eur J Pain* 11:139–152.

Cockayne DA, Hamilton SG, Zhu QM, Dunn PM, Zhong Y, Novakovic S, Malmberg AB, et al. (2000) Urinary bladder hyporeflexia and reduced pain-related behaviour in P2X3-deficient mice. *Nature* 407:1011–1015.

Collins SM, Barbara G, Vallance B (1999) Stress, inflammation and the irritable bowel syndrome. *Can J Gastroenterol* 13 Suppl A:47A–49A.

Corey DP, Garcia-Anoveros J, Holt JR, Kwan KY, Lin SY, Vollrath MA, Amalfitano A, et al. (2004) TRPA1 is a candidate for the mechanosensitive transduction channel of vertebrate hair cells. *Nature* 432:723–730.

Cruz-Orengo L, Dhaka A, Heuermann RJ, Young TJ, Montana MC, Cavanaugh EJ, Kim D, Story GM (2008) Cutaneous nociception evoked by 15-delta PGJ2 via activation of ion channel TRPA1. *Mol Pain* 4:30.

Daly D, Rong W, Chess-Williams R, Chapple C, Grundy D (2007) Bladder afferent sensitivity in wild-type and TRPV1 knockout mice. *J Physiol* 583:663–674.

Dang K, Bielfeldt K, Lamb K, Gebhart GF (2005) Gastric ulcers evoke hyperexcitability and enhance P2X receptor function in rat gastric sensory neurons. *J Neurophysiol* 93:3112–3119.

de Groat WC (1987) Neuropeptides in pelvic afferent pathways. *Experientia* 43:801–813.

DeBerry J, Ness TJ, Robbins MT, Randich A (2007) Inflammation-induced enhancement of the visceromotor reflex to urinary bladder distention: modulation by endogenous opioids and the effects of early-in-life experience with bladder inflammation. *J Pain* 8:914–923.

Eisenberg ER, Moldwin RM (2003) Etiology: where does prostatitis stop and interstitial cystitis begin? *World J Urol* 21:64–69.

Elitt CM, McIlwrath SL, Lawson JJ, Malin SA, Molliver DC, Cornuet PK, Koerber HR, Davis BM, Albers KM (2006) Artemin overexpression in skin enhances expression of TRPV1 and TRPA1 in cutaneous sensory neurons and leads to behavioral sensitivity to heat and cold. *J Neurosci* 26:8578–8587.

Fasanella KE, Christianson JA, Chanthaphavong RS, Davis BM (2008) Distribution and neurochemical identification of pancreatic afferents in the mouse. *J Comp Neurol* 509:42–52.

Fitzgerald M (1995) Developmental biology of inflammatory pain. *Br J Anaesth* 75:177–185.

Fitzgerald M, Beggs S (2001) The neurobiology of pain: developmental aspects. *Neuroscientist* 7:246–257.

Fitzpatrick CC, DeLancey JO, Elkins TE, McGuire EJ (1993) Vulvar vestibulitis and interstitial cystitis: a disorder of urogenital sinus-derived epithelium? *Obstet Gynecol* 81:860–862.

Gildenberg PL, Hirshberg RM (1984) Limited myelotomy for the treatment of intractable cancer pain. *J Neurol Neurosurg Psychiatry* 47:94–96.

Green T, Dockray GJ (1987) Calcitonin gene-related peptide and substance P in afferents to the upper gastrointestinal tract in the rat. *Neurosci Lett* 76:151–156.

Green T, Dockray GJ (1988) Characterization of the peptidergic afferent innervation of the stomach in the rat, mouse and guinea-pig. *Neuroscience* 25:181–193.

Guo A, Vulchanova L, Wang J, Li X, Elde R (1999) Immunocytochemical localization of the vanilloid receptor 1 (VR1): relationship to neuropeptides, the P2X3 purinoceptor and IB4 binding sites. *Eur J Neurosci* 11:946–958.

Gwee KA, Graham JC, McKendrick MW, Collins SM, Marshall JS, Walters SJ, Read NW (1996) Psychometric scores and persistence of irritable bowel after infectious diarrhoea. *Lancet* 347:150–153.

Hartel M, di Mola FF, Selvaggi F, Mascetta G, Wente MN, Felix K, Giese NA, et al. (2006) Vanilloids in pancreatic cancer: potential for chemotherapy and pain management. *Gut* 55:519–528.

Hoogerwerf WA, Shenoy M, Winston JH, Xiao SY, He Z, Pasricha PJ (2004) Trypsin mediates nociception via the proteinase-activated receptor 2: a potentially novel role in pancreatic pain. *Gastroenterology* 127:883–891.

Hoogerwerf WA, Zou L, Shenoy M, Sun D, Micci MA, Lee-Hellmich H, Xiao SY, Winston JH, Pasricha PJ (2001) The proteinase-activated receptor 2 is involved in nociception. *J Neurosci* 21:9036–9042.

Houghton AK, Wang CC, Westlund KN (2001) Do nociceptive signals from the pancreas travel in the dorsal column? *Pain* 89:207–220.

Hughes PA, Brierley SM, Young RL, Blackshaw LA (2007) Localization and comparative analysis of acid-sensing ion channel (ASIC1, 2, and 3) mRNA expression in mouse colonic sensory neurons within thoracolumbar dorsal root ganglia. *J Comp* Neurol 500:863–875.

Ishikura H, Nishimura S, Matsunami M, Tsujiuchi T, Ishiki T, Sekiguchi F, Naruse M, Nakatani T, Kamanaka Y, Kawabata A (2007) The proteinase inhibitor camostat mesilate suppresses pancreatic pain in rodents. *Life Sci* 80:1999–2004.

Ito Y, Okada Y, Sato M, Sawai H, Funahashi H, Murase T, Hayakawa T, Manabe T (2005) Expression of glial cell line-derived neurotrophic factor family members and their receptors in pancreatic cancers. *Surgery* 138:788–794.

Jones CA, Nyberg L (1997) Epidemiology of interstitial cystitis. *Urology* 49:2–9.

Jones RC, 3rd, Otsuka E, Wagstrom E, Jensen CS, Price MP, Gebhart GF (2007) Short-term sensitization of colon mechanoreceptors is associated with long-term hypersensitivity to colon distention in the mouse. *Gastroenterology* 133:184–194.

Jones RC, 3rd, Xu L, Gebhart GF (2005) The mechanosensitivity of mouse colon afferent fibers and their sensitization by inflammatory mediators require transient receptor potential vanilloid 1 and acid-sensing ion channel 3. *J Neurosci* 25:10981–10989.

Kapural L, Narouze SN, Janicki TI, Mekhail N (2006) Spinal cord stimulation is an effective treatment for the chronic intractable visceral pelvic pain. *Pain Med* 7:440–443.

Kawabata A, Matsunami M, Tsutsumi M, Ishiki T, Fukushima O, Sekiguchi F, Kawao N, Minami T, Kanke T, Saito N (2006) Suppression of pancreatitis-related allodynia/hyperalgesia by proteinase-activated receptor-2 in mice. *Br J Pharmacol* 148:54–60.

Keast JR, de Groat WC (1992) Segmental distribution and peptide content of primary afferent neurons innervating the urogenital organs and colon of male rats. *J Comp Neurol* 319:615–623.

Kim YS, Kwon SJ (2000) High thoracic midline dorsal column myelotomy for severe visceral pain due to advanced stomach cancer. *Neurosurgery* 46:85–90; discussion 90-82.

Kwong K, Kollarik M, Nassenstein C, Ru F, Undem BJ (2008) P2X2 receptors differentiate placodal vs. neural crest C-fiber phenotypes innervating guinea pig lungs and esophagus. *Am J Physiol Lung Cell Mol Physiol* 295:L858–865.

Laukkarinen JM, Weiss ER, van Acker GJ, Steer ML, Perides G (2008) Protease-activated receptor-2 exerts contrasting model-specific effects on acute experimental pancreatitis. *J Biol Chem* 283:20703–20712.

Lawson JJ, McIlwrath SL, Woodbury CJ, Davis BM, Koerber HR (2004) Mouse cutaneous C-fibers containing TRPV1 are responsive to heat, but mechanically insensitive. *Soc Neurosci Abstr* 288.5.

Lidow MS (2002) Long-term effects of neonatal pain on nociceptive systems. *Pain* 99:377–383.

Lin C, Al-Chaer ED (2003) Long-term sensitization of primary afferents in adult rats exposed to neonatal colon pain. *Brain Res* 971:73–82.

Lindsay TH, Halvorson KG, Peters CM, Ghilardi JR, Kuskowski MA, Wong GY, Mantyh PW (2006) A quantitative analysis of the sensory and sympathetic innervation of the mouse pancreas. *Neuroscience* 137:1417–1426.

Lynn PA, Blackshaw LA (1999) In vitro recordings of afferent fibres with receptive fields in the serosa, muscle and mucosa of rat colon. *J Physiol* 518 (Pt 1):271–282.

Ma J, Jiang Y, Jiang Y, Sun Y, Zhao X (2008) Expression of nerve growth factor and tyrosine kinase receptor A and correlation with perineural invasion in pancreatic cancer. *J Gastroenterol Hepatol* 23:1852–1859.

Malin SA, Christianson JA, Bielefeldt K, Davis BM (2009) TRPV1 expression defines functionally distinct pelvic colon afferents. *J Neurosci* 29:743–752.

Malin SA, Molliver DC, Koerber HR, Cornuet P, Frye R, Albers KM, Davis BM (2006) Glial cell line-derived neurotrophic factor family members sensitize nociceptors in vitro and produce thermal hyperalgesia in vivo. *J Neurosci* 26:8588–8599.

Malykhina AP, Qin C, Foreman RD, Akbarali HI (2004) Colonic inflammation increases Na+ currents in bladder sensory neurons. *Neuroreport* 15:2601–2605.

Malykhina AP, Qin C, Greenwood-van Meerveld B, Foreman RD, Lupu F, Akbarali HI (2006) Hyperexcitability of convergent colon and bladder dorsal root ganglion neurons after colonic inflammation: mechanism for pelvic organ cross-talk. *Neurogastroenterol Motil* 18:936–948.

Mitsui T, Kakizaki H, Matsuura S, Ameda K, Yoshioka M, Koyanagi T (2001) Afferent fibers of the hypogastric nerves are involved in the facilitating effects of chemical bladder irritation in rats. *J Neurophysiol* 86:2276–2284.

Morrison JF (1973) Splanchnic slowly adapting mechanoreceptors with punctate receptive fields in the mesentery and gastrointestinal tract of the cat. *J Physiol* 233:349–361.

Namkung W, Yoon JS, Kim KH, Lee MG (2008) PAR2 exerts local protection against acute pancreatitis via modulation of MAP kinase and MAP kinase phosphatase signaling. *Am J Physiol Gastrointest Liver Physiol* 295:G886–894.

Nathan JD, Peng RY, Wang Y, McVey DC, Vigna SR, Liddle RA (2002) Primary sensory neurons: a common final pathway for inflammation in experimental pancreatitis in rats. *Am J Physiol Gastrointest Liver Physiol* 283:G938–946.

Ness TJ, Gebhart GF (1988) Characterization of neurons responsive to noxious colorectal distension in the T13–L2 spinal cord of the rat. *J Neurophysiol* 60:1419–1438.

Noble MD, Romac J, Wang Y, Hsu J, Humphrey JE, Liddle RA (2006) Local disruption of the celiac ganglion inhibits substance P release and ameliorates caerulein-induced pancreatitis in rats. *Am J Physiol Gastrointest Liver Physiol* 291:G128–134.

Okada Y, Eibl G, Guha S, Duffy JP, Reber HA, Hines OJ (2004) Nerve growth factor stimulates MMP-2 expression and activity and increases invasion by human pancreatic cancer cells. *Clin Exp Metastasis* 21:285–292.

Orozco OE, Walus L, Sah DW, Pepinsky RB, Sanicola M (2001) GFRalpha3 is expressed predominantly in nociceptive sensory neurons. *Eur J Neurosci* 13:2177–2182.

Ozaki N, Gebhart GF (2001) Characterization of mechanosensitive splanchnic nerve afferent fibers innervating the rat stomach. *Am J Physiol Gastrointest Liver Physiol* 281:G1449–1459.

Page AJ, Blackshaw LA (1998) An in vitro study of the properties of vagal afferent fibres innervating the ferret oesophagus and stomach. *J Physiol* 512 (Pt 3):907–916.

Page AJ, Brierley SM, Martin CM, Price MP, Symonds E, Butler R, Wemmie JA, Blackshaw LA (2005) Different contributions of ASIC channels 1a, 2, and 3 in gastrointestinal mechanosensory function. *Gut* 54:1408–1415.

Palecek J (2004) The role of dorsal columns pathway in visceral pain. *Physiol Res* 53 Suppl 1:S125–130.

Pezzone MA, Liang R, Fraser MO (2005) A model of neural cross-talk and irritation in the pelvis: implications for the overlap of chronic pelvic pain disorders. *Gastroenterology* 128:1953–1964.

Price MP, McIlwrath SL, Xie J, Cheng C, Qiao J, Tarr DE, Sluka KA, Brennan TJ, Lewin GR, Welsh MJ (2001) The DRASIC cation channel contributes to the detection of cutaneous touch and acid stimuli in mice. *Neuron* 32:1071–1083.

Prior A, Wilson K, Whorwell PJ, Faragher EB (1989) Irritable bowel syndrome in the gynecological clinic. Survey of 798 new referrals. *Dig Dis Sci* 34:1820–1824.

Qin C, Chen JD, Zhang J, Foreman RD (2007) Characterization of T9–T10 spinal neurons with duodenal input and modulation by gastric electrical stimulation in rats. *Brain Res* 1152:75–86.

Randich A, Uzzell T, DeBerry JJ, Ness TJ (2006) Neonatal urinary bladder inflammation produces adult bladder hypersensitivity. *J Pain* 7:469–479.

Razavi R, Chan Y, Afifiyan FN, Liu XJ, Wan X, Yantha J, Tsui H, et al. (2006) TRPV1+ sensory neurons control beta cell stress and islet inflammation in autoimmune diabetes. *Cell* 127:1123–1135.

Reiter RC (1990) A profile of women with chronic pelvic pain. *Clin Obstet Gynecol* 33:130–136.

Ricco MM, Kummer W, Biglari B, Myers AC, Undem BJ (1996) Interganglionic segregation of distinct vagal afferent fibre phenotypes in guinea-pig airways. *J Physiol* 496 (Pt 2):521–530.

Robinson DR, McNaughton PA, Evans ML, Hicks GA (2004) Characterization of the primary spinal afferent innervation of the mouse colon using retrograde labelling. *Neurogastroenterol Motil* 16:113–124.

Romac JM, McCall SJ, Humphrey JE, Heo J, Liddle RA (2008) Pharmacologic disruption of TRPV1-expressing primary sensory neurons but not genetic deletion of TRPV1 protects mice against pancreatitis. *Pancreas* 36:394–401.

Rong W, Spyer KM, Burnstock G (2002) Activation and sensitisation of low and high threshold afferent fibres mediated by P2X receptors in the mouse urinary bladder. *J Physiol* 541:591–600.

Ruda MA, Ling QD, Hohmann AG, Peng YB, Tachibana T (2000) Altered nociceptive neuronal circuits after neonatal peripheral inflammation. *Science* 289:628–631.

Sandner-Kiesling A, Pan HL, Chen SR, James RL, DeHaven-Hudkins DL, Dewan DM, Eisenach JC (2002) Effect of kappa opioid agonists on visceral nociception induced by uterine cervical distension in rats. *Pain* 96:13–22.

Sawada Y, Hosokawa H, Matsumura K, Kobayashi S (2008) Activation of transient receptor potential ankyrin 1 by hydrogen peroxide. *Eur J Neurosci* 27:1131–1142.

Schneider G, Hamacher R, Eser S, Friess H, Schmid RM, Saur D (2008) Molecular biology of pancreatic cancer—new aspects and targets. *Anticancer Res* 28:1541–1550.

Sengupta JN, Gebhart GF (1994a) Characterization of mechanosensitive pelvic nerve afferent fibers innervating the colon of the rat. *J Neurophysiol* 71:2046–2060.

Sengupta JN, Gebhart GF (1994b) Mechanosensitive properties of pelvic nerve afferent fibers innervating the urinary bladder of the rat. *J Neurophysiol* 72:2420–2430.

Sharkey KA, Williams RG (1983) Extrinsic innervation of the rat pancreas: demonstration of vagal sensory neurones in the rat by retrograde tracing. *Neurosci Lett* 42:131–135.

Sharkey KA, Williams RG, Dockray GJ (1984) Sensory substance P innervation of the stomach and pancreas. Demonstration of capsaicin-sensitive sensory neurons in the rat by combined immunohistochemistry and retrograde tracing. *Gastroenterology* 87:914–921.

Silverman JD, Kruger L (1988) Lectin and neuropeptide labeling of separate populations of dorsal root ganglion neurons and associated "nociceptor" thin axons in rat testis and cornea whole-mount preparations. *Somatosens Res* 5:259–267.

Simons SH, van Dijk M, Anand KS, Roofthooft D, van Lingen RA, Tibboel D (2003) Do we still hurt newborn babies? A prospective study of procedural pain and analgesia in neonates. *Arch Pediatr Adolesc Med* 157:1058–1064.

Sipe WE, Brierley SM, Martin CM, Phillis BD, Cruz FB, Grady EF, Liedtke W, et al. (2008) Transient receptor potential vanilloid 4 mediates protease activated receptor 2-induced sensitization of colonic afferent nerves and visceral hyperalgesia. *Am J Physiol Gastrointest Liver Physiol* 294:G1288–1298.

Slater R, Cantarella A, Gallella S, Worley A, Boyd S, Meek J, Fitzgerald M (2006) Cortical pain responses in human infants. *J Neurosci* 26:3662–3666.

Stander S, Moormann C, Schumacher M, Buddenkotte J, Artuc M, Shpacovitch V, Brzoska T, et al. (2004) Expression of vanilloid receptor subtype 1 in cutaneous sensory nerve fibers, mast cells, and epithelial cells of appendage structures. *Exp Dermatol* 13:129–139.

Starowicz K, Nigam S, Di Marzo V (2007) Biochemistry and pharmacology of endovanilloids. *Pharmacol Ther* 114:13–33.

Sternini C, De Giorgio R, Furness JB (1992) Calcitonin gene-related peptide neurons innervating the canine digestive system. *Regul Pept* 42:15–26.

Stevens B, Yamada J, Beyene J, Gibbins S, Petryshen P, Stinson J, Narciso J (2005) Consistent management of repeated procedural pain with sucrose in preterm neonates: Is it effective and safe for repeated use over time? *Clin J Pain* 21:543–548.

Sugiura T, Dang K, Lamb K, Bielefeldt K, Gebhart GF (2005) Acid-sensing properties in rat gastric sensory neurons from normal and ulcerated stomach. *J Neurosci* 25:2617–2627.

Suh YG, Oh U (2005) Activation and activators of TRPV1 and their pharmaceutical implication. *Curr Pharm Des* 11:2687–2698.

Tan LL, Bornstein JC, Anderson CR (2008) Distinct chemical classes of medium-sized transient receptor potential channel vanilloid 1-immunoreactive dorsal root ganglion neurons innervate the adult mouse jejunum and colon. *Neuroscience* 156:334–343.

Traub RJ (2000) Evidence for thoracolumbar spinal cord processing of inflammatory, but not acute colonic pain. *Neuroreport* 11:2113–2116.

Traub RJ, Hutchcroft K, Gebhart GF (1999) The peptide content of colonic afferents decreases following colonic inflammation. *Peptides* 20:267–273.

Traub RJ, Murphy A (2002) Colonic inflammation induces fos expression in the thoracolumbar spinal cord increasing activity in the spinoparabrachial pathway. *Pain* 95:93–102.

Trevisani M, Siemens J, Materazzi S, Bautista DM, Nassini R, Campi B, Imamachi N, Andre E, Patacchini R, Cottrell GS, Gatti R, Basbaum AI, Bunnett NW, Julius D, Geppetti P (2007). 4-Hydroxynonenal, an endogenous aldehyde, causes pain and neurogenic inflammation through activation of the irritant receptor TRPA1. *Proc Natl Acad Sci USA* 104:13519–13524.

Undem BJ, Chuaychoo B, Lee MG, Weinreich D, Myers AC, Kollarik M (2004) Subtypes of vagal afferent C-fibres in guinea-pig lungs. *J Physiol* 556:905–917.

Van Der Stelt M, Di Marzo V (2004) Endovanilloids. Putative endogenous ligands of transient receptor potential vanilloid 1 channels. *Eur J Biochem* 271:1827–1834.

Walker SM (2008) Pain in children: recent advances and ongoing challenges. *Br J Anaesth* 101:101–110.

Wang HF, Shortland P, Park MJ, Grant G (1998) Retrograde and transganglionic transport of horseradish peroxidase-conjugated cholera toxin B subunit, wheatgerm agglutinin and isolectin B4 from Griffonia simplicifolia I in primary afferent neurons innervating the rat urinary bladder. *Neuroscience* 87:275–288.

Whitfield MF, Grunau RE (2000) Behavior, pain perception, and the extremely low-birth weight survivor. *Clin Perinatol* 27:363–379.

Willis WD, Al-Chaer ED, Quast MJ, Westlund KN (1999) A visceral pain pathway in the dorsal column of the spinal cord. *Proc Natl Acad Sci U S A* 96:7675–7679.

Willis WD, Jr., Westlund KN (2001) The role of the dorsal column pathway in visceral nociception. *Curr Pain Headache Rep* 5:20–26.

Winston J, Shenoy M, Medley D, Naniwadekar A, Pasricha PJ (2007) The vanilloid receptor initiates and maintains colonic hypersensitivity induced by neonatal colon irritation in rats. *Gastroenterology* 132:615–627.

Won MH, Park HS, Jeong YG, Park HJ (1998) Afferent innervation of the rat pancreas: retrograde tracing and immunohistochemistry in the dorsal root ganglia. *Pancreas* 16:80–87.

Woo DH, Jung SJ, Zhu MH, Park CK, Kim YH, Oh SB, Lee CJ (2008) Direct activation of transient receptor potential vanilloid 1(TRPV1) by diacylglycerol (DAG). *Mol Pain* 4:42.

Wood JD (2002) Neuropathophysiology of irritable bowel syndrome. *J Clin Gastroenterol* 35:S11–22.

Xu GY, Shenoy M, Winston JH, Mittal S, Pasricha PJ (2008) P2X receptor-mediated visceral hyperalgesia in a rat model of chronic visceral hypersensitivity. *Gut.*

Xu GY, Winston JH, Shenoy M, Yin H, Pendyala S, Pasricha PJ (2007) Transient receptor potential vanilloid 1 mediates hyperalgesia and is up-regulated in rats with chronic pancreatitis. *Gastroenterology* 133:1282–1292.

Xu L, Gebhart GF (2008) Characterization of mouse lumbar splanchnic and pelvic nerve urinary bladder mechanosensory afferents. *J Neurophysiol* 99:244–253.

Yiangou Y, Facer P, Baecker PA, Ford AP, Knowles CH, Chan CL, Williams NS, Anand P (2001) ATP-gated ion channel P2X(3) is increased in human inflammatory bowel disease. *Neurogastroenterol Motil* 13:365–369.

Yu S, Undem BJ, Kollarik M (2005) Vagal afferent nerves with nociceptive properties in guinea-pig oesophagus. *J Physiol* 563:831–842.

Zhang L, Jones S, Brody K, Costa M, Brookes SJ (2004) Thermosensitive transient receptor potential channels in vagal afferent neurons of the mouse. *Am J Physiol Gastrointest Liver Physiol* 286:G983–991.

Zhong F, Christianson JA, Davis BM, Bielefeldt K (2008) Dichotomizing axons in spinal and vagal afferents of the mouse stomach. *Dig Dis Sci* 53:194–203.

Zhu Z, Friess H, diMola FF, Zimmermann A, Graber HU, Korc M, Buchler MW (1999) Nerve growth factor expression correlates with perineural invasion and pain in human pancreatic cancer. *J Clin Oncol* 17:2419–2428.

Zhu ZW, Friess H, Wang L, Bogardus T, Korc M, Kleeff J, Buchler MW (2001) Nerve growth factor exerts differential effects on the growth of human pancreatic cancer cells. *Clin Cancer Res* 7:105–112.

Zwick M, Davis BM, Woodbury CJ, Burkett JN, Koerber HR, Simpson JF, Albers KM (2002) Glial cell line-derived neurotrophic factor is a survival factor for isolectin B4-positive, but not vanilloid receptor 1-positive, neurons in the mouse. *J Neurosci* 22:4057–4065.

4 Cancer Pain
From the Development of Mouse Models to Human Clinical Trials

Juan Miguel Jimenez Andrade and Patrick Mantyh

CONTENTS

4.1 INTRODUCTION

Cancer-associated pain can be present at any time during the course of the disease, but the frequency and intensity of cancer pain tends to increase with advancing stages of cancer. In patients with advanced cancer, 62%–86% experience significant pain, which is described as moderate to severe in approximately 40%–50% and as very severe in 25%–30% (van den Beuken-van Everdingen et al. 2007). Bone cancer pain is the most common pain in patients with advanced cancer; two-thirds of patients with metastatic bone disease experience severe pain (Coleman 2006; Mercadante and Fulfaro 2007). Although bone is not a vital organ, many of the most common tumors (breast, prostate, thyroid, kidney, and lung) have a strong predilection for bone metastasis (Figure 4.1). Tumor metastases to the skeleton are major contributors to morbidity and mortality in metastatic cancer. Tumor growth in bone results in pain, hypercalcemia, anemia, increased susceptibility to infection, skeletal fractures,

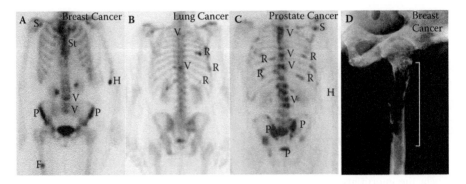

FIGURE 4.1 *Bone metastases* in patients with advanced breast, lung, and prostate cancer. [99m] Technetium-methylene diphosphonate bone scintigraphy demonstrating multifocal increased uptake due to skeletal metastases from a breast (A), lung (B), and prostate carcinoma (C). Note the high prevalence of metastases in vertebra (V), pelvis (P), and femur of breast and prostate carcinoma. Bone metastases ultimately induce skeletal-related events such as fracture, which is seen in the proximal right femur of a 52-year-old woman with bone metastases from a breast carcinoma (D), which negatively influence the functional status, quality of life, and survival of the patient. S=scapule; St=sternum; H=humerus; R=rib. (With permission from Garcia et al. 2003. *European Journal of Nuclear Medicine and Molecular Imaging.*)

compression of the spinal cord, spinal instability, and decreased mobility—all of which compromise the patient's survival and quality of life (Coleman 2006, 2008). Once tumor cells have metastasized to the skeleton, tumor-induced bone pain is usually described as dull in character, constant in presentation, and gradually increasing in intensity with time (Dy et al. 2008). As tumor-induced bone remodeling progresses, severe *incident pain* frequently occurs (Mercadante 1997), and given that the onset of this pain is both acute and unpredictable, this component of bone cancer pain can be highly debilitating to the patient's functional status and quality of life (Coleman 1997; Mercadante 1997). Incident or *breakthrough pain*, which is defined as an intermittent episode of extreme pain, can occur spontaneously, or more commonly is induced by either movement of or weight-bearing on the tumor-bearing bone(s) (Mercadante et al. 2004).

Currently, the treatment of pain from bone metastases involves the use of multiple complementary approaches, including radiotherapy applied to the painful area, surgery, chemotherapy, bisphosphonates, calcitonin, and analgesics (Mercadante 1997; Mercadante and Fulfaro 2007). However, bone cancer pain is one of the most difficult of all persistent pains to fully control (Mercadante 1997) because the metastases are generally not limited to a single site and the analgesics that are most commonly used to treat bone cancer pain—nonsteroidal anti-inflammatory drugs (NSAIDs) (Mercadante 1997) and opioids (Mercadante 1997; Cherny 2000; Lussier et al. 2004; Reid et al. 2008)—are limited by significant adverse side effects (Weber and Huber 1999; Harris 2008).

Over the last decade a major focus of our lab has been to investigate the mechanisms that drive cancer pain (mostly due to bone cancer) and to develop mechanism-based therapies to attenuate the pain and reduce disease progression (Honore et al. 2000; Halvorson et al. 2005; Sevcik et al. 2006). These efforts have provided a new understanding of the factors that drive bone cancer pain and disease progression and have provided the preclinical data that has resulted in several clinical trials of these mechanism-based therapies.

4.2 DEVELOPMENT OF A MURINE MODEL OF BONE CANCER: TUMOR GROWTH, SKELETAL REMODELING, AND PAIN BEHAVIORS

Given the enormous consequences in terms of human suffering that cancer pain can cause, when we first began exploring the mechanisms that drive cancer pain, it was surprising to find that there was simply not a well-established animal model for studying any form of cancer pain. However, there were two commonly used *in vivo* models to study the tumor-induced bone destruction. In the first model, tumor cells are injected into the left ventricle of the heart and then spread to multiple sites including the bone marrow, where they multiply, grow, and destroy the surrounding bone (Arguello et al. 1988; Yoneda et al. 1994). While this model replicates the observation that most tumor cells metastasize to multiple sites including bone, a major problem with this model is the animal-to-animal variability in the sites, size, and extent of the metastasis. Since the tumors frequently metastasize to vital organs such as the lung or liver, the general health of the animal is also variable, making behavioral assessment difficult. Additionally, as the tumors frequently metastasize to bone in the vertebral column, tumor growth in the vertebrae can result in collapse of the vertebral column and compression of the spinal cord with resultant spinal dysfunction and paralysis. Given these problems, the development of a model of bone cancer pain using intracardiac injection proved difficult at best.

The second major model used to study tumor-induced bone destruction involves the direct injection of lytic sarcoma cells into the intramedullary space of the mouse tibia or femur. The major problem with this model was that as the injection site could not be plugged using conventional sealing agents (because it is a wet, bony surface) and the tumor cells rapidly escaped and avidly grew in nearby skin and joints, resulting in large extraskeletal tumor masses that not only interfered with behavioral analysis but also destroyed nerves passing though these sites and generated a neuropathic pain state. We chose to adapt and modify this model by plugging the injection hole with a dental amalgam, which tightly binds and seals the injection hole in the distal head of the femur. This plugging of the injection site allowed us to contain the tumor cells within the intramedullary space and prevented tumor invasion into surrounding soft tissue (Figure 4.2) (Honore et al. 2000). This advance, along with the techniques where we could simultaneously measure bone cancer-induced pain behaviors, tumor growth, and tumor-induced bone remodeling, has provided us with the first preclinical cancer pain model, which we then used to define the mechanisms that generate and maintain bone cancer pain.

Using primarily osteolytic 2472 murine osteosarcoma tumor cells that are injected and confined to the intramedullary space of the mouse femur, these tumor cells

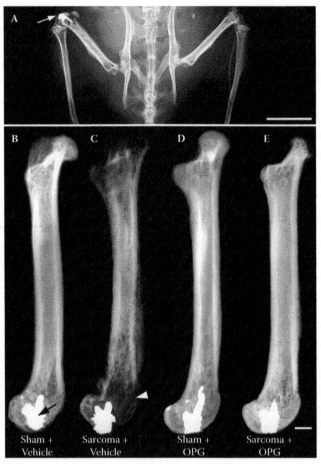

FIGURE 4.2 Osteoprotegerin (OPG) attenuates sarcoma-induced bone destruction in a mouse model of bone cancer pain. A: Lower power frontal radiograph of mouse pelvis and hind limbs following unilateral injection of 2472 murine osteosarcoma cells into the distal end of the femur and closure of the injection with a dental amalgam plug (arrow). The amalgam plug was used to prevent tumor cells from growing outside the bone. High-resolution radiographs of sham-injected (B and D) and sarcoma-injected (C and E) femurs from mice that received vehicle (B and C) or OPG (D and E). Note that at day 17 after the injection of the osteosarcoma cells, there is significant bone destruction in the distal femur without OPG (C; white arrowhead), whereas tumor-induced bone destruction is not evident in sarcoma-injected mouse that received OPG (E). Scale bars represent 10 mm (*a*) and 0.5 mm (*b–e*; bottom panel). (Reprinted with permission from Honore P, Luger NM, Sabino MA, Schwei MJ, Rogers SD, Mach DB, O'Keefe P F, Ramnaraine ML, Clohisy DR, Mantyh PW (2000) Osteoprotegerin blocks bone cancer-induced skeletal destruction, skeletal pain and pain-related neurochemical reorganization of the spinal cord. *Nat Med* 6:521–528.)

grow in a highly reproducible fashion as they proliferate, replacing the hematopoetic cells in the bone marrow (Schwei et al. 1999; Sabino et al. 2002). Eventually, the entire marrow space is homogeneously filled with tumor cells and tumor-associated inflammatory/immune cells. In terms of bone remodeling, injection of osteosarcoma cells to the femur induces a dramatic proliferation and hypertrophy of osteoclasts at the tumor–bone interface, with significant bone destruction in both the proximal and distal heads of the femur (Figure 4.2). In the osteosarcoma model, ongoing pain and movement-evoked pain-related behaviors increased in severity with time. These pain-related behaviors correlated with the tumor growth and progressive tumor-induced bone destruction, which mirrors what occurs in patients with primary or metastatic bone cancer. While the sarcoma cells constituted the tumor used in the first bone cancer pain model, we have since developed other bone cancer pain models using prostate, breast, melanoma, colon, and lung tumors, all of which provided insight into the similarities and differences by which different tumors drive bone cancer pain (Sabino et al. 2003).

4.3 MECHANISMS UNDERLYING BONE CANCER PAIN

4.3.1 TUMOR AND OSTEOCLAST-INDUCED ACIDOSIS AND BONE CANCER PAIN

Recent reports in both murine and human bone cancer pain have suggested that osteoclasts play an essential role in cancer-induced bone loss and that osteoclasts contribute to the etiology of bone cancer pain (Luger et al. 2001; Sabino et al. 2002). Osteoclasts are terminally differentiated, multinucleated, monocyte lineage cells that resorb bone by maintaining an extracellular microenvironment of acidic pH (4.0–5.0) at the osteoclast-mineralized bone interface (Delaisse and Vaes 1992). Tumor-induced release of protons and acidosis may be particularly important in the generation of bone cancer pain. Both osteolytic (bone destroying) and osteoblastic (bone forming) cancers are characterized by osteoclast proliferation and hypertrophy (Clohisy et al. 2000).

Bisphosphonates, a class of anti-resorptive compounds that induce osteoclast apoptosis, have also been reported to reduce pain in patients with osteolytic (primarily bone destroying) and osteoblastic (primarily bone forming) skeletal metastases (Fulfaro et al. 1998; Major et al. 2000; Berenson et al. 2001). Bisphosphonates are pyrophosphate analogues that display high affinity for calcium ions, causing them to rapidly target the mineralized matrix of bone (Rogers et al. 2000). These drugs have been reported to act directly on osteoclasts, inducing their apoptosis by impairing either the synthesis of adenosinetriphosphate or cholesterol, both of which are necessary for cell survival (Gatti and Adami 1999; Rodan and Martin 2000). Osteoclasts treated with bisphosphonates undergo morphologic changes including cell shrinkage, chromatin condensation, nuclear fragmentation, and loss of the ruffled border that are indicative of apoptosis (Rogers et al. 2000). Studies in both clinical (Fulfaro et al. 1998; Major et al. 2000; Berenson et al. 2001) and animal (Sasaki et al. 1995; Hiraga 1996; Yoneda et al. 2000) models of bone cancer have reported anti-resorptive effects of bisphosphonate therapy. The effect bisphosphonates have on tumor growth and long-term survival rate remains controversial.

In a recent study of the bisphosphonate alendronate in the murine 2472 sarcoma model (Sevcik et al. 2004), a reduction in the number of osteoclasts and osteoclast activity was observed. In this model, alendronate also attenuates ongoing and movement-evoked bone cancer pain and the neurochemical reorganization of the peripheral and central nervous system while also promoting both tumor growth and tumor necrosis. These results suggest that in bone cancer, alendronate can simultaneously modulate pain, bone destruction, tumor growth, and tumor necrosis and that administration of alendronate along with a tumoricidal agent may synergistically improve the survival and quality of life of patients with bone cancer pain. Recent studies in humans with ibandronate, a novel nitrogen containing bisphosphonate, has shown remarkable effects in being able to rapidly induce long-lasting relief of bone cancer pain (Tripathy et al. 2004).

Although bisphosphonates are currently being used to reduce tumor-induced bone destruction and bone cancer pain induced by both primarily osteolytic and osteoblastic tumors, the use of osteoprotegerin (OPG) or antibodies that have OPG-like activities (for instance: sequestering antibody AMG-162, which is also known as denosumab) holds significant promise for alleviating bone cancer pain. OPG is a secreted, soluble receptor that is a member of the tumor necrosis factor receptor (TNFR) family (Simonet et al. 1997). This decoy receptor prevents the activation and proliferation of osteoclasts by binding to and sequestering OPG ligand (OPGL; also known as receptor for activator of NFκB ligand, RANKL) (Simonet et al. 1997; Yasuda et al. 1998; Rodan and Martin 2000). While OPG has been shown to decrease pain behaviors in the murine sarcoma model of bone cancer (Figure 4.3) (Luger et al. 2001), a monoclonal antibody (AMG-162, denosumab) that also blocks the interaction of OPGL and

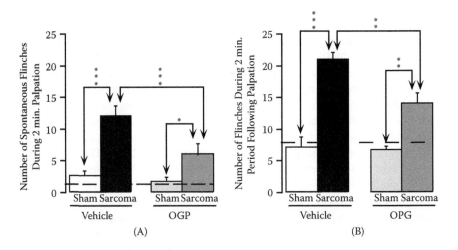

FIGURE 4.3 Osteoprotegerin (OPG) attenuates both spontaneous and movement-evoked bone cancer–related pain behaviors. The number of spontaneous flinching behaviors in a 2-minute observation period (A) and the number of flinching behaviors in a 2-minute observation period after the completion of a normally non-noxious palpation (B) 17 days after sham or sarcoma injection into the femora of mice that subsequently received vehicle or OPG. Daily treatment with OPG significantly reduces spontaneous and movement-evoked nociceptive behaviors indicative of pain. Data represent mean +SEM. Dashed lines, baseline values. *, $P < 0.05$; **, $P < 0.01$; and ***, $P < 0.001$; one-way ANOVA and Fisher PLSD (brackets and downward arrows indicate groups being compared). (With permission from Honore et al. 2002. *Nature Medicine*.)

RANK (Figure 4.4) is being evaluated in late clinical trials for use in attenuating bone loss associated with breast and multiple myeloma cancer (Figure 4.5). Additionally, denosumab is being evaluated for its potential to reduce the skeletal-related events (pain, fracture) due to the spread of cancer to the bone in multiple myeloma and multiple solid tumors and for its potential to delay bone metastases in prostate cancer. These

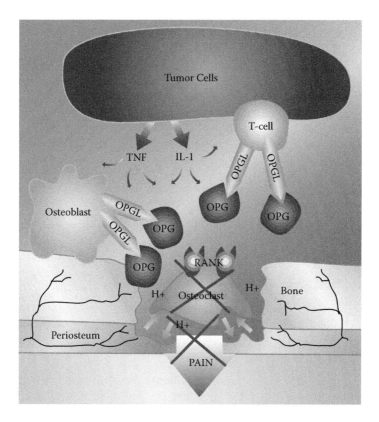

FIGURE 4.4 Proposed mechanism by which osteoprotegerin (OPG) and monoclonal antibodies to RANKL block bone cancer pain. Bone destruction (osteolysis) induced by growth of tumor cancer cells in the bone is accomplished by an increase in the number and activity of osteoclasts whose activity is dependent on osteoprotegerin ligand (OPGL) produced by osteoblasts and activated T cells. The receptor for OPGL, receptor activator of nuclear factor κB (RANK), is expressed in mature osteoclasts. The osteoclasts' activity and induction of bone resorption is dependent on the binding of OPGL (which is also know as RANK ligand; RANKL) to RANK. Malignant tumor cells secrete cytokines and growth factors that activate T cells and osteoclasts. Bone resorption is accompanied by an osteoclast-induced low pH at the tumor–bone interface. Bone cancer pain occurs as the result of activation of acid-sensing ion channels that are expressed by nociceptors, the release of inflammatory mediators, and the activation and sensitization of nociceptive sensory neurons in the richly innervated bone marrow and mineralized bone, which results in bone pain. OPG is a secreted "decoy" TNFR-related receptor that binds and sequesters OPGL/RANKL, thus preventing RANK activation, which is thought to break the bone resorption–pain cycle. (Modified with permission from Thomson and Tonge. 2000. *Nature Medicine*.)

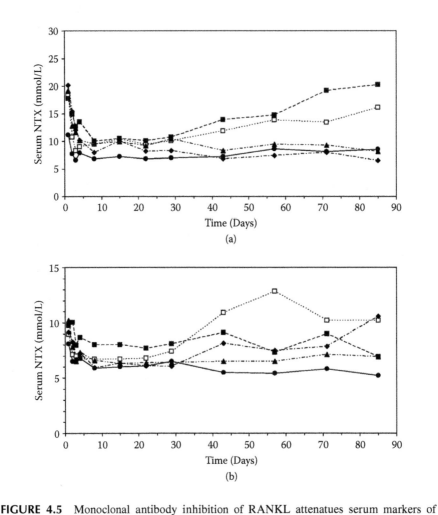

FIGURE 4.5 Monoclonal antibody inhibition of RANKL attenatues serum markers of bone resorption in human patients with breast cancer and multiple myeloma. A randomized, double-blind, double-dummy, active-controlled, multicenter study to determine the safety and efficacy of denosumab in 25 patients with multiple myeloma and 29 patients with breast cancer with bone metastases. Denosumab is a fully human monoclonal antibody that binds with high affinity to and inhibits the activity of human OPGL/RANKL, a key mediator of osteoclast activity. The pharmacodynamic effects of denosumab treatment on bone resorption: absolute median values of second-morning-void urinary NTX/creatinine in patients with breast cancer (A) and multiple myeloma (B). BCE, bone collagen equivalent. □, pamidronate 90 mg i.v.; ■, denosumab 0.1 mg/kg s.c.; ◆, denosumab 0.3 mg/kg s.c.;▲, denosumab 1.0 mg/kg s.c.; ●, denosumab 3.0 mg/kg s.c. Note that not only does denosumab reduce the bone resorption markers following a single subcutaneous (s.c.) dose but also the duration of bone resorption suppression was dose dependent and persisted for up to 84 days after a single s.c. dose of 1.0 or 3.0 mg/kg. (With permission from Body et al. 2006. *Clin Cancer Res.*)

results suggest that a substantial part of the actions of OPG results from inhibition of tumor-induced bone destruction via a reduction in osteoclast function (Figure 4.4). This reduction of osteoclast function in turn inhibits the neurochemical changes associated with peripheral and central sensitization that are thought to be involved in the generation and maintenance of cancer pain.

The finding that sensory neurons can be directly excited by protons or acid originating from cells such as osteoclasts in bone has generated intense clinical interest in pain research (Sutherland et al. 2000). Studies have shown that subsets of sensory neurons express different acid-sensing ion channels (Olson et al. 1998; Julius and Basbaum 2001). Two acid-sensing ion channels expressed by nociceptors are the transient receptor potential vanilloid-1 (TRPV1) (Tominaga et al. 1998; Caterina et al. 2000) and the acid-sensing ion channel-3 (ASIC-3) (Bassilana et al. 1997; Olson et al. 1998; Sutherland et al. 2000). Both of these channels are sensitized and excited by a decrease in pH. The tumor stroma (Griffiths 1991) and areas of the tumor that are necrotic typically exhibit lower extracellular pH than surrounding normal tissues. As inflammatory and immune cells invade the tumor stroma, these cells also release protons that generate a local acidosis. As tumor cells outgrow their vascular supply, tumor apoptosis that occurs in the tumor environment may also contribute to the development of an acidic environment.

It has been shown that TRPV1 is expressed by a subset of sensory nerve fibers that innervate the mouse femur, and that in an *in vivo* model of bone cancer pain, acute or chronic administration of a TRPV1 antagonist or disruption of the TRPV1 gene results in a significant attenuation of both ongoing and movement-evoked nocifensive behaviors (Ghilardi et al. 2005). In addition, previous studies have also shown that in a sarcoma model of bone cancer pain, administration of a TRPV1 antagonist retains its efficacy at early, middle, and late stages of tumor growth (Ghilardi et al. 2005). The ability of a TRPV1 antagonist to maintain its analgesic potency with disease progression is probably influenced by the fact that sensory nerve fibers innervating the tumor-bearing mouse femur maintain their expression of TRPV1 even as tumor growth and tumor-induced bone destruction progresses. These results suggest that the TRPV1 channel plays a role in the integration of nociceptive signaling in a severe pain state and that antagonists of TRPV1 may be effective in attenuating difficult-to-treat mixed chronic pain states, such as that encountered in patients with bone cancer pain. In addition to acidosis, previous studies have suggested that a major source of bone pain is mechanical distortion of the periosteum (Mercadante 1997; Mundy 1999; Alder 2000). Thus, following fracture due to tumor-induced bone remodeling, the pain associated with the fracture is partially relieved if the bone and periosteum are repositioned and stabilized in their normal orientation (Rubert and Malawer 2000). Both osteolytic and osteoblastic tumors induce a loss of the mechanical strength and stability of mineralized bone, so that normally innocuous mechanical stress now results in distortion of the periosteum. Thus, as bisphosphonates and OPG-like molecules reduce the rate of bone remodeling, and in doing so preserve the mechanical strength of bone, and antagonists of acid-sensing channels expressed by nociceptors should reduce the activation and sensitization of nociceptors innervating the bone, these therapies may prove particularly useful in reducing bone cancer pain.

4.3.2   TUMOR-DERIVED PRODUCTS IN GENERATION OF BONE CANCER PAIN

In most cancers the tumor mass is composed of tumor and tumor stromal cells, the latter of which includes macrophages, neutrophils, T-lymphocytes, fibroblasts, and endothelial cells. Tumor and/or tumor stromal cells have been shown to secrete a variety of factors that sensitize or directly excite primary afferent neurons, such

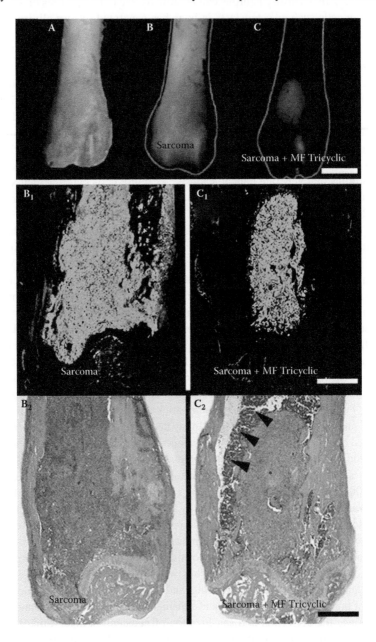

as prostaglandins (Nielsen et al. 1991; Galasko 1995), tumor necrosis factor alpha (Watkins and Maier 1999; Nadler et al. 2000; DeLeo and Yezierski 2001; Khatami 2008), endothelins (Nelson and Carducci 2000; Davar 2001), interleukins -1 and -6 (Watkins et al. 1995; Opree and Kress 2000; DeLeo and Yezierski 2001), epidermal growth factor (Purow et al. 2008), transforming growth factor-beta (Poon et al. 2001; Roman et al. 2001), platelet-derived growth factor (Radinsky 1991; Kuhnert et al. 2008; Lin et al. 2008; Ono 2008), and nerve growth factor. Receptors for many of these factors are expressed by primary afferent neurons.

4.3.2.1 Prostaglandins

Cancer cells and tumor-associated macrophages have both been shown to express high levels of cyclooxygenase (COX) isoenzymes, leading to high levels of prostaglandins (Kundu et al. 2001; Ohno et al. 2001; Shappell et al. 2001; Farooqui et al. 2007; Wang and Dubois 2008). Prostaglandins are lipid-derived eicosanoids that are synthesized from arachidonic acid by COX isoenzymes COX-1 and COX-2. Prostaglandins have been shown to be involved in the sensitization and/or direct excitation of nociceptors by binding to several prostanoid receptors expressed by nociceptors (Baba et al. 2001).

Studies have shown in the sarcoma model of bone cancer pain that chronic inhibition of COX-2 activity with selective COX-2 inhibitors resulted in significant attenuation of bone cancer pain behaviors as well as many of the neurochemical changes suggestive of both peripheral and central sensitization (Sabino et al. 2002). In addition, prostaglandins have been shown to be involved in tumor growth, cell survival, and angiogenesis (Harris et al. 2000; Masferrer et al. 2000; Reddy et al. 2000; Williams et al. 2000; Lal et al. 2001; Sonoshita et al. 2001; Yaqub et al. 2008). Therefore, as well as blocking cancer pain, COX-2 inhibitors are also capable of retarding tumor growth within bone (Figure 4.6) (Sabino et al. 2002). Chronic administration of a selective COX-2 inhibitor significantly reduced tumor burden in sarcoma-bearing

FIGURE 4.6 Sustained administration of a COX-2 inhibitor reduces tumor burden in mice with bone cancer as evaluated by epi-fluorescence microscopy. Low-power photomicrographs of the anterior aspect of a sarcoma-bearing femur 14 days after sarcoma injections (A). When the same bone in A was illuminated with a light source and band pass filters to visualize sarcoma cells transfected with green fluorescent protein (GFP) (B), GFP-expressing tumor cells had completely filled the intramedullary space. Femora in C were identical to B, except these animals received chronic treatment of a COX-2 inhibitor from day 6 to day 14 (C). Note the reduction in GFP fluorescence after chronic COX-2 inhibition (B versus C). Tumor burden was also visualized and quantified as is shown in the overlapping confocal immunofluorescence picture (B_1). Seven-micrometer-thick tissue sections of the distal femur from sarcoma-bearing mice that received regular diet (B_1) or diet containing a COX-2 inhibitor (C_1) were stained with an antibody raised against GFP to label sarcoma cells. Serial 7-μm-thick sections were stained with hematoxylin and eosin to additionally define and confirm the presence of tumor cells within the intramedullary space (B_2, C_2,*). The presence of normal marrow cells in animals treated with the COX-2 inhibitor are delineated by arrowheads. Note that animals that received chronic administration of a COX-2 inhibitor demonstrated a reduction in tumor burden (C_1,C_2) as compared with vehicle-treated mice (B_1,B_2). Scale bar = 1.5 mm in A–C and 750 μm in B_1, B_2, C_1, and C_2. (Reprinted with permission from Sabino MA, Ghilardi JR, Jongen JL, et al. (2002) Simultaneous reduction in cancer pain, bone destruction, and tumor growth by selective inhibition of cyclooxygenase-2. *Cancer Res* 62:7343–7349.)

bones (Figure 4.6), which may, in turn, reduce factors released by tumor cells capable of exciting primary afferent fibers (Davar 2001). Acute or chronic administration of a selective COX-2 inhibitor significantly attenuated both ongoing and movement-evoked pain. Whereas acute administration of a COX-2 inhibitor presumably reduces prostaglandins capable of activating sensory or spinal cord neurons, chronic inhibition of COX-2 also appears to simultaneously reduce osteoclastogenesis, bone resorption, and tumor burden. Together, suppression of prostaglandin synthesis and release at multiple sites by selective inhibition of COX-2 may synergistically improve the survival and quality of life of patients with bone cancer pain.

4.3.2.2   Endothelins

Endothelins (ET-1, -2, and -3) are a family of vasoactive peptides that are expressed at high levels by several types of tumors, including those that arise from the prostate (Nelson and Carducci 2000). Clinical studies have shown a correlation between the severity of the pain and plasma levels of endothelins in prostate cancer patients (Smollich and Wulfing 2008). Endothelins could contribute to cancer pain by directly sensitizing or exciting nociceptors, as a subset of small unmyelinated primary afferent neurons express endothelin A receptors (Pomonis et al. 2001). Furthermore, direct application of endothelin to peripheral nerves induces activation of primary afferent fibers and an induction of pain-related behaviors (Yuyama et al. 2004). Like prostaglandins, endothelins that are produced by cancer cells are also thought to be involved in regulating angiogenesis (Boldrini et al. 2006) and tumor growth (Asham et al. 1998; Herrmann et al. 2006).

In the sarcoma model, acute or chronic administration of the endothelin A receptor (ET_AR) selective antagonist ABT-627 significantly attenuated ongoing and movement-evoked bone cancer pain. Chronic administration of ABT-627 also reduced several neurochemical indexes of peripheral and central sensitization without influencing tumor growth or bone destruction (Peters et al. 2004). As tumor expression and release of ET-1 has been shown to be regulated by the local environment, location-specific expression and release of ET-1 by tumor cells may provide insight into the mechanisms that underlie the heterogeneity of bone cancer pain that is frequently observed in humans with multiple skeletal metastases.

4.3.2.3   Kinins

Previous studies have shown that bradykinin and related kinins are released in response to tissue injury and that these kinins play a significant role in driving acute and chronic inflammatory pain (Couture et al. 2001). The action of bradykinin is mediated by two receptors, B_1 and B_2. Whereas the B_2 receptor is constitutively expressed at high levels by sensory neurons, the B_1 receptor is normally expressed at low but detectable levels by sensory neurons and these B_1 receptors are significantly upregulated following peripheral inflammation and/or tissue injury (Fox et al. 2003). Tumor metastases to the skeleton induce significant bone remodeling with accompanying tissue injury, which presumably induces the release of bradykinin. It has been demonstrated that both bone cancer–induced ongoing and movement-evoked nocifensive behaviors were reduced following the pharmacologic blockade of the B_1 receptor (Sevcik et al. 2005a).

4.3.2.4 Nerve Growth Factor

One important concept that has emerged over the past decade is that in addition to nerve growth factor (NGF) being able to directly activate sensory neurons that express the trkA receptor, NGF modulates expression and function of a wide variety of molecules and proteins expressed by sensory neurons that express the trkA or p75 receptor. Some of these molecules and proteins include neurotransmitters (substance P and CGRP), receptors (bradykinin R), channels (P2X3, TRPV1, ASIC-3, and sodium channels), transcription factors (ATF-3), and structural molecules (neurofilaments and the sodium channel anchoring molecule p11) (Hefti et al. 2006; Pezet and McMahon 2006) (Figure 4.7). Additionally, NGF has been shown to modulate the trafficking and insertion of sodium channels such as Nav 1.8 (Gould et al. 2000) and TRPV1 (Ji et al. 2002) in the sensory neurons as well as modulating the expression profile of supporting cells in the dorsal root ganglia (DRG) and peripheral nerves, such as non-myelinating Schwann cells and macrophages (Brown et al. 2004; Chen et al. 2007). Therefore, anti-NGF antibody therapy may be particularly effective in blocking bone cancer pain because NGF appears to be integrally involved in the upregulation, sensitization, and disinhibition of multiple neurotransmitters, ion channels, and receptors in the primary afferent nerve and DRG fibers that synergistically increase nociceptive signals originating from the tumor-bearing bone.

To test the hypothesis that an anti-NGF therapy would be efficacious in reducing bone cancer pain, we examined the analgesic efficacy of a murine anti-NGF monoclonal antibody in tumor-induced bone pain using the primarily osteolytic 2472 murine osteosarcoma, which expresses high levels of NGF, and the primarily osteoblastic canine ACE-1 prostate, where NGF expression is undetectable (Figure 4.8). In both of these models of tumor-induced bone cancer pain, it was demonstrated that administration of an anti-NGF antibody was highly efficacious in reducing both early- and late-stage bone cancer pain–related behaviors, but that this reduction in pain-related behaviors was greater than that achieved with acute administration of 10 mg/kg of morphine sulfate (Figure 4.8) (Halvorson et al. 2005; Sevcik et al. 2005b). These data suggest that therapeutic targeting of NGF or its cognate receptor TrkA may be useful in blocking bone cancer pain whether or not the tumor that has metastasized to bone expresses NGF. Presumably, in the case where the tumor cells themselves do not express NGF, it is the tumor stromal cells that are expressing and secreting NGF, as these tumor stromal cells have been shown to comprise 2%–60% of the total tumor mass.

4.3.3 Neuropathic Component of Bone Cancer Pain

Numerous studies have demonstrated that the periosteum is densely innervated by both sensory and sympathetic fibers (Asmus et al. 2000; Irie et al. 2002; Martin et al. 2007). Using a combination of minimal decalcification techniques and antigen amplification techniques, it has also been demonstrated that the bone marrow and mineralized bone also receive a significant innervation by both sensory and sympathetic nerve fibers (Bjurholm et al. 1988a, b; Tabarowski et al. 1996). Since sensory

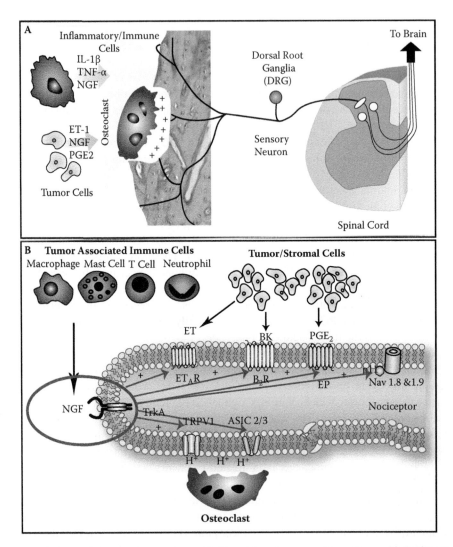

FIGURE 4.7 Schematic showing factors in bone (A) and receptors/channels expressed by nociceptors that innervate the skeleton (B) that drive bone cancer pain. A variety of cells (tumor cells, and stomal cells including inflammatory/immune cells, and osteoclasts) drive bone cancer pain (A). Nociceptors that innervate the bone use several different types of receptors to detect and transmit noxious stimuli that are produced by cancer cells, tumor-associated immune cells, or other aspects of the tumor microenvironment. There are multiple factors that may contribute to the pain associated with cancer (B). The transient receptor potential vanilloid receptor-1 (TRPV1) and acid-sensing ion channels (ASICs) detect extracellular protons produced by tumor-induced tissue damage or abnormal osteoclast mediated bone resorption. Tumor cells and associated inflammatory (immune) cells produce a variety of chemical mediators including prostaglandins (PGE2), nerve growth factor (NGF), endothelins, bradykinin, and extracellular ATP. Several of these proinflammatory mediators have receptors on peripheral terminals and can directly activate or sensitize nociceptors. NGF and its cognate receptor TrkA may serve as a master regulator of bone cancer pain by modulating the sensitivity or increasing the expression of several receptors and ion channels contributing to increased excitability of nociceptors in the vicinity of the tumor.

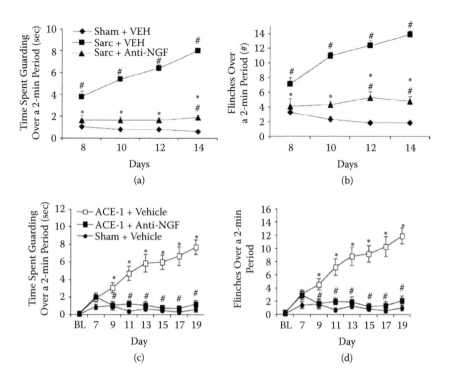

FIGURE 4.8 Anti-NGF attenuates spontaneous bone cancer pain in a model where the tumor cells do (mouse sarcoma cells) and do not (canine prostate cells) express NGF. Anti-NGF treatment (10 mg/kg, i.p., given on days 7, 12, and 17 post-tumor injection) attenuated ongoing bone cancer pain behavior on days 7 to 19 post-tumor injection. In these experiments, 2472 mouse sarcoma cells (A,B) or canine prostate carcinoma (ACE-1) cells (C,D) were injected into the femur of adult male mice. The time spent guarding and number of spontaneous flinches of the afflicted limb over a 2-minute observation period were used as measures of ongoing pain. Anti-NGF significantly reduced ongoing pain behaviors in tumor-injected mice as compared with sarcoma + vehicle (A, B) or ACE-1 + vehicle (C, D). Note that whereas the 2472 cells express relatively high levels of NGF mRNA in vitro, ACE-1 cells in vitro express undetectabe levels of NGF mRNA or protein, suggesting that NGF could be released mainly from tumor associated macrophage and immune cells. Bars represent mean S.E.M. *, $P < 0.05$ versus sham + vehicle; #, $P < 0.05$ versus ACE-1 + vehicle. (With permission from Sevcik et al. 2004, *Pain* and Halvorson et al. 2005, *Cancer Res.*)

and sympathetic neurons are present within the bone marrow, mineralized bone and periosteum, these nerve fibers are ultimately impacted by fractures, ischemia, or the presence of tumor cells. Sensory fibers in any of these compartments may play a role in the generation and maintenance of bone cancer pain.

In examining the changes in the sensory innervation of bone induced by the primarily osteolytic sarcoma cells, sensory fibers were observed at and within the leading edge of the tumor in the deep stromal regions of the tumor. Additionally, these sensory nerve fibers displayed a discontinuous and fragmented appearance, suggesting that following initial activation by the osteolytic tumor cells, the distal processes of the sensory fibers were ultimately injured by the invading tumor cells. In contrast, in examining the sensory innervation of bone following injection of the primarily osteoblastic prostate cancer cells suggests that there is simultaneous injury and sprouting of sensory fibers in the bone (Halvorson et al. 2006).

In the sarcoma-injected animals, there was expression of activating transcription factor-3 (ATF-3) in the nucleus of sensory neurons that innervate the femur. ATF-3 is a member of the ATF/CREB family of transcription factors, which is not expressed at detectable levels in normal sensory neurons or in sensory neurons following peripheral inflammation, but is strongly expressed in sensory neurons following injury to peripheral nerves in neuropathic pain models (Tsujino et al. 2000). It is likely that the expression of ATF-3 in sensory neurons of tumor-bearing animals is a result of the tumor-induced injury and destruction of the distal tips of the sensory and sympathetic nerve fibers that innervate the bone.

This tumor-induced injury of sensory nerve fibers in the sarcoma model was also accompanied by an increase in ongoing and movement-evoked pain behaviors, an up-regulation of galanin by sensory neurons that innervate the tumor-bearing femur, up-regulation of glial fibrillary acidic protein and hypertrophy of satellite cells surrounding sensory neuron cell bodies within the ipsilateral DRG, and macrophage infiltration of the DRG ipsilateral to the tumor-bearing femur (Peters et al. 2005; Sevcik et al. 2005b). Similar neurochemical changes have been described following peripheral nerve injury and in other non-cancerous neuropathic pain states (Obata et al. 2003). Chronic treatment with gabapentin in the sarcoma model also did not influence tumor growth, tumor-induced bone destruction or the tumor-induced neurochemical reorganization that occurs in sensory neurons or the spinal cord, but did attenuate both ongoing and movement-evoked bone cancer–related pain behaviors (Peters et al. 2005). These results suggest that even when the tumor is confined within the bone, a component of bone cancer pain is due to tumor-induced injury to primary afferent nerve fibers that normally innervate the tumor-bearing bone.

4.4 CONCLUSIONS

For the first time, animal models of cancer pain are now available and effectively mirror the clinical picture observed in humans with bone cancer pain. Information generated from these models has begun to provide insight into the mechanisms that generate and maintain bone cancer pain and helped target potential mechanism-based therapies to treat this chronic pain state. It is noteworthy that in these models analgesics such as a bisphosphonate, inhibitors of RANK, and a cyclooxygenase-2 inhibitor there appears to be an influence disease progression in the tumor-bearing bone. These and other studies using models of bone cancer suggest that it may be possible to develop novel mechanism-based therapies that not only reduce tumor-induced bone pain but may provide added benefit in synergistically reducing disease progression. Successful development and clinical use of these therapies has the potential to positively impact the patient's functional status, quality of life, and survival.

REFERENCES

Alder C-P (2000) *Bone diseases*. Berlin: Springer.

Arguello F, Baggs RB, Frantz CN (1988) A murine model of experimental metastasis to bone and bone marrow. *Cancer Res* 48:6876–6881.

Asham EH, Loizidou M, Taylor I (1998) Endothelin-1 and tumour development. *Eur J Surg Oncol* 24:57–60.

Asmus SE, Parsons S, Landis SC (2000) Developmental changes in the transmitter properties of sympathetic neurons that innervate the periosteum. *J Neurosci* 20:1495–1504.

Baba H, Kohno T, Moore KA, Woolf CJ (2001) Direct activation of rat spinal dorsal horn neurons by prostaglandin E2. *J Neurosci* 21:1750–1756.

Bassilana F, Champigny G, Waldmann R, de Weille JR, Heurteaux C, Lazdunski M (1997) The acid-sensitive ionic channel subunit ASIC and the mammalian degenerin MDEG form a hetero-multimeric H+-gated Na+ channel with novel properties. *J Biol Chem* 272:28819–28822.

Berenson JR, Rosen LS, Howell A, Porter L, Coleman RE, Morley W, Dreicer R, Kuross SA, Lipton A, Seaman JJ (2001) Zoledronic acid reduces skeletal-related events in patients with osteolytic metastases. *Cancer* 91:1191–1200.

Bjurholm A, Kreicbergs A, Brodin E, Schultzberg M (1988a) Substance P- and CGRP-immunoreactive nerves in bone. *Peptides* 9:165–171.

Bjurholm A, Kreicbergs A, Terenius L, Goldstein M, Schultzberg M (1988b) Neuropeptide Y-, tyrosine hydroxylase- and vasoactive intestinal polypeptide-immunoreactive nerves in bone and surrounding tissues. *J Auton Nerv Syst* 25:119–125.

Boldrini L, Pistolesi S, Gisfredi S, Ursino S, Ali G, Pieracci N, Basolo F, Parenti G, Fontanini G (2006) Expression of endothelin 1 and its angiogenic role in meningiomas. *Virchows Arch* 449:546–553.

Brown A, Ricci MJ, Weaver LC (2004) NGF message and protein distribution in the injured rat spinal cord. *Exp Neurol* 188:115–127.

Caterina MJ, Leffler A, Malmberg AB, Martin WJ, Trafton J, Petersen-Zeitz KR, Koltzenburg M, Basbaum AI, Julius D (2000) Impaired nociception and pain sensation in mice lacking the capsaicin receptor. *Science* 288:306–313.

Chen ZL, Yu WM, Strickland S (2007) Peripheral regeneration. *Annu Rev Neurosci* 30:209–233.

Cherny N (2000) New strategies in opioid therapy for cancer pain. *J Oncol Manag* 9:8–15.

Clohisy DR, Perkins SL, Ramnaraine ML (2000) Review of cellular mechanisms of tumor osteolysis. *Clin Orthop Rel Res* 373:104–114.

Coleman RE (1997) Skeletal complications of malignancy. *Cancer* 80:1588–1594.

Coleman RE (2006) Clinical features of metastatic bone disease and risk of skeletal morbidity. *Clin Cancer Res* 12:6243s–6249s.

Coleman RE (2008) Risks and benefits of bisphosphonates. *Br J Cancer* 98:1736–1740.

Couture R, Harrisson M, Vianna RM, Cloutier F (2001) Kinin receptors in pain and inflammation. *Eur J Pharmacol* 429:161–176.

Davar G (2001) Endothelin-1 and metastatic cancer pain. *Pain Medicine* 2:24–27.

Delaisse JM, Vaes G (eds.) (1992) *Mechanism of mineral solubilization and matrix degradation in osteoclastic bone resorption*. Ann Arbor: CRC Press.

DeLeo JA, Yezierski RP (2001) The role of neuroinflammation and neuroimmune activation in persistent pain. *Pain* 90:1–6.

Dy SM, Asch SM, Naeim A, Sanati H, Walling A, Lorenz KA (2008) Evidence-based standards for cancer pain management. *J Clin Oncol* 26:3879–3885.

Farooqui M, Li Y, Rogers T, Poonawala T, Griffin RJ, Song CW, Gupta K (2007) COX-2 inhibitor celecoxib prevents chronic morphine-induced promotion of angiogenesis, tumour growth, metastasis and mortality, without compromising analgesia. *Br J Cancer* 97:1523–1531.

Fox A, Wotherspoon G, McNair K, Hudson L, Patel S, Gentry C, Winter J (2003) Regulation and function of spinal and peripheral neuronal B1 bradykinin receptors in inflammatory mechanical hyperalgesia. *Pain* 104:683–691.

Fulfaro F, Casuccio A, Ticozzi C, Ripamonti C (1998) The role of bisphosphonates in the treatment of painful metastatic bone disease: a review of phase III trials. *Pain* 78:157–169.

Galasko CS (1995) Diagnosis of skeletal metastases and assessment of response to treatment. *Clin Ortho Relat Res* 64–75.

Gatti D, Adami S (1999) New bisphosphonates in the treatment of bone diseases. *Drugs & Aging* 15:285–296.

Ghilardi JR, Rohrich H, Lindsay TH, et al. (2005) Selective blockade of the capsaicin receptor TRPV1 attenuates bone cancer pain. *J Neurosci* 25:3126–3131.

Gould HJ, 3rd, Gould TN, England JD, Paul D, Liu ZP, Levinson SR (2000) A possible role for nerve growth factor in the augmentation of sodium channels in models of chronic pain. *Brain Res* 854:19–29.

Griffiths JR (1991) Are cancer cells acidic? *Br J Cancer* 64:425–427.

Halvorson KG, Kubota K, Sevcik MA, Lindsay TH, Sotillo JE, Ghilardi JR, Rosol TJ, Boustany L, Shelton DL, Mantyh PW (2005) A blocking antibody to nerve growth factor attenuates skeletal pain induced by prostate tumor cells growing in bone. *Cancer Res* 65:9426–9435.

Halvorson KG, Sevcik MA, Ghilardi JR, Rosol TJ, Mantyh PW (2006) Similarities and differences in tumor growth, skeletal remodeling and pain in an osteolytic and osteoblastic model of bone cancer. *Clin J Pain* 22:587–600.

Harris JD (2008) Management of expected and unexpected opioid-related side effects. *Clin J Pain* 24 Suppl 10:S8–S13.

Harris RE, Alshafie GA, Abou-Issa H, Seibert K (2000) Chemoprevention of breast cancer in rats by celecoxib, a cyclooxygenase 2 inhibitor. *Cancer Res* 60:2101–2103.

Hefti FF, Rosenthal A, Walicke PA, Wyatt S, Vergara G, Shelton DL, Davies AM (2006) Novel class of pain drugs based on antagonism of NGF. *Trends Pharmacol* Sci 27:85–91.

Herrmann E, Bogemann M, Bierer S, Eltze E, Hertle L, Wulfing C (2006) The endothelin axis in urologic tumors: mechanisms of tumor biology and therapeutic implications. *Expert Rev Anticancer Ther* 6:73–81.

Hiraga T, Tanaka, S., Yamamoto, M., Nakajima, T., Ozawa, H. (1996) Inhibitory effects of bisphosphonate YM175 on bone resorption induced by metastic bone tumor. *Bone* 18:1–7.

Honore P, Luger NM, Sabino MA, Schwei MJ, Rogers SD, Mach DB, O'Keefe P F, Ramnaraine ML, Clohisy DR, Mantyh PW (2000) Osteoprotegerin blocks bone cancer-induced skeletal destruction, skeletal pain and pain-related neurochemical reorganization of the spinal cord. *Nat Med* 6:521–528.

Irie K, Hara-Irie F, Ozawa H, Yajima T (2002) Calcitonin gene-related peptide (CGRP)-containing nerve fibers in bone tissue and their involvement in bone remodeling. *Microsc Res Tech* 58:85–90.

Ji RR, Samad TA, Jin SX, Schmoll R, Woolf CJ (2002) p38 MAPK activation by NGF in primary sensory neurons after inflammation increases TRPV1 levels and maintains heat hyperalgesia. *Neuron* 36:57–68.

Julius D, Basbaum AI (2001) Molecular mechanisms of nociception. *Nature* 413:203–210.

Khatami M (2008) 'Yin and Yang' in inflammation: duality in innate immune cell function and tumorigenesis. *Expert Opin Biol Ther* 8:1461–1472.

Kuhnert F, Tam BY, Sennino B, Gray JT, Yuan J, Jocson A, Nayak NR, Mulligan RC, McDonald DM, Kuo CJ (2008) Soluble receptor-mediated selective inhibition of VEGFR and PDGFR beta signaling during physiologic and tumor angiogenesis. *Proc Natl Acad Sci U S A* 105:10185–10190.

Kundu N, Yang QY, Dorsey R, Fulton AM (2001) Increased cyclooxygenase-2 (cox-2) expression and activity in a murine model of metastatic breast cancer. *Int J Cancer* 93:681–686.

Lal G, Ash C, Hay K, Redston M, Kwong E, Hancock B, Mak T, Kargman S, Evans JF, Gallinger S (2001) Suppression of intestinal polyps in Msh2-deficient and non-Msh2-deficient multiple intestinal neoplasia mice by a specific cyclooxygenase-2 inhibitor and by a dual cyclooxygenase-1/2 inhibitor. *Cancer Res* 61:6131–6136.

Lin SY, Yang J, Everett AD, Clevenger CV, Koneru M, Mishra PJ, Kamen B, Banerjee D, Glod J (2008) The isolation of novel mesenchymal stromal cell chemotactic factors from the conditioned medium of tumor cells. *Exp Cell Res.*

Luger NM, Honore P, Sabino MA, Schwei MJ, Rogers SD, Mach DB, Clohisy DR, Mantyh PW (2001) Osteoprotegerin diminishes advanced bone cancer pain. *Cancer Res* 61:4038–4047.

Lussier D, Huskey AG, Portenoy RK (2004) Adjuvant analgesics in cancer pain management. *Oncologist* 9:571–591.

Major PP, Lipton A, Berenson J, Hortobagyi G (2000) Oral bisphosphonates: a review of clinical use in patients with bone metastases. *Cancer* 88:6–14.

Martin CD, Jimenez-Andrade JM, Ghilardi JR, Mantyh PW (2007) Organization of a unique net-like meshwork of CGRP+ sensory fibers in the mouse periosteum: implications for the generation and maintenance of bone fracture pain. *Neurosci Lett* 427:148–152.

Masferrer JL, Leahy KM, Koki AT, Zweifel BS, Settle SL, Woerner BM, Edwards DA, Flickinger AG, Moore RJ, Seibert K (2000) Antiangiogenic and antitumor activities of cyclooxygenase-2 inhibitors. *Cancer Res* 60:1306–1311.

Mercadante S (1997) Malignant bone pain: pathophysiology and treatment. *Pain* 69:1–18.

Mercadante S, Fulfaro F (2007) Management of painful bone metastases. *Curr Opin Oncol* 19:308–314.

Mercadante S, Villari P, Ferrera P, Casuccio A (2004) Optimization of opioid therapy for preventing incident pain associated with bone metastases. *J Pain Symptom Manage* 28:505–510.

Mundy GR (1999) *Bone remodeling and its disorders.* London: Taylor & Francis.

Nadler RB, Koch AE, Calhoun EA, Campbell PL, Pruden DL, Bennett CL, Yarnold PR, Schaeffer AJ (2000) IL-1-beta and TNF-alpha in prostatic secretions are indicators in the evaluation of men with chronic prostatitis. *J Urol* 164:214–218.

Nelson JB, Carducci MA (2000) The role of endothelin-1 and endothelin receptor antagonists in prostate cancer. *BJU International* 85:45–48.

Nielsen OS, Munro AJ, Tannock IF (1991) Bone metastases: pathophysiology and management policy. *J Clin Oncol* 9:509–524.

Obata K, Yamanaka H, Fukuoka T, Yi D, Tokunaga A, Hashimoto N, Yoshikawa H, Noguchi K (2003) Contribution of injured and uninjured dorsal root ganglion neurons to pain behavior and the changes in gene expression following chronic constriction injury of the sciatic nerve in rats. *Pain* 101:65–77.

Ohno R, Yoshinaga K, Fujita T, Hasegawa K, Iseki H, Tsunozaki H, Ichikawa W, Nihei Z, Sugihara K (2001) Depth of invasion parallels increased cyclooxygenase-2 levels in patients with gastric carcinoma. *Cancer* 91:1876–1881.

Olson TH, Riedl MS, Vulchanova L, Ortiz-Gonzalez XR, Elde R (1998) An acid sensing ion channel (ASIC) localizes to small primary afferent neurons in rats. *Neuroreport* 9:1109–1113.

Ono M (2008) Molecular links between tumor angiogenesis and inflammation: inflammatory stimuli of macrophages and cancer cells as targets for therapeutic strategy. *Cancer Sci* 99:1501–1506.

Opree A, Kress M (2000) Involvement of the proinflammatory cytokines tumor necrosis fac-
    tor-alpha, IL-1 beta, and IL-6 but not IL-8 in the development of heat hyperalgesia:
    effects on heat-evoked calcitonin gene-related peptide release from rat skin. *J Neurosci*
    20:6289–6293.
Peters CM, Ghilardi JR, Keyser CP, Kubota K, Lindsay TH, Luger NM, Mach DB, Schwei
    MJ, Sevcik MA, Mantyh PW (2005) Tumor-induced injury of primary afferent sensory
    nerve fibers in bone cancer pain. *Exp Neurol* 193:85–100.
Peters CM, Lindsay TH, Pomonis JD, Luger NM, Ghilardi JR, Sevcik MA, Mantyh PW
    (2004) Endothelin and the tumorigenic component of bone cancer pain. *Neuroscience*
    126:1043–1052.
Pezet S, McMahon SB (2006) Neurotrophins: mediators and modulators of pain. *Annu Rev
    Neurosci* 29:507–538.
Pomonis JD, Rogers SD, Peters CM, Ghilardhi JR, Mantyh PW (2001) Expression and local-
    ization of endothelin receptors: implication for the involvement of peripheral glia in
    nociception. *J Neurosci* 21:999–1006.
Poon RT, Fan ST, Wong J (2001) Clinical implications of circulating angiogenic factors in
    cancer patients. *J Clin Oncol* 19:1207–1225.
Purow BW, Sundaresan TK, Burdick MJ, Kefas BA, Comeau LD, Hawkinson MP, Su Q,
    Kotliarov Y, Lee J, Zhang W, Fine HA (2008) Notch-1 regulates transcription of the
    epidermal growth factor receptor through p53. *Carcinogenesis* 29:918–925.
Radinsky R (1991) Growth factors and their receptors in metastasis. *Seminars in Cancer
    Biology* 2:169–177.
Reddy BS, Hirose Y, Lubet R, Steele V, Kelloff G, Paulson S, Seibert K, Rao CV (2000)
    Chemoprevention of colon cancer by specific cyclooxygenase-2 inhibitor, celecoxib,
    administered during different stages of carcinogenesis. *Cancer Res* 60:293–297.
Reid CM, Gooberman-Hill R, Hanks GW (2008) Opioid analgesics for cancer pain: symptom
    control for the living or comfort for the dying? A qualitative study to investigate the fac-
    tors influencing the decision to accept morphine for pain caused by cancer. *Ann Oncol*
    19:44–48.
Rodan G, Martin T (2000) Therapeutic approaches to bone disease. *Science* 289:1508–1514.
Rogers MJ, Gordon S, Benford HL, Coxon FP, Luckman SP, Monkkonen J, Frith JC
    (2000) Cellular and molecular mechanisms of action of bisphosphonates. *Cancer*
    88:2961–2978.
Roman C, Saha D, Beauchamp R (2001) TGF-beta and colorectal carcinogenesis. *Microsc Res
    Tech* 52:450–457.
Rubert CHR, Malawer M (2000) Orthopedic management of skeletal metastases. In: *Tumor
    bone disease and osteoporsis in cancer patients* (Body, J.-J., ed), pp 305–356. New
    York: Marcel Dekker.
Sabino M, Luger N, Mach D, Rogers S, Schwei M, Feia K, Mantyh P (2003) Different tumors
    in bone each give rise to a distinct pattern of skeletal destruction, bone cancer-related
    pain behaviors and neurochemical changes in the central nervous system. *Int J Cancer*
    104:550–558.
Sabino MA, Ghilardi JR, Jongen JL, et al. (2002) Simultaneous reduction in cancer pain, bone
    destruction, and tumor growth by selective inhibition of cyclooxygenase-2. *Cancer Res*
    62:7343–7349.
Sasaki A, Boyce B, Story B, et al. (1995) Bisphosphonate risedronate reduces metastic human
    breast cancer burden in bone in nude mice. *Cancer Res* 77:279–285.
Schwei MJ, Honore P, Rogers SD, Salak-Johnson JL, Finke MP, Ramnaraine ML, Clohisy
    DR, Mantyh PW (1999) Neurochemical and cellular reorganization of the spinal cord in
    a murine model of bone cancer pain. *J Neurosci* 19:10886–10897.

Sevcik M, Jonas B, Lindsay T, Halvorson K, Ghilardi J, Kuskowski M, Mukherjee P, Maggio J, Mantyh P (2006) Endogenous opioids inhibit early stage pancreatic pain in a mouse model of pancreatic cancer. *Gastroenterolgy* 131:900–910.

Sevcik MA, Ghilardi JR, Halvorson KG, Lindsay TH, Kubota K, Mantyh PW (2005a) Analgesic efficacy of bradykinin B1 antagonists in a murine bone cancer pain model. *J Pain* 6:771–775.

Sevcik MA, Ghilardi JR, Peters CM, et al. (2005b) Anti-NGF therapy profoundly reduces bone cancer pain and the accompanying increase in markers of peripheral and central sensitization. *Pain* 115:128–141.

Sevcik MA, Luger NM, Mach DB, et al. (2004) Bone cancer pain: the effects of the bisphosphonate alendronate on pain, skeletal remodeling, tumor growth and tumor necrosis. *Pain* 111:169–180.

Shappell SB, Manning S, Boeglin WE, et al. (2001) Alterations in lipoxygenase and cyclooxygenase-2 catalytic activity and mRNA expression in prostate carcinoma. *Neoplasia* 3:287–303.

Simonet WS, Lacey DL, Dunstan CR, et al. (1997) Osteoprotegerin—a novel secreted protein involved in the regulation of bone density. *Cell* 89:309–319.

Smollich M, Wulfing P (2008) Targeting the endothelin system: novel therapeutic options in gynecological, urological and breast cancers. *Expert Rev Anticancer Ther* 8:1481–1493.

Sonoshita M, Takaku K, Sasaki N, Sugimoto Y, Ushikubi F, Narumiya S, Oshima M, Taketo MM (2001) Acceleration of intestinal polyposis through prostaglandin receptor EP2 in Apc(Delta 716) knockout mice. *Nature Medicine* 7:1048–1051.

Sutherland S, Cook S, EW M (2000) Chemical mediators of pain due to tissue damage and ischemia. *Prog Brain Res* 129:21–38.

Tabarowski Z, Gibson-Berry K, Felten SY (1996) Noradrenergic and peptidergic innervation of the mouse femur bone marrow. *Acta Histochem* 98:453–457.

Tominaga M, Caterina MJ, Malmberg AB, Rosen TA, Gilbert H, Skinner K, Raumann BE, Basbaum AI, Julius D (1998) The cloned capsaicin receptor integrates multiple pain-producing stimuli. *Neuron* 21:531–543.

Tripathy D, Body JJ, Bergstrom B (2004) Review of ibandronate in the treatment of metastatic bone disease: experience from phase III trials. *Clin Ther* 26:1947–1959.

Tsujino H, Kondo E, Fukuoka T, Dai Y, Tokunaga A, Miki K, Yonenobu K, Ochi T, Noguchi K (2000) Activating transcription factor 3 (ATF3) induction by axotomy in sensory and motoneurons: a novel neuronal marker of nerve injury. *Mol Cell Neurosc* 15:170–182.

van den Beuken-van Everdingen M, de Rijke J, Kessels A, Schouten H, van Kleef M, Patijn J (2007) Prevalence of pain in patients with cancer: a systematic review of the past 40 years. *Ann Oncol* 18:1437–1449.

Wang D, Dubois RN (2008) Pro-inflammatory prostaglandins and progression of colorectal cancer. *Cancer Lett* 267:197–203.

Watkins LR, Goehler LE, Relton J, Brewer MT, Maier SF (1995) Mechanisms of tumor necrosis factor-alpha (TNF-alpha) hyperalgesia. *Brain Research* 692:244–250.

Watkins LR, Maier SF (1999) Implications of immune-to-brain communication for sickness and pain. *Proc Natl Acad Sci U S A* 96:7710–7713.

Weber M, Huber C (1999) Documentation of severe pain, opioid doses, and opioid-related side effects in outpatients with cancer: a retrospective study. *J Pain Symptom Manage* 17:49–54.

Williams CS, Tsujii M, Reese J, Dey SK, DuBois RN (2000) Host cyclooxygenase-2 modulates carcinoma growth. *J Clin Invest* 105:1589–1594.

Yaqub S, Henjum K, Mahic M, Jahnsen FL, Aandahl EM, Bjornbeth BA, Tasken K (2008) Regulatory T cells in colorectal cancer patients suppress anti-tumor immune activity in a COX-2 dependent manner. *Cancer Immunol Immunother* 57:813–821.

Yasuda H, Shima N, Nakagawa N, et al. (1998) Identity of osteoclastogenesis inhibitory factor
    (Ocif) and osteoprotegerin (Opg)—a mechanism by which Opg/Ocif inhibits osteoclas-
    togenesis in vitro. *Endocrinology* 139:1329–1337.
Yoneda T, Sasaki A, Mundy GR (1994) Osteolytic bone metastasis in breast cancer. *Breast
    Cancer Res Treat* 32:73–84.
Yoneda T, Williams PJ, Myoi A, Michigami T, Mbalaviele G (2000) Cellular and molecular
    mechanisms of development of skeletal metastases. In: *Tumor bone diseases and osteo-
    porosis in cancer patients* (Body, J.-J., ed). New York: Marcel Dekker.
Yuyama H, Koakutsu A, Fujiyasu N, et al. (2004) Effects of selective endothelin ET(A) recep-
    tor antagonists on endothelin-1-induced potentiation of cancer pain. *Eur J Pharmacol*
    492:177–182.

5 Therapeutic Targeting of Peripheral Cannabinoid Receptors in Inflammatory and Neuropathic Pain States

Igor Spigelman

CONTENTS

5.1 INTRODUCTION

Synthetic and naturally occurring cannabinoids are a focus of strong social, legal, and medical controversy concerning their therapeutic utility, yet studies show that cannabinoids reduce the hyperalgesia and allodynia associated with persistent pain of inflammatory and neuropathic origin in humans and animals. Furthermore, cannabinoids are effective in alleviating chronic pain symptoms after prolonged repeated treatment, unlike opioids, which have only limited effectiveness. A major impediment to the widespread use of cannabinoid analgesics has been their centrally mediated psychotropic side effects. In addition, there are various other conditions where selective activation (or blockade) of peripheral cannabinoid receptors could prove to be of clinical benefit. This chapter will contrast between the peripheral and central actions of cannabinoids to build a case for selective targeting of the peripheral cannabinoid receptors for therapeutic gain.

5.2 CANNABINOID RECEPTORS

The targets of the antinociceptive cannabinoids may be defined by the distribution of two cloned subtypes of cannabinoid receptors, CB1R and CB2R (Matsuda et al. 1990; Munro et al. 1993). These are members of the superfamily of G protein-coupled receptors (GPCRs); both CB1R and CB2R are coupled to $G_{i/o}$ proteins (Howlett et al. 2002). Another recent addition to the cannabinoid receptor family is the G protein-coupled receptor, GPR55, which couples go $G_{\alpha11-13}$ (Begg et al. 2005; Lauckner et al. 2008; Ryberg et al. 2005) (Figure 5.1). CB_1R is actually the most

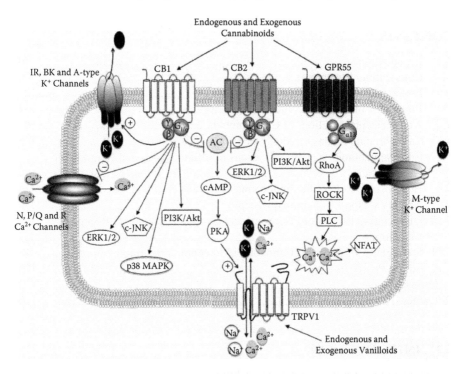

FIGURE 5.1 Schematic of the four G-protein coupled receptors targeted by endogenous and exogenous cannabinoids/vanilloids and some of their cellular targets. Abbreviations: AC, adenylyl cyclase; Akt, serine/threonine protein kinase; BK, large conductance Ca^{2+}-activated K^+ channel; cAMP, cyclic adenosine 5'-monophosphate; ERK, extracellular signal-regulated protein kinase; IR, inwardly rectifying; c-JNK, c-Jun N-terminal kinase; NFAT, nuclear factor of activated T cell; PI3K, phosphatidylinositol 3-kinase; PKA, protein kinase A; PLC, phospholipase C; ROCK, Rho-associated protein kinase.

abundant central nervous system (CNS) GPCR expressed at high levels in the hippocampus, cortex, cerebellum, and basal ganglia (Herkenham et al. 1990; Mackie 2005; Matsuda et al. 1990; Tsou et al. 1998). CB1R activation leads to inhibition of adenylyl cyclase (Howlett 1984), blockade of several voltage-gated Ca^{2+}-channels (Brown et al. 2004; Guo and Ikeda 2004), and activation of several K^+-channels (Deadwyler et al. 1995; Felder et al. 1995; Stumpff et al. 2005) (Figure 5.1). There is also some evidence that CB1R activation can block the K^+ M-current in central neurons (Schweitzer 2000). Central CB_1Rs are also localized in regions involved in pain transmission and modulation, specifically in the spinal dorsal horn and periaqueductal gray (Farquhar-Smith et al. 2000; Lichtman et al. 1996; Tsou et al. 1998). In the forebrain, ultrastructural studies have demonstrated a high degree of CB1R localization to presynaptic terminals of cholecystokinin-containing inhibitory interneurons, consistent with the ability of CB1R agonists to decrease evoked release of γ-aminobutyric acid (GABA) (Katona et al. 1999; Tsou et al. 1999). Later studies revealed that presynaptic terminals of glutamatergic fibers in the hippocampus and cerebellum also express CB_1Rs, albeit at much lower levels than in GABAergic neurons (Kawamura et al. 2006). These studies also provided the much needed anatomical basis for the well-known ability of CB1R agonists to decrease excitatory glutamatergic neurotransmission in these brain regions (Ameri et al. 1999; Hoffman et al. 2003; Kreitzer and Regehr 2001; Maejima et al. 2001). In the basal ganglia, CB1Rs are produced in and transported to the terminals of GABAergic medium-sized spiny neurons of the dorsal and ventral striatum (Julian et al. 2003; Matsuda et al. 1993), resulting in a dense CB1R-positive innervation of pallidal and nigral structures (Egertová and Elphick 2000; Katona et al. 1999; Tsou et al. 1998). CB1Rs were also localized to the glutamatergic terminals of corticostriatal neurons (Rodriguez et al. 2001), and functional studies demonstrated that their activation leads to decreased glutamate release from corticostriatal inputs (Gerdeman and Lovinger 2001; Huang et al. 2001). In the brainstem, CB1Rs are expressed at relatively low levels within medullary respiratory control centers (Glass et al. 1997; Herkenham et al. 1990), but are highly expressed in axon terminals within medullary nuclei which control emesis, such as the area postrema (Van Sickle et al. 2001). This distribution is in agreement with the relative lack of respiratory effects and the potent antiemetic actions of cannabinoids (Pertwee 2005a). Since the chemoreceptor trigger zone of the area postrema lies outside the blood–brain barrier, activation of CB1Rs in this area by cannabinoids that do not penetrate the CNS should produce antiemetic actions without CNS side effects.

In the peripheral nervous system, CB1Rs have been detected in dorsal root ganglion (DRG) neurons of heterogeneous size (Salio et al. 2002; Sanudo-Pena et al. 1999), with variable degrees of CB1R mRNA and protein localization to different sensory neuron subtypes. Thus, several groups reported predominant CB_1R localization to large-diameter non-nociceptive neurons (Bridges et al. 2003; Hohmann and Herkenham 1999b; Price et al. 2003), and others localized CB1Rs primarily to small-diameter nociceptors (Ahluwalia et al. 2000; Ahluwalia et al. 2002; Binzen et al. 2006). By contrast, we detected CB1Rs in the majority (89%) of DRG sensory neurons with similar degree of localization in nociceptor and non-nociceptor populations (Mitrirattanakul et al. 2006). Axoplasmic flow of CB1Rs has been demonstrated in peripheral sensory axons, implying transport to terminals where cannabinoids are presumed to produce their antinociceptive effects (Hohmann and Herkenham 1999a). Immunohistochemical studies also revealed CB1R immunoreactivity in both small unmyelinated and large myelinated nerve fiber bundles in the human skin (Ständer et al. 2005). These studies also demonstrated CB1Rs in human macrophages, mast cells, sebaceous cells, and keratinocytes (Ständer et al. 2005). The localization of CB1Rs on the central terminals of primary afferents was controversial for many years, in part because earlier ultrastructural studies failed to detect CB1Rs on these terminals in rats and primates (Farquhar-Smith et al. 2000; Ong and Mackie 1999). However, after detection of CB1Rs on glutamatergic terminals in the hippocampus and cerebellum (Katona et al. 2006; Kawamura et al. 2006), a "re-examination" of the excitatory (glutamatergic) terminals of Aδ- and C-fiber primary afferents in the spinal cord did reveal the presence of CB1Rs on these terminals (Nyilas et al. 2009). These presynaptic CB1Rs likely account for the ability of cannabinoid agonists to decrease the frequency of excitatory postsynaptic currents recorded in spinal cord neurons, thus contributing to the modulation of spinal nociceptive neurotransmission (Morisset et al. 2001).

CB1Rs have also been localized to various neurons of the gastrointestinal (G-I) tract in different species, including humans (Izzo et al. 2001b; Izzo and Coutts 2005). Virtually all cholinergic sensory neurons, interneurons, and motoneurons in myenteric ganglia express CB1R in close association with synaptic protein labeling (Coutts et al. 2002). This differential distribution agrees with the inhibitory actions of cannabinoids on G-I motility and secretion (Izzo et al. 2001b; Izzo and Coutts 2005). Pharmacological studies have also localized CB1Rs to presynaptic terminals of postganglionic sympathetic neurons, where they are thought to mediate depressant effects on sympathetic outflow by inhibiting noradrenaline release (Ishac et al. 1996; Kurz et al. 2008; Niederhoffer et al. 2003; Schultheiss et al. 2005). The presence of CB1Rs was detected in endothelial cells of various vascular beds (Golech et al. 2004a; Rajesh et al. 2007; Sugiura et al. 1998); these likely contribute to the vasodilatory actions of CB1R agonists (Wagner et al. 1998). CB1Rs are expressed in various structures of the human eye (Porcella et al. 2000; Straiker et al. 1999), with particularly high levels of expression in the ciliary body (Porcella et al. 2000). Selective activation of CB1Rs, but not CB2Rs, decreases intraocular pressure (Laine et al. 2003; Oltmanns et al. 2008; Pate et al. 1998; Song and Slowey 2000). However, the precise mechanisms by which cannabinoids decrease intraocular pressure have yet to be elucidated (Tomida et al. 2004). CB1R transcripts are also detected in human spleen, tonsils, and peripheral blood leukocytes, although at levels much lower than those found in the brain (Bouaboula et al. 1993).

By contrast, CB2Rs were initially localized and are most highly expressed by immunocompetent cells of the spleen, thymus, and various circulating immune cell populations (Galiegue et al. 1995; Lynn and Herkenham 1994). While not without controversy, pharmacological studies have suggested, and molecular and immunocytochemical studies have later confirmed, localization of CB2R in both peripheral and CNS neurons (Duncan et al. 2008; Gong et al. 2006; Griffin et al. 1997; Onaivi et al. 2006; Pertwee et al. 1995; Skaper et al. 1996; Ständer et al. 2005; Van Sickle et al. 2005). However, CB2R transcripts in the normal brain are present at much lower levels than CB1R transcripts. It is noteworthy that in contrast to the predominant presynaptic axon terminal location of CB1Rs, CB2Rs appear to localize to the cell bodies and dendrites of central (Gong et al. 2006) and peripheral (Duncan et al. 2008) neurons. Although CB2Rs couple to $G_{i/o}$ proteins and inhibit adenylyl cyclase, they do not couple to inhibition of voltage-gated Ca^{2+}-channels or activation of K^+-channels (Felder et al. 1995) (Figure 5.1); this may account for the lack of significant psychotropic effects upon administration of CB2R-selective agonists (Hanus et al. 1999; Malan, Jr. et al. 2001). The physiological role of CB2Rs in central neurons is presently unclear; however, administration of CB2R-selective ligands or direct intracerebroventricular administration of CB2R antisense oligonucleotides does modify behavior (Onaivi et al. 2006, 2008). CB2Rs were also localized with CB1Rs to the endothelial cells of human brain capillaries where they were proposed to play a role in regulation of cerebrovascular blood flow and blood–brain barrier permeability (Golech et al. 2004b). In keeping with the immunomodulatory role of CB2Rs, brain and spinal cord microglia (the only hemopoietic lineage cell type in the CNS) are endowed with CB2Rs (Nunez et al. 2004). In addition, GPR55, a novel cannabinoid receptor which is present both in brain and the periphery, may account for some of the actions of cannabinoids by activating signaling pathways quite distinct from those used by CB1/CB2Rs (Ross 2009). The broad expression of the three cannabinoid receptors in the CNS and various visceral organs implies involvement in a variety of physiological processes that are the subject of intensive investigations.

5.3 ENDOCANNABINOIDS AND THEIR METABOLISM

The endogenous lipid cannabinoids that bind to their receptors cannot be sequestered in vesicles and are therefore synthesized on demand and immediately released by neuronal tissues (Di Marzo et al. 1994; Stella et al. 1997). For example, N-arachidonoylethanolamine (anandamide, AEA) is mainly produced by a two-step enzymatic pathway involving calcium-dependent transacylase and phospholipase D (Cadas et al. 1997; Okamoto et al. 2004; Sugiura et al. 1996). Then, AEA either diffuses (Glaser et al. 2003) or is actively transported into cells (Patricelli and Cravatt 2001) and is rapidly degraded by the membrane-bound fatty acid amide hydrolase (FAAH) to arachidonic acid. Another endocannabinoid, 2-arachidonoyl glycerol (2-AG) is synthesized via the diacylglycerol lipase (DAGL)-mediated hydrolysis of diacylglycerol and metabolized primarily by monoacylglycerol lipase (MAGL) (Dinh et al. 2002). There is also evidence that FAAH and two recently characterized serine hydrolases (ABHD6 and ABHD12) may contribute to 2-AG metabolism (Blankman et al. 2007). Interestingly, FAAH is mainly a postsynaptic enzyme, whereas MAGL is localized to presynaptic axon terminals, suggesting possible differences in the functional roles for AEA and 2-AG (Gulyas et al. 2004). The brain levels of 2-AG are at least two orders of magnitude higher than AEA (Stella et al. 1997). Both AEA and 2-AG are cleared by a high-affinity, selective transporter, which has been characterized biochemically but not molecularly (Hillard et al. 2007; Moore et al. 2005). The biochemistry and metabolism of AEA and 2-AG, as well as other less-well-studied endocannabinoids, have been the subject of excellent reviews (Bisogno et al. 2005; Cravatt and Lichtman 2003; Di Marzo et al. 1999; Hillard 2000).

5.4 ENDOCANNABINOIDS AND SYNAPTIC PLASTICITY

The postsynaptic localization of the endocannabinoid production and transport machinery versus the presynaptic location of CB1Rs led to the current and widely accepted view that brain endocannabinoids are synthesized following excitatory activation in postsynaptic neurons yet act as retrograde messengers at presynaptic terminals to decrease the release of various neurotransmitters (Kreitzer and Regehr 2001; Ohno-Shosaku et al. 2001; Wilson and Nicoll 2001). The endocannabinoid-mediated plasticity involves at least four different types of transient and long-lasting synaptic depression, which is found at both excitatory and inhibitory synapses in many different brain regions (Chevaleyre et al. 2006). In addition, endocannabinoids can modify the induction of non-endocannabinoid-mediated forms of synaptic plasticity (Chevaleyre et al. 2006). The widespread involvement of the endocannabinoid system in synaptic plasticity implies a major role in learning and memory and, consequently, behavior. Future studies will be needed to determine specifically how endocannabinoid-mediated synaptic plasticity contributes to modification of behavior.

5.5 ANTINOCICEPTIVE ACTIONS OF CANNABINOIDS

Endocannabinoids such as AEA, naturally occurring Δ^9–tetrahydrocannabinol (Δ^9–THC), and synthetic cannabinoids such as WIN 55,212-2 or CP 55,940 inhibit responses to noxious thermal and mechanical stimulation in a variety of tests (Fride and Mechoulam 1993; Lichtman and Martin 1991; Martin et al. 1996; Smith et al. 1994; Sofia et al. 1973; Welch et al. 1998). Blockade of peripheral or central CB1Rs leads to hyperalgesia, suggesting tonic activation of CB1Rs by endocannabinoids (Brusberg et al. 2009; Calignano et al. 1998; Richardson et al. 1997; Strangman et al. 1998). Other studies determined that cannabinoids are also effective in reducing thermal and mechanical hyperalgesia and mechanical allodynia induced by peripheral inflammation (Martin et al. 1999; Richardson et al. 1998a; Richardson et al. 1998b) and peripheral nerve injury in rodents (Herzberg et al. 1997). Similarly, chronic neuropathic pain symptoms in humans are alleviated by cannabinoids (Abrams et al. 2007; Berman et al. 2004; Karst et al. 2003; Notcutt et al. 2004; Nurmikko et al. 2007; Wilsey et al. 2008). Cannabinoids are effective in alleviating neuropathic pain symptoms after prolonged, repeated treatment (Bridges et al. 2001; Costa et al. 2004), unlike opioids, which have only limited effectiveness (Mao et al. 1995; Ossipov et al. 1995; Rashid et al. 2004). The antinociceptive (analgesic) and antihyperalgesic effects of cannabinoids were initially thought to be mediated largely by CB1Rs because they are blocked by the selective CB1R antagonist SR 141716A (Calignano et al. 1998; Lichtman and Martin 1997; Richardson et al. 1998b; Rinaldi-Carmona et al. 1994).

5.6 CANNABINOID ACTIONS ON MOTOR CONTROL AND COGNITION

In addition to analgesic effects, acute consumption of cannabis reversibly impairs a variety of cognitive and performance tasks, including memory and learning (Hampson and Deadwyler 1999). Activation of central CB1Rs by cannabinoids such as Δ^9–THC or AEA produces other complex effects on behavior unique to this class of compounds: at low doses a mixture of stimulatory and depressant effects is observed, and at higher doses central depression predominates (Dewey 1986; Pertwee 1988). Catalepsy, motor deficits, and hypothermia are among the effects observed after administration of centrally acting cannabinoids (Dewey 1986; Maccarrone and Wenger 2005; Pertwee 1988). These properties, which appear in large part mediated by CB1Rs, have greatly limited the clinical use of cannabinoids for the treatment of chronic pain states. At the same time, the increased understanding of CB1R function in the CNS has prompted the evaluation of CB1R ligands for treatment in various other disorders, including psychoses, obesity, multiple sclerosis, stroke, brain trauma, drug addiction, and movement disorders (Basavarajappa and Hungund 2005; Di Marzo and Petrocellis 2006; Mechoulam et al. 2002; Robson 2001).

The synthesis of CB2R-selective agonists such as HU308 and AM1241 was a major development because these compounds produced antinociceptive and antihyperalgesic effects in transient and persistent pain states by activating what was at the time thought to be essentially a peripheral cannabinoid receptor (Hanus et al. 1999; Ibrahim et al. 2003). Importantly, these compounds lack the type of side effects associated with activation of central CB1Rs, which effectively renewed interest in the development of peripherally active cannabinoid-based analgesics.

5.7 ENDOCANNABINOID SYSTEM ALTERATIONS IN INFLAMMATORY AND NEUROPATHIC PAIN STATES

A decade ago it was demonstrated that the antihyperalgesic effectiveness of centrally administered synthetic cannabinoid WIN 55,212-2 is greater after induction of rodent hindpaw inflammation than its antinociceptive effects in non-inflamed hindpaws (Martin et al. 1999). Another study showed that endocannabinoid (AEA) levels in the periaqueductal gray increase in response to peripheral inflammation (Walker et al. 1999). Since then many studies have demonstrated that both CB1Rs and CB2Rs undergo increased expression during inflammation and after development of peripheral nerve injury–induced painful neuropathies (Table 5.1). Such transcription-driven increases in cannabinoid receptors were demonstrated both in the peripheral tissues and the CNS. Increases in the tissue levels of endocannabinoids have also been demonstrated in inflammatory and neuropathic pain states. For example, one study revealed spinal nerve injury–induced increases in the levels of endocannabinoids within sensory ganglia (Mitrirattanakul et al. 2006). Presently it is unknown whether increased endocannabinoid levels are due to their increased "on-demand" synthesis by hyperexcitable neurons or to decreases in the activity of enzymes that metabolize endocannabinoids.

TABLE 5.1

Pathology or Experimentally Induced Alterations in the Endocannabinoid System

Experimental/Clinical Pathology	Animal Species, Anatomical Location of Measurements	Endocannabinoid System Alteration	Reference
Formalin-induced hindpaw inflammation	Rat, hindpaw skin	Unchanged AEA, 2-AG, PEA	(Beaulieu et al. 2000)
CFA-induced hindpaw inflammation	Rat, DRG neurons and dermal nerve fibers	↑ CB1R expression	(Amaya et al. 2006)
CFA-induced hindpaw inflammation	Rat, DRG neurons	Unchanged FAAH distribution	(Lever et al. 2009)
Carrageenan-induced hindpaw inflammation	Rat, hindpaw skin	↓ AEA, ↓ PEA, unchanged FAAH and MAGL activity	(Jhaveri et al. 2008)
Gingivitis and periodontitis	Human, gingival tissue and crevicular fluid	↑ CB1R mRNA and protein, ↑ CB2R mRNA and protein	(Nakajima et al. 2006)
Osteoarthritis and rheumatoid arthritis	Human, knee synovial fluid and tissue	↑ CB1R mRNA and protein, ↑ CB2R mRNA and protein, ↑ AEA and 2-AG	(Richardson et al. 2008)
Oxazolone-induced contact dermatitis	Mouse, ear	↑ CB2R mRNA, ↑ 2-AG	(Oka et al. 2006)
DNFB-induced contact dermatitis	Mouse, ear	↓ CB1R mRNA, ↑ CB2R mRNA, ↑ AEA and 2-AG	(Karsak et al. 2007)
Endometrial inflammation	Human, endometrium	↑ CB2R mRNA and protein expression (but not CB1R)	(Iuvone et al. 2008)
Mustard oil-induced colitis, DSS-induced colitis	Mouse, colon	↑ CB1R expression (neurons and endothelium), ↑ CB2R expression (immune cells)	(Kimball et al. 2006)
Cerulein-induced and acute pancreatitis	Mouse/human, pancreas	↑ CB1R expression, ↑ CB2R expression, ↑ AEA (but not 2-AG)	(Michalski et al. 2007)
Croton oil-induced intestinal inflammation	Mouse, small intestine	↑ CB1R mRNA and protein expression, unchanged AEA and 2-AG, ↑ FAAH activity	(Izzo et al. 2001a)

TABLE 5.1 (Continued)

Pathology or Experimentally Induced Alterations in the Endocannabinoid System

Experimental/Clinical Pathology	Animal Species, Anatomical Location of Measurements	Endocannabinoid System Alteration	Reference
Acrolein-induced bladder inflammation	Rat, bladder and spinal cord	↑ CB2R mRNA and protein expression in bladder (unchanged CB1R), unchanged CB1R and CB2R expression in spinal cord	(Merriam et al. 2008)
LPS-induced sepsis	Mouse, intestine	↑ CB1R (but not CB2R) expression, ↑ FAAH expression	(De Filippis et al. 2008)
LPS-enhanced GI transit	Rat, ileum and spleen	Unchanged CB2R expression	(Duncan et al. 2008)
Acetic acid-induced ileus	Mouse, small intestine	↑ CB1R expression, ↑ AEA (but not 2-AG), unchanged FAAH activity	(Mascolo et al. 2002)
Toxin A-induced inflammation	Rat, ileum	↑ AEA and 2-AG	(McVey et al. 2003)
DNB- and DSS-induced colitis	Mouse, colon	↑ CB1R expression	(Massa et al. 2004)
Ulcerative colitis, DNBS and TNBS-induced colitis	Human, rat, and mouse, colon	↑ AEA (but not 2-AG)	(D'argenio et al. 2006)
Cholera toxin-induced diarrhea	Mouse, small intestine	↑ CB1R mRNA, ↑ AEA (but not 2-AG), unchanged FAAH activity	(Izzo et al. 2003)
Painful neuroma	Human, skin and nerve	↑ CB2R expression	(Anand et al. 2008)
L5 spinal nerve ligation–induced neuropathy or sciatic nerve section	Rat and mouse, peripheral nerve, DRG and spinal cord	↑ CB2R expression	(Wotherspoon et al. 2005)
L5-L6 spinal nerve ligation-induced neuropathy	Rat, spinal cord and DRG	↑ CB2R mRNA in spinal cord	(Beltramo et al. 2006)
L5 spinal nerve ligation-induced neuropathy	Rat, L4 DRG neurons	↑ CB1R expression, ↑ AEA and 2-AG	(Mitrirattanakul et al. 2006)

TABLE 5.1 (Continued)
Pathology or Experimentally Induced Alterations in the Endocannabinoid System

Experimental/Clinical Pathology	Animal Species, Anatomical Location of Measurements	Endocannabinoid System Alteration	Reference
Sciatic nerve constriction- or L5-L6 spinal nerve ligation-induced neuropathy	Rat, spinal cord	↑ CB1R mRNA	(Zhang et al. 2003)
Infraorbital nerve constriction-induced	Rat, brainstem	↑ CB1R expression	(Liang et al. 2007)
Sciatic nerve constriction-induced neuropathy	Rat, spinal cord	↑ CB1R expression	(Lim et al. 2003)
Sciatic nerve constriction-induced neuropathy	Mouse, spinal cord	↑ CB2R expression	(Racz et al. 2008)
Tibial nerve axotomy-induced neuropathy	Rat, thalamus	↑ CB1R expression in contralateral thalamus	(Siegling et al. 2001)
Partial saphenous nerve ligation-induced neuropathy	Rat and mouse, skin, DRG and spinal cord	↑ CB1R expression, ↑ CB2R expression	(Walczak et al. 2005, 2006)
Sciatic nerve axotomy	Rat, L5 DRG neurons	↑ FAAH localization in large neurons	(Lever et al. 2009)
L5 spinal nerve section	Rat, L5 DRG neurons	↑ FAAH expression, ↑ FAAH localization in large neurons	(Lever et al. 2009)
Exercise	Human, plasma	↑ AEA (but not 2-AG)	(Sparling et al. 2003)
Fasting	Rat, small intestine, stomach, and brain	↑ AEA in small intestine (but not in stomach or brain)	(Gomez et al. 2002)
Fasting	Rat brain	↑ AEA and 2-AG in limbic forebrain and hypothalamus (but not cerebellum)	(Kirkham et al. 2002)
Feeding	Rat brain	↓ 2-AG in hypothalamus	(Kirkham et al. 2002)
Zucker rat obesity model	Rat, hypothalamus	↑ 2-AG (but not AEA or PEA)	(Di Marzo et al. 2001b)
Leptin receptor-deficient mouse obesity model	Mouse, hypothalamus and cerebellum	↑ AEA and 2-AG (but not PEA) in hypothalamus (but not cerebellum)	(Di Marzo et al. 2001b)

TABLE 5.1 (Continued)
Pathology or Experimentally Induced Alterations in the Endocannabinoid System

Experimental/Clinical Pathology	Animal Species, Anatomical Location of Measurements	Endocannabinoid System Alteration	Reference
Leptin-null mouse obesity model	Mouse, hypothalamus	↑ 2-AG (but not AEA)	(Di Marzo et al. 2001b)
Zucker rat obesity model	Rat, adipose tissue	↑ CB1R mRNA	(Bensaid et al. 2003)
Leptin-null mouse obesity model	Mouse, uterus	Unchanged CB1R binding, ↑ AEA and 2-AG (but not PEA), ↓ FAAH and MAGL activity, ↑ DAGL, ↓ endocannabinoid transporter activity	(Maccarone et al. 2005)
Alcohol-induced liver cirrhosis	Human/rat, hepatic vascular endothelial cells/circulating monocytes	↑ CB1R expression, ↑ AEA	(Batkai et al. 2001)
CCl$_4$-induced liver cirrhosis	Rat, circulating monocytes	↑ AEA	(Ros et al. 2002)
Alcohol- or hepatitis C-induced liver cirrhosis	Human, plasma	↑ AEA	(Fernandez-Rodriguez et al. 2004)
CCl$_4$-induced liver cirrhosis	Rat, mesenteric arteries	↑ CB1R mRNA and protein expression	(Domenicali et al. 2005)
Liver cirrhosis of various origin	Human, liver biopsy	↑ CB2R expression	(Julien et al. 2005)
Liver cirrhosis of various origin	Human, liver biopsy	↑ CB1R expression	(Teixeira-Clerc et al. 2006)
Bile duct ligation-induced cirrhosis	Rat, superior mesenteric artery	↑ CB1R mRNA and protein expression	(Moezi et al. 2006)
Hepatic ischemia/reperfusion (15 or 60 min)	Rat, plasma	↑ AEA (15 min), ↑ 2-AG (15 and 60 min)	(Kurabayashi et al. 2005)
Hemorrhagic shock	Rat, macrophages	↑ AEA	(Wagner et al. 1997)
Acute myocardial infarction	Rat, monocytes and platelets	↑ AEA and 2-AG	(Wagner et al. 2001)
LPS-induced septic shock/hypotension	Rat, macrophages and platelets	↑ AEA (macrophages), ↑ 2-AG (platelets and macrophages	(Varga et al. 1998)
LPS-induced septic shock/hypotension	Mouse, macrophages	↑ AEA (but not 2 AG), ↑ FAAH mRNA and activity	(Liu et al. 2003)

TABLE 5.1 (Continued)

Pathology or Experimentally Induced Alterations in the Endocannabinoid System

Experimental/Clinical Pathology	Animal Species, Anatomical Location of Measurements	Endocannabinoid System Alteration	Reference
LPS treatment of lymphocytes	Human, lymphocytes	Unchanged CB1R and CB2R cDNA and binding, ↑ AEA , ↓ FAAH expression, cDNA and activity	(Maccarrone et al. 2001b)
Endotoxic shock	Human, serum	↑ AEA and 2-AG	(Wang et al. 2001)
Bone cement implantation syndrome	Human, blood	↑ AEA and 2-AG	(Motobe et al. 2004)
Miscarriage	Human, lymphocytes	↓ FAAH activity and expression	(Maccarrone et al. 2000)
Miscarriage	Human, lymphocytes	Unchanged CB1R binding, ↓ FAAH activity and expression, unchanged AMT	(Maccarrone et al. 2001c)
Embryo implantation failure	Human, blood and lymphocytes	Unchanged CB1R/CB2R binding, ↑ AEA (but not 2-AG), ↓ FAAH activity and expression	(Maccarrone et al. 2002)
Miscarriage	Human, placenta	↑ FAAH expression	(Chamley et al. 2008)
Ectopic pregnancy	Human, Fallopian tube and endometrium	↑ CB1R mRNA and protein expression	(Horne et al. 2008)
Spontaneously hypertensive rats	Rat, myocardium and aortic endothelium	↑ CB1R expression, ↑ AEA (but not 2-AG), ↑ FAAH expression	(Batkai et al. 2004)
Atherosclerosis	Human/mouse, atherosclerotic plaques	Undetected CB1R expression, ↑ CB2R expression	(Steffens et al. 2005)
Hepatic ischemia/reperfusion	Mouse, liver	↑ CB2R expression, ↑ AEA and 2-AG	(Batkai et al. 2007)
High IOP-induced ischemia/reperfusion	Rat, retina	↓ CB1R expression, ↓ CB1R binding, ↓ AEA, ↑ FAAH expression, unchanged AEA transporter activity	(Nucci et al. 2007)
Stroke	Human, cortex	↑ AEA, OEA and PEA	(Schabitz et al. 2002)
MCA occlusion	Rat, neocortex	↑ CB1R expression	(Jin et al. 2000)

TABLE 5.1 (Continued)

Pathology or Experimentally Induced Alterations in the Endocannabinoid System

Experimental/Clinical Pathology	Animal Species, Anatomical Location of Measurements	Endocannabinoid System Alteration	Reference
MCA occlusion	Rat, whole brain	↑ AEA (but not 2-AG)	(Muthian et al. 2004)
Brain hypoxia/ischemia, MCA occlusion	Rat, microglia	↑ CB2R expression	(Ashton et al. 2007)
Traumatic brain injury	Mouse, brain	↑ 2-AG	(Panikashvili et al. 2001)
Traumatic brain injury, intracerebral NMDA injection,	Rat (neonatal), brain	↑ CB1R mRNA, ↑ AEA (but not 2-AG)	(Hansen et al. 2001)
Picrotoxinin-induced seizures	Rat, whole brain	↑ 2-AG	(Sugiura et al. 2000)
Kainic acid-induced excitotoxicity	Rat, hippocampus	↑ AEA (but not 2-AG)	(Marsicano et al. 2003)
Febrile seizures	Rat (adult), hippocampus	↑ CB1R expression, unchanged AEA and 2-AG, unchanged FAAH and MAGL activity	(Chen et al. 2003)
Pilocarpine-induced temporal lobe epilepsy	Rat, hippocampus	↑ CB1R expression	(Wallace et al. 2003)
Depression/suicide	Human, prefrontal cortex	↑ CB1R expression	(Hungund et al. 2004)
Schizophrenia	Human, cerebrospinal fluid	↑ AEA	(Giuffrida et al. 2004; Leweke et al. 1999)
Schizophrenia	Human, blood	↑ AEA	(De Marchi et al. 2003)
Schizophrenia	Human, dorsolateral prefrontal cortex, anterior cingulate cortex	↑ CB1R binding (region-specific)	(Dean et al. 2001; Zavitsanou et al. 2004)
Schizophrenia	Human, anterior cingulate cortex	Unchanged CB1R expression	(Koethe et al. 2007)
Major depression	Human, anterior cingulate cortex	↓ CB1R immunopositive glial cells	(Koethe et al. 2007)
Alcoholism with depression and suicide	Human, prefrontal cortex	↑ CB1R expression, ↑ AEA and 2-AG	(Vinod et al. 2005)
Alcohol withdrawal/dependence	Rat, limbic forebrain and midbrain	Region- and time-dependent changes in AEA and 2-AG	(Gonzalez et al. 2004)

TABLE 5.1 (Continued)
Pathology or Experimentally Induced Alterations in the Endocannabinoid System

Experimental/Clinical Pathology	Animal Species, Anatomical Location of Measurements	Endocannabinoid System Alteration	Reference
Alcohol withdrawal/dependence	Rat, hippocampus	↑ CB1R expression, ↑ AEA and 2-AG	(Mitrirattanakul et al. 2007)
Alcohol intoxication	Rat, prefrontal cortex, amygdala, striatum, hypothalamus	↓ AEA and 2-AG	(Rubio et al. 2007)
Chronic unpredictable stress	Rat, hippocampus and limbic forebrain	↓ CB1R expression and binding, ↓ 2-AG (hippocampus but not limbic forebrain)	(Hill et al. 2005)
Acute restraint stress	Mouse, amygdala, limbic forebrain and cerebellum	↓ AEA (amygdala only)	(Patel et al. 2005)
Repeated restraint stress	Mouse, amygdala and limbic forebrain	↓ AEA, ↑ 2-AG (amygdala and limbic forebrain)	(Patel et al. 2005)
Extinction of aversive memory	Mouse, amygdala	↑ AEA and 2-AG	(Marsicano et al. 2002)
Reserpine-induced model of Parkinson's disease	Rat, globus pallidus	Unchanged AEA, ↑ 2-AG	(Di Marzo et al. 2000b)
6-HD-induced model of Parkinson's disease	Rat, striatum	Unchanged CB1R binding, ↑ AEA (but not 2-AG), ↓ FAAH and AMT	(Gubellini et al. 2002; Maccarrone et al. 2003)
6-HD-induced model of Parkinson's disease	Rat, striatum	↓ FAAH activity, unchanged N-Acyltransferase activity	(Fernandez-Espejo et al. 2004)
6-HD-induced model of Parkinson's disease	Rat, striatum	↓ AEA (but not 2-AG), unchanged FAAH activity	(Ferrer et al. 2003)
6-HD-induced model of Parkinson's disease	Rat, striatum and substantia nigra	Unchanged CB1R expression and coupling efficiency	(Herkenham et al. 1991; Romero et al. 2000)
6-HD-induced model of Parkinson's disease	Rat, striatum	↑ CB1R mRNA	(Mailleux and Vanderhaeghen 1993; Romero et al. 2000; Zeng et al. 1999)
Parkinson's disease, MPTP-induced lesions	Human and marmoset, striatum	↑ CB1R binding	(Lastres-Becker et al. 2001a)

TABLE 5.1 (Continued)
Pathology or Experimentally Induced Alterations in the Endocannabinoid System

Experimental/Clinical Pathology	Animal Species, Anatomical Location of Measurements	Endocannabinoid System Alteration	Reference
Parkinson's disease	Human, cerebrospinal fluid	↑ AEA	(Pisani et al. 2005)
Huntington's disease, Huntington's disease transgenic mouse model	Human and mouse, striatum	↓ CB1R binding	(Glass et al. 1993, 2004; Richfield and Herkenham 1994)
Huntington's disease transgenic mouse model	Mouse, striatum	↓ CB1R mRNA and binding	(Lastres-Becker et al. 2002a)
3-NP induced striatum lesions	Rat, basal ganglia	↓ CB1R mRNA and binding, ↓ AEA and 2-AG (striatum), ↑ AEA (substantia nigra)	(Lastres-Becker et al. 2001b, 2002b)
Alzheimer's disease	Human, neuritic plaque-associated glia	↑ CB2R (but not CB1R) expression in microglia, ↑ FAAH expression in astrocytes	(Benito et al. 2003)
Alzheimer's disease and β-amyloid-induced toxicity	Human and rat, neurons and neuritic plaque-associated glia	↓ CB1R expression, ↑ CB2R expression in microglia	(Ramirez et al. 2005)
Down's syndrome	Human, brain microglia and astrocytes	↑ CB2R expression in microglia, ↑ FAAH expression in astrocytes	(Nunez et al. 2008)
Simian immunodeficiency virus–induced encephalitis	Macaque, brain microglia and astrocytes	↑ CB2R expression in microglia, ↑ FAAH expression in astrocytes	(Benito et al. 2005)
HIV gp120-induced neurotoxicity	Rat, neocortex	Unchanged CB1R binding, ↓ AEA, ↑ FAAH expression and activity	(Maccarrone et al. 2004)
Experimental autoimmune encephalomyelitis	Mouse, brain and spinal cord	↑ AEA, 2-AG and PEA	(Baker et al. 2001)
Experimental autoimmune encephalomyelitis	Mouse, brain and spinal cord, brain microglia and peripheral macrophages	↑ CB2R mRNA in CNS, activated microglia and peripheral macrophages, unchanged 2-AG levels	(Maresz et al. 2005)

TABLE 5.1 (Continued)
Pathology or Experimentally Induced Alterations in the Endocannabinoid System

Experimental/Clinical Pathology	Animal Species, Anatomical Location of Measurements	Endocannabinoid System Alteration	Reference
Experimental autoimmune encephalomyelitis	Mouse, brain	↓ CB1R expression in caudate- putamen, globus pallidus and cerebellum	(Cabranes et al. 2006)
Multiple sclerosis and amyotrophic lateral sclerosis	Human, spinal cord	↑ CB2R expression in microglia	(Yiangou et al. 2006)
Intracerebroventricular LPS-induced inflammation	Rat, brain	↑ CB2R expression	(Mukhopadhyay et al. 2006)
HOSCC-induced hindpaw tumor	Mouse, L5 DRG neurons	↑ CB1R expression	(Guerrero et al. 2008)
Colorectal cancer	Human, colon	Unchanged CB1R, CB2R and FAAH expression, ↑ AEA and 2-AG	(Ligresti et al. 2003)
Colorectal cancer	Human, colon	↓ CB1R expression	(Wang et al. 2008)
Colorectal cancer	Human, colon	↓ CB1R expression, ↑ CB2R expression	(Cianchi et al. 2008)
Osteolytic fibrosarcoma–induced bone cancer pain	Mouse, hindpaw skin, DRG and tibial nerve	↑ CB1R mRNA and protein, ↓ AEA, ↑ FAAH mRNA and activity,	(Khasabova et al. 2008)
Prostate cancer and benign prostate hyperplasia	Human, prostate	↑ CB1R mRNA and protein expression	(Czifra et al. 2009)
Mantle cell lymphoma	Human, lymphoid tissues	↑ CB1R mRNA, ↑ CB2R mRNA	(Islam et al. 2003)
Non-Hodgkin lymphomas of the B-cell type	Human, lymphoid tissues	↑ CB1R mRNA and protein expression, ↑ CB2R mRNA	(Gustafsson et al. 2008)
Non-Hodgkin lymphomas of the B- and T-cell type	Human, lymphoid tissues	↑ CB1R expression	(Rayman et al. 2007)
Pancreatic ductal adenocarcinoma	Human, pancreas	Unchanged AEA and 2-AG, variable CB1R, CB2R, FAAH and MAGL mRNA expression	(Michalski et al. 2008)
Hepatic carcinoma	Human, liver	↑ CB1R expression, ↑ CB2R expression,	(Xu et al. 2006)

TABLE 5.1 (Continued)
Pathology or Experimentally Induced Alterations in the Endocannabinoid System

Experimental/Clinical Pathology	Animal Species, Anatomical Location of Measurements	Endocannabinoid System Alteration	Reference
Prostate cancer	Human, cell lines	↑ CB1R mRNA and protein expression, ↑ AEA and 2-AG	(Sarfaraz et al. 2005)
Estrogen-induced pituitary hyperplasia	Rat, pituitary	↓ CB1R mRNA and protein expression	(Gonzalez et al. 2000)
Pituitary adenomas	Human, pituitary	↑ CB1R mRNA and protein expression, ↑ AEA and 2-AG	(Pagotto et al. 2001)
Glioblastoma and meningiomas	Human, resected brain tissue	↓ AEA, OEA, PEA and SEA (but not 2-AG)	(Maccarone et al. 2001a)
Glioblastomas	Human, resected brain tissue	↑ AEA (but not 2-AG), ↓ PEA, ↓ FAAH and NAPE-PLD	(Petersen et al. 2005)
Meningiomas	Human, resected brain tissue	↑ AEA and 2-AG, ↓ PEA, ↓ FAAH and NAPE-PLD	(Petersen et al. 2005)
Glioblastomas, astrocytomas, medulloblastomas	Human adult and pediatric brain tissue	↑ CB2R expression (not in medulloblastomas)	(Ellert-Miklaszewska et al. 2007)
Astroglioma	Human, brain	Unchanged CB1R and CB2R mRNA	(Held-Feindt et al. 2006)

3-NP, 3-nitropropionic acid; AMT, anandamide membrane transporter; CFA, complete Freund's adjuvant; IOP, intraocular pressure; MCA, middle cerebral artery; HOSCC, human oral squamous cell carcinoma; DNB, dinitrobenzene; DNBS, 2,4-dinitrobenzene sulfonic acid; DSS, dextran sulfate sodium; LPS, lipopolysaccharide; NAPE-PLD, N-acylphosphatidylethanolamine-hydrolysing phospholipase D; NMDA, N-methyl, D-aspartate; OEA, N-oleylethanolamide; PEA, N-palmitoylethanolamide; SEA, N-stearoylethanolamine; TNBS, 2,4,6-trinitrobenzene sulfonic acid.

The increases in cannabinoid receptor expression result in increased potency or efficacy of the exogenously applied cannabinoids, depending on whether the ligand is a full (e.g., WIN 55,212-2) or a partial agonist (e.g., Δ^9–THC) (Pertwee 2008). It is also likely that increased cannabinoid receptor expression contributes to the effectiveness of cannabinoids in providing relief from painful neuropathy symptoms after repeated administration. By contrast, several studies demonstrated that morphine has only limited effectiveness in alleviating peripheral neuropathy symptoms, possibly due to the decreased expression of peripheral opioid receptors (Mao et al. 1995; Ossipov et al. 1995; Rashid et al. 2004). In addition, chronic opioid treatment leads to considerable analgesic tolerance and development of hyperalgesic effects, which together with the well-known respiratory depressant effects have led to the failure of opioid therapy to successfully treat chronic pain populations (see Horvath et al., this volume). By contrast, recent clinical studies have reaffirmed that long-term treatment with cannabinoids for symptomatic relief of peripheral neuropathy symptoms does not result in any appreciable decrement in clinical effectiveness after long-term administration (Nurmikko et al. 2007).

5.8   HOMEOSTATIC ROLE OF THE ENDOCANNABINOID SYSTEM

In addition to inflammatory and neuropathic pain states, it appears that any pathological condition that involves an inflammatory response, whether generated from an injury, a foreign organism, or the autoimmune system, results in the up-regulation of the endocannabinoid system. Thus, alterations (usually increases) in cannabinoid receptors and/or their endogenous ligands have been observed in temporal lobe epilepsy (Wallace et al. 2003), alcohol withdrawal and dependence (Mitrirattanakul et al. 2007), brain ischemia (Jin et al. 2000), endometritis (Iuvone et al. 2008), pancreatitis (Michalski et al. 2007), and various other types of injury (Table 5.1). Also, injury symptomatology is exacerbated in the presence of CB1R or CB2R antagonists or in mice with genetic deletions of CB1R and CB2R (Baker et al. 2000; Batkai et al. 2007; Kimball et al. 2006; Massa et al. 2004; Palazuelos et al. 2008; Panikashvili et al. 2005; Schuelert and McDougall 2008). While the specific alterations appear to depend on the type of injury/pathology, an overall emerging view is that the up-regulation of the endocannabinoid system is the organism's compensatory mechanism designed to alleviate the negative consequences of tissue injury and facilitate repair (Di Marzo and Petrocellis 2006; Mechoulam and Shohami 2007; Pertwee 2005b). Nevertheless, there are several clinically relevant situations where increases in CB1R expression may have an adverse effect. In these conditions, selective blockade of peripheral CB1Rs could prove to be of clinical benefit (Kunos et al. 2009). Also, mice with a genetic deletion of GPR55 were recently shown not to develop either inflammatory- or nerve injury–induced hyperalgesia, suggesting that selective GPR55 antagonists may also be of utility for treating inflammatory or neuropathic pain (Staton et al. 2008).

5.9   ACTIONS AT NON-CANNABINOID RECEPTORS

In addition to the diverse physiological effects of cannabinoid receptor activation, certain cannabinoids have effects at other targets. For example, anandamide administration in CB1R -/- mice still produces cannabinomimetic effects in various behavioral tests (Baskfield et al. 2004; Di Marzo et al. 2000a). Although some of these effects could be ascribed to actions at CB2Rs or GPR55Rs, others may be due to activation of non-cannabinoid receptors or to receptor-independent interactions with membrane ion channels and intracellular second-messenger systems (Oz 2006). In particular, several endogenous and synthetic cannabinoids have demonstrated effects at transient receptor potential (TRP) receptors. Certain TRP receptors (e.g., TRPV1) are highly expressed in nociceptors where they play an important role in detection of nociceptive signals and nociceptor sensitization in inflammatory and neuropathic pain states (Tominaga and Caterina 2004).

Several studies demonstrated that near physiological concentrations of AEA produce local vasodilation (Zygmunt et al. 1999), vas deferens relaxation (Ross et al. 2001), and excitation of the central terminals of sensory afferents (Tognetto et al. 2001), all via TRPV1 receptor activation. Such studies led to the idea that endocannabinoids acting via TRPV1 may contribute to nociception and hyperalgesia, reviewed by Di Marzo et al. (2001a) and Szolcsanyi (2000). Indeed, AEA was implicated in the inflammatory response of certain tissues. Thus, toxin A-induced inflammation and edema of the ileum was shown to be dependent on activation of TRPV1 receptors by endogenous AEA (McVey et al. 2003). Similar findings were obtained in cyclophosphamide-induced bladder hyperreflexia and cystitis (Dinis et al. 2004). It was also shown that inflammatory mediators can convert anandamide into a potent activator of TRPV1 receptors, possibly via receptor sensitization (Singh Tahim et al. 2005). On the other hand, many studies showed that cannabinoids require micromolar levels to activate TRP receptors, whereas activation of antinociceptive cannabinoid receptors occurs at nanomolar levels. Thus, with the possible exception of N-arachidonoyl dopamine (nm activation of TRPV1 receptors [Huang et al. 2002]), both endogenous and exogenous cannabinoids that possess TRP activity act as partial agonists at TRP receptors (Akopian et al. 2008; Nemeth et al. 2003; Roberts et al. 2002; Ross 2003). In addition, cannabinoid receptor-mediated activation of calcineurin results in TRPV1 receptor desensitization (Jeske et al. 2006; Patwardhan et al. 2006). Recent studies have also demonstrated that selective activation of CB2Rs on human sensory neurons blocks capsaicin-induced inward currents and cytoplasmic Ca^{2+} elevation via inhibition of adenylyl cyclase (Anand et al. 2008). Collectively, these studies help explain why exogenous cannabinoids produce analgesic and antihyperalgesic effects rather than pronociceptive effects after local peripheral or systemic administration.

5.10 DISSOCIATING EFFECTS OF PERIPHERAL AND CENTRAL CANNABINOID RECEPTOR ACTIVATION

Early studies have assumed a central action of cannabinoids based on the high degree of CB1R expression in the brain, including various sites associated with pain signal transmission and modulation. Subsequently, multiple studies with genetically engineered mice lacking CB1Rs have confirmed their role in cannabinoid-induced analgesia (Ledent et al. 1999; Zimmer et al. 1999) but did not localize their actions to peripheral or central receptors. Evidence for important peripheral sites of cannabinoid analgesic effects came from studies where local administration of cannabinoids into inflamed tissue attenuated hyperalgesia and allodynia via peripheral CB1Rs, at doses that produced minimal centrally mediated side effects (Amaya et al. 2006; Gutierrez et al. 2007; Richardson et al. 1998b). Peripheral CB1R activation was also shown to reduce mechanical activation of A-δ nociceptors from inflamed skin but not from non-inflamed skin (Potenzieri et al. 2008a). Similarly, local activation of peripheral CB1Rs attenuates hyperalgesia produced by thermal injury (Johanek and Simone 2004), nerve injury (Fox et al. 2001), and cancer (Guerrero et al. 2008; Potenzieri et al. 2008b). However, the crucial role of peripheral cannabinoid receptors in the antihyperalgesic actions of systemically administered cannabinoids was demonstrated only recently using conditional deletion of CB1Rs located on nociceptive primary afferent neurons (Agarwal et al. 2007). In these conditional peripheral CB1R knockout mice, the antihyperalgesic effects of systemically administered cannabinoids were nearly completely lost in models of carrageenan-induced inflammation and sciatic nerve injury–induced neuropathy. By contrast, the effects of central CB1R activation were retained in the conditional knockouts but lost in the global CB1R-null mice (Agarwal et al. 2007).

5.11 STRATEGIES FOR PERIPHERAL CANNABINOID RECEPTOR TARGETING

Considerable experimental and clinical evidence points to the homeostatic role of the endocannabinoid system in ameliorating the negative consequences of tissue injury. Therapeutic targeting of the peripheral cannabinoid receptors could provide relief of injury symptoms and speed up tissue repair, while minimizing the side-effects associated with activation of central cannabinoid receptors.

One approach already taken was the development of CB2R-selective ligands. Given the recent demonstrations of CB2Rs on human sensory nerve fibers (Anand et al. 2008; Ständer et al. 2005) and the increased expression of CB2Rs within human and rat sensory neurons after inflammation and peripheral nerve injury (Anand et al. 2008; Beltramo et al. 2006; Wotherspoon et al. 2005), CB2R-selective agonists promise to become an important treatment option for inflammatory and neuropathic pain states (Guindon and Hohmann 2008). CB2R-selective agonists are also being considered for the treatment of myocardial ischemia and atherosclerosis (Pacher et al. 2008). In addition, CB2R-selective antagonists may be of value in the treatment of certain degenerative bone diseases such as rheumatoid arthritis (Lunn et al. 2007). Many CB2R-selective ligands have been developed (Huffman 2000; Huffman et al. 2002), although brain-impermeant analogues are not being emphasized because of the limited localization of CB2Rs in the CNS under normal conditions (Ibrahim et al. 2003), as well as their increased central expression in neuropathic pain states (Beltramo et al. 2006; Zhang et al. 2003) and in autoimmune disorders (Benito et al. 2008). A potential concern with administration of CB2R agonists for the treatment of chronic pain symptoms is excessive suppression of the immune system, which could make them unsuitable as therapeutics in patients with compromised immune systems.

One alternative strategy might be to develop selective CB1R agonists that do not penetrate the blood–brain barrier, thereby providing pain relief without the side effects associated with central CB1R activation. Indeed, one compound with a dual CB1R/CB2R agonist profile (~170-fold preference for CB1R over CB2R) and restricted CNS permeability was recently shown to possess antihyperalgesic properties without appreciable central side effects (Dziadulewicz et al. 2007). Another study demonstrated this compound's effectiveness against colorectal distention–induced visceral pain; this action was blocked by CB1R but not CB2R antagonists

(Brusberg et al. 2009). Several other CB1R ligands were suggested to exhibit limited brain penetration and few psychotropic side effects (Fride et al. 2004). However, some derivatives turned out to have little activity at CB1Rs or CB2Rs; their effects appear to be mediated through other mechanisms yet to be defined (Pertwee et al. 2005). Other new derivatives may bind CB1Rs but may exhibit antagonistic activity at these receptors (Ben-Shabat et al. 2006). Thus, development of peripherally acting CB1R-selective agonists continues to represent an important goal. Such brain-impermeant analgesics would still be expected to produce side effects of peripheral CB1R activation such as constipation, hypotension, and possibly weight gain. There is also a potential concern for the development of tolerance to CB1R agonists during prolonged treatment. However, in recent clinical trials of cannabis preparations for neuropathic pain treatment, such side effects were well tolerated and there was no evidence for development of analgesic tolerance with long-term treatment (Ellis et al. 2009; Nurmikko et al. 2007; Wilsey et al. 2008).

Peripherally acting CB1R-selective analgesics are unlikely to replace non-steroidal anti-inflammatory analgesics (NSAIAs) and opioids as the mainstay treatment for acute or postoperative pain. Indeed, several studies demonstrated the relatively poor response of cannabinoids in postoperative pain relief (Beaulieu 2006; Buggy et al. 2003). However, CB1R-selective analgesics may be a panacea for the treatment of various types of chronic pain in situations where NSAIAs or opioids may be contraindicated. For example, patients with G-I ulcers treated with CB1R agonists could benefit from the demonstrated antiulcer effects of cannabinoids (Izzo et al. 2001b). Similarly, asthmatic patients could benefit from the bronchodilator properties of cannabinoids (Gong, Jr. et al. 1984), which are not dependent on prostaglandins (Laviolette and Belanger 1986).

Other therapeutic applications where selective activation of peripheral CB1Rs could prove useful include (1) decreasing intraocular pressure in glaucoma resistant to conventional therapies (Porcella et al. 2001); (2) antiemetic actions via CB1R activation in area postrema, which are located outside the blood–brain barrier (Machado Rocha et al. 2008); (3) antidiarrheal actions (Esfandyari et al. 2007); (4) antitumorigenic actions (Ligresti et al. 2006); and (5) treatment of bone diseases associated with accelerated osteoclastic bone resorption including osteoporosis, rheumatoid arthritis, and bone metastasis (Idris 2008).

Alternatively, there are several conditions where selective blockade of peripheral CB1Rs would be desirable to prevent the anxiety and depression symptoms associated with blockade of central CB1Rs (Christensen et al. 2007). These conditions, which have been the subject of recent reviews, include metabolic and vascular regulation in obesity, hepatic steatosis of various origins, and treatment of dyslipidimeas and insulin resistance (Kunos et al. 2009; Magen et al. 2008; Sarzani 2008). Also, development of peripherally acting selective blockers of GPR55 may prove useful for the treatment of chronic pain states given the recent demonstration of a non-hyperalgesic phenotype of GPR55-null mice after peripheral nerve injury or inflammation (Staton et al. 2008).

Another option is to develop peripherally acting selective inhibitors of endocannabinoid metabolism to elevate endocannabinoid levels, which would result in increased activation of both CB1Rs and CB2Rs. An important advantage of such metabolism inhibitors over CB1R or CB2R agonists is that increases in endocannabinoid levels would be achieved at the physiological sites of endocannabinoid production and release. Indeed, selective FAAH inhibitors have already been developed (Kathuria et al. 2003) and demonstrated to ameliorate neuropathic pain symptoms (Lichtman et al. 2004). Inhibitors of MAGL have also been developed (Ghafouri et al. 2004; Quistad et al. 2006). To our knowledge, these FAAH and MAGL inhibitors are all brain-permeable (Bisogno et al. 2006; Ghafouri et al. 2004; Karbarz et al. 2009; Kathuria et al. 2003; Lichtman et al. 2004; Makara et al. 2005; Minkkila et al. 2009; Myllymaki et al. 2007; Quistad et al. 2006). It is noteworthy that despite CNS permeability, these inhibitors do not appear to exhibit the adverse side effects of central CB1R activation (Comelli et al. 2007; Esfandyari et al. 2007; Kathuria et al. 2003). Also, new selective inhibitors of anandamide uptake have recently been synthesized (Ortar et al. 2008). Unlike their parent compound (Moore et al. 2005), these inhibitors have no appreciable effects on FAAH or MAGL activity (Ortar et al. 2008). The physiological consequences of selective AEA uptake inhibition have yet to be determined, but if it results in elevated AEA levels, it could prove to be an important addition to cannabinoid-based therapeutics.

5.12   FUTURE PROSPECTS

Selective activation of peripheral CB1Rs brings the promise of completely dissociating the antinociceptive and antihyperalgesic effects of cannabinoids from their centrally mediated psychotropic side effects. Similarly, restricting CB1R blockers to the periphery would dissociate anxiogenic and depressant side effects of such blockers from their therapeutic actions. Various academic institutions and the pharmaceutical industry are currently pursuing the development of orally bioavailable, brain-impermeable cannabinoid receptor ligands. Even if such centrally impermeant drugs do not become a mainstay of clinical pharmacology, they should be useful tools in unraveling the complexities of the endlessly fascinating endocannabinoid system.

ACKNOWLEDGMENT

Support contributed by NIH grants AA016100 and DA023163.

REFERENCES

Abrams, D. I., C. A. Jay, S. B. Shade et al. 2007. Cannabis in painful HIV-associated sensory neuropathy: a randomized placebo-controlled trial. *Neurology* 68:515–21.
Agarwal, N., P. Pacher, I. Tegeder et al. 2007. Cannabinoids mediate analgesia largely via peripheral type 1 cannabinoid receptors in nociceptors. *Nat. Neurosci.* 10:870–9.
Ahluwalia, J., L. Urban, S. Bevan, M. Capogna, and I. Nagy. 2002. Cannabinoid 1 receptors are expressed by nerve growth factor- and glial cell-derived neurotrophic factor-responsive primary sensory neurones. *Neuroscience* 110:747–53.
Ahluwalia, J., L. Urban, M. Capogna, S. Bevan, and I. Nagy. 2000. Cannabinoid 1 receptors are expressed in nociceptive primary sensory neurons. *Neuroscience* 100:685–8.
Akopian, A. N., N. B. Ruparel, A. Patwardhan, and K. M. Hargreaves. 2008. Cannabinoids desensitize capsaicin and mustard oil responses in sensory neurons via TRPA1 activation. *J. Neurosci.* 28:1064–75.

Amaya, F., G. Shimosato, Y. Kawasaki et al. 2006. Induction of CB1 cannabinoid receptor by inflammation in primary afferent neurons facilitates antihyperalgesic effect of peripheral CB1 agonist. *Pain* 124:175–83.

Ameri, A., A. Wilhelm, and T. Simmet. 1999. Effects of the endogeneous cannabinoid, anandamide, on neuronal activity in rat hippocampal slices. *Br. J. Pharmacol.* 126:1831–9.

Anand, U., W. R. Otto, D. Sanchez-Herrera et al. 2008. Cannabinoid receptor CB2 localisation and agonist-mediated inhibition of capsaicin responses in human sensory neurons. *Pain* 138:667–80.

Ashton, J. C., R. M. Rahman, S. M. Nair et al. 2007. Cerebral hypoxia-ischemia and middle cerebral artery occlusion induce expression of the cannabinoid CB2 receptor in the brain. *Neurosci. Lett.* 412:114–7.

Baker, D., G. Pryce, J. L. Croxford et al. 2000. Cannabinoids control spasticity and tremor in a multiple sclerosis model. *Nature* 404:84–7.

Baker, D., G. Pryce, J. L. Croxford et al. 2001. Endocannabinoids control spasticity in a multiple sclerosis model. *FASEB J.* 15:300–2.

Basavarajappa, B. S. and B. L. Hungund. 2005. Role of the endocannabinoid system in the development of tolerance to alcohol. *Alcohol.* 40:15–24.

Baskfield, C. Y., B. R. Martin, and J. L. Wiley. 2004. Differential effects of delta9-tetrahydrocannabinol and methanandamide in CB1 knockout and wild-type mice. *J. Pharmacol. Exp. Ther.* 309:86–91.

Batkai, S., Z. Jarai, J. A. Wagner et al. 2001. Endocannabinoids acting at vascular CB1 receptors mediate the vasodilated state in advanced liver cirrhosis. *Nat. Med.* 7:827–32.

Batkai, S., D. Osei-Hyiaman, H. Pan et al. 2007. Cannabinoid-2 receptor mediates protection against hepatic ischemia/reperfusion injury. *FASEB J.* 21:1788–800.

Batkai, S., P. Pacher, D. Osei-Hyiaman et al. 2004. Endocannabinoids acting at cannabinoid-1 receptors regulate cardiovascular function in hypertension. *Circulation* 110:1996–2002.

Beaulieu, P. 2006. Effects of nabilone, a synthetic cannabinoid, on postoperative pain. *Can. J. Anaesth.* 53:769–75.

Beaulieu, P., T. Bisogno, S. Punwar et al. 2000. Role of the endogenous cannabinoid system in the formalin test of persistent pain in the rat. *Eur. J. Pharmacol.* 396:85–92.

Begg, M., P. Pacher, S. Batkai et al. 2005. Evidence for novel cannabinoid receptors. *Pharmacol. Ther.* 106:133–45.

Beltramo, M., N. Bernardini, R. Bertorelli et al. 2006. CB2 receptor-mediated antihyperalgesia: possible direct involvement of neural mechanisms. *Eur. J. Neurosci.* 23:1530–8.

Ben-Shabat, S., L. O. Hanus, G. Katzavian, and R. Gallily. 2006. New cannabidiol derivatives: synthesis, binding to cannabinoid receptor, and evaluation of their antiinflammatory activity. *J. Med. Chem.* 49:1113–7.

Benito, C., W. K. Kim, I. Chavarria et al. 2005. A glial endogenous cannabinoid system is upregulated in the brains of macaques with simian immunodeficiency virus-induced encephalitis. *J. Neurosci.* 25:2530–6.

Benito, C., E. Nunez, R. M. Tolon et al. 2003. Cannabinoid CB2 receptors and fatty acid amide hydrolase are selectively overexpressed in neuritic plaque-associated glia in Alzheimer's disease brains. *J. Neurosci.* 23:11136–41.

Benito, C., R. M. Tolon, M. R. Pazos et al. 2008. Cannabinoid CB2 receptors in human brain inflammation. *Br. J. Pharmacol.* 153:277–85.

Bensaid, M., M. Gary-Bobo, A. Esclangon et al. 2003. The cannabinoid CB1 receptor antagonist SR141716 increases Acrp30 mRNA expression in adipose tissue of obese fa/fa rats and in cultured adipocyte cells. *Mol. Pharmacol.* 63:908–14.

Berman, J. S., C. Symonds, and R. Birch. 2004. Efficacy of two cannabis based medicinal extracts for relief of central neuropathic pain from brachial plexus avulsion: results of a randomised controlled trial. *Pain* 112:299–306.

Binzen, U., W. Greffrath, S. Hennessy et al. 2006. Co-expression of the voltage-gated potassium channel Kv1.4 with transient receptor potential channels (TRPV1 and TRPV2) and the cannabinoid receptor CB1 in rat dorsal root ganglion neurons. *Neuroscience* 142:527–39.

Bisogno, T., M. G. Cascio, B. Saha et al. 2006. Development of the first potent and specific inhibitors of endocannabinoid biosynthesis. *Biochim. Biophys. Acta* 1761:205–12.

Bisogno, T., A. Ligresti, and Marzo Di, V. 2005. The endocannabinoid signalling system: biochemical aspects. *Pharmacol. Biochem. Behav.* 81:224–38.

Blankman, J. L., G. M. Simon, and B. F. Cravatt. 2007. A comprehensive profile of brain enzymes that hydrolyze the endocannabinoid 2-arachidonoylglycerol. *Chem. Biol.* 14:1347–56.

Bouaboula, M., M. Rinaldi, P. Carayon et al. 1993. Cannabinoid-receptor expression in human leukocytes. *Eur. J. Biochem.* 214:173–80.

Bridges, D., K. Ahmad, and A. S. Rice. 2001. The synthetic cannabinoid WIN55,212-2 attenuates hyperalgesia and allodynia in a rat model of neuropathic pain. *Br. J. Pharmacol.* 133:586–94.

Bridges, D., A. S. Rice, M. Egertova et al. 2003. Localisation of cannabinoid receptor 1 in rat dorsal root ganglion using in situ hybridisation and immunohistochemistry. *Neuroscience* 119:803–12.

Brown, S. P., P. K. Safo, and W. G. Regehr. 2004. Endocannabinoids inhibit transmission at granule cell to Purkinje cell synapses by modulating three types of presynaptic calcium channels. *J. Neurosci.* 24:5623–31.

Brusberg, M., S. Arvidsson, D. Kang et al. 2009. CB_1 receptors mediate the analgesic effects of cannabinoids on colorectal distension-induced visceral pain in rodents. *J. Neurosci.* 29:1554–64.

Buggy, D. J., L. Toogood, S. Maric et al. 2003. Lack of analgesic efficacy of oral delta-9-tetrahydrocannabinol in postoperative pain. *Pain* 106:169–72.

Cabranes, A., G. Pryce, D. Baker, and J. Fernandez-Ruiz. 2006. Changes in CB1 receptors in motor-related brain structures of chronic relapsing experimental allergic encephalomyelitis mice. *Brain Res.* 1107:199–205.

Cadas, H., E. di Tomaso, and D. Piomelli. 1997. Occurrence and biosynthesis of endogenous cannabinoid precursor, N-arachidonoyl phosphatidylethanolamine, in rat brain. *J. Neurosci.* 17:1226–42.

Calignano, A., G. La Rana, A. Giuffrida, and D. Piomelli. 1998. Control of pain initiation by endogenous cannabinoids. *Nature* 394:277–81.

Chamley, L. W., A. Bhalla, P. R. Stone et al. 2008. Nuclear localisation of the endocannabinoid metabolizing enzyme fatty acid amide hydrolase (FAAH) in invasive trophoblasts and an association with recurrent miscarriage. *Placenta* 29:970–5.

Chen, K., A. Ratzliff, L. Hilgenberg et al. 2003. Long-term plasticity of endocannabinoid signaling induced by developmental febrile seizures. *Neuron* 39:599–611.

Chevaleyre, V., K. A. Takahashi, and P. E. Castillo. 2006. Endocannabinoid-mediated synaptic plasticity in the CNS. *Annu. Rev. Neurosci.* 29:37–76.

Christensen, R., P. K. Kristensen, E. M. Bartels, H. Bliddal, and A. Astrup. 2007. Efficacy and safety of the weight-loss drug rimonabant: a meta-analysis of randomised trials. *Lancet* 370:1706–13.

Cianchi, F., L. Papucci, N. Schiavone et al. 2008. Cannabinoid receptor activation induces apoptosis through tumor necrosis factor alpha-mediated ceramide de novo synthesis in colon cancer cells. *Clin. Cancer Res.* 14:7691–700.

Comelli, F., G. Giagnoni, I. Bettoni, M. Colleoni, and B. Costa. 2007. The inhibition of monoacylglycerol lipase by URB602 showed an anti-inflammatory and anti-nociceptive effect in a murine model of acute inflammation. *Br. J. Pharmacol.* 152:787–94.

Costa, B., M. Colleoni, S. Conti et al. 2004. Repeated treatment with the synthetic cannabinoid WIN 55,212-2 reduces both hyperalgesia and production of pronociceptive mediators in a rat model of neuropathic pain. *Br. J. Pharmacol.* 141:4–8.

Coutts, A. A., A. J. Irving, K. Mackie, R. G. Pertwee, and S. Anavi-Goffer. 2002. Localisation of cannabinoid CB_1 receptor immunoreactivity in the guinea pig and rat myenteric plexus. *J. Comp. Neurol.* 448:410–22.

Cravatt, B. F. and A. H. Lichtman. 2003. Fatty acid amide hydrolase: an emerging therapeutic target in the endocannabinoid system. *Curr. Opin. Chem. Biol.* 7:469–75.

Czifra, G., A. Varga, K. Nyeste et al. 2009. Increased expressions of cannabinoid receptor-1 and transient receptor potential vanilloid-1 in human prostate carcinoma. *J. Cancer Res. Clin. Oncol.* 135:507–14.

D'argenio, G., M. Valenti, G. Scaglione et al. 2006. Up-regulation of anandamide levels as an endogenous mechanism and a pharmacological strategy to limit colon inflammation. *FASEB J.* 20:568–70.

De Filippis, D., T. Iuvone, A. D'Amico et al. 2008. Effect of cannabidiol on sepsis-induced motility disturbances in mice: involvement of CB receptors and fatty acid amide hydrolase. *Neurogastroenterol. Motil.* 20:919–27.

De Marchi, N., L. De Petrocellis, P. Orlando et al. 2003. Endocannabinoid signalling in the blood of patients with schizophrenia. *Lipids Health Dis.* 2:5.

Deadwyler, S. A., R. E. Hampson, J. Mu, A. Whyte, and S. Childers. 1995. Cannabinoids modulate voltage sensitive potassium A-current in hippocampal neurons via a cAMP-dependent process. *J. Pharmacol. Exp. Ther.* 273:734–43.

Dean, B., S. Sundram, R. Bradbury, E. Scarr, and D. Copolov. 2001. Studies on [3H]CP-55940 binding in the human central nervous system: regional specific changes in density of cannabinoid-1 receptors associated with schizophrenia and cannabis use. *Neuroscience* 103:9–15.

Dewey, W. L. 1986. Cannabinoid pharmacology. *Pharmacol. Rev.* 38:151–78.

Di Marzo, V., T. Bisogno, and L. De Petrocellis. 2001a. Anandamide: some like it hot. *Trends Pharmacol. Sci.* 22:346–9.

Di Marzo, V., T. Bisogno, L. De Petrocellis, D. Melck, and B. R. Martin. 1999. Cannabimimetic fatty acid derivatives: the anandamide family and other endocannabinoids. *Curr. Med. Chem.* 6:721–44.

Di Marzo, V., C. S. Breivogel, Q. Tao et al. 2000a. Levels, metabolism, and pharmacological activity of anandamide in CB_1 cannabinoid receptor knockout mice: evidence for non-CB_1, non-CB_2 receptor-mediated actions of anandamide in mouse brain. *J. Neurochem.* 75:2434–44.

Di Marzo, V., A. Fontana, H. Cadas et al. 1994. Formation and inactivation of endogenous cannabinoid anandamide in central neurons. *Nature* 372:686–91.

Di Marzo, V., S. K. Goparaju, L. Wang et al. 2001b. Leptin-regulated endocannabinoids are involved in maintaining food intake. *Nature* 410:822–5.

Di Marzo, V., M. P. Hill, T. Bisogno, A. R. Crossman, and J. M. Brotchie. 2000b. Enhanced levels of endogenous cannabinoids in the globus pallidus are associated with a reduction in movement in an animal model of Parkinson's disease. *FASEB J.* 14:1432–8.

Di Marzo, V. and L. D. Petrocellis. 2006. Plant, synthetic, and endogenous cannabinoids in medicine. *Annu. Rev. Med.* 57:553–74.

Dinh, T. P., D. Carpenter, F. M. Leslie et al. 2002. Brain monoglyceride lipase participating in endocannabinoid inactivation. *Proc. Natl. Acad. Sci. U.S.A.* 99:10819–24.

Dinis, P., A. Charrua, A. Avelino et al. 2004. Anandamide-evoked activation of vanilloid receptor 1 contributes to the development of bladder hyperreflexia and nociceptive transmission to spinal dorsal horn neurons in cystitis. *J. Neurosci.* 24:11253–63.

Domenicali, M., J. Ros, G. Fernandez-Varo et al. 2005. Increased anandamide induced relaxation in mesenteric arteries of cirrhotic rats: role of cannabinoid and vanilloid receptors. *Gut* 54:522–7.

Duncan, M., A. Mouihate, K. Mackie et al. 2008. Cannabinoid CB2 receptors in the enteric nervous system modulate gastrointestinal contractility in lipopolysaccharide-treated rats. *Am. J. Physiol Gastrointest. Liver Physiol.* 295:G78–G87.

Dziadulewicz, E. K., S. J. Bevan, C. T. Brain et al. 2007. Naphthalen-1-yl-(4-pentyloxynaphthalen-1-yl)methanone: a potent, orally bioavailable human CB1/CB2 dual agonist with antihyperalgesic properties and restricted central nervous system penetration. *J. Med. Chem.* 50:3851–6.

Egertová, M. and M. R. Elphick. 2000. Localisation of cannabinoid receptors in the rat brain using antibodies to the intracellular C-terminal tail of CB. *J. Comp. Neurol.* 422:159–71.

Ellert-Miklaszewska, A., W. Grajkowska, K. Gabrusiewicz, B. Kaminska, and L. Konarska. 2007. Distinctive pattern of cannabinoid receptor type II (CB2) expression in adult and pediatric brain tumors. *Brain Res.* 1137:161–9.

Ellis, R. J., W. Toperoff, F. Vaida et al. 2009. Smoked medicinal cannabis for neuropathic pain in HIV: a randomized, crossover clinical trial. *Neuropsychopharmacology* 34:672–80.

Esfandyari, T., M. Camilleri, I. Busciglio et al. 2007. Effects of a cannabinoid receptor agonist on colonic motor and sensory functions in humans: a randomized, placebo-controlled study. *Am. J. Physiol. Gastrointest. Liver Physiol.* 293:G137–G145.

Farquhar-Smith, W. P., M. Egertova, E. J. Bradbury et al. 2000. Cannabinoid CB$_1$ receptor expression in rat spinal cord. *Mol. Cell. Neurosci.* 15:510–21.

Felder, C. C., K. E. Joyce, E. M. Briley et al. 1995. Comparison of the pharmacology and signal transduction of the human cannabinoid CB$_1$ and CB$_2$ receptors. *Mol. Pharmacol.* 48:443–50.

Fernandez-Espejo, E., I. Caraballo, Fonseca F. Rodriguez De et al. 2004. Experimental parkinsonism alters anandamide precursor synthesis, and functional deficits are improved by AM404: a modulator of endocannabinoid function. *Neuropsychopharmacology* 29:1134–42.

Fernandez-Rodriguez, C. M., J. Romero, T. J. Petros et al. 2004. Circulating endogenous cannabinoid anandamide and portal, systemic and renal hemodynamics in cirrhosis. *Liver Int.* 24:477–83.

Ferrer, B., N. Asbrock, S. Kathuria, D. Piomelli, and A. Giuffrida. 2003. Effects of levodopa on endocannabinoid levels in rat basal ganglia: implications for the treatment of levodopa-induced dyskinesias. *Eur. J. Neurosci.* 18:1607–14.

Fox, A., A. Kesingland, C. Gentry et al. 2001. The role of central and peripheral Cannabinoid$_1$ receptors in the antihyperalgesic activity of cannabinoids in a model of neuropathic pain. *Pain* 92:91–100.

Fride, E., C. Feigin, D. E. Ponde et al. 2004. (+)-Cannabidiol analogues which bind cannabinoid receptors but exert peripheral activity only. *Eur. J. Pharmacol.* 506:179–88.

Fride, E., and R. Mechoulam. 1993. Pharmacological activity of the cannabinoid receptor agonist, anandamide, a brain constituent. *Eur. J. Pharmacol.* 231:313–4.

Galiegue, S., S. Mary, J. Marchand et al. 1995. Expression of central and peripheral cannabinoid receptors in human immune tissues and leukocyte subpopulations. *Eur. J. Biochem.* 232:54–61.

Gerdeman, G., and D. M. Lovinger. 2001. CB1 cannabinoid receptor inhibits synaptic release of glutamate in rat dorsolateral striatum. *J. Neurophysiol.* 85:468–71.

Ghafouri, N., G. Tiger, R. K. Razdan et al. 2004. Inhibition of monoacylglycerol lipase and fatty acid amide hydrolase by analogues of 2-arachidonoylglycerol. *Br. J. Pharmacol.* 143:774–84.

Giuffrida, A., F. M. Leweke, C. W. Gerth et al. 2004. Cerebrospinal anandamide levels are elevated in acute schizophrenia and are inversely correlated with psychotic symptoms. *Neuropsychopharmacology* 29:2108–14.

Glaser, S. T., N. A. Abumrad, F. Fatade et al. 2003. Evidence against the presence of an anandamide transporter. *Proc. Natl. Acad. Sci. U.S.A.* 100:4269–74.

Glass, M., M. Dragunow, and R. L. Faull. 1997. Cannabinoid receptors in the human brain: a detailed anatomical and quantitative autoradiographic study in the fetal, neonatal and adult human brain. *Neuroscience* 77:299–318.

Glass, M., R. L. Faull, and M. Dragunow. 1993. Loss of cannabinoid receptors in the substantia nigra in Huntington's disease. *Neuroscience* 56:523–7.

Glass, M., Dellen A. van, C. Blakemore, A. J. Hannan, and R. L. Faull. 2004. Delayed onset of Huntington's disease in mice in an enriched environment correlates with delayed loss of cannabinoid CB1 receptors. *Neuroscience* 123:207–12.

Golech, S. A., R. M. McCarron, Y. Chen et al. 2004a. Human brain endothelium: coexpression and function of vanilloid and endocannabinoid receptors. *Brain Res. Mol. Brain Res.* 132:87–92.

Gomez, R., M. Navarro, B. Ferrer et al. 2002. A peripheral mechanism for CB1 cannabinoid receptor-dependent modulation of feeding. *J. Neurosci.* 22:9612–7.

Gong, H., Jr., D. P. Tashkin, M. S. Simmons, B. Calvarese, and B. J. Shapiro. 1984. Acute and subacute bronchial effects of oral cannabinoids. *Clin. Pharmacol. Ther.* 35:26–32.

Gong, J. P., E. S. Onaivi, H. Ishiguro et al. 2006. Cannabinoid CB2 receptors: immunohistochemical localization in rat brain. *Brain Res.* 1071:10–23.

Gonzalez, S., G. Mauriello-Romanazzi, F. Berrendero et al. 2000. Decreased cannabinoid CB1 receptor mRNA levels and immunoreactivity in pituitary hyperplasia induced by prolonged exposure to estrogens. *Pituitary* 3:221–6.

Gonzalez, S., M. Valenti, R. de Miguel et al. 2004. Changes in endocannabinoid contents in reward-related brain regions of alcohol-exposed rats, and their possible relevance to alcohol relapse. *Br. J. Pharmacol.* 143:455–64.

Griffin, G., S. R. Fernando, R. A. Ross et al. 1997. Evidence for the presence of CB2-like cannabinoid receptors on peripheral nerve terminals. *Eur. J. Pharmacol.* 339:53–61.

Gubellini, P., B. Picconi, M. Bari et al. 2002. Experimental parkinsonism alters endocannabinoid degradation: implications for striatal glutamatergic transmission. *J. Neurosci.* 22:6900–7.

Guerrero, A. V., P. Quang, N. Dekker, R. C. Jordan, and B. L. Schmidt. 2008. Peripheral cannabinoids attenuate carcinoma-induced nociception in mice. *Neurosci. Lett.* 433:77–81.

Guindon, J. and A. G. Hohmann. 2008. Cannabinoid CB2 receptors: a therapeutic target for the treatment of inflammatory and neuropathic pain. *Br. J. Pharmacol.* 153:319–34.

Gulyas, A. I., B. F. Cravatt, M. H. Bracey et al. 2004. Segregation of two endocannabinoid-hydrolyzing enzymes into pre- and postsynaptic compartments in the rat hippocampus, cerebellum and amygdala. *Eur. J. Neurosci.* 20:441–58.

Guo, J. and S. R. Ikeda. 2004. Endocannabinoids modulate N-type calcium channels and G-protein-coupled inwardly rectifying potassium channels via CB1 cannabinoid receptors heterologously expressed in mammalian neurons. *Mol. Pharmacol.* 65:665–74.

Gustafsson, K., X. Wang, D. Severa et al. 2008. Expression of cannabinoid receptors type 1 and type 2 in non-Hodgkin lymphoma: growth inhibition by receptor activation. *Int. J. Cancer.* 123:1025–33.

Gutierrez, T., J. N. Farthing, A. M. Zvonok, A. Makriyannis, and A. G. Hohmann. 2007. Activation of peripheral cannabinoid CB1 and CB2 receptors suppresses the maintenance of inflammatory nociception: a comparative analysis. *Br. J. Pharmacol.* 150:153–63.

Hampson, R. E. and S. A. Deadwyler. 1999. Cannabinoids, hippocampal function and memory. *Life Sci.* 65:715–23.

Hansen, H. H., P. C. Schmid, P. Bittigau et al. 2001. Anandamide, but not 2-arachidonoylglycerol, accumulates during in vivo neurodegeneration. *J. Neurochem.* 78:1415–27.

Hanus, L., A. Breuer, S. Tchilibon et al. 1999. HU-308: a specific agonist for CB(2), a peripheral cannabinoid receptor. *Proc. Natl. Acad. Sci. U.S.A.* 96:14228–33.

Held-Feindt, J., L. Dorner, G. Sahan, H. M. Mehdorn, and R. Mentlein. 2006. Cannabinoid receptors in human astroglial tumors. *J. Neurochem.* 98:886–93.

Herkenham, M., A. B. Lynn, B. R. de Costa, and E. K. Richfield. 1991. Neuronal localization of cannabinoid receptors in the basal ganglia of the rat. *Brain Res.* 547:267–74.

Herkenham, M., A. B. Lynn, M. D. Little et al. 1990. Cannabinoid receptor localization in brain. *Proc. Natl. Acad. Sci. U.S.A.* 87:1932–6.

Herzberg, U., E. Eliav, G. J. Bennett, and I. J. Kopin. 1997. The analgesic effects of R(+)-WIN 55,212-2 mesylate, a high affinity cannabinoid agonist, in a rat model of neuropathic pain. *Neurosci. Lett.* 221:157–60.

Hill, M. N., S. Patel, E. J. Carrier et al. 2005. Downregulation of endocannabinoid signaling in the hippocampus following chronic unpredictable stress. *Neuropsychopharmacology* 30:508–15.

Hillard, C. J. 2000. Biochemistry and pharmacology of the endocannabinoids arachidonyle-thanolamide and 2-arachidonylglycerol. *Prostaglandins Other Lipid Mediat.* 61:3–18.

Hillard, C. J., L. Shi, V. R. Tuniki, J. R. Falck, and W. B. Campbell. 2007. Studies of anand-amide accumulation inhibitors in cerebellar granule neurons: comparison to inhibition of fatty acid amide hydrolase. *J. Mol. Neurosci.* 33:18–24.

Hoffman, A. F., A. C. Riegel, and C. R. Lupica. 2003. Functional localization of cannabinoid receptors and endogenous cannabinoid production in distinct neuron populations of the hippocampus. *Eur. J. Neurosci.* 18:524–34.

Hohmann, A. G. and M. Herkenham. 1999a. Cannabinoid receptors undergo axonal flow in sensory nerves. *Neuroscience* 92:1171–5.

Hohmann, A. G. and M. Herkenham. 1999b. Localization of central cannabinoid CB_1 receptor messenger RNA in neuronal subpopulations of rat dorsal root ganglia: a double-label in situ hybridization study. *Neuroscience* 90:923–31.

Horne, A. W., J. A. Phillips, III, N. Kane et al. 2008. CB1 expression is attenuated in Fallopian tube and decidua of women with ectopic pregnancy. *PLoS.ONE.* 3:e3969.

Howlett, A. C. 1984. Inhibition of neuroblastoma adenylate cyclase by cannabinoid and nant-radol compounds. *Life Sci.* 35:1803–10.

Howlett, A. C., F. Barth, T. I. Bonner et al. 2002. International Union of Pharmacology. XXVII. Classification of cannabinoid receptors. *Pharmacol. Rev.* 54:161–202.

Huang, C. C., S. W. Lo, and K. S. Hsu. 2001. Presynaptic mechanisms underlying cannabi-noid inhibition of excitatory synaptic transmission in rat striatal neurons. *J. Physiol.* 532:731–48.

Huang, S. M., T. Bisogno, M. Trevisani et al. 2002. An endogenous capsaicin-like substance with high potency at recombinant and native vanilloid VR1 receptors. *Proc. Natl. Acad. Sci. U.S.A.* 99:8400–5.

Huffman, J. W. 2000. The search for selective ligands for the CB_2 receptor. *Curr. Pharm. Des* 6:1323–37.

Huffman, J. W., S. M. Bushell, J. R. Miller, J. L. Wiley, and B. R. Martin. 2002. 1-Methoxy-, 1-deoxy-11-hydroxy- and 11-hydroxy-1-methoxy-Delta(8)-tetrahydrocannabinols: new selective ligands for the CB_2 receptor. *Bioorg. Med. Chem.* 10:4119–29.

Hungund, B. L., K. Y. Vinod, S. A. Kassir et al. 2004. Upregulation of CB1 receptors and agonist-stimulated [^{35}S]GTPgammaS binding in the prefrontal cortex of depressed sui-cide victims. *Mol. Psychiatry.* 9:184–90.

Ibrahim, M. M., H. Deng, A. Zvonok et al. 2003. Activation of CB_2 cannabinoid receptors by AM1241 inhibits experimental neuropathic pain: pain inhibition by receptors not pres-ent in the CNS. *Proc. Natl. Acad. Sci. U.S.A.* 100:10529–33.

Idris, A. I. 2008. Role of cannabinoid receptors in bone disorders: alternatives for treatment. *Drug News Perspect.* 21:533–40.

Ishac, E. J., L. Jiang, K. D. Lake et al. 1996. Inhibition of exocytotic noradrenaline release by presynaptic cannabinoid CB1 receptors on peripheral sympathetic nerves. *Br. J. Pharmacol.* 118:2023–8.

Islam, T. C., A. C. Asplund, J. M. Lindvall et al. 2003. High level of cannabinoid receptor 1, absence of regulator of G protein signalling 13 and differential expression of Cyclin D1 in mantle cell lymphoma. *Leukemia.* 17:1880–90.

Iuvone, T., Filippis D. De, Sardo A. Di Spiezio et al. 2008. Selective CB2 up-regulation in women affected by endometrial inflammation. *J. Cell Mol. Med.* 12:661–70.

Izzo, A. A., F. Capasso, A. Costagliola et al. 2003. An endogenous cannabinoid tone attenuates cholera toxin-induced fluid accumulation in mice. *Gastroenterology* 125:765–74.

Izzo, A. A. and A. A. Coutts. 2005. Cannabinoids and the digestive tract. *Handb. Exp. Pharmacol.* :573–98.

Izzo, A. A., F. Fezza, R. Capasso et al. 2001a. Cannabinoid CB1-receptor mediated regulation of gastrointestinal motility in mice in a model of intestinal inflammation. *Br. J. Pharmacol.* 134:563–70.

Izzo, A. A., N. Mascolo, and F. Capasso. 2001b. The gastrointestinal pharmacology of cannabinoids. *Curr. Opin. Pharmacol.* 1:597–603.

Jeske, N. A., A. M. Patwardhan, N. Gamper et al. 2006. Cannabinoid WIN 55,212-2 regulates TRPV1 phosphorylation in sensory neurons. *J. Biol. Chem.* 281:32879–90.

Jhaveri, M. D., D. Richardson, I. Robinson et al. 2008. Inhibition of fatty acid amide hydrolase and cyclooxygenase-2 increases levels of endocannabinoid related molecules and produces analgesia via peroxisome proliferator-activated receptor-alpha in a model of inflammatory pain. *Neuropharmacology* 55:85–93.

Jin, K. L., X. O. Mao, P. C. Goldsmith, and D. A. Greenberg. 2000. CB1 cannabinoid receptor induction in experimental stroke. *Ann. Neurol.* 48:257–61.

Johanek, L. M. and D. A. Simone. 2004. Activation of peripheral cannabinoid receptors attenuates cutaneous hyperalgesia produced by a heat injury. *Pain* 109:432–42.

Julian, M. D., A. B. Martin, B. Cuellar et al. 2003. Neuroanatomical relationship between type 1 cannabinoid receptors and dopaminergic systems in the rat basal ganglia. *Neuroscience* 119:309–18.

Julien, B., P. Grenard, F. Teixeira-Clerc et al. 2005. Antifibrogenic role of the cannabinoid receptor CB2 in the liver. *Gastroenterology* 128:742–55.

Karbarz, M. J., L. Luo, L. Chang et al. 2009. Biochemical and biological properties of 4-(3-phenyl-[1,2,4] thiadiazol-5-yl)-piperazine-1-carboxylic acid phenylamide, a mechanism-based inhibitor of fatty acid amide hydrolase. *Anesth. Analg.* 108:316–29.

Karsak, M., E. Gaffal, R. Date et al. 2007. Attenuation of allergic contact dermatitis through the endocannabinoid system. *Science.* 316:1494–7.

Karst, M., K. Salim, S. Burstein et al. 2003. Analgesic effect of the synthetic cannabinoid CT-3 on chronic neuropathic pain: a randomized controlled trial. *JAMA* 290:1757–62.

Kathuria, S., S. Gaetani, D. Fegley et al. 2003. Modulation of anxiety through blockade of anandamide hydrolysis. *Nat. Med.* 9:76–81.

Katona, I., B. Sperlagh, A. Sik et al. 1999. Presynaptically located CB_1 cannabinoid receptors regulate GABA release from axon terminals of specific hippocampal interneurons. *J. Neurosci.* 19:4544–58.

Kawamura, Y., M. Fukaya, T. Maejima et al. 2006. The CB1 cannabinoid receptor is the major cannabinoid receptor at excitatory presynaptic sites in the hippocampus and cerebellum. *J. Neurosci.* 26:2991–3001.

Khasabova, I. A., S. G. Khasabov, C. Harding-Rose et al. 2008. A decrease in anandamide signaling contributes to the maintenance of cutaneous mechanical hyperalgesia in a model of bone cancer pain. *J. Neurosci.* 28:11141–52.

Kimball, E. S., C. R. Schneider, N. H. Wallace, and P. J. Hornby. 2006. Agonists of cannabi-noid receptor 1 and 2 inhibit experimental colitis induced by oil of mustard and by dex-tran sulfate sodium. *Am. J. Physiol Gastrointest. Liver Physiol.* 291:G364–G371.

Kirkham, T. C., C. M. Williams, F. Fezza, and Marzo Di, V. 2002. Endocannabinoid levels in rat limbic forebrain and hypothalamus in relation to fasting, feeding and satiation: stimulation of eating by 2-arachidonoyl glycerol. *Br. J. Pharmacol.* 136:550–7.

Koethe, D., I. C. Llenos, J. R. Dulay et al. 2007. Expression of CB1 cannabinoid receptor in the anterior cingulate cortex in schizophrenia, bipolar disorder, and major depression. *J. Neural Transm.* 114:1055–63.

Kreitzer, A. C. and W. G. Regehr. 2001. Retrograde inhibition of presynaptic calcium influx by endogenous cannabinoids at excitatory synapses onto Purkinje cells. *Neuron* 29:717–27.

Kunos, G., D. Osei-Hyiaman, S. Batkai, K. A. Sharkey, and A. Makriyannis. 2009. Should peripheral CB_1 cannabinoid receptors be selectively targeted for therapeutic gain? *Trends Pharmacol. Sci.* 30:1–7.

Kurabayashi, M., I. Takeyoshi, D. Yoshinari et al. 2005. 2-Arachidonoylglycerol increases in ischemia-reperfusion injury of the rat liver. *J. Invest Surg.* 18:25–31.

Kurz, C. M., C. Gottschalk, E. Schlicker, and M. Kathmann. 2008. Identification of a presyn-aptic cannabinoid CB1 receptor in the guinea-pig atrium and sequencing of the guinea-pig CB1 receptor. *J. Physiol. Pharmacol.* 59:3–15.

Laine, K., K. Jarvinen, and T. Jarvinen. 2003. Topically administered CB(2)-receptor agonist, JWH-133, does not decrease intraocular pressure (IOP) in normotensive rabbits. *Life Sci.* 72:837–42.

Lastres-Becker, I., F. Berrendero, J. J. Lucas et al. 2002a. Loss of mRNA levels, binding and activation of GTP-binding proteins for cannabinoid CB1 receptors in the basal ganglia of a transgenic model of Huntington's disease. *Brain Res.* 929:236–42.

Lastres-Becker, I., M. Cebeira, M. L. de Ceballos et al. 2001a. Increased cannabinoid CB1 receptor binding and activation of GTP-binding proteins in the basal ganglia of patients with Parkinson's syndrome and of MPTP-treated marmosets. *Eur. J. Neurosci.* 14:1827–32.

Lastres-Becker, I., F. Fezza, M. Cebeira et al. 2001b. Changes in endocannabinoid transmission in the basal ganglia in a rat model of Huntington's disease. *Neuroreport* 12:2125–9.

Lastres-Becker, I., M. Gomez, R. de Miguel, J. A. Ramos, and J. Fernandez-Ruiz. 2002b. Loss of cannabinoid CB(1) receptors in the basal ganglia in the late akinetic phase of rats with experimental Huntington's disease. *Neurotox. Res.* 4:601–8.

Lauckner, J. E., J. B. Jensen, H. Y. Chen et al. 2008. GPR55 is a cannabinoid receptor that increases intracellular calcium and inhibits M current. *Proc. Natl. Acad. Sci. U.S.A.* 105:2699–704.

Laviolette, M. and J. Belanger. 1986. Role of prostaglandins in marihuana-induced broncho-dilation. *Respiration.* 49:10–5.

Ledent, C., O. Valverde, G. Cossu et al. 1999. Unresponsiveness to cannabinoids and reduced addictive effects of opiates in CB1 receptor knockout mice. *Science.* 283:401–4.

Lever, I. J., M. Robinson, M. Cibelli et al. 2009. Localization of the endocannabinoid-degrad-ing enzyme fatty acid amide hydrolase in rat dorsal root ganglion cells and its regulation after peripheral nerve injury. *J. Neurosci.* 29:3766–80.

Leweke, F. M., A. Giuffrida, U. Wurster, H. M. Emrich, and D. Piomelli. 1999. Elevated endogenous cannabinoids in schizophrenia. *Neuroreport.* 10:1665–9.

Liang, Y. C., C. C. Huang, and K. S. Hsu. 2007. The synthetic cannabinoids attenuate allo-dynia and hyperalgesia in a rat model of trigeminal neuropathic pain. *Neuropharmacol.* 53:169–77.

Lichtman, A. H., S. A. Cook, and B. R. Martin. 1996. Investigation of brain sites mediating cannabinoid-induced antinociception in rats: evidence supporting periaqueductal gray involvement. *J. Pharmacol. Exp. Ther.* 276:585–93.

Lichtman, A. H., D. Leung, C. C. Shelton et al. 2004. Reversible inhibitors of fatty acid amide hydrolase that promote analgesia: evidence for an unprecedented combination of potency and selectivity. *J. Pharmacol. Exp. Ther.* 311:441–8.

Lichtman, A. H. and B. R. Martin. 1991. Spinal and supraspinal components of cannabinoid-induced antinociception. *J. Pharmacol. Exp. Ther.* 258:517–23.

Lichtman, A. H. and B. R. Martin. 1997. The selective cannabinoid antagonist SR 141716A blocks cannabinoid-induced antinociception in rats. *Pharmacol Biochem. Behav.* 57:7–12.

Ligresti, A., T. Bisogno, I. Matias et al. 2003. Possible endocannabinoid control of colorectal cancer growth. *Gastroenterology.* 125:677–87.

Ligresti, A., A. S. Moriello, K. Starowicz et al. 2006. Antitumor activity of plant cannabinoids with emphasis on the effect of cannabidiol on human breast carcinoma. *J. Pharmacol. Exp. Ther.* 318:1375–87.

Lim, G., B. Sung, R. R. Ji, and J. Mao. 2003. Upregulation of spinal cannabinoid-1-receptors following nerve injury enhances the effects of Win 55,212-2 on neuropathic pain behaviors in rats. *Pain* 105:275–83.

Liu, J., S. Batkai, P. Pacher et al. 2003. Lipopolysaccharide induces anandamide synthesis in macrophages via CD14/MAPK/phosphoinositide 3-kinase/NF-kappaB independently of platelet-activating factor. *J. Biol. Chem.* 278:45034–9.

Lunn, C. A., J. Fine, A. Rojas-Triana et al. 2007. Cannabinoid CB_2-selective inverse agonist protects against antigen-induced bone loss. *Immunopharmacol.Immunotoxicol.* 29:387–401.

Lynn, A. B. and M. Herkenham. 1994. Localization of cannabinoid receptors and nonsaturable high-density cannabinoid binding sites in peripheral tissues of the rat: implications for receptor-mediated immune modulation by cannabinoids. *J. Pharmacol. Exp.Ther.* 268:1612–23.

Maccarrone, M., M. Attina, A. Cartoni, M. Bari, and A. Finazzi-Agro. 2001a. Gas chromatography-mass spectrometry analysis of endogenous cannabinoids in healthy and tumoral human brain and human cells in culture. *J. Neurochem.* 76:594–601.

Maccarrone, M., T. Bisogno, H. Valensise et al. 2002. Low fatty acid amide hydrolase and high anandamide levels are associated with failure to achieve an ongoing pregnancy after IVF and embryo transfer. *Mol. Hum. Reprod.* 8:188–95.

Maccarrone, M., L. De Petrocellis, M. Bari et al. 2001b. Lipopolysaccharide downregulates fatty acid amide hydrolase expression and increases anandamide levels in human peripheral lymphocytes. *Arch. Biochem. Biophys.* 393:321–8.

Maccarrone, M., E. Fride, T. Bisogno et al. 2005. Up-regulation of the endocannabinoid system in the uterus of leptin knockout (ob/ob) mice and implications for fertility. *Mol. Hum. Reprod.* 11:21–8.

Maccarrone, M., P. Gubellini, M. Bari et al. 2003. Levodopa treatment reverses endocannabinoid system abnormalities in experimental parkinsonism. *J. Neurochem.* 85:1018–25.

Maccarrone, M., S. Piccirilli, N. Battista et al. 2004. Enhanced anandamide degradation is associated with neuronal apoptosis induced by the HIV-1 coat glycoprotein gp120 in the rat neocortex. *J. Neurochem.* 89:1293–300.

Maccarrone, M., H. Valensise, M. Bari et al. 2000. Relation between decreased anandamide hydrolase concentrations in human lymphocytes and miscarriage. *Lancet* 355:1326–9.

Maccarrone, M., H. Valensise, M. Bari et al. 2001c. Progesterone up-regulates anandamide hydrolase in human lymphocytes: role of cytokines and implications for fertility. *J. Immunol.* 166:7183–9.

Maccarrone, M. and T. Wenger. 2005. Effects of cannabinoids on hypothalamic and reproductive function. *Handb. Exp. Pharmacol.* 168:555–71.

Machado Rocha, F. C., S. C. Stefano, Haiek R. De Cassia, L. M. Rosa Oliveira, and D. X. Da Silveira. 2008. Therapeutic use of Cannabis sativa on chemotherapy-induced nausea and vomiting among cancer patients: systematic review and meta-analysis. *Eur. J. Cancer Care (Engl.)* 17:431–43.

Mackie, K. 2005. Distribution of cannabinoid receptors in the central and peripheral nervous system. *Handb. Exp. Pharmacol.* 168:299–325.

Maejima, T., K. Hashimoto, T. Yoshida, A. Aiba, and M. Kano. 2001. Presynaptic inhibition caused by retrograde signal from metabotropic glutamate to cannabinoid receptors. *Neuron* 31:463–75.

Magen, I., Y. Avraham, E. Berry, and R. Mechoulam. 2008. Endocannabinoids in liver disease and hepatic encephalopathy. *Curr. Pharm. Des.* 14:2362–9.

Mailleux, P., and J. J. Vanderhaeghen. 1993. Dopaminergic regulation of cannabinoid receptor mRNA levels in the rat caudate-putamen: an in situ hybridization study. *J. Neurochem.* 61:1705–12.

Makara, J. K., M. Mor, D. Fegley et al. 2005. Selective inhibition of 2-AG hydrolysis enhances endocannabinoid signaling in hippocampus. *Nat. Neurosci.* 8:1139–41.

Malan, T. P., Jr., M. M. Ibrahim, H. Deng et al. 2001. CB$_2$ cannabinoid receptor-mediated peripheral antinociception. *Pain* 93:239–45.

Mao, J., D. D. Price, and D. J. Mayer. 1995. Experimental mononeuropathy reduces the anti-nociceptive effects of morphine: implications for common intracellular mechanisms involved in morphine tolerance and neuropathic pain. *Pain* 61:353–64.

Maresz, K., E. J. Carrier, E. D. Ponomarev, C. J. Hillard, and B. N. Dittel. 2005. Modulation of the cannabinoid CB2 receptor in microglial cells in response to inflammatory stimuli. *J. Neurochem.* 95:437–45.

Marsicano, G., S. Goodenough, K. Monory et al. 2003. CB1 cannabinoid receptors and on-demand defense against excitotoxicity. *Science* 302:84–8.

Marsicano, G., C. T. Wotjak, S. C. Azad et al. 2002. The endogenous cannabinoid system controls extinction of aversive memories. *Nature* 418:530–4.

Martin, W. J., A. G. Hohmann, and J. M. Walker. 1996. Suppression of noxious stimulus-evoked activity in the ventral posterolateral nucleus of the thalamus by a cannabi-noid agonist: correlation between electrophysiological and antinociceptive effects. *J. Neurosci.* 16:6601–11.

Martin, W. J., C. M. Loo, and A. I. Basbaum. 1999. Spinal cannabinoids are anti-allodynic in rats with persistent inflammation. *Pain* 82:199–205.

Mascolo, N., A. A. Izzo, A. Ligresti et al. 2002. The endocannabinoid system and the molecu-lar basis of paralytic ileus in mice. *FASEB J.* 16:1973–5.

Massa, F., G. Marsicano, H. Hermann et al. 2004. The endogenous cannabinoid system pro-tects against colonic inflammation. *J. Clin. Invest.* 113:1202–9.

Matsuda, L. A., T. I. Bonner, and S. J. Lolait. 1993. Localization of cannabinoid receptor mRNA in rat brain. *J. Comp Neurol.* 327:535–50.

Matsuda, L. A., S. J. Lolait, M. J. Brownstein, A. C. Young, and T. I. Bonner. 1990. Structure of a cannabinoid receptor and functional expression of the cloned cDNA. *Nature* 346:561–4.

McVey, D. C., P. C. Schmid, H. H. Schmid, and S. R. Vigna. 2003. Endocannabinoids induce ileitis in rats via the capsaicin receptor (VR1). *J. Pharmacol. Exp. Ther.* 304:713–22.

Mechoulam, R. and E. Shohami. 2007. Endocannabinoids and traumatic brain injury. *Mol. Neurobiol.* 36:68–74.

Mechoulam, R., M. Spatz, and E. Shohami. 2002. Endocannabinoids and neuroprotection. *Sci. STKE.* 2002:RE5.

Merriam, F. V., Z. Y. Wang, S. D. Guerios, and D. E. Bjorling. 2008. Cannabinoid recep-tor 2 is increased in acutely and chronically inflamed bladder of rats. *Neurosci. Lett.* 445:130–4.

Michalski, C. W., T. Laukert, D. Sauliunaite et al. 2007. Cannabinoids ameliorate pain and reduce disease pathology in cerulein-induced acute pancreatitis. *Gastroenterology* 132:1968–78.

Michalski, C. W., F. E. Oti, M. Erkan et al. 2008. Cannabinoids in pancreatic cancer: correlation with survival and pain. *Int. J. Cancer.* 122:742–50.

Minkkila, A., M. J. Myllymaki, S. M. Saario et al. 2009. The synthesis and biological evaluation of para-substituted phenolic N-alkyl carbamates as endocannabinoid hydrolyzing enzyme inhibitors. *Eur. J. Med. Chem.* 44(7):2994–3008.

Mitrirattanakul, S., H. López-Valdés, J. Liang et al. 2007. Bi-directional alterations of hippocampal cannabinoid 1 receptors and their endogenous ligands in a rat model of alcohol withdrawal and dependence. *Alcohol. Clin. Exp. Res.* 31:855–67.

Mitrirattanakul, S., N. Ramakul, A. V. Guerrero et al. 2006. Site-specific increases in peripheral cannabinoid receptors and their endogenous ligands in a model of neuropathic pain. *Pain* 126:102–14.

Moezi, L., S. A. Gaskari, H. Liu et al. 2006. Anandamide mediates hyperdynamic circulation in cirrhotic rats via CB(1) and VR(1) receptors. *Br. J. Pharmacol.* 149:898–908.

Moore, S. A., G. G. Nomikos, A. K. Ckason-Chesterfield et al. 2005. Identification of a high-affinity binding site involved in the transport of endocannabinoids. *Proc. Natl. Acad.Sci .U.S.A.* 102:17852–7.

Morisset, V., J. Ahluwalia, I. Nagy, and L. Urban. 2001. Possible mechanisms of cannabinoid-induced antinociception in the spinal cord. *Eur. J Pharmacol* 429:93–100.

Motobe, T., T. Hashiguchi, T. Uchimura et al. 2004. Endogenous cannabinoids are candidates for lipid mediators of bone cement implantation syndrome. *Shock* 21:8–12.

Mukhopadhyay, S., S. Das, E. A. Williams et al. 2006. Lipopolysaccharide and cyclic AMP regulation of CB(2) cannabinoid receptor levels in rat brain and mouse RAW 264.7 macrophages. *J. Neuroimmunol.* 181:82–92.

Munro, S., K. L. Thomas, and M. Abu-Shaar. 1993. Molecular characterization of a peripheral receptor for cannabinoids. *Nature* 365:61–5.

Muthian, S., D. J. Rademacher, C. T. Roelke, G. J. Gross, and C. J. Hillard. 2004. Anandamide content is increased and CB1 cannabinoid receptor blockade is protective during transient, focal cerebral ischemia. *Neuroscience* 129:743–50.

Myllymaki, M. J., S. M. Saario, A. O. Kataja et al. 2007. Design, synthesis, and in vitro evaluation of carbamate derivatives of 2-benzoxazolyl- and 2-benzothiazolyl-(3-hydroxyphenyl)-methanones as novel fatty acid amide hydrolase inhibitors. *J. Med. Chem.* 50:4236–42.

Nakajima, Y., Y. Furuichi, K. K. Biswas et al. 2006. Endocannabinoid, anandamide in gingival tissue regulates the periodontal inflammation through NF-kappaB pathway inhibition. *FEBS Lett.* 580:613–9.

Nemeth, J., Z. Helyes, M. Than et al. 2003. Concentration-dependent dual effect of anandamide on sensory neuropeptide release from isolated rat tracheae. *Neurosci. Lett.* 336:89–92.

Niederhoffer, N., K. Schmid, and B. Szabo. 2003. The peripheral sympathetic nervous system is the major target of cannabinoids in eliciting cardiovascular depression. *Naunyn Schmiedebergs Arch. Pharmacol.* 367:434–43.

Notcutt, W., M. Price, R. Miller et al. 2004. Initial experiences with medicinal extracts of cannabis for chronic pain: results from 34 'N of 1' studies. *Anaesthesia* 59:440–52.

Nucci, C., V. Gasperi, R. Tartaglione et al. 2007. Involvement of the endocannabinoid system in retinal damage after high intraocular pressure-induced ischemia in rats. *Invest Ophthalmol. Vis. Sci.* 48:2997–3004.

Nunez, E., C. Benito, M. R. Pazos et al. 2004. Cannabinoid CB2 receptors are expressed by perivascular microglial cells in the human brain: an immunohistochemical study. *Synapse* 53:208–13.

Nunez, E., C. Benito, R. M. Tolon et al. 2008. Glial expression of cannabinoid CB(2) receptors and fatty acid amide hydrolase are beta amyloid-linked events in Down's syndrome. *Neuroscience* 151:104–10.

Nurmikko, T. J., M. G. Serpell, B. Hoggart et al. 2007. Sativex successfully treats neuropathic pain characterised by allodynia: a randomised, double-blind, placebo-controlled clinical trial. *Pain* 133:210–20.

Nyilas, R., L. C. Gregg, K. Mackie et al. 2009. Molecular architecture of endocannabinoid signaling at nociceptive synapses mediating analgesia. *Eur. J. Neurosci.* 29:1964–78.

Ohno-Shosaku, T., T. Maejima, and M. Kano. 2001. Endogenous cannabinoids mediate retrograde signals from depolarized postsynaptic neurons to presynaptic terminals. *Neuron* 29:729–38.

Oka, S., J. Wakui, S. Ikeda et al. 2006. Involvement of the cannabinoid CB2 receptor and its endogenous ligand 2-arachidonoylglycerol in oxazolone-induced contact dermatitis in mice. *J. Immunol.* 177:8796–805.

Okamoto, Y., J. Morishita, K. Tsuboi, T. Tonai, and N. Ueda. 2004. Molecular characterization of a phospholipase D generating anandamide and its congeners. *J. Biol. Chem.* 279:5298–305.

Oltmanns, M. H., S. S. Samudre, I. G. Castillo et al. 2008. Topical WIN55212-2 alleviates intraocular hypertension in rats through a CB1 receptor mediated mechanism of action. *J. Ocul. Pharmacol. Ther.* 24:104–15.

Onaivi, E. S., H. Ishiguro, J. P. Gong et al. 2008. Functional expression of brain neuronal CB2 cannabinoid receptors are involved in the effects of drugs of abuse and in depression. *Ann. N.Y. Acad. Sci.* 1139:434–49.

Onaivi, E. S., H. Ishiguro, J. P. Gong et al. 2006. Discovery of the presence and functional expression of cannabinoid CB2 receptors in brain. *Ann. N.Y. Acad. Sci.* 1074:514–36.

Ong, W. Y. and K. Mackie. 1999. A light and electron microscopic study of the CB1 cannabinoid receptor in the primate spinal cord. *J. Neurocytol.* 28:39–45.

Ortar, G., Moriello A. Schiano, M. G. Cascio et al. 2008. New tetrazole-based selective anandamide uptake inhibitors. *Bioorg. Med. Chem. Lett.* 18:2820–4.

Ossipov, M. H., Y. Lopez, M. L. Nichols, D. Bian, and F. Porreca. 1995. The loss of antinociceptive efficacy of spinal morphine in rats with nerve ligation injury is prevented by reducing spinal afferent drive. *Neurosci. Lett.* 199:87–90.

Oz, M. 2006. Receptor-independent actions of cannabinoids on cell membranes: focus on endocannabinoids. *Pharmacol. Ther.* 111:114–44.

Pacher, P., P. Mukhopadhyay, R. Mohanraj et al. 2008. Modulation of the endocannabinoid system in cardiovascular disease: therapeutic potential and limitations. *Hypertension* 52:601–7.

Pagotto, U., G. Marsicano, F. Fezza et al. 2001. Normal human pituitary gland and pituitary adenomas express cannabinoid receptor type 1 and synthesize endogenous cannabinoids: first evidence for a direct role of cannabinoids on hormone modulation at the human pituitary level. *J .Clin. Endocrinol. Metab.* 86:2687–96.

Palazuelos, J., N. Davoust, B. Julien et al. 2008. The CB(2) cannabinoid receptor controls myeloid progenitor trafficking: involvement in the pathogenesis of an animal model of multiple sclerosis. *J. Biol. Chem.* 283:13320–9.

Panikashvili, D., R. Mechoulam, S. M. Beni, A. Alexandrovich, and E. Shohami. 2005. CB1 cannabinoid receptors are involved in neuroprotection via NF-kappa B inhibition. *J. Cereb. Blood Flow Metab.* 25:477–84.

Panikashvili, D., C. Simeonidou, S. Ben Shabat et al. 2001. An endogenous cannabinoid (2-AG) is neuroprotective after brain injury. *Nature* 413:527–31.

Pate, D. W., K. Jarvinen, A. Urtti, V. Mahadevan, and T. Jarvinen. 1998. Effect of the CB1 receptor antagonist, SR141716A, on cannabinoid-induced ocular hypotension in normotensive rabbits. *Life Sci.* 63:2181–8.

Patel, S., C. T. Roelke, D. J. Rademacher, and C. J. Hillard. 2005. Inhibition of restraint stress-induced neural and behavioural activation by endogenous cannabinoid signalling. *Eur. J. Neurosci.* 21:1057–69.

Patricelli, M. P. and B. F. Cravatt. 2001. Proteins regulating the biosynthesis and inactivation of neuromodulatory fatty acid amides. *Vitam. Horm.* 62:95–131.

Patwardhan, A. M., N. A. Jeske, T. J. Price et al. 2006. The cannabinoid WIN 55,212-2 inhibits transient receptor potential vanilloid 1 (TRPV1) and evokes peripheral antihyperalgesia via calcineurin. *Proc. Natl. Acad. Sci. U.S.A.* 103:11393–8.

Pertwee, R., G. Griffin, S. Fernando et al. 1995. AM630, a competitive cannabinoid receptor antagonist. *Life Sci.* 56:1949–55.

Pertwee, R. G. 1988. The central neuropharmacology of psychotropic cannabinoids. *Pharmacol. Ther.* 36:189–261.

Pertwee, R. G. 2005a. Pharmacological actions of cannabinoids. *Handb. Exp. Pharmacol.* :1–51.

Pertwee, R. G. 2005b. The therapeutic potential of drugs that target cannabinoid receptors or modulate the tissue levels or actions of endocannabinoids. *AAPS. J.* 7:E625–E654.

Pertwee, R. G. 2008. The diverse CB1 and CB2 receptor pharmacology of three plant cannabinoids: delta9-tetrahydrocannabinol, cannabidiol and delta9-tetrahydrocannabivarin. *Br. J. Pharmacol.* 153:199–215.

Pertwee, R. G., A. Thomas, L. A. Stevenson, Y. Maor, and R. Mechoulam. 2005. Evidence that (-)-7-hydroxy-4'-dimethylheptyl-cannabidiol activates a non-CB(1), non-CB(2), non-TRPV1 target in the mouse vas deferens. *Neuropharmacology* 48:1139–46.

Petersen, G., B. Moesgaard, P. C. Schmid et al. 2005. Endocannabinoid metabolism in human glioblastomas and meningiomas compared to human non-tumour brain tissue. *J. Neurochem.* 93:299–309.

Pisani, A., F. Fezza, S. Galati et al. 2005. High endogenous cannabinoid levels in the cerebrospinal fluid of untreated Parkinson's disease patients. *Ann. Neurol.* 57:777–9.

Porcella, A., C. Maxia, G. L. Gessa, and L. Pani. 2000. The human eye expresses high levels of CB1 cannabinoid receptor mRNA and protein. *Eur. J. Neurosci.* 12:1123–7.

Porcella, A., C. Maxia, G. L. Gessa, and L. Pani. 2001. The synthetic cannabinoid WIN55212-2 decreases the intraocular pressure in human glaucoma resistant to conventional therapies. *Eur. J. Neurosci.* 13:409–12.

Potenzieri, C., T. S. Brink, C. Pacharinsak, and D. A. Simone. 2008a. Cannabinoid modulation of cutaneous Adelta nociceptors during inflammation. *J. Neurophysiol.* 100:2794–806.

Potenzieri, C., C. Harding-Rose, and D. A. Simone. 2008b. The cannabinoid receptor agonist, WIN 55, 212-2, attenuates tumor-evoked hyperalgesia through peripheral mechanisms. *Brain Res.* 1215:69–75. Epub@2008 Apr 6:69–75.

Price, T. J., G. Helesic, D. Parghi, K. M. Hargreaves, and C. M. Flores. 2003. The neuronal distribution of cannabinoid receptor type 1 in the trigeminal ganglion of the rat. *Neuroscience* 120:155–62.

Quistad, G. B., R. Klintenberg, P. Caboni, S. N. Liang, and J. E. Casida. 2006. Monoacylglycerol lipase inhibition by organophosphorus compounds leads to elevation of brain 2-arachidonoylglycerol and the associated hypomotility in mice. *Trends Neurosci.* 211:78–83.

Racz, I., X. Nadal, J. Alferink et al. 2008. Crucial role of CB_2 cannabinoid receptor in the regulation of central immune responses during neuropathic pain. *J. Neurosci.* 28:12125–35.

Rajesh, M., P. Mukhopadhyay, S. Batkai et al. 2007. CB2-receptor stimulation attenuates TNF-alpha-induced human endothelial cell activation, transendothelial migration of monocytes, and monocyte-endothelial adhesion. *Am. J. Physiol Heart Circ. Physiol.* 293:H2210–H2218.

Ramirez, B. G., C. Blazquez, del Pulgar T.G., M. Guzman, and M. L. de Ceballos. 2005. Prevention of Alzheimer's disease pathology by cannabinoids: neuroprotection mediated by blockade of microglial activation. *J. Neurosci.* 25:1904–13.

Rashid, M. H., M. Inoue, K. Toda, and H. Ueda. 2004. Loss of peripheral morphine analgesia contributes to the reduced effectiveness of systemic morphine in neuropathic pain. *J. Pharmacol. Exp. Ther.* 309:380–7.

Rayman, N., K. H. Lam, J. Van Leeuwen et al. 2007. The expression of the peripheral cannabinoid receptor on cells of the immune system and non-Hodgkin's lymphomas. *Leuk. Lymphoma.* 48:1389–99.

Richardson, D., R. G. Pearson, N. Kurian et al. 2008. Characterisation of the cannabinoid receptor system in synovial tissue and fluid in patients with osteoarthritis and rheumatoid arthritis. *Arthritis Res. Ther.* 10:R43.

Richardson, J. D., L. Aanonsen, and K. M. Hargreaves. 1997. SR 141716A, a cannabinoid receptor antagonist, produces hyperalgesia in untreated mice. *Eur. J. Pharmacol.* 319:R3–R4.

Richardson, J. D., L. Aanonsen, and K. M. Hargreaves. 1998a. Antihyperalgesic effects of spinal cannabinoids. *Eur. J Pharmacol* 345:145–53.

Richardson, J. D., S. Kilo, and K. M. Hargreaves. 1998b. Cannabinoids reduce hyperalgesia and inflammation via interaction with peripheral CB1 receptors. *Pain* 75:111–9.

Richfield, E. K. and M. Herkenham. 1994. Selective vulnerability in Huntington's disease: preferential loss of cannabinoid receptors in lateral globus pallidus. *Ann. Neurol.* 36:577–84.

Rinaldi-Carmona, M., F. Barth, M. Heaulme et al. 1994. SR141716A, a potent and selective antagonist of the brain cannabinoid receptor. *FEBS Lett.* 350:240–4.

Roberts, L. A., M. J. Christie, and M. Connor. 2002. Anandamide is a partial agonist at native vanilloid receptors in acutely isolated mouse trigeminal sensory neurons. *Br. J. Pharmacol.* 137:421–8.

Robson, P. 2001. Therapeutic aspects of cannabis and cannabinoids. *Br. J. Psychiatry* 178:107–15.

Rodriguez, J. J., K. Mackie, and V. M. Pickel. 2001. Ultrastructural localization of the CB1 cannabinoid receptor in mu-opioid receptor patches of the rat Caudate putamen nucleus. *J. Neurosci.* 21:823–33.

Romero, J., F. Berrendero, A. Perez-Rosado et al. 2000. Unilateral 6-hydroxydopamine lesions of nigrostriatal dopaminergic neurons increased CB1 receptor mRNA levels in the caudate-putamen. *Life Sci.* 66:485–94.

Ros, J., J. Claria, J. To-Figueras et al. 2002. Endogenous cannabinoids: a new system involved in the homeostasis of arterial pressure in experimental cirrhosis in the rat. *Gastroenterology* 122:85–93.

Ross, R. A. 2003. Anandamide and vanilloid TRPV1 receptors. *Br. J. Pharmacol.* 140:790–801.

Ross, R. A. 2009. The enigmatic pharmacology of GPR55. *Trends Pharmacol. Sci.* 30:156–63.

Ross, R. A., T. M. Gibson, H. C. Brockie et al. 2001. Structure-activity relationship for the endogenous cannabinoid, anandamide, and certain of its analogues at vanilloid receptors in transfected cells and vas deferens. *Br. J. Pharmacol.* 132:631–40.

Rubio, M., D. McHugh, J. Fernandez-Ruiz, H. Bradshaw, and J. M. Walker. 2007. Short-term exposure to alcohol in rats affects brain levels of anandamide, other N-acylethanolamines and 2-arachidonoyl-glycerol. *Neurosci. Lett.* 421(3): 270–4.

Ryberg, E., H. K. Vu, N. Larsson et al. 2005. Identification and characterisation of a novel splice variant of the human CB1 receptor. *FEBS Lett.* 579:259–64.

Salio, C., J. Fischer, M. F. Franzoni, and M. Conrath. 2002. Pre- and postsynaptic localizations of the CB_1 cannabinoid receptor in the dorsal horn of the rat spinal cord. *Neuroscience* 110:755–64.

Sanudo-Pena, M. C., N. M. Strangman, K. Mackie, J. M. Walker, and K. Tsou. 1999. CB1 receptor localization in rat spinal cord and roots, dorsal root ganglion, and peripheral nerve. *Acta Pharmacol. Sin.* 20:1115–20.

Sarfaraz, S., F. Afaq, V. M. Adhami, and H. Mukhtar. 2005. Cannabinoid receptor as a novel target for the treatment of prostate cancer. *Cancer Res.* 65:1635–41.

Sarzani, R. 2008. Endocannabinoids, blood pressure and the human heart. *J. Neuroendocrinol.* 20 Suppl 1:58–62.

Schabitz, W. R., A. Giuffrida, C. Berger et al. 2002. Release of fatty acid amides in a patient with hemispheric stroke: a microdialysis study. *Stroke* 33:2112–4.

Schuelert, N. and J. J. McDougall. 2008. Cannabinoid-mediated antinociception is enhanced in rat osteoarthritic knees. *Arthritis Rheum.* 58:145–53.

Schultheiss, T., K. Flau, M. Kathmann, M. Gothert, and E. Schlicker. 2005. Cannabinoid CB1 receptor-mediated inhibition of noradrenaline release in guinea-pig vessels, but not in rat and mouse aorta. *Naunyn Schmiedebergs Arch. Pharmacol.* 372:139–46.

Schweitzer, P. 2000. Cannabinoids decrease the K^+ M-current in hippocampal CA1 neurons. *J. Neurosci.* 20:51–8.

Siegling, A., H. A. Hofmann, D. Denzer, F. Mauler, and Vry J. De. 2001. Cannabinoid CB(1) receptor upregulation in a rat model of chronic neuropathic pain. *Eur. J. Pharmacol.* 415:R5–R7.

Singh Tahim A., P. Santha, and I. Nagy. 2005. Inflammatory mediators convert anandamide into a potent activator of the vanilloid type 1 transient receptor potential receptor in nociceptive primary sensory neurons. *Neuroscience* 136:539–48.

Skaper, S. D., A. Buriani, Toso R. Dal et al. 1996. The ALIAmide palmitoylethanolamide and cannabinoids, but not anandamide, are protective in a delayed postglutamate paradigm of excitotoxic death in cerebellar granule neurons. *Proc. Natl. Acad.S ci. U.S.A.* 93:3984–9.

Smith, P. B., D. R. Compton, S. P. Welch et al. 1994. The pharmacological activity of anandamide, a putative endogenous cannabinoid, in mice. *J. Pharmacol. Exp. Ther.* 270:219–27.

Sofia, R. D., S. D. Nalepa, J. J. Harakal, and H. B. Vassar. 1973. Anti-edema and analgesic properties of delta9-tetrahydrocannabinol (THC). *J Pharmacol Exp Ther* 186:646–55.

Song, Z. H. and C. A. Slowey. 2000. Involvement of cannabinoid receptors in the intraocular pressure-lowering effects of WIN55212-2. *J. Pharmacol. Exp. Ther.* 292:136–9.

Sparling, P. B., A. Giuffrida, D. Piomelli, L. Rosskopf, and A. Dietrich. 2003. Exercise activates the endocannabinoid system. *Neuroreport.* 14:2209–11.

Ständer, S., M. Schmelz, D. Metze, T. Luger, and R. Rukwied. 2005. Distribution of cannabinoid receptor 1 (CB1) and 2 (CB2) on sensory nerve fibers and adnexal structures in human skin. *J. Dermatol. Sci.* 38:177–88.

Staton, P. C., J. P. Hatcher, D. J. Walker et al. 2008. The putative cannabinoid receptor GPR55 plays a role in mechanical hyperalgesia associated with inflammatory and neuropathic pain. *Pain* 139:225–36.

Steffens, S., N. R. Veillard, C. Arnaud et al. 2005. Low dose oral cannabinoid therapy reduces progression of atherosclerosis in mice. *Nature* 434:782–6.

Stella, N., P. Schweitzer, and D. Piomelli. 1997. A second endogenous cannabinoid that modulates long-term potentiation. *Nature* 388:773–8.

Straiker, A. J., G. Maguire, K. Mackie, and J. Lindsey. 1999. Localization of cannabinoid CB1 receptors in the human anterior eye and retina. *Invest Ophthalmol. Vis. Sci.* 40:2442–8.

Strangman, N. M., S. L. Patrick, A. G. Hohmann, K. Tsou, and J. M. Walker. 1998. Evidence for a role of endogenous cannabinoids in the modulation of acute and tonic pain sensitivity. *Brain Res.* 813:323–8.

Stumpff, F., M. Boxberger, A. Krauss et al. 2005. Stimulation of cannabinoid (CB1) and prostanoid (EP2) receptors opens BKCa channels and relaxes ocular trabecular meshwork. *Exp. Eye Res.* 80:697–708.

Sugiura, T., T. Kodaka, S. Nakane et al. 1998. Detection of an endogenous cannabimimetic molecule, 2-arachidonoylglycerol, and cannabinoid CB1 receptor mRNA in human vascular cells: is 2-arachidonoylglycerol a possible vasomodulator? *Biochem. Biophys. Res. Commun.* 243:838–43.

Sugiura, T., S. Kondo, A. Sukagawa et al. 1996. Transacylase-mediated and phosphodiesterase-mediated synthesis of N-arachidonoylethanolamine, an endogenous cannabinoid-receptor ligand, in rat brain microsomes. Comparison with synthesis from free arachidonic acid and ethanolamine. *Eur. J. Biochem.* 240:53–62.

Sugiura, T., N. Yoshinaga, S. Kondo, K. Waku, and Y. Ishima. 2000. Generation of 2-arachidonoylglycerol, an endogenous cannabinoid receptor ligand, in picrotoxinin-administered rat brain. *Biochem. Biophys. Res. Commun.* 271:654–8.

Szolcsanyi, J. 2000. Anandamide and the question of its functional role for activation of capsaicin receptors. *Trends Pharmacol. Sci.* 21:203–4.

Teixeira-Clerc, F., B. Julien, P. Grenard et al. 2006. CB1 cannabinoid receptor antagonism: a new strategy for the treatment of liver fibrosis. *Nat. Med.* 12:671–6.

Tognetto, M., S. Amadesi, S. Harrison et al. 2001. Anandamide excites central terminals of dorsal root ganglion neurons via vanilloid receptor-1 activation. *J. Neurosci.* 21:1104–9.

Tomida, I., R. G. Pertwee, and A. Azuara-Blanco. 2004. Cannabinoids and glaucoma. *Br. J. Ophthalmol.* 88:708–13.

Tominaga, M. and M. J. Caterina. 2004. Thermosensation and pain. *J. Neurobiol.* 61:3–12.

Tsou, K., S. Brown, M. C. Sanudo-Pena, K. Mackie, and J. M. Walker. 1998. Immunohistochemical distribution of cannabinoid CB1 receptors in the rat central nervous system. *Neuroscience* 83:393–411.

Tsou, K., K. Mackie, M. C. Sanudo-Pena, and J. M. Walker. 1999. Cannabinoid CB1 receptors are localized primarily on cholecystokinin-containing GABAergic interneurons in the rat hippocampal formation. *Neuroscience* 93:969–75.

Van Sickle, M. D., M. Duncan, P. J. Kingsley et al. 2005. Identification and functional characterization of brainstem cannabinoid CB2 receptors. *Science* 310:329–32.

Van Sickle, M. D., L. D. Oland, W. Ho et al. 2001. Cannabinoids inhibit emesis through CB1 receptors in the brainstem of the ferret. *Gastroenterology* 121:767–74.

Varga, K., J. A. Wagner, D. T. Bridgen, and G. Kunos. 1998. Platelet- and macrophage-derived endogenous cannabinoids are involved in endotoxin-induced hypotension. *FASEB J.* 12:1035–44.

Vinod, K. Y., V. Arango, S. Xie et al. 2005. Elevated levels of endocannabinoids and CB_1 receptor-mediated G-protein signaling in the prefrontal cortex of alcoholic suicide victims. *Biol. Psychiatry* 57:480–6.

Wagner, J. A., K. Hu, J. Bauersachs et al. 2001. Endogenous cannabinoids mediate hypotension after experimental myocardial infarction. *J. Am. Coll. Cardiol.* 38:2048–54.

Wagner, J. A., K. Varga, E. F. Ellis et al. 1997. Activation of peripheral CB1 cannabinoid receptors in haemorrhagic shock. *Nature* 390:518–21.

Wagner, J. A., K. Varga, and G. Kunos. 1998. Cardiovascular actions of cannabinoids and their generation during shock. *J. Mol. Med.* 76:824–36.

Walczak, J. S., V. Pichette, F. Leblond, K. Desbiens, and P. Beaulieu. 2005. Behavioral, pharmacological and molecular characterization of the saphenous nerve partial ligation: a new model of neuropathic pain. *Neuroscience* 132:1093–102.

Walczak, J. S., V. Pichette, F. Leblond, K. Desbiens, and P. Beaulieu. 2006. Characterization of chronic constriction of the saphenous nerve, a model of neuropathic pain in mice showing rapid molecular and electrophysiological changes. *J. Neurosci. Res.* 83:1310–22.

Walker, J. M., S. M. Huang, N. M. Strangman, K. Tsou, and M. C. Sanudo-Pena. 1999. Pain modulation by release of the endogenous cannabinoid anandamide. *Proc. Natl. Acad. Sci. U.S.A.* 96:12198–203.

Wallace, M. J., R. E. Blair, K. W. Falenski, B. R. Martin, and R. J. DeLorenzo. 2003. The endogenous cannabinoid system regulates seizure frequency and duration in a model of temporal lobe epilepsy. *J. Pharmacol. Exp. Ther.* 307:129–37.

Wang, D., H. Wang, W. Ning et al. 2008. Loss of cannabinoid receptor 1 accelerates intestinal tumor growth. *Cancer Res.* 68:6468–76.

Wang, Y., Y. Liu, Y. Ito et al. 2001. Simultaneous measurement of anandamide and 2-arachidonoylglycerol by polymyxin B-selective adsorption and subsequent high-performance liquid chromatography analysis: increase in endogenous cannabinoids in the sera of patients with endotoxic shock. *Anal. Biochem.* 294:73–82.

Welch, S. P., J. W. Huffman, and J. Lowe. 1998. Differential blockade of the antinociceptive effects of centrally administered cannabinoids by SR141716A. *J. Pharmacol. Exp. Ther.* 286:1301–8.

Wilsey, B., T. Marcotte, A. Tsodikov et al. 2008. A randomized, placebo-controlled, crossover trial of cannabis cigarettes in neuropathic pain. *J. Pain.* 9:506–21.

Wilson, R. I. and R. A. Nicoll. 2001. Endogenous cannabinoids mediate retrograde signalling at hippocampal synapses. *Nature* 410:588–92.

Wotherspoon, G., A. Fox, P. McIntyre et al. 2005. Peripheral nerve injury induces cannabinoid receptor 2 protein expression in rat sensory neurons. *Neuroscience* 135:235–45.

Xu, X., Y. Liu, S. Huang et al. 2006. Overexpression of cannabinoid receptors CB1 and CB2 correlates with improved prognosis of patients with hepatocellular carcinoma. *Cancer Genet. Cytogenet.* 171:31–8.

Yiangou, Y., P. Facer, P. Durrenberger et al. 2006. COX-2, CB2 and P2X7-immunoreactivities are increased in activated microglial cells/macrophages of multiple sclerosis and amyotrophic lateral sclerosis spinal cord. *BMC. Neurol.* 6:12.

Zavitsanou, K., T. Garrick, and X. F. Huang. 2004. Selective antagonist [3H]SR141716A binding to cannabinoid CB1 receptors is increased in the anterior cingulate cortex in schizophrenia. *Prog. Neuropsychopharmacol. Biol. Psychiatry* 28:355–60.

Zeng, B. Y., B. Dass, A. Owen et al. 1999. Chronic L-DOPA treatment increases striatal cannabinoid CB1 receptor mRNA expression in 6-hydroxydopamine-lesioned rats. *Neurosci. Lett.* 276:71–4.

Zhang, J., C. Hoffert, H. K. Vu et al. 2003. Induction of CB2 receptor expression in the rat spinal cord of neuropathic but not inflammatory chronic pain models. *Eur. J. Neurosci.* 17:2750–4.

Zimmer, A., A. M. Zimmer, A. G. Hohmann, M. Herkenham, and T. I. Bonner. 1999. Increased mortality, hypoactivity, and hypoalgesia in cannabinoid CB_1 receptor knockout mice. *Proc. Natl. Acad. Sci. U.S.A.* 96:5780–5.

Zygmunt, P. M., J. Petersson, D. A. Andersson et al. 1999. Vanilloid receptors on sensory nerves mediate the vasodilator action of anandamide. *Nature* 400:452–7.

6 Molecular Strategies for Therapeutic Targeting of Primary Sensory Neurons in Chronic Pain Syndromes

Ichiro Nishimura, Devang Thakor, Audrey Lin, Supanigar Ruangsri, and Igor Spigelman

CONTENTS

6.1 INTRODUCTION

Pain is one of the primitive behaviors that is highly conserved among species and is critical for survival when facing environmental stresses. In an extremely rare occurrence, families in a remote village of Pakistan were discovered to completely lack the capability for nociception (Cox et al. 2006). The resulting lack of pain appeared to manifest with frequent injuries ranging from bone fractures and extensive burns to accidental death. Whole genome linkage analysis revealed candidate single nucleotide polymorphisms (SNPs) in the SCN9A allele, which encodes the tetrodotoxin-sensitive voltage-gated sodium channel NaV1.7. SNPs in these families (Cox et al. 2006), as well as in other families from Canada exhibiting a similar phenotype (Goldberg et al. 2007; Ahmad et al. 2007), result in truncation of the NaV1.7 molecule that leads to loss of function. Furthermore, SNPs causing protein substitutions in the SCN9A allele have been found to lead to gain of function mutations in patients with primary erythermalgia (Drenth et al. 2005; Han et al. 2006; Yang et al. 2004) and paroxysmal extreme pain disorder (Fertleman et al. 2006; Yiangou et al. 2007), both of which manifest with localized severe pain episodes. The information gained from patients has been particularly valuable for understanding the role of NaV1.7 in inheritable disorders with inflammation-mediated pain.

Neuropathic pain, on the other hand, is a chronic disability commonly caused by an initial primary lesion to or dysfunction of the peripheral nervous system (PNS) (Chen et al. 2004). Even after the initial causes are eliminated, neuropathic pain continues; although the precise pathophysiological mechanisms have yet to be elucidated, a change in constitutive characteristics of the PNS has been postulated to play an important role. A large amount of literature has described alterations in the expression or local distribution of various proteins in the affected PNS (Zimmermann 2001; Niederberger and Geisslinger 2008; Hains and Waxman 2007; Wood et al. 2004). For example, altered expression of sodium channels might explain the observed decreases in nociceptive thresholds to mechanical stimuli. Mouse models with null mutations of the NaV1.3, NaV1.6, NaV1.7, NaV1.8, or NaV1.9 sodium channels have been generated; however, these animal models have failed to express the anticipated neuropathic pain behavioral modifications (Nassar et al. 2006; Cummins et al. 2005; Nassar et al. 2004; Stirling et al. 2005; Laird et al. 2002; Roza et al. 2003; Hillsley et al. 2006), leaving challenges for researchers to understand disease etiology and define molecular targets.

Investigation on the molecular pathophysiological mechanisms of neuropathic pain has been limited in part due to the lack of available experimental tools that allow the modulation of target molecules *in vitro* and *in vivo*. The scope of this chapter is to provide information regarding the available molecular tools and to discuss potential challenges. Our recent attempts in devising new gene transfer vectors for the PNS are also summarized.

6.2 CHALLENGES ASSOCIATED WITH GENE TRANSFER TO NEURONAL CELLS *IN VITRO*

Indispensable experimental systems in molecular biology include *in vitro* cell culture combined with strategies for introducing exogenous genes for independent promoter-driven over-expression or RNA interference (RNAi)-mediated suppression of target genes. The purpose of such experimental systems is to increase or decrease the function of the target gene, respectively, thus allowing the elucidation of the cellular consequences of altered gene expression. To accommodate this fundamental experimental system, it is important to establish a stable method for shuttling exogenous DNA into cells. Features associated with neuronal cells often present specific challenges to the accomplishment of this task. In this section, the major challenges are indicated and possible solutions discussed.

6.2.1 Challenge 1: Gene Delivery to Non-Dividing Neuronal Cells

6.2.1.1 Non-Viral Vectors

Currently, cationic gene transfer agents are most commonly used for gene transfer *in vitro*, yielding 50% to 80% transfection rates in actively proliferating cells such as NIH3T3 mouse fibroblasts, CHO (Chinese hamster ovary) cells, and HEK293 (human embryonic kidney) cells. Cationic non-viral gene delivery systems offer a number of advantages including ease of production and use, stability, low immunogenicity and toxicity, and capacity to deliver larger DNA payloads than those of viral delivery systems. Acute dissociation of dorsal root ganglia (DRG) has been established as an *in vitro* model of sensory neurons and has been frequently used as a standard experimental model for studies of the PNS function and pathophysiology (Spigelman, Gold, and Light 2001). However, relatively poor transfection efficiency in non-dividing neuronal cells has reduced the utility of non-viral gene transfer for investigating sensory disorders.

Cationic gene transfer agents include cationic lipids as well as cationic polymers such as polyethylenimine (PEI) and cationized gelatin (CG) (Kim et al. 2004). It has been postulated that cationic agents condense DNA molecules and generate polyplexes of less than 100 nm in diameter as well as with ζ potential (surface charge) over +10 mV (Anderson et al. 2005; Akinc et al. 2008). Positively charged nanoparticles and polyplexes are thought to interact with negatively charged cell membrane components. The resulting non-specific adsorption facilitates internalization via clathrin-dependent endocytosis. Polyplex-containing endosomes are thought to be osmotically lysed by a "proton sponge effect," eventually resulting in the release of DNA molecule for transgene expression (Figure 6.1). However, this process is relatively inefficient, and endosomolysis remains a major barrier to non-viral gene delivery by cationic agents. Although it has not been demonstrated in neuronal cells, cationic agent-assisted gene transfer can be improved by incorporation of endosome destabilizing agents (Zhou et al. 2007; Boeckle, Wagner, and Ogris 2005; Kloeckner et al. 2006).

Neurons present a particular challenge to successful non-viral gene transfer primarily due to their non-proliferative status; the maintenance of nuclear envelope integrity in non-dividing cells impedes nuclear vector penetration, which is thought to be required for efficient transgene transcription (Berry et al. 2001). A number of attempts have been made to enhance nuclear gene delivery of non-viral vectors by attachment of nuclear localization elements to either plasmid DNA or the vector itself (Escriou et al. 2003; Rolland 2006; van der Aa et al. 2006). However, these approaches have typically attained only limited success while requiring special

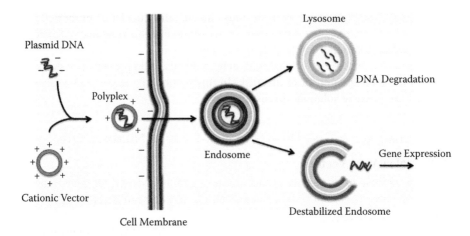

FIGURE 6.1 Diagram of cationic vector gene transfer. Negatively charged plasmid DNA molecules are complexed with cationic vector and form polyplexes of less than 100 nm in diameter as well as with ζ potential (surface charge) over +10 mV. The positive surface charge of polyplexes is believed to interact with the negatively charged cell membrane, initiating internalization. The excess cationic agent can destabilize the endosome and release plasmid DNA resulting in the transgene expression. However, when the endosome matures to become a lysosome, as often seen in non-dividing cells, plasmid DNA is degraded and transgene expression does not occur.

chemical conjugation/purification steps as well as costly peptide synthesis reactions (Rolland 2006; van der Aa et al. 2006). Thus, an alternative approach involves the development of simple vectors that can be more efficiently internalized by the plasma membrane, given the hypothesis that increased overall uptake should lead to an increase in the absolute amount of DNA delivered to the nucleus even in the absence of specific nuclear targeting.

Recently, a new cationic vector was generated by the conjugation of the naturally occurring polysaccharide pullulan and polyamine spermine; this vector allowed efficient *in vitro* gene expression in rapidly proliferating tumor cells (Kanatani et al. 2006; Jo, Ikai, Okazaki, Nagane, et al. 2007; Jo, Ikai, Okazaki, Yamamoto, et al. 2007). In terms of safety, the individual use of these materials is advantageous, as pullulan is already commercially used in oral healthcare and pharmaceutical coating applications (Leathers 2003) and spermine is ubiquitously found at high concentrations in human cells, where it is thought to naturally interact with nucleic acids (Tabor and Tabor 1984).

Thakor et al. (2009) reported that pullulan-spermine/DNA complexes could be used to transfect dissociated DRG somata (Thakor, Teng, and Tabata 2009). Surprisingly, the best transfection efficiency for DRG neuronal cultures was found at a pullulan amine:DNA phosphate ratio of 1:1; particles formed under these conditions exhibited a diameter and ζ potential of approximately 350 nm and -40 mV, respectively, and were termed "anioplexes." Successful transfection by such anioplexes was counterintuitive, as cationic polymer-mediated transfection has been traditionally thought to rely on electrostatic interactions with negatively charged membrane components.

Pullulan-spermine anioplex transfection was instead suggested to involve specific binding to glycoproteins and uptake through lipid rafts. The negative surface charge of these particles resulted in unique properties not usually observed in cationic polymer-based non-viral vectors, including serum compatibility and a lack of observable cytotoxicity, even at high DNA concentrations and long exposure times (Thakor, Teng, and Tabata 2009). Thus, the use of anionic vectors for neuronal cell transfection may present a promising alternative for future studies.

6.2.1.2 Viral Vectors

Engineered viral vectors that permit gene transduction to relatively non-proliferative cells have recently become more efficient with reproducible results. Adenovirus (Ad), adeno-associated virus (AAV), herpes simplex virus (HSV), and lentivirus have all been applied to DRG cells *in vitro* (Table 6.1).

TABLE 6.1

Gene Transfer Viral Vectors Used for Investigation of Sensory Neurons

Vectors	Particle Size	Viral Genome	Transgene	Receptor	*In Vitro*	*In Vivo*
Adenovirus	70-90 nm, naked	35 Kb (Double-stranded DNA)	~8 Kb	CAR, heparan sulfate, CD46, CD80, CD86, integrins	DRG neurons, retinal cells, cochlear cells, neuronal precursor cells, taste bud cells	Trigeminal and olfactory sensory neurons through nostril application or injection to the jaw
Adeno-associated virus	20 nm, naked	4.7 Kb (Single-stranded DNA)	30-35 Kb	Heparan sulfate	DRG neurons, retinal cells, cochlear cells	DRG neurons through injection into DRG, and sciatic nerve
Herpes simplex virus	200 nm, enveloped	Amplicon plasmid DNA	~150 Kb	HVEM, nectin1, nectin2, heparan sulfate	DRG neurons, fetal neuron cells, retinal cells	DRG neurons through injection into DRG, sciatic nerve and skin
Lentivirus	80-100 nm, enveloped	9.2 Kb (Single-stranded RNA)	~8 Kb	CD4, CCR5, CSCR4 (Neuronal receptor unknown)	DRG neurons, photoreceptor cells, olfactory sensory neurons	Trigeminal and DRG neurons through injection into DRG, sciatic nerve or temporo-mandibular joint

Ad belongs to a family of viruses containing linear double-stranded DNA. The virion is non-enveloped with a size of 70 to 90 nm, and the viral DNA is 35 Kb with the ability to incorporate insert gene of up to 8 Kb. Ad infects non-replicating cells and has been shown to transfer genes to DRG explant cultures (Dijkhuizen et al. 1997).

AAV is also a non-enveloped virus, but of much smaller size (20 nm), that contains 4.7 Kb of single-stranded DNA. However, AAV can accommodate a much larger DNA insert of about 35 Kb. Out of 11 AAV serotypes, serotype 2 (AAV2) has been extensively characterized for gene transfer and has been shown to be effective for cultured DRG neurons (Fleming et al. 2001).

Herpes simplex virus (HSV), a human neurotropic virus, naturally infects sensory neurons. HSV type 1 (HSV-1) has been modified for gene transfer, essentially resulting in a defective HSV containing a plasmid (amplicon) composed of the transgene and two essential cis-acting HSV genomic DNA sequences. The amplicon-based HSV-1 vector does not contain viral genes and thus does not replicate. Furthermore, a single virion can contain multiple copies of the transgene. The amplicon-based HSV-1 vector has been used for both cultured (Goins et al. 1999) and *in vivo* (Walwyn et al. 2006) DRG neuronal transduction.

Lentiviral vectors are derived from human immunodeficiency virus (HIV), a type of retrovirus containing a single-stranded RNA genome. With the third-generation lentiviral vector, critical additional safety features are incorporated. First, self-inactivating (SIN) vectors have been incorporated; this involves a deletion in the enhancer region of the 3' U3 of the long terminal repeat (LTR). Thus, this vector cannot be converted into a full-length RNA. The second modification is the removal of the *tat* gene from the packaging vector. The Tat protein has been shown to be dispensable when the 5' LTR is replaced with a heterologous constitutive promoter. The resulting third-generation lentiviral vectors have provided a safer system (Dull et al. 1998; Stripecke et al. 2003; Koya et al. 2002) and have been applied to DRG neuronal cell culture systems (Mikami and Yang 2005) (Figure 6.2).

6.2.2 CHALLENGE 2: DRG NEURONS AND CELL MEMBRANE ACTIVITY

Several viruses are known to affect sensory neuron physiology. For example, while the Sendai virus may influence specific membrane properties such as calcium channel activation several days prior to general structural and functional degeneration of sensory neurons (Maehlen et al. 1991), HSV can alter electrophysiological properties of sensory neurons recorded with sharp intracellular electrodes (Fukuda and Kurata 1981). However, replication-defective HSV-1 does not appear to alter electrophysiological properties of transfected neurons (Farkas, Nakajima, and Nakajima 1994).

HIV type-1 virus has been shown to induce neuronal dysfunction that may be caused by non-structural protein Tat and envelope glycoprotein gp120. It has been shown that Tat protein applied to human fetal neurons *in vitro* elicits dose-dependent depolarization in the absence and presence of tetrodotoxin (Cheng et al. 1998). The envelope glycoprotein gp120 has been shown to induce neuronal apoptosis and axonal degeneration in DRG cultures (Melli et al. 2006), which has been postulated to cause HIV-related sensory neuropathy. The third-generation lentiviral vector

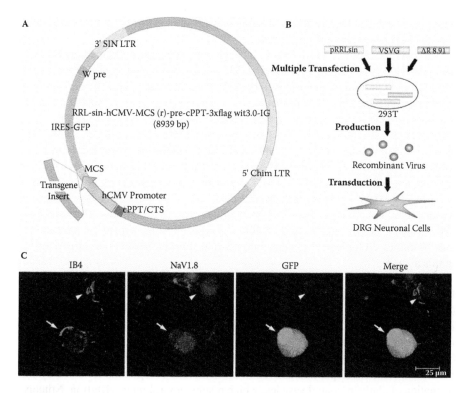

FIGURE 6.2 (see color insert following page 166) Cultured DRG neurons transduced with GFP-expressing lentivirus vector. **A:** Third-generation lentiviral vector map (pRRLsin-hCMV). Central polypurine tract (cPPT) and the central termination sequence (CTS) are located upstream of the CMV promoter, which enhances the nuclear import efficiency of the DNA plasmid. In this vector, immediately downstream of the Multiple Cloning Site (MCS) are the Internal Ribosome Entry Site (IRES) and the Green Fluorescent Protein (GFP) gene, which allows the co-expression of GFP with transgene. The promoter and enhancer elements of the lentiviral genome has been deleted at the U3 region of the long terminal repeat (LTR), therefore becoming self-inactivating (SIN). **B:** Flow chart of lentivirus production. **C:** Rat DRG culture was treated with lentivirus vector. The transduction rate assessed by GFP generally yields 50%-60% for IB4 positive small diameter neuronal cells. The expression pattern of endogenous molecules such as NaV1.8 was not affected by lentivirus vectors.

lacks *tat* gene but requires gp120. In fact, after lentivirus transduction, the host cells appear to maintain the immunoreactive gp120 (Pham et al. 2004).

Although there is one report on patch clamp recording of DRG neurons after lentiviral transduction (Mikami and Yang 2005), we have experienced a persistent inability to obtain tight (GΩ) seals using patch clamp electrodes on cultured DRG neurons after lentiviral transduction (A. Lin, F. Schweizer, S. Ruangsri, and I. Nishimura, unpublished observations). Transduced neurons were identified by their green fluorescent protein (GFP) and appeared to have normal features of DRG neurons including isolectin B4 (IB4) binding and NaV1.8 immunoreactivity (Figure 6.2). Therefore, it is conceivable that lentiviral transduction may result in yet-to-be-defined structural changes of the neuronal plasma membrane.

The passage of many viral vectors through cell membranes requires host receptor interactions that may affect neuronal cell membrane characteristics. HSV vector contains a gD protein, which binds to the gD receptor on the host cell and mediates HSV entry into the cell through the herpes virus entry mediator (HVEM). HVEM, a member of the tumor necrosis factor receptor family, and nectin-1 are the major mediators of cell entry for HSV-1 (Akhtar et al. 2008). HIV associates with CD4 as its primary receptor together with CCR5 and CXCR4 as co-receptors in macrophages and T cells, respectively. However, lentiviral entry mechanisms in neuronal cells are not fully understood. Whereas treatment with a neutralizing anti-CD4 antibody has been shown to block lentiviral entry into CD4+ lymphocytes, a similar treatment did not attenuate lentiviral entry into human DRG neurons (Harouse et al. 1989; Kunsch, Hartle, and Wigdahl 1989). It is postulated that a different mechanism may be employed for lentiviral entry into neurons; future investigations are required in this area.

In contrast, Ad appears not to interfere with sensory neuron function (Smith and Romero 1999). Ad is primarily recognized by Coxsackie virus and Ad receptor (CAR). Recently, several other primary receptors have been found for Ad, including cell membrane-associated heparan sulfate, CD46, CD80, CD86, sialic acid, and integrins aMb2 and aLb2. AAV-2 has been found to use cell membrane-associated heparan sulfate proteoglycans (Summerford and Samulski 1998). Ad and AAV do not have a viral envelope, and activation of membrane receptors is followed by the formation of clathrin-coated vesicles, which results in viral internalization. Notably, HSV and lentivirus are enveloped viruses. We speculate that the remaining viral envelope might significantly affect neuronal membrane properties (Figure 6.3).

It may be possible that viral glycoproteins left in the cell membrane might continue to destabilize the membrane. Usually such membrane destabilization by specific peptides in viral coat glycoproteins is the initial event involved in the fusion of the viral membrane with the cell membrane (Teissier and Pecheur 2007). The compromised membrane might be simply unable to withstand the forces involved in making the tight seal needed for patch clamp recordings.

The use of vesicular stomatitis virus glycoprotein (VSV-G) to pseudotype retroviral vectors enabled high-titer production vectors broad tropism (Burns et al. 1993). The lentiviral vector is one of the VSV-G–pseudotyped vectors that has been considerably used in experimental settings (Farley et al. 2007). It has been demonstrated that VSV-G, a component of the viral envelope that remains in the membrane of transfected cells, is essential for the induction of the specific antiviral toll-like receptor 4 (TLR4) signaling cascade, which leads to the type I interferon (IFN) response (Georgel et al. 2007). Although TLR4 is best known as a sensor of microbial lipopolysaccharide, it is also well placed to respond to viral infection. Because neurons have been shown to express several TLRs, neuronal cells might be capable of innate immune responses to viral envelopes, including IFN expression. It has been shown that structural changes in the plasma membrane occur in cultured cells following IFN treatment (Chang et al. 1978; Chang, Jay, and Friedman 1978); this may also result in difficulty in obtaining patch clamp recordings. Furthermore, nucleic

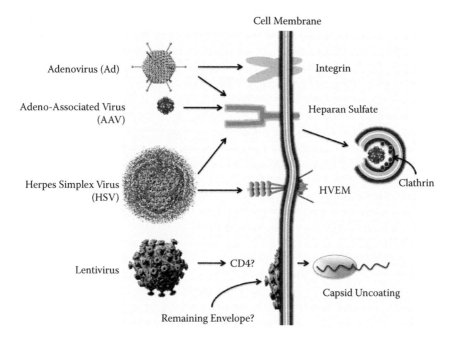

FIGURE 6.3   (see color insert following page 166) Diagram of viral vector entry into neuronal cells. Viral vector entry requires putative host receptors. Adenovirus and adeno-associated virus undergo clathrin-coated vesicular endocytosis, whereas herpes simplex virus and lentivirus release capsid into host cells. It appears that the remaining viral envelope glycoprotein on the neuronal cell membrane may negatively affect the ability to make electrophysiological recordings and disrupt normal membrane function.

acid from the viral genome, once endocytosed, can also activate TLR7 and 8 in the endosome (Beutler et al. 2007).

Thus, these problems of viral vector transduction may need to be addressed to permit electrophysiological assessment of target genes, and neuronal functional data obtained through any transmembrane gene transfer systems may need careful interpretation.

6.3   ADMINISTRATION ROUTES FOR *IN VIVO* GENE TRANSFER TO PRIMARY SENSORY NEURONS

Loosely placed chromic gut ligatures around the rat sciatic nerve have been shown to induce reductions of nociceptive thresholds that are representative of tactile or thermal allodynia and hyperalgesia, in addition to nocifensive behaviors that are representative of dysesthesia (Bennett and Xie 1988). These are all common symptoms of human neuropathy. Like most animal neuropathy models, this chronic constriction injury (CCI) shows sustained abnormal hyperexcitability and ectopic burst discharge of primary nociceptors, implicating these cells in neuropathic pathophysiology (Kajander and Bennett 1992). The CCI model requires minimal intervention to the nerve itself and thus may closely represent peripheral somatic nerve injuries.

However, variability in the degree of injury to the sciatic nerve has been reported in the CCI model, probably due to varying tightness of the ligatures and possible immune reaction to the chromic gut sutures. An alternative CCI model has been developed using fixed-diameter polyethlene cuffs that were loosely placed around the sciatic nerve (Mosconi and Kruger 1996). This sciatic nerve entrapment (SNE) model can induce more consistent morphometric changes to the affected nerves than the original CCI model, in addition to reproducing all behavioral symptomology. Varying the diameter of the cuffs can produce a wide range of changes in the fiber spectrum of the sciatic nerve distal to the injury site. However, the severity of change in the sciatic nerve fiber spectrum does not necessarily correlate with the presence of induced pain symptoms. Therefore, it has been postulated that there is a peripheral mechanism underlying neuropathic pain involving physical and/or biochemical alterations in the local microenvironment at the site of injury (Chung and Chung 2002; Wood et al. 2004; Hains and Waxman 2007).

Animal models have served to identify candidate targets for neuropathic pain treatment. For effective validation of molecular targets and mechanistic elucidation of the disease process, versatile methodologies for selective manipulation of sensory neuron gene expression *in vivo* must play a significant role. *In vivo* gene delivery to peripheral sensory neurons can be attained from four access points: the DRG itself, the sciatic nerve, the intrathecal space, or peripherally innervated tissue such as muscle and skin (Figure 6.4).

6.3.1 CHALLENGE 1: GENE TRANSFER TO DRG

The direct injection of viral vectors into DRG (Xu et al. 2003), sciatic nerve (Gu et al. 2005), or spinal cord (Maeda et al. 2008) has been shown to yield transgene

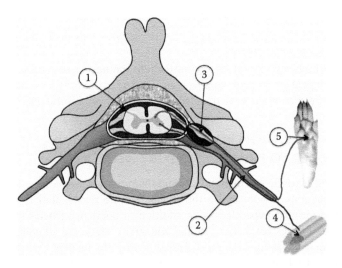

FIGURE 6.4 Diagram of *in vivo* gene transfer routes into DRG neurons: (1) intrathecal injection; (2) direct injection into sciatic nerve; (3) direct injection into DRG; (4) intramuscular injection; (5) subcutaneous injection.

expression in DRG neurons. For example, the injection of solutions containing viral vectors into the DRG is achieved through surgical exposure of lumbar L4 and L5 DRG by removing part of the vertebral bone. A small volume of viral solution is infused for approximately 20 minutes through a glass micropipette connected to a Hamilton syringe. Transduction efficiencies of AAV through direct injection to DRG or sciatic nerve are approximately 30% and 10%, respectively, in a DRG section at 1 week after injection; after 3 weeks, they reach 70% and 35%, respectively (Xu et al. 2003).

Intrathecal injection, especially by minimally invasive lumbar puncture, some-what ameliorates the potential problems associated with surgical exposure of target neuronal tissues (Beutler et al. 2005; Fairbanks 2003), while permitting effective delivery of various therapeutic agents to DRG neurons (Wang et al. 2005; Aldskogius and Kozlova 2002; Lai et al. 2002). However, transgene expression may occur in multiple spinal cord tissues due to spreading of injectate (Shi et al. 2003; Hocking and Wildsmith 2004; Stienstra and Veering 1998); specific transgene expression effi-ciency in DRG neurons has not been reported with this methodology. Intrathecal catheterization, either through the cervical or lumbar approach, has been frequently used for delivery to DRG or spinal cord neurons *in vivo* (Beutler et al. 2005; Xu et al. 2003). It must be noted, however, that atlanto-occipital catheterization in rodents has been associated with significant postsurgical mortality (3%–10%) (Storkson et al. 1996) and neurological morbidity (10%–20%) (Kristensen et al. 1993; Tsang et al. 1997).

6.3.2 CHALLENGE 2: RETROGRADE GENE TRANSFER THROUGH NEURONS USING VIRAL VECTORS

Retrograde delivery by peripheral intramuscular or subcutaneous injection presents the simplest and least invasive option, because it does not require surgery or body cavity penetration. It also offers the unique advantage of specific access to neurons innervating a particular muscle or region of skin (Katada et al. 2006; Glorioso, Mata, and Fink 2003; Puigdellivol-Sanchez et al. 2002).

Ad and AAV have both been used to retrogradely target spinal cord motoneu-rons after intramuscular injection (Baumgartner and Shine 1998; Pirozzi et al. 2006; Kaspar et al. 2003). However, intramuscular delivery to sensory neurons remains questionable, with no reports for AAV as well as one reported success (Ghadge et al. 1995) and one failure (Glatzel et al. 2000) for Ad. Poliovirus is an enterovirus with an innate affinity for motoneurons (Jackson et al. 2003). Intramuscular inoculation with a poliovirus vector gave rise to transient reporter gene expression in both spinal cord motoneurons and DRG sensory neurons that was partially attenuated within 5 days and undetectable within 2 weeks (Jackson et al. 2003). Gene delivery was less effective in adult animals than juveniles, most likely because of reduced poliovirus receptor expression in adult neurons (Jackson et al. 2003).

Lentiviral vectors are capable of stable gene expression in non-dividing cells but have little or no propensity for retrograde axonal transport (Mazarakis et al. 2001). To overcome this problem, equine infectious anemia virus (EIAV) and HIV have been pseudotyped with envelope glycoproteins of the rabies virus, which undergoes efficient retrograde axonal transport (Mazarakis et al. 2001; Mentis et al. 2006). Intramuscular injection of these vectors has resulted in effective motoneuron transduction (Mazarakis et al. 2001; Mentis et al. 2006); however, sensory neuron transduction has not been reported.

HSV, a human neurotropic virus that naturally infects sensory neurons after peripheral virion uptake and retrograde axonal transport, is so far the only viral vector to clearly show stable and efficient DRG gene delivery after non-invasive peripheral administration (Kennedy 1997). After inoculation of replication-defective HSV-1 on abraded rat hindpaw skin, reporter protein was detected in L3 DRG sensory neurons at 4 days (Wilson and Yeomans 2002).

6.3.3 Challenge 3: Non-Viral Vector Gene Transfer to the PNS *in Vivo*

DNA complexed with linear PEI has been reported to form polyplexes with an average diameter of 132 nm and a ζ-potential of 30 mV (Wightman et al. 2001). Wang et al. (2001) injected PEI/plasmid DNA polyplexes into the mouse tongue and observed transgene expression in hypoglossal motor neurons in the brain stem (Wang et al. 2001). This study demonstrated the possibility of retrograde gene transfer using a non-viral vector. We have investigated this possibility for the PNS and in particular for sensory neuron transfection. PEI/DNA polyplexes were injected subcutaneously into the rat hindpaw. We have observed transgene expression in the DRG, suggesting that PEI/DNA polyplexes could retrogradely transfer genes (Thakor et al. 2007). However, optimization of this vector failed due to substantial toxicity. Therefore, we investigated CG/DNA polyplex for gene transfer to DRG neurons in rats. The larger size and lower ζ-potential of the CG/DNA polyplexes suggested that CG/DNA polyplexes are less tightly packed than PEI/DNA polyplexes, which may allow easier disengagement of DNA from the carrier once inside the cell. Six days after CG/DNA injection into the hindpaw, reporter gene expression was determined in L4 and L5 DRG. The total number of GFP+ cells in L4 and L5 DRG was 1164 ± 176. It has been reported that the L4 and L5 DRG collectively contain between 10,500 and 11,000 ± 2,000 neurons (Swett et al. 1991). Therefore, subcutaneous CG/DNA injection is thought to have resulted in gene transfer to approximately 10% of L4 and L5 DRG neurons. However, it may be more relevant to calculate the CG/DNA transfection efficiency in DRG neurons specifically innervating the hindpaw injection site (Puigdellivol-Sanchez et al. 2002).

6.4 "LOSS- OR GAIN-OF-FUNCTION" EXPERIMENTS IN PERIPHERAL NEUROPATHY ANIMAL MODELS

To dissect the molecular mechanism of peripheral neuropathy, the function of candidate molecules may be specifically manipulated, and relevant outcomes such as the alleviation of neuropathy symptoms may be critically evaluated. Loss- or gainof-function models can be established at the genetic, post-transcriptional and posttranslational levels, thus providing novel opportunities to investigate the role of selected molecules in experimentally induced neuropathic pain models. Genetically modified mouse models have been commonly used for loss- or gain-of-function studies, either via interfering with the proper transcription of target genes by homologous recombination of mutated alleles, or by increasing the transcription of target genes with promoter-driven DNA inserts, respectively. Transcriptional regulation and dysregulation is an attractive hypothesis for disease onset; however, mutant mouse models have not yet fully explained the pathogenesis of peripheral neuropathy to date, in part because pathogenesis may also involve post-transcriptional and post-translational regulations.

6.4.1 CHALLENGE 1: POST-TRANSCRIPTIONAL REGULATION

After gene transcription, heterogeneous nuclear RNA undergoes a series of post-transcriptional modifications to become mature mRNA. Loss-of-function studies may be accomplished by inducing the degradation of target mRNA species using antisense oligonucleotide or RNAi. Antisense oligonucleotides comprise a short-length single-stranded DNA containing a sequence that is reverse and complementary to that of the target mRNA. When the oligonucleotide DNA hybridizes to the mRNA, RNaseH can degrade the DNA/RNA, contributing to the reduction of mRNA copies. Honore et al. (2002) applied antisense oligonucleotides against the P2X(3) receptor using continuous infusion via a mini-osmotic pump into the intrathecal space (Honore et al. 2002). This resulted in the reduction of complete Freund's adjuvant- and SNE-induced thermal hyperalgesia, but did not decrease carrageenan-induced hyperalgesia. Liu et al. (2005) used antisense oligonucleotides against NaV1.8 through intrathecal injection (twice a day for 5 days), which resulted in a 40% reduction of NaV1.8 mRNA levels in DRG (Liu et al. 2005). Mechanical allodynia induced by CCI was gradually decreased after antisense oligonucleotide treatment and reached significance at day 13. Christoph et al. (2007) applied antisense oligonucleotides against vanilloid receptor (TRPV1) mRNA to CCI-treated rats and found that after 5 days of continuous intrathecal injection, the protein level of TRPV1 was not altered in DRG and spinal cord tissues (Christoph et al. 2007). However, TRPV1 antisense oligonucleotide treatment reduced CCI-induced mechanical allodynia, as well as capsaicin-induced visceral pain symptoms.

RNAi is an endogenous system involving short, double-stranded RNAs that primarily regulates post-transcriptional gene silencing. Two species of such RNA molecules have been identified (small interfering RNA [siRNA] and micro RNA [miRNA]) that can initiate RNA-dependent gene silencing controlled by RNA-induced silencing complexes. Exogenously applied siRNA has been shown to induce targeted mRNA knockdown. DRG injection of cationic lipid/ephrinB2 siRNA complexes has been reported to decrease mechanical allodynia in rats with crushed spinal nerves (Kobayashi et al. 2007). Intrathecal application of siRNAs targeting P2X(3) (Dorn et al. 2004), TRPV1 (Christoph et al. 2006), and NaV1.8 (Dong et al. 2007) has also been shown to reduce neuropathic pain symptoms in rats.

Although most antisense oligonucleotides and siRNAs have been delivered through intrathecal injection, we have recently achieved siRNA-derived gene knockdown in DRG using non-invasive retrograde gene transfer in rats. We constructed a plasmid DNA containing U6 promoter-driven siRNA targeting NaV1.8 and CMV promoter-driven GFP. This plasmid DNA complexed with CG can be delivered by subcutaneous injection to the hindpaw, which is innervated by the peripheral axon terminals of L4/L5 DRG neurons (S. Ruangsri, A. Lin, I. Spigelman, and I. Nishimura,

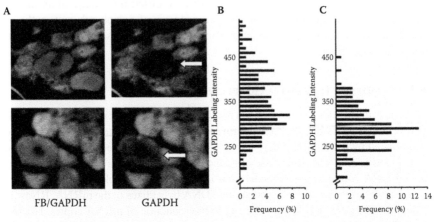

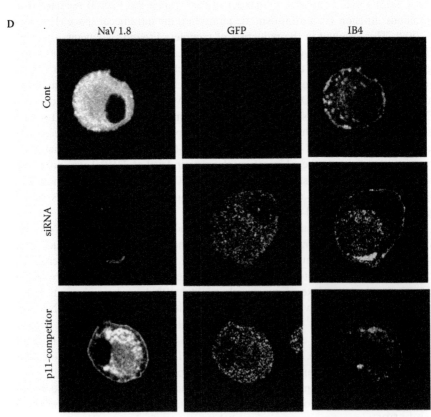

unpublished data). Acutely dissociated L4/L5 DRG neurons expressing GFP showed a significant reduction of NaV1.8 immunoreactivity (Figure 6.5).

Several investigations involving antisense oligonucleotides and siRNAs have shown a puzzling result; these strategies reduced neuropathic pain symptoms, but the level of targeted mRNA species and putative proteins within the DRG itself was not significantly affected (Christoph et al. 2007; Luo et al. 2005; Dong et al. 2007). These siRNAs were functionally active as they silenced the target mRNAs *in vitro*; thus it is possible that the observed *in vivo* effects may have occurred outside of the DRG itself and/or involved system-level post-transcriptional mechanisms such as active peripheral mRNA transport. Primary sensory neurons of DRG possess some of the longest axons among the neuraxis (Thakor et al. 2009). Neuronal protein synthesis has been believed to occur primarily in the somata; however, recent data increasingly support the hypothesis that axonal sites are capable of autonomous protein synthesis (Alvarez, Giuditta, and Koenig 2000; Giuditta et al. 2002). To facilitate this axonal protein translation, it is essential that mRNA synthesized in the nuclei of DRG somata be transported through the axon (Willis et al. 2005; Price et al. 2006; Bi et al. 2006); thus mRNAs transported in peripheral axons may represent additional targets for RNAi strategies. Future studies may examine post-transcriptional regulation including mRNA axonal transport as a potential pathological mechanism of peripheral neuropathy.

6.4.2 CHALLENGE 2: POST-TRANSLATIONAL REGULATION

After translation from mRNA, proteins may undergo post-translational modifications such as ubiquitin-associated degradation, phosphorylation, and intracellular trafficking. For example, NaV proteins bind to the Nedd4 family of protein ligases, which are subjected to endoplasmic reticulum-associated degradation (van Bemmelen et al. 2004; Fotia et al. 2004). This may result in the reduction of functional expression. After inflammation or after repeated noxious stimuli, the activation thresholds of peripheral terminals are decreased. This peripheral sensitization may be supplemented by phosphorylation of membrane-bound ion channels. For example, inflammatory cytokines are known to activate protein kinase A or C, leading to phosphorylation of NaV1.8 (Gold, Levine, and Correa 1998). However, the pathological role of post-translational modifications in neuropathic pain has yet to be fully established.

FIGURE 6.5 (see color insert following page 166) *In vivo* retrograde transfection of cationized gelatin-plasmid DNA polyplex. **A**: Representative L4 DRG sections 60 hours after unilateral injection of CG/pSilencer1.0U6-GAPDH polyplexes and fast blue (FB) injection. Left panels show sections immunolabeled for GAPDH (green); FB+ cells appear blue. Right panels show identical sections without FB overlay. ⇦ : FB+ cells showing GAPDH suppression. **B, C**: GAPDH immunoreactivity in individual FB+ cells of L4 DRG ipsilateral (**B**: n = 126 cells) and contralateral (**C**: n = 139 cells) to CG/pSilencer1.0U6-GAPDH injection. The data show a significant downward shift in GAPDH labeling intensity in ipsilateral FB+ cells after CG/pSilencer1.0U6-GAPDH injection (p = 0.028). (A–C are adapted from Thakor et al. 2007.) **D**: Acutely dissociated L4/L5 DRG neurons after injection of CG/NaV1.8 siRNA-GFP plasmid DNA polyplexes or CG/p11 competitor-GFP plasmid DNA polyplexes. (**A**, **B**, and **C**: Reprinted with permission from Thakor et al., 2007 *Molecular Therapy*, 15(12): 2124–31.)

Functional expression of ion channels further requires molecular assembly as well as intracellular trafficking. Annexin A2 light chain p11 has been shown to facilitate such intracellular trafficking for acid-sensing ion channel 1a (Donier et al. 2005), acid-sensitive potassium channel 1 (Girard et al. 2002), and NaV1.8 (Okuse et al. 2002). We have designed a short peptide that has a sequence with high affinity to p11 and constructed a CMV-promoter-driven expression vector with IRES GFP. The experimental peptide-expressing vector was retrogradely transfected *in vivo* to rat DRG using the CG/DNA system (A. Lin, S. Ruangsri, I. Spigelman, and I. Nishimura, unpublished data). GFP-positive cells were stained for NaV1.8 and showed an unusual cytoplasmic aggregation (Figure 6.5). This result suggests that designer peptides may interfere with functional expression of the target molecule at the post-translational level; this may present a novel therapeutic opportunity.

6.5 CONCLUSIONS

Molecular biological methods have been increasingly applied in investigations of peripheral neuropathy mechanisms. A major bottleneck has been the relative inefficiency of gene transfer to neuronal cells *in vitro* and *in vivo*. However, recently, new arrays of materials have been developed and applied to studies of the PNS. Putative roles of candidate molecules in the pathogenesis of peripheral neuropathy may be elucidated through these new experimental modalities.

ACKNOWLEDGMENTS

The authors' investigations were supported in part by NIH grants DE014573, NS049137, and DA023153; NSF IGERT DGE9972802; the Japan Society for the Promotion of Science; the Takanawakai; and the Clinical Implant Society of Japan. The investigation was conducted in part in a facility constructed with support from Research Facilities Improvement Program grant number C06 RR014529 from NCRR/NIH.

REFERENCES

Ahmad, S., L. Dahllund, A. B. Eriksson, D. Hellgren, U. Karlsson, P. E. Lund, I. A. Meijer, et al. 2007. A stop codon mutation in SCN9A causes lack of pain sensation. *Hum Mol Genet* 16 (17):2114–21.

Akhtar, J., V. Tiwari, M. J. Oh, M. Kovacs, A. Jani, S. K. Kovacs, T. Valyi-Nagy, and D. Shukla. 2008. HVEM and nectin-1 are the major mediators of herpes simplex virus 1 (HSV-1) entry into human conjunctival epithelium. *Invest Ophthalmol Vis Sci* 49 (9):4026–35.

Akinc, A., A. Zumbuehl, M. Goldberg, E. S. Leshchiner, V. Busini, N. Hossain, S. A. Bacallado, et al. 2008. A combinatorial library of lipid-like materials for delivery of RNAi therapeutics. *Nat Biotechnol* 26 (5):561–9.

Aldskogius, H., and E. N. Kozlova. 2002. Strategies for repair of the deafferented spinal cord. *Brain Res Brain Res Rev* 40 (1–3):301–8.

Alvarez, J., A. Giuditta, and E. Koenig. 2000. Protein synthesis in axons and terminals: significance for maintenance, plasticity and regulation of phenotype. With a critique of slow transport theory. *Prog Neurobiol* 62 (1):1–62.

Anderson, D. G., A. Akinc, N. Hossain, and R. Langer. 2005. Structure/property studies of polymeric gene delivery using a library of poly(beta-amino esters). *Mol Ther* 11 (3):426–34.

Baumgartner, B. J., and H. D. Shine. 1998. Neuroprotection of spinal motoneurons following targeted transduction with an adenoviral vector carrying the gene for glial cell line–derived neurotrophic factor. *Exp Neurol* 153 (1):102–12.

Bennett, G. J., and Y. K. Xie. 1988. A peripheral mononeuropathy in rat that produces disorders of pain sensation like those seen in man. *Pain* 33 (1):87–107.

Berry, M., L. Barrett, L. Seymour, A. Baird, and A. Logan. 2001. Gene therapy for central nervous system repair. *Curr Opin Mol Ther* 3 (4):338–49.

Beutler, A. S., M. S. Banck, C. E. Walsh, and E. D. Milligan. 2005. Intrathecal gene transfer by adeno-associated virus for pain. *Curr Opin Mol Ther* 7 (5):431–9.

Beutler, B., C. Eidenschenk, K. Crozat, J. L. Imler, O. Takeuchi, J. A. Hoffmann, and S. Akira. 2007. Genetic analysis of resistance to viral infection. *Nat Rev Immunol* 7 (10):753–66.

Bi, J., N. P. Tsai, Y. P. Lin, H. H. Loh, and L. N. Wei. 2006. Axonal mRNA transport and localized translational regulation of kappa-opioid receptor in primary neurons of dorsal root ganglia. *Proc Natl Acad Sci U S A* 103 (52):19919–24.

Boeckle, S., E. Wagner, and M. Ogris. 2005. C- versus N-terminally linked melittin-polyethylenimine conjugates: the site of linkage strongly influences activity of DNA polyplexes. *J Gene Med* 7 (10):1335–47.

Burns, J. C., T. Friedmann, W. Driever, M. Burrascano, and J. K. Yee. 1993. Vesicular stomatitis virus G glycoprotein pseudotyped retroviral vectors: concentration to very high titer and efficient gene transfer into mammalian and nonmammalian cells. *Proc Natl Acad Sci U S A* 90 (17):8033–7.

Chang, E. H., E. F. Grollman, F. T. Jay, G. Lee, L. D. Kohn, and R. M. Friedman. 1978. Membrane alterations following interferon treatment. *Adv Exp Med Biol* 110:85–99.

Chang, E. H., F. T. Jay, and R. M. Friedman. 1978. Physical, morphological, and biochemical alterations in the membrane of AKR mouse cells after interferon treatment. *Proc Natl Acad Sci U S A* 75 (4):1859–63.

Chen, H., T. J. Lamer, R. H. Rho, K. A. Marshall, B. T. Sitzman, S. M. Ghazi, and R. P. Brewer. 2004. Contemporary management of neuropathic pain for the primary care physician. *Mayo Clin Proc* 79 (12):1533–45.

Cheng, J., A. Nath, B. Knudsen, S. Hochman, J. D. Geiger, M. Ma, and D. S. Magnuson. 1998. Neuronal excitatory properties of human immunodeficiency virus type 1 Tat protein. *Neuroscience* 82 (1):97–106.

Christoph, T., C. Gillen, J. Mika, A. Grunweller, M. K. Schafer, K. Schiene, R. Frank, et al. 2007. Antinociceptive effect of antisense oligonucleotides against the vanilloid receptor VR1/TRPV1. *Neurochem Int* 50 (1):281–90.

Christoph, T., A. Grunweller, J. Mika, M. K. Schafer, E. J. Wade, E. Weihe, V. A. Erdmann, R. Frank, C. Gillen, and J. Kurreck. 2006. Silencing of vanilloid receptor TRPV1 by RNAi reduces neuropathic and visceral pain in vivo. *Biochem Biophys Res Commun* 350 (1):238–43.

Chung, J. M., and K. Chung. 2002. Importance of hyperexcitability of DRG neurons in neuropathic pain. *Pain Pract* 2 (2):87–97.

Cox, J. J., F. Reimann, A. K. Nicholas, G. Thornton, E. Roberts, K. Springell, G. Karbani, et al. 2006. An SCN9A channelopathy causes congenital inability to experience pain. *Nature* 444 (7121):894–8.

Cummins, T. R., S. D. Dib-Hajj, R. I. Herzog, and S. G. Waxman. 2005. Nav1.6 channels generate resurgent sodium currents in spinal sensory neurons. *FEBS Lett* 579 (10):2166–70.

Dijkhuizen, P. A., W. T. Hermens, M. A. Teunis, and J. Verhaagen. 1997. Adenoviral vector–directed expression of neurotrophin-3 in rat dorsal root ganglion explants results in a robust neurite outgrowth response. *J Neurobiol* 33 (2):172–84.

Dong, X. W., S. Goregoaker, H. Engler, X. Zhou, L. Mark, J. Crona, R. Terry, J. Hunter, and T. Priestley. 2007. Small interfering RNA-mediated selective knockdown of Na(V)1.8 tetrodotoxin-resistant sodium channel reverses mechanical allodynia in neuropathic rats. *Neuroscience* 146 (2):812–21.

Donier, E., F. Rugiero, K. Okuse, and J. N. Wood. 2005. Annexin II light chain p11 promotes functional expression of acid-sensing ion channel ASIC1a. *J Biol Chem* 280 (46):38666–72.

Dorn, G., S. Patel, G. Wotherspoon, M. Hemmings-Mieszczak, J. Barclay, F. J. Natt, P. Martin, et al. 2004. siRNA relieves chronic neuropathic pain. *Nucleic Acids Res* 32 (5):e49.

Drenth, J. P., R. H. te Morsche, G. Guillet, A. Taieb, R. L. Kirby, and J. B. Jansen. 2005. SCN9A mutations define primary erythermalgia as a neuropathic disorder of voltage gated sodium channels. *J Invest Dermatol* 124 (6):1333–8.

Dull, T., R. Zufferey, M. Kelly, R. J. Mandel, M. Nguyen, D. Trono, and L. Naldini. 1998. A third-generation lentivirus vector with a conditional packaging system. *J Virol* 72 (11):8463–71.

Escriou, V., M. Carriere, D. Scherman, and P. Wils. 2003. NLS bioconjugates for targeting therapeutic genes to the nucleus. *Adv Drug Deliv Rev* 55 (2):295–306.

Fairbanks, C. A. 2003. Spinal delivery of analgesics in experimental models of pain and analgesia. *Adv Drug Deliv Rev* 55 (8):1007–41.

Farkas, R. H., S. Nakajima, and Y. Nakajima. 1994. Cultured neurons infected with an HSV-1-derived vector remain electrically excitable and responsive to neurotransmitter. *Neurosci Lett* 165 (1–2):153–6.

Farley, D. C., S. Iqball, J. C. Smith, J. E. Miskin, S. M. Kingsman, and K. A. Mitrophanous. 2007. Factors that influence VSV-G pseudotyping and transduction efficiency of lentiviral vectors—in vitro and in vivo implications. *J Gene Med* 9 (5):345–56.

Fertleman, C. R., M. D. Baker, K. A. Parker, S. Moffatt, F. V. Elmslie, B. Abrahamsen, J. Ostman, et al. 2006. SCN9A mutations in paroxysmal extreme pain disorder: allelic variants underlie distinct channel defects and phenotypes. *Neuron* 52 (5):767–74.

Fleming, J., S. L. Ginn, R. P. Weinberger, T. N. Trahair, J. A. Smythe, and I. E. Alexander. 2001. Adeno-associated virus and lentivirus vectors mediate efficient and sustained transduction of cultured mouse and human dorsal root ganglia sensory neurons. *Hum Gene Ther* 12 (1):77–86.

Fotia, A. B., J. Ekberg, D. J. Adams, D. I. Cook, P. Poronnik, and S. Kumar. 2004. Regulation of neuronal voltage-gated sodium channels by the ubiquitin-protein ligases Nedd4 and Nedd4-2. *J Biol Chem* 279 (28):28930–5.

Fukuda, J., and T. Kurata. 1981. Loss of membrane excitability after herpes simplex virus infection in tissue-cultured nerve cells from adult mammals. *Brain Res* 211 (1):235–41.

Georgel, P., Z. Jiang, S. Kunz, E. Janssen, J. Mols, K. Hoebe, S. Bahram, M. B. Oldstone, and B. Beutler. 2007. Vesicular stomatitis virus glycoprotein G activates a specific antiviral Toll-like receptor 4-dependent pathway. *Virology* 362 (2):304–13.

Ghadge, G. D., R. P. Roos, U. J. Kang, R. Wollmann, P. S. Fishman, A. M. Kalynych, E. Barr, and J. M. Leiden. 1995. CNS gene delivery by retrograde transport of recombinant replication-defective adenoviruses. *Gene Ther* 2 (2):132–7.

Girard, C., N. Tinel, C. Terrenoire, G. Romey, M. Lazdunski, and M. Borsotto. 2002. p11, an annexin II subunit, an auxiliary protein associated with the background K+ channel, TASK-1. *Embo J* 21 (17):4439–48.

Giuditta, A., B. B. Kaplan, J. van Minnen, J. Alvarez, and E. Koenig. 2002. Axonal and presynaptic protein synthesis: new insights into the biology of the neuron. *Trends Neurosci* 25 (8):400–4.

Glatzel, M., E. Flechsig, B. Navarro, M. A. Klein, J. C. Paterna, H. Bueler, and A. Aguzzi. 2000. Adenoviral and adeno-associated viral transfer of genes to the peripheral nervous system. *Proc Natl Acad Sci U S A* 97 (1):442–7.

Glorioso, J. C., M. Mata, and D. J. Fink. 2003. Exploiting the neurotherapeutic potential of peptides: targeted delivery using HSV vectors. *Expert Opin Biol Ther* 3 (8):1233–9.

Goins, W. F., K. A. Lee, J. D. Cavalcoli, M. E. O'Malley, S. T. DeKosky, D. J. Fink, and J. C. Glorioso. 1999. Herpes simplex virus type 1 vector-mediated expression of nerve growth factor protects dorsal root ganglion neurons from peroxide toxicity. *J Virol* 73 (1):519–32.

Gold, M. S., J. D. Levine, and A. M. Correa. 1998. Modulation of TTX-R INa by PKC and PKA and their role in PGE2-induced sensitization of rat sensory neurons in vitro. *J Neurosci* 18 (24):10345–55.

Goldberg, Y. P., J. MacFarlane, M. L. MacDonald, J. Thompson, M. P. Dube, M. Mattice, R. Fraser, et al. 2007. Loss-of-function mutations in the Nav1.7 gene underlie congenital indifference to pain in multiple human populations. *Clin Genet* 71 (4):311–9.

Gu, Y., Y. Xu, G. W. Li, and L. Y. Huang. 2005. Remote nerve injection of mu opioid receptor adeno-associated viral vector increases antinociception of intrathecal morphine. *J Pain* 6 (7):447–54.

Hains, B. C., and S. G. Waxman. 2007. Sodium channel expression and the molecular pathophysiology of pain after SCI. *Prog Brain Res* 161:195–203.

Han, C., A. M. Rush, S. D. Dib-Hajj, S. Li, Z. Xu, Y. Wang, L. Tyrrell, X. Wang, Y. Yang, and S. G. Waxman. 2006. Sporadic onset of erythermalgia: a gain-of-function mutation in Nav1.7. *Ann Neurol* 59 (3):553–8.

Harouse, J. M., C. Kunsch, H. T. Hartle, M. A. Laughlin, J. A. Hoxie, B. Wigdahl, and F. Gonzalez-Scarano. 1989. CD4-independent infection of human neural cells by human immunodeficiency virus type 1. *J Virol* 63 (6):2527–33.

Hillsley, K., J. H. Lin, A. Stanisz, D. Grundy, J. Aerssens, P. J. Peeters, D. Moechars, B. Coulie, and R. H. Stead. 2006. Dissecting the role of sodium currents in visceral sensory neurons in a model of chronic hyperexcitability using Nav1.8 and Nav1.9 null mice. *J Physiol* 576 (Pt 1):257–67.

Hocking, G., and J. A. Wildsmith. 2004. Intrathecal drug spread. *Br J Anaesth* 93 (4):568–78.

Honore, P., K. Kage, J. Mikusa, A. T. Watt, J. F. Johnston, J. R. Wyatt, C. R. Faltynek, M. F. Jarvis, and K. Lynch. 2002. Analgesic profile of intrathecal P2X(3) antisense oligonucleotide treatment in chronic inflammatory and neuropathic pain states in rats. *Pain* 99 (1–2):11–9.

Jackson, C. A., J. Messinger, M. T. Palmer, J. D. Peduzzi, and C. D. Morrow. 2003. Gene expression in the muscle and central nervous system following intramuscular inoculation of encapsidated or naked poliovirus replicons. *Virology* 314 (1):45–61.

Jo, J., T. Ikai, A. Okazaki, K. Nagane, M. Yamamoto, Y. Hirano, and Y. Tabata. 2007. Expression profile of plasmid DNA obtained using spermine derivatives of pullulan with different molecular weights. *J Biomater Sci Polym Ed* 18 (7):883–99.

Jo, J., T. Ikai, A. Okazaki, M. Yamamoto, Y. Hirano, and Y. Tabata. 2007. Expression profile of plasmid DNA by spermine derivatives of pullulan with different extents of spermine introduced. *J Control Release* 118 (3):389–98.

Kajander, K. C., and G. J. Bennett. 1992. Onset of a painful peripheral neuropathy in rat: a partial and differential deafferentation and spontaneous discharge in A beta and A delta primary afferent neurons. *J Neurophysiol* 68 (3):734–44.

Kanatani, I., T. Ikai, A. Okazaki, J. Jo, M. Yamamoto, M. Imamura, A. Kanematsu, et al. 2006. Efficient gene transfer by pullulan-spermine occurs through both clathrin- and raft/caveolae-dependent mechanisms. *J Control Release* 116 (1):75–82.

Kaspar, B. K., J. Llado, N. Sherkat, J. D. Rothstein, and F. H. Gage. 2003. Retrograde viral delivery of IGF-1 prolongs survival in a mouse ALS model. *Science* 301 (5634):839–42.

Katada, A., J. D. Vos, B. B. Swelstad, and D. L. Zealear. 2006. A sequential double labeling technique for studying changes in motoneuronal projections to muscle following nerve injury and reinnervation. *J Neurosci Methods* 155 (1):20–7.

Kennedy, P. G. 1997. Potential use of herpes simplex virus (HSV) vectors for gene therapy of neurological disorders. *Brain* 120 (Pt 7):1245–59.

Kim, S. W., T. Ogawa, Y. Tabata, and I. Nishimura. 2004. Efficacy and cytotoxicity of cationic-agent-mediated nonviral gene transfer into osteoblasts. *J Biomed Mater Res A* 71 (2):308–15.

Kloeckner, J., S. Boeckle, D. Persson, W. Roedl, M. Ogris, K. Berg, and E. Wagner. 2006. DNA polyplexes based on degradable oligoethylenimine-derivatives: combination with EGF receptor targeting and endosomal release functions. *J Control Release* 116 (2):115–22.

Kobayashi, H., T. Kitamura, M. Sekiguchi, M. K. Homma, Y. Kabuyama, S. Konno, S. Kikuchi, and Y. Homma. 2007. Involvement of EphB1 receptor/EphrinB2 ligand in neuropathic pain. *Spine* 32 (15):1592–8.

Koya, R. C., N. Kasahara, V. Pullarkat, A. M. Levine, and R. Stripecke. 2002. Transduction of acute myeloid leukemia cells with third generation self-inactivating lentiviral vectors expressing CD80 and GM-CSF: effects on proliferation, differentiation, and stimulation of allogeneic and autologous anti-leukemia immune responses. *Leukemia* 16 (9):1645–54.

Kristensen, J. D., C. Post, T. Gordh, Jr., and B. A. Svensson. 1993. Spinal cord morphology and antinociception after chronic intrathecal administration of excitatory amino acid antagonists in the rat. *Pain* 54 (3):309–16.

Kunsch, C., H. T. Hartle, and B. Wigdahl. 1989. Infection of human fetal dorsal root ganglion glial cells with human immunodeficiency virus type 1 involves an entry mechanism independent of the CD4 T4A epitope. *J Virol* 63 (12):5054–61.

Lai, J., M. S. Gold, C. S. Kim, D. Bian, M. H. Ossipov, J. C. Hunter, and F. Porreca. 2002. Inhibition of neuropathic pain by decreased expression of the tetrodotoxin-resistant sodium channel, NaV1.8. *Pain* 95 (1–2):143–52.

Laird, J. M., V. Souslova, J. N. Wood, and F. Cervero. 2002. Deficits in visceral pain and referred hyperalgesia in Nav1.8 (SNS/PN3)-null mice. *J Neurosci* 22 (19):8352–6.

Leathers, T. D. 2003. Biotechnological production and applications of pullulan. *Appl Microbiol Biotechnol* 62 (5–6):468–73.

Liu, Y., S. Yao, W. Song, Y. Wang, D. Liu, and L. Zen. 2005. Effects of intrathecally administerd NaV1.8 antisense oligonucleotide on the expression of sodium channel mRNA in dorsal root ganglion. *J Huazhong Univ Sci Technolog Med Sci* 25 (6):696–9.

Luo, M. C., D. Q. Zhang, S. W. Ma, Y. Y. Huang, S. J. Shuster, F. Porreca, and J. Lai. 2005. An efficient intrathecal delivery of small interfering RNA to the spinal cord and peripheral neurons. *Mol Pain* 1:29.

Maeda, S., A. Kawamoto, Y. Yatani, H. Shirakawa, T. Nakagawa, and S. Kaneko. 2008. Gene transfer of GLT-1, a glial glutamate transporter, into the spinal cord by recombinant adenovirus attenuates inflammatory and neuropathic pain in rats. *Mol Pain* 4:65.

Maehlen, J., P. Wallen, A. Love, E. Norrby, and K. Kristensson. 1991. Paramyxovirus infections alter certain functional properties in cultured sensory neurons. *Brain Res* 540 (1–2):123–30.

Mazarakis, N. D., M. Azzouz, J. B. Rohll, F. M. Ellard, F. J. Wilkes, A. L. Olsen, E. E. Carter, et al. 2001. Rabies virus glycoprotein pseudotyping of lentiviral vectors enables retrograde axonal transport and access to the nervous system after peripheral delivery. *Hum Mol Genet* 10 (19):2109–21.

Melli, G., S. C. Keswani, A. Fischer, W. Chen, and A. Hoke. 2006. Spatially distinct and functionally independent mechanisms of axonal degeneration in a model of HIV-associated sensory neuropathy. *Brain* 129 (Pt 5):1330–8.

Mentis, G. Z., M. Gravell, R. Hamilton, N. A. Shneider, M. J. O'Donovan, and M. Schubert. 2006. Transduction of motor neurons and muscle fibers by intramuscular injection of HIV-1-based vectors pseudotyped with select rabies virus glycoproteins. *J Neurosci Methods* 157 (2):208–17.

Mikami, M., and J. Yang. 2005. Short hairpin RNA-mediated selective knockdown of NaV1.8 tetrodotoxin-resistant voltage-gated sodium channel in dorsal root ganglion neurons. *Anesthesiology* 103 (4):828–36.

Mosconi, T., and L. Kruger. 1996. Fixed-diameter polyethylene cuffs applied to the rat sciatic nerve induce a painful neuropathy: ultrastructural morphometric analysis of axonal alterations. *Pain* 64 (1):37–57.

Nassar, M. A., M. D. Baker, A. Levato, R. Ingram, G. Mallucci, S. B. McMahon, and J. N. Wood. 2006. Nerve injury induces robust allodynia and ectopic discharges in Nav1.3 null mutant mice. *Mol Pain* 2:33.

Nassar, M. A., L. C. Stirling, G. Forlani, M. D. Baker, E. A. Matthews, A. H. Dickenson, and J. N. Wood. 2004. Nociceptor-specific gene deletion reveals a major role for Nav1.7 (PN1) in acute and inflammatory pain. *Proc Natl Acad Sci U S A* 101 (34):12706–11.

Niederberger, E., and G. Geisslinger. 2008. Proteomics in neuropathic pain research. *Anesthesiology* 108 (2):314–23.

Okuse, K., M. Malik-Hall, M. D. Baker, W. Y. Poon, H. Kong, M. V. Chao, and J. N. Wood. 2002. Annexin II light chain regulates sensory neuron-specific sodium channel expression. *Nature* 417 (6889):653–6.

Pham, H. M., E. R. Arganaraz, B. Groschel, D. Trono, and J. Lama. 2004. Lentiviral vectors interfering with virus-induced CD4 down-modulation potently block human immunodeficiency virus type 1 replication in primary lymphocytes. *J Virol* 78 (23):13072–81.

Pirozzi, M., A. Quattrini, G. Andolfi, G. Dina, M. C. Malaguti, A. Auricchio, and E. I. Rugarli. 2006. Intramuscular viral delivery of paraplegin rescues peripheral axonopathy in a model of hereditary spastic paraplegia. *J Clin Invest* 116 (1):202–8.

Price, T. J., C. M. Flores, F. Cervero, and K. M. Hargreaves. 2006. The RNA binding and transport proteins staufen and fragile X mental retardation protein are expressed by rat primary afferent neurons and localize to peripheral and central axons. *Neuroscience* 141 (4):2107–16.

Puigdellivol-Sanchez, A., A. Valero-Cabre, A. Prats-Galino, X. Navarro, and C. Molander. 2002. On the use of fast blue, fluoro-gold and diamidino yellow for retrograde tracing after peripheral nerve injury: uptake, fading, dye interactions, and toxicity. *J Neurosci Methods* 115 (2):115–27.

Rolland, A. 2006. Nuclear gene delivery: the Trojan horse approach. *Expert Opin Drug Deliv* 3 (1):1–10.

Roza, C., J. M. Laird, V. Souslova, J. N. Wood, and F. Cervero. 2003. The tetrodotoxin-resistant Na+ channel Nav1.8 is essential for the expression of spontaneous activity in damaged sensory axons of mice. *J Physiol* 550 (Pt 3):921–6.

Shi, L., G. P. Tang, S. J. Gao, Y. X. Ma, B. H. Liu, Y. Li, J. M. Zeng, Y. K. Ng, K. W. Leong, and S. Wang. 2003. Repeated intrathecal administration of plasmid DNA complexed with polyethylene glycol-grafted polyethylenimine led to prolonged transgene expression in the spinal cord. *Gene Ther* 10 (14):1179–88.

Smith, G. M., and M. I. Romero. 1999. Adenoviral-mediated gene transfer to enhance neuronal survival, growth, and regeneration. *J Neurosci Res* 55 (2):147–57.

Spigelman, I., M. S. Gold, and A.R. Light. 2001. Electrophysiological recording techniques in pain research. In *Methods in pain research*, edited by L. Kruger. New York: CRC Press.

Stienstra, R., and B. T. Veering. 1998. Intrathecal drug spread: is it controllable? *Reg Anesth Pain Med* 23 (4):347–51; discussion 384–7.

Stirling, L. C., G. Forlani, M. D. Baker, J. N. Wood, E. A. Matthews, A. H. Dickenson, and M. A. Nassar. 2005. Nociceptor-specific gene deletion using heterozygous NaV1.8-Cre recombinase mice. *Pain* 113 (1–2):27–36.

Storkson, R. V., A. Kjorsvik, A. Tjolsen, and K. Hole. 1996. Lumbar catheterization of the spinal subarachnoid space in the rat. *J Neurosci Methods* 65 (2):167–72.

Stripecke, R., R. C. Koya, H. Q. Ta, N. Kasahara, and A. M. Levine. 2003. The use of lentiviral vectors in gene therapy of leukemia: combinatorial gene delivery of immunomodulators into leukemia cells by state-of-the-art vectors. *Blood Cells Mol Dis* 31 (1):28–37.

Summerford, C., and R. J. Samulski. 1998. Membrane-associated heparan sulfate proteoglycan is a receptor for adeno-associated virus type 2 virions. *J Virol* 72 (2):1438–45.

Swett, J. E., Y. Torigoe, V. R. Elie, C. M. Bourassa, and P. G. Miller. 1991. Sensory neurons of the rat sciatic nerve. *Exp Neurol* 114 (1):82–103.

Tabor, C.W., and H. Tabor. 1984. Polyamines. *Annu Rev Biochem* 53:749–90.

Teissier, E., and E. I. Pecheur. 2007. Lipids as modulators of membrane fusion mediated by viral fusion proteins. *Eur Biophys J* 36 (8):887–99.

Thakor, D. K., A. Lin, Y. Matsuka, E. M. Meyer, S. Ruangsri, I. Nishimura, and I. Spigelman. 2009. Increased peripheral nerve excitability and local NaV1.8 mRNA upregulation in painful neuropathy. *Mol Pain* In press.

Thakor, D. K., Y. D. Teng, and Y. Tabata. 2009. Neuronal gene delivery by negatively charged pullulan-spermine/DNA anioplexes. *Biomaterials* 30 (9):1815–26.

Thakor, D., I. Spigelman, Y. Tabata, and I. Nishimura. 2007. Subcutaneous peripheral injection of cationized gelatin/DNA polyplexes as a platform for non-viral gene transfer to sensory neurons. *Mol Ther* 15 (12):2124–31.

Tsang, B. K., Z. He, T. Ma, I. K. Ho, and J. H. Eichhorn. 1997. Decreased paralysis and better motor coordination with microspinal versus PE10 intrathecal catheters in pain study rats. *Anesth Analg* 84 (3):591–4.

van Bemmelen, M. X., J. S. Rougier, B. Gavillet, F. Apotheloz, D. Daidie, M. Tateyama, I. Rivolta, et al. Cardiac voltage-gated sodium channel Nav1.5 is regulated by Nedd4-2 mediated ubiquitination. *Circ Res* 95 (3):284–91.

van der Aa, M. A., E. Mastrobattista, R. S. Oosting, W. E. Hennink, G. A. Koning, and D. J. Crommelin. 2006. The nuclear pore complex: the gateway to successful nonviral gene delivery. *Pharm Res* 23 (3):447–59.

Walwyn, W. M., Y. Matsuka, D. Arai, D. C. Bloom, H. Lam, C. Tran, I. Spigelman, and N. T. Maidment. 2006. HSV-1-mediated NGF delivery delays nociceptive deficits in a genetic model of diabetic neuropathy. *Exp Neurol* 198 (1):260–70.

Wang, S., N. Ma, S. J. Gao, H. Yu, and K. W. Leong. 2001. Transgene expression in the brain stem effected by intramuscular injection of polyethylenimine/DNA complexes. *Mol Ther* 3 (5 Pt 1):658–64.

Wang, X., C. Wang, J. Zeng, X. Xu, P. Y. Hwang, W. C. Yee, Y. K. Ng, and S. Wang. 2005. Gene transfer to dorsal root ganglia by intrathecal injection: effects on regeneration of peripheral nerves. *Mol Ther* 12 (2):314–20.

Wightman, L., R. Kircheis, V. Rossler, S. Carotta, R. Ruzicka, M. Kursa, and E. Wagner. 2001. Different behavior of branched and linear polyethylenimine for gene delivery in vitro and in vivo. *J Gene Med* 3 (4):362–72.

Willis, D., K. W. Li, J. Q. Zheng, J. H. Chang, A. Smit, T. Kelly, T. T. Merianda, J. Sylvester, J. van Minnen, and J. L. Twiss. 2005. Differential transport and local translation of cytoskeletal, injury-response, and neurodegeneration protein mRNAs in axons. *J Neurosci* 25 (4):778–91.

Wilson, S. P., and D. C. Yeomans. 2002. Virally mediated delivery of enkephalin and other neuropeptide transgenes in experimental pain models. *Ann N Y Acad Sci* 971:515–21.

Wood, J. N., J. P. Boorman, K. Okuse, and M. D. Baker. 2004. Voltage-gated sodium channels and pain pathways. *J Neurobiol* 61 (1):55–71.

Xu, Y., Y. Gu, P. Wu, G. W. Li, and L. Y. Huang. 2003. Efficiencies of transgene expression in nociceptive neurons through different routes of delivery of adeno-associated viral vectors. *Hum Gene Ther* 14 (9):897–906.

Yang, Y., Y. Wang, S. Li, Z. Xu, H. Li, L. Ma, J. Fan, et al. 2004. Mutations in SCN9A, encoding a sodium channel alpha subunit, in patients with primary erythermalgia. *J Med Genet* 41 (3):171–4.

Yiangou, Y., P. Facer, I. P. Chessell, C. Bountra, C. Chan, C. Fertleman, V. Smith, and P. Anand. 2007. Voltage-gated ion channel Nav1.7 innervation in patients with idiopathic rectal hypersensitivity and paroxysmal extreme pain disorder (familial rectal pain). *Neurosci Lett* 427 (2):77–82.

Zhou, J., J. W. Yockman, S. W. Kim, and S. E. Kern. 2007. Intracellular kinetics of non-viral gene delivery using polyethylenimine carriers. *Pharm Res* 24 (6):1079–87.

Zimmermann, M. 2001. Pathobiology of neuropathic pain. *Eur J Pharmacol* 429 (1–3):23–37.

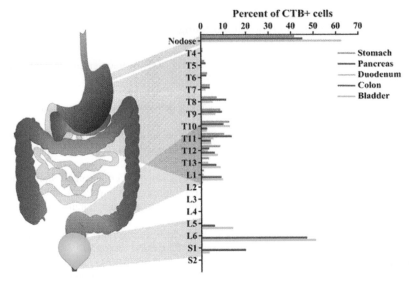

FIGURE 3.1 Vertebral distribution of visceral afferents innervating different organs. Sensory neurons innervating visceral structures in the mouse were retrogradely labeled using Alexa fluor-conjugated cholera toxin β (CTB); and the distribution of afferents innervating the stomach (Zhong et al. 2008), pancreas (Fasanella et al. 2008), duodenum (Zhong et al. 2008), colon (Christianson et al. 2006a, 2007), or bladder (Christianson et al. 2007) in nodose or dorsal root ganglia (DRG) in studies from our laboratories are expressed as a percentage of total CTB-positive afferents in the above graph.

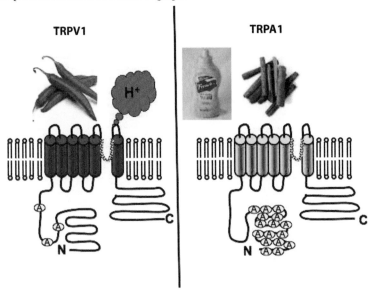

FIGURE 3.3 Structure and ligands for TRPV1 and TRPA1. Both TRPA1 and TRPV1 are non-selective cation channels and have six transmembrane domains with the pore-forming region contained in a loop between the fifth and sixth transmembrane domain. TRPV1 respond to exogenous compounds like the active ingredient in chili peppers (capsaicin) and endogenous molecules like protons. TRPA1 binds a number of exogenous pungent compounds like isothiocyanate (mustard oil) and cinnamonaldehyde (found in cinnamon). It also binds a large number of derivatives of membrane lipids (see text).

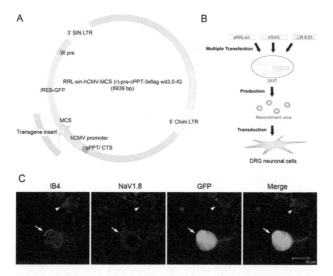

FIGURE 6.2 Cultured DRG neurons transduced with GFP-expressing lentivirus vector. **A:** Third-generation lentiviral vector map (pRRLsinhCMV). Central polypurine tract (cPPT) and the central termination sequence (CTS) are located upstream of the CMV promoter, which enhances the nuclear import efficiency of the DNA plasmid. In this vector, immediately downstream of the Multiple Cloning Site (MCS) are the Internal Ribosome Entry Site (IRES) and the Green Fluorescent Protein (GFP) gene, which allows the co-expression of GFP with transgene. The promoter and enhancer elements of the lentiviral genome has been deleted at the U3 region of the long terminal repeat (LTR), therefore becoming self-inactivating (SIN). **B:** Flow chart of lentivirus production. **C:** Rat DRG culture was treated with lentivirus vector. The transduction rate assessed by GFP generally yields 50%~60% for IB4 positive small diameter neuronal cells. The expression pattern of endogenous molecules such as NaV1.8 was not affected by lentivirus vectors.

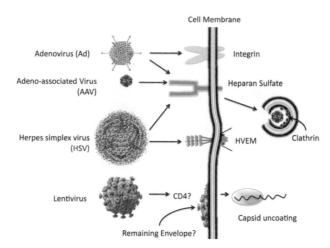

FIGURE 6.3 Diagram of viral vector entry into neuronal cells. Viral vector entry requires putative host receptors. Adenovirus and adeno-associated virus undergo clathrin-coated vesicular endocytosis, whereas herpes simplex virus and lentivirus release capsid into host cells. It appears that the remaining viral envelope glycoprotein on the neuronal cell membrane may negatively affect the ability to make electrophysiological recordings and disrupt normal membrane function.

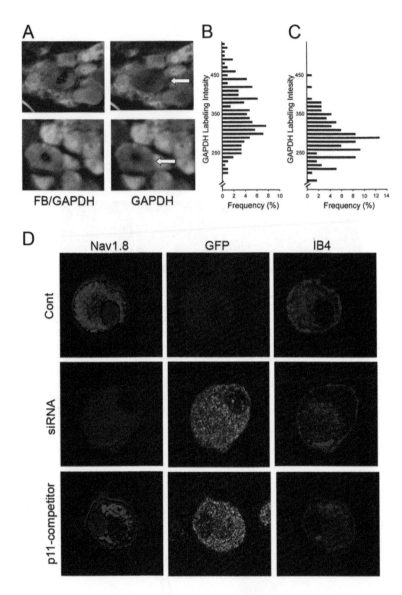

FIGURE 6.5 *In vivo* retrograde transfection of cationized gelatin-plasmid DNA polyplex. **A**: Representative L4 DRG sections 60 hours after unilateral injection of CG/pSilencer1.0U6-GAPDH polyplexes and fast blue (FB) injection. Left panels show sections immunolabeled for GAPDH (green); FB+ cells appear blue. Right panels show identical sections without FB overlay. ⇐ : FB+ cells showing GAPDH suppression. **B, C**: GAPDH immunoreactivity in individual FB+ cells of L4 DRG ipsilateral (**B**: n = 126 cells) and contralateral (**C**: n = 139 cells) to CG/pSilencer1.0U6-GAPDH injection. The data show a significant downward shift in GAPDH labeling intensity in ipsilateral FB+ cells after CG/pSilencer1.0U6-GAPDH injection (p=0.028). A–C are adapted from (Thakor et al. 2007). **D**: Acutely dissociated L4/L5 DRG neurons after injection of CG/Nav1.8 siRNA-GFP plasmid DNA polyplexes or CG/p11 competitor-GFP plasmid DNA polyplexes. (**A**, **B**, and **C**: Reprinted with permission from Thakor et al., Molecular Therapy, 2007;15(12): 2124–31.)

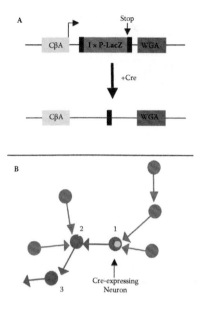

FIGURE 7.1 Genetic approach to establish Cre-dependent expression of the transneuronal tracer wheat germ agglutinin (WGA) in neurochemically distinct neuronal subsets. See text for explanations.

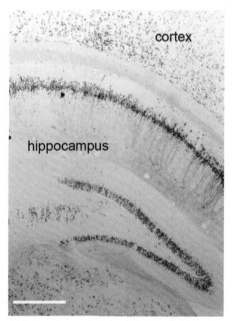

FIGURE 7.2 The ZW transgene is expressed throughout the CNS. This section is stained for expression of the floxed lacZ transgene (blue). Although we anticipated ubiquitous expression of the transgene, because we used the strong, chicken beta-actin promoter, random insertion of the ZW transgene nevertheless still resulted in a mosaic expression in the CNS. Scale bar 100 μm.

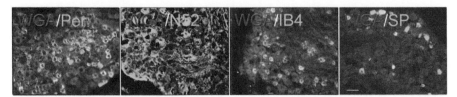

FIGURE 7.3 Preferential expression of WGA in dorsal root ganglion cells at the origin of unmyelinated (C) fibers of the nonpeptidergic class. We crossed the ZW mice with others that express Cre in nociceptors that express the Nav1.8 subtype of voltage-gated Na channel (Stirling et al. 2005). In double transgenic (N-ZW) animals, most WGA-positive DRG neurons co-stain for the neurofilament peripherin (Per) and bind the isolectin IB4 (yellow denotes double-labeled neurons). In contrast, very few DRG neurons express N52 (a marker of neurons with myelinated axons) or immunostaining with antibodies against the peptide, substance P (SP). Scale bar 100 μm.

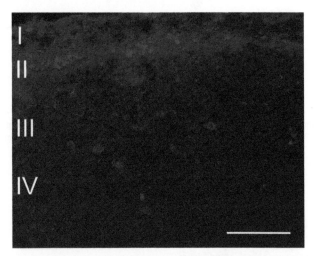

FIGURE 7.4 Anterograde and transneuronal transfer of WGA to the dorsal horn of the spinal cord in the adult N-ZW mouse. In double transgenic N-ZW animals, in addition to a dense band of labeled terminals in lamina IIi, we detected the WGA tracer (red) in large numbers of postsynaptic neurons located in laminae III–V of the dorsal horn. Scale bar 100 μm.

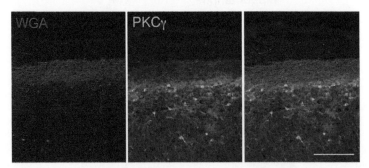

FIGURE 7.5 PKCγ-expressing interneurons of inner lamina II do not receive inputs from nonpeptidergic primary sensory neurons. In double-transgenic N-ZW animals, we found that most WGA+ afferents terminate (red) dorsal to the band of PKCγ interneurons (green), overlapping the termination zone of the IB4 positive afferents. Scale bar 100 μm.

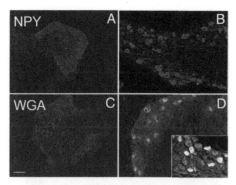

FIGURE 7.7 Peripheral nerve injury–induced expression of the transneuronal tracer in DRG neurons. We crossed the ZW mice with others that express the Cre recombinase in neurons that express neuropeptide Y (NPY-Cre; DeFalco et al. 2001). Under normal conditions, there is no expression of NPY (A) or WGA (C) in DRG of double transgenic NPY-ZW mice. However, a sciatic nerve transection induces expression of NPY (and thus Cre) (B), which in turn induces expression of WGA (D) in myelinated (N52+; yellow) sensory neurons (inset in D). Scale bar 100 μm.

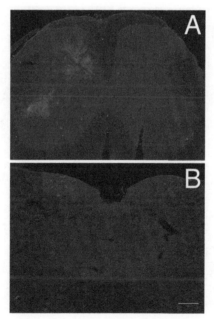

FIGURE 7.8 Anterograde and transneuronal transfer of WGA to postsynaptic neurons located in deep laminae of the spinal cord (A) and in the dorsal column nuclei (B) in NPY-ZW mice. In NPY-ZW animals, most WGA+ afferents terminate in the medial part of deep laminae of the spinal cord and in the nucleus gracilis ipsilateral to the sciatic nerve section, which is consistent with the projection pattern of myelinated sensory neurons. Even after long survival times, we did not see an expansion of the WGA terminal field into the superficial dorsal horn, indicating that there is limited if any sprouting of these afferents after nerve injury. Scale bar 100 μm in A, 50 μm in B.

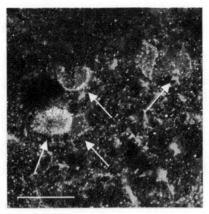

FIGURE 7.9 PKCγ-expressing interneurons of inner lamina II receive inputs from myelinated sensory neurons (transneuronal transport in the NPY-ZW mouse). The presence of WGA (red) in PKCγ-expressing interneurons (green) indicates that these interneurons have taken up the tracer after its release from myelinated primary afferents. Yellow fluorescence illustrates double-labeled neurons. Scale bar 30 μm.

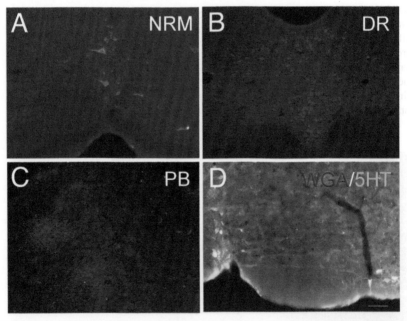

FIGURE 7.10 Triggering WGA expression in bulbospinal serotonergic neurons. We crossed the ZW mice with others that express the Cre recombinase exclusively in 5HT neurons of the brainstem. This was achieved using the ePet1-Cre mice (Scott et al. 2005). In double-transgenic mice (ePet-ZW), the WGA (red) is expressed in all raphe nuclei, including the nucleus raphe magnus (NRM, A) and the dorsal raphe (DR, B). Anterograde transport of WGA was also detected in axonal terminals in various brainstem nuclei, including the parabrachial nuclei (PB, C). Both 5HT (green) and non-5HT neurons within the raphe magnus and pallidus (D) contain the WGA tracer. Scale bar 100 μm in A, B and D; 50 μm in C.

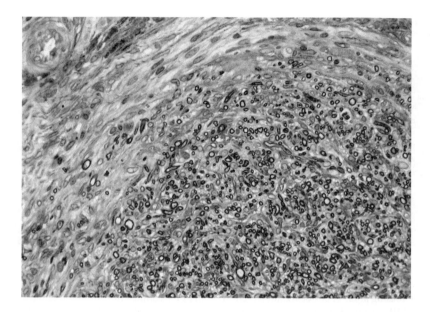

FIGURE 8.1 Sciatic nerve following transection injury. Plastic-embedded section of rat sciatic nerve stained with Methylene Blue Azure II. Note the lack of a defining perineurial sheath surrounding the nerve fibers and the disorganized interstitial structure. This abnormal environment is a consequence of the tissue injury, which has compromised the integrity of the endoneurial compartment and in so doing has altered the chemistry of the fluids bathing the nerve fibers. Tissue injury produces an inflammatory environment that can be modeled as an "inflammatory soup" of chemicals that are in part regulated by cytokines, and that cause spontaneous neurophysiological activity that activates nociceptive pathways.

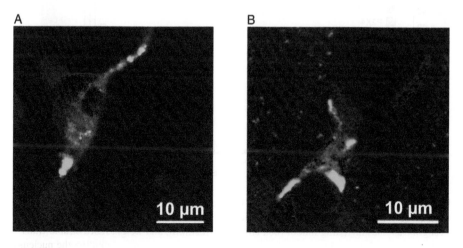

FIGURE 9.3 CBR2 are expressed in perivascular microglia (A, ED2/CD163) and microglia (B, Iba-1). Representative confocal images of CBR2 (red), ED2/CD163 (green in A), Iba-1 (green in B) and DAPI (blue) staining of the dorsal horn L5 spinal cord ipsilateral to L5 nerve transection (4 days after surgery).

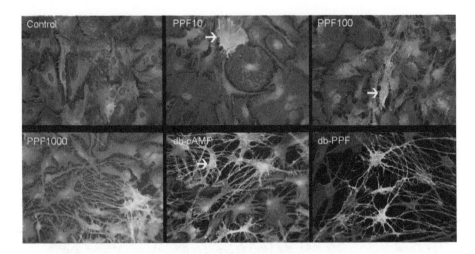

FIGURE 9.4 Seven-day treatment with PPF enhances GLT-1 expression in cultured astro-cytes. Double-label immunocytochemical localization of GLT-1 (green) and GFAP (red). Note the increasing areas of co-localization (yellow) with PPF treatment that mimics that seen with db-cAMP. Arrows identify astrocytes that are low GFAP, high GLT-1 expressing. PPF10, 100, 1000: PPF-treated 10, 100, or 1000 M; db-cAMP: 250 M db-cAMP; db-PPF: 250 M db-cAMP plus 1000 M PPF. (Reprinted with permission from *Glia* (2006) 54(3):193–203.)

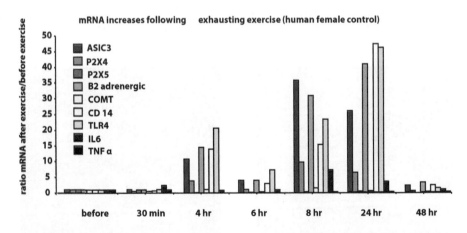

FIGURE 11.9 mRNA increases in human leukocytes at times indicated after 45 minutes of exhausting exercise. All data are from one female subject. Exercise consisted of aerobic and strength exercises including indoor rock climbing, treadmill, and weight machines.

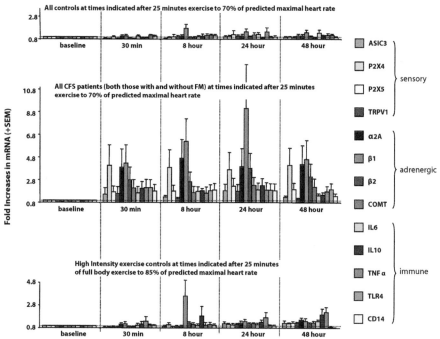

FIGURE 11.10 mRNA increases in CFS patients after 25 minutes of moderate exercise (70% of age adjusted maximum heart rate) (middle) but not in control subjects (top). Much more intense exercise (85% of age-adjusted maximum heart rate) in control subjects (bottom) increases mRNA for only a few of the genes. Faint horizontal line near the bottom of each panel indicates the baseline levels.

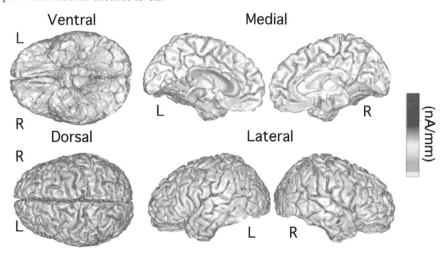

FIGURE 13.2 Example of localization of theta activity in a patient with brachial plexus avulsion pain. Activity is localized to the contralateral somatosensory cortex and bilaterally to the mesial orbitofrontal cortices. (Modified from Schulman J.J., Zonenshayn M., Ramirez R.R., Ribary U., and Llinas R. 2005. Thalamocortical dysrhythmia syndrome: MEG imaging of neuropathic pain. *Thalamus and Related Systems* 3: 33–39.

7 Transgenic Mouse Models for the Tracing of "Pain" Pathways

Allan I. Basbaum and João M. Bráz

CONTENTS

7.1 INTRODUCTION

The traditional, textbook view of the "pain" pathway illustrates an unmyelinated primary afferent C-fiber, the nociceptor, contacting a second-order dorsal horn neuron at the origin of the spinothalamic and spinoreticular pathways. Although the ultimate cortical target of these different pathways is unclear, there is no question that a better understanding of the mechanisms through which noxious stimuli produce pain requires a better understanding of these circuits. The limitations of our knowledge, of course, go beyond the need to identify cortical targets. We recognize now that there are neurochemically and physiologically distinct populations of afferents, projection neurons, and diverse central nervous system (CNS) targets.

In fact, even the classification of nociceptors into *peptidergic* and *nonpeptidergic* categories is oversimplified (Snider and McMahon 1998). Thus, an array of transient receptor potential (TRP) channels, which respond to different temperatures, natural products, or environmental irritants, establishes subcategories of nociceptors, as do the various Na+ channel subtypes (Caterina and Julius 1999; McCleskey and Gold 1999; Cummins et al. 2007). Even the rather broad categorization of myelinated versus unmyelinated nociceptor is but a first approximation to the diversity of afferent fibers that transmit "pain" messages (Talavera et al. 2008). Furthermore, the spinal cord is also far more complicated and contains various classes of projection neurons (Todd 2002; Morris et al. 2004; Klop et al. 2005), which not only are differentially distributed in the gray matter (e.g., laminae I, V, VII, and X) but also differ in the selectivity of their responses to non-noxious and noxious stimuli, in their receptive field sizes, and in their central targets.

What is still not clear, however, is the extent to which there are unique functional correlates of these neurochemically distinct populations of neurons along the pain pathway. For example, it is still not clear to what extent distinct classes of afferents differ in the types of pain provoked by their activation. Of particular interest is the differential contribution of neurons of laminae I and V to nociceptive processing. Some groups argue that only the lamina I neurons are essential for the highly selective discriminative aspect of the pain experience, and that the wide-dynamic-range neurons of lamina V are primarily contributors to sensorimotor integration (Craig 2004). Others argue for an essential contribution of the lamina V neurons (Price et al. 2003; Martin et al. 2004; Mazario and Basbaum 2007). Also unknown is the extent to which subpopulations of dorsal root and trigeminal ganglion (DRG and TG) neurons feed into these sensory-discriminative and limbic/emotional processing regions of the brain, especially in light of the relatively recent discovery of major spinohypothalamic (Burstein et al. 1987; Giesler et al. 1994) and spinoparabrachial-amygdala pathways (Bernard et al. 1989; Bernard and Besson 1990; Jasmin et al. 1997) in addition to the more traditional spinothalamic and spinoreticulothalamic systems.

That neurochemically distinct populations of nociceptors indeed access different central circuits is illustrated by the demonstration that the major classes of nociceptors differ in their patterns of axon termination in the spinal cord dorsal horn. The peptide population terminates almost exclusively in the outer laminae of the superficial dorsal horn (laminae I and outer II), targeting projection neurons that transmit nociceptive messages to brainstem and/or thalamus; by contrast, the IB4 population primarily targets interneurons of the inner part of lamina II, a region just dorsal to a distinct subset of interneurons that synthesize the gamma isoform of protein kinase C (PKCγ) (Malmberg et al. 1997). Finally, myelinated neurons project primarily to deeper laminae (III–VII) of the spinal cord (and to a smaller extent to lamina I). These observations support the view that different classes of nociceptors indeed wire to different CNS circuits. Based on the remarkable electrophysiological specificity of afferents and their neurochemical distinctiveness, there is now a general consensus for specificity (i.e., labeled line) features to the afferent, at least with respect to response properties. But whether these afferents converge upon functionally distinct but related populations of projection neuron, resulting in functionally segregated ascending circuits, or whether there is convergence upon populations of projection neurons with common functional properties remains to be determined.

Unfortunately, the information about these circuits is extremely limited. Not only is the identity of the neuron immediately postsynaptic to the different nociceptors inadequately specified, but the neurons and circuits that lie downstream of the first synapse in the dorsal horn are also largely uncharacterized. In some cases, the identity of postsynaptic neurons has been determined by electrophysiological analyses, and synapses have been characterized at the electron microscopic level (Westlund et al. 1992; Alvarez et al. 2004; Hwang et al. 2004; Shields et al. 2007; Neumann et al. 2008), but the sample from which the information is derived is extremely small. Studies that monitor Fos expression provide a much more extensive picture of populations of neurons activated by noxious stimuli (Menétrey et al. 1989; Abbadie et al. 1997; Neumann et al. 2008), but there is no information about the circuits that underlie Fos activation. Also unclear are the third-order neurons to which the laminae I and V neurons project. With some exceptions, the map of the intervening circuits is, in fact, largely unknown.

7.2 TRADITIONAL ANATOMICAL TRACING APPROACHES

The limitation in our knowledge reflects the limitation of existing neuroanatomical tract tracing technology. Classically, circuit organization has been established using a variety of anterograde and retrograde tracers that define regional associations between neurons or by electrophysiological recordings that include antidromic activation of projection neurons. In recent years, the number of neuronal tracers has dramatically increased and now includes a host of fluorescent tracers, plant lectins, viruses, toxins, and various fusion proteins (e.g., tau-lacZ/GFP) (Kuypers and Huisman 1984; Kobbert et al. 2000; Vercelli et al. 2000). These traditional tracers have proven to be immensely useful, but they have some inherent limitations. For example, most approaches are hampered by the inevitable spread of the tracer at the site of injection, making it difficult to determine which neurons carried the tracer and to identify the population of neurons that are the targets of the labeled cells. An injection of the retrograde tracer fluorogold into the spinal cord, for example, will inevitably result in uptake by terminals that target interneurons as well as projection neurons. Intracellular or axonal injection of tracers can avoid some of the problems associated with poorly defined injection sites, but they are difficult, and generally limited to analyzing the axonal arborization of single neurons. Finally, the majority of tracers provide information only about the connections made by a single neuron or by a population of neurons, with neurons immediately downstream (or upstream in the case of retrograde tracers) of the injection site.

The development of transneuronal tracers that can be transferred from neuron to neuron, across synapses, was a big step forward. Among the earliest successful examples of this approach were radioisotope-labeled amino acids (Kristensson and Olsson 1973; Su and Polak 1987) or carbohydrates (which were largely limited to the visual system), bacterial toxins (Kissa et al. 2002; Maskos et al. 2002), plant lectins (Borges et al. 1982; Sawchenko and Gerfen 1985; Cabot et al. 1991), and neurotropic viruses. The latter, which include herpes simplex (HSV-1 and -2; Martin and Dolivo

1983; Norgren et al. 1992), rabies (Wickersham 2007a; Ugolini 2008), and pseudo-rabies (PRV; Jasmin et al. 1997; Card and Enquist 2001; Braz et al. 2009), are ideal as transneuronal tracers because they produce infections that spread within chains of synaptically linked networks of neurons, in both the anterograde and retrograde directions. Because the viruses replicate after infecting a neuron, they are powerful self-amplifying markers (Card 1998; Loewy 1998). But viral transport is not without its limitations. For example, labeling depends on viral concentration: low titer can result in minimal uptake and transport, but high titers increase cell lysis. In fact, the lytic nature of the virus literally often obliterates the injection site, making it difficult to determine the specific cells of origin of the transport (Jasmin et al. 1997). Other virus limitations include the selective tropism of some strains, which can restrict replication competency of the virus to subsets of neurons. In fact, not all neurons are permissive to infection, possibly because they lack specific viral receptors (Sik et al. 2006). For example, PRV does not produce transneuronal labeling in the olfactory system (Babic et al. 1994) and is not robust enough in DRG cells to be visualized by immunohistochemistry (Rotto-Percelay et al. 1992). In addition, although viral tracers are very effective tools for transneuronal retrograde transport and have been used beautifully to examine the CNS circuits that likely regulate different somatic and visceral peripheral structures, when they are injected in the CNS, it is impossible to know from which neuronal population the transport originated. This limitation results from the virus's lack of specificity for a neuronal subtype. Viral tracers with conditional replication properties (DeFalco et al. 2001; Wickersham et al. 2007b) are very powerful and promising tools to overcome this problem, but here the analysis is restricted to a single or very small subset of neurons into which the genetically directed retrograde tracer is targeted. The problem is exacerbated if one wishes to study the circuits that influence, rather than the immediate input to, a particular cell population.

7.3 GENETIC TRACING WITH TRANSNEURONAL TRACERS

For all of these reasons, new approaches to circuit mapping have been sought, and in recent years transgenic expression of tracer molecules has revolutionized this analysis. In this approach, the tracer is induced and sustained for the duration of the experiment. Thus, the tracer accumulates over long periods of time, allowing large quantities to be produced and transferred transneuronally. This considerably facilitates subsequent detection of the marker and allows for high-resolution mapping. This approach to inducing the tracer not only obviates the need for invasive surgery, which by its nature might destroy part of the neural circuitry that is to be analyzed, but it also eliminates the problem of identifying the "injection site." A few particular advantages of the genetic approach over conventional tracing techniques are the reproducibility of labeling between animals, the avoidance of uptake by axons of passage, which almost inevitably occurs when the tracer is microinjected into the tissue, and the uniform labeling of cells that can be produced.

Building on the ability of plant lectins to transport transneuronally, Yoshihara et al. (1999) developed a transgenic mouse in which the lectin wheat germ agglutinin (WGA) is expressed in a subset of CNS neurons. In their first study, the authors drove WGA expression off a Purkinje cell-specific promoter, L7. In these mice, the lectin was not only strongly expressed in Purkinje cells but was also detectable in second- and third-order neurons downstream of the Purkinje cells. The tracer was revealed by immunocytochemistry following the transneuronal anterograde transport of the WGA. In a similar approach, Zou et al. (2001) used transneuronal tracing with selective expression of barley lectin to study the central projection of subsets of primary olfactory neurons.

7.4 A GENETIC APPROACH TO THE ANATOMICAL MAPPING OF COMPLEX CIRCUITS: THE ZW MOUSE

Despite its simplicity and elegance, this new approach still has some limitations. Because the tracer expression in these mice is under the control of neuron-specific promoters, the circuits that are revealed are system specific. In other words, the utility of those transgenic mice is limited to the CNS circuit influenced by the particular promoter. This problem inspired us to generate a more powerful transgenic mouse line (the ZW mouse) in which transneuronal labeling of circuits originating from any region of the brain or spinal cord can be induced (Braz et al. 2002). In the remainder of this chapter, we describe this new mouse line and illustrate how this powerful new genetic tracing system can help us to better characterize the anatomy of the local and long-distance circuits that are engaged by different classes of neurons, including nociceptors (Braz et al. 2005; Neumann et al. 2008). We also illustrate how the mice can be used to study brainstem serotonergic descending control systems (Braz and Basbaum 2008). Finally, we describe how this approach can be used to better understand how different "pain" transmission circuits are altered in the setting of injury, a process that we believe is a major contributor to the pain produced after nerve injury (Braz and Basbaum, 2009).

The ZW mouse is a transgenic mouse strain that expresses a neuronal tracer (WGA) that is both temporally and spatially inducible. The Z refers to the lacZ gene, which is constitutively expressed in CNS neurons of the ZW mouse. The W refers to the WGA, which we induce in subsets of neurons. To produce the mouse, we combined the Cre/loxP site-specific recombination system with the transgenic approach of Yoshihara et al. (1999). Figure 7.1A illustrates the construct (pCZW) used to generate the ZW transgenic mice. The expression vector includes a loxP-flanked lacZ gene inserted upstream of a WGA cDNA. Both genes are driven off a cytomegalovirus (CMV) enhancer/chicken beta-actin (CβA) promoter. In this arrangement, the cells constitutively produce β-galactosidase (β-gal), which is the protein product of the lacZ gene, but there is no constitutive WGA expression.

To induce expression of the WGA, one must first excise the lacZ gene. This is accomplished by taking advantage of a bacterial enzyme (Cre recombinase), which recognizes specific sequences that surround the lacZ gene (loxP sites) and excises anything in between. Thus, in the presence of Cre, the lacZ gene is removed, which allows for transcription of the WGA gene. The WGA mRNA is then translated, resulting in synthesis and eventually transport of the tracer. Figure 7.1B schematically illustrates a hypothetical cluster of neurons in a transgenic mouse carrying the

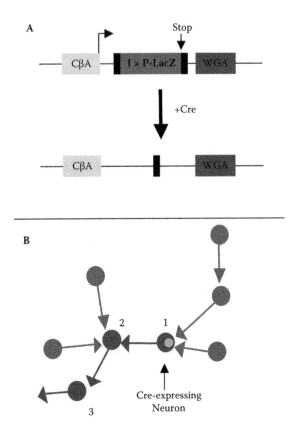

FIGURE 7.1 (see color insert following page 166) Genetic approach to establish Cre-dependent expression of the transneuronal tracer wheat germ agglutinin (WGA) in neurochemically distinct neuronal subsets. See text for explanations.

ZW construct. All neurons are blue as they express the lacZ gene. At the appropriate time and in the desired region of the CNS, the Cre is induced and then it translocates to the nucleus (green nucleus of neuron 1, Figure 7.1B) where it cuts out the lacZ gene. (Note that Cre can be provided by crossing the ZW mouse with others that express the Cre recombinase or by injecting the ZW mouse with a Cre-expressing viral vector). The Cre-mediated recombination event initiates expression of WGA in the Cre-expressing cell only (red neuron 1, Figure 7.1B). Subsequent transneuronal transport of the tracer (red neurons 2 and 3, Figure 7.1B) reveals the relevant circuit from within the large population of neurons that are not part of the circuit.

In the adult ZW transgenic mouse, we detected β-gal throughout the CNS and peripheral nerve system (Figure 7.2). However, even though we used a supposed ubiquitous promoter to drive expression of the ZW transgene, we found that only about one-half of the neurons have the potential to express the transgene. This incomplete expression is, in fact, common and results from a mosaic expression of the transgene, undoubtedly related to the site of insertion of the plasmid into the host genome.

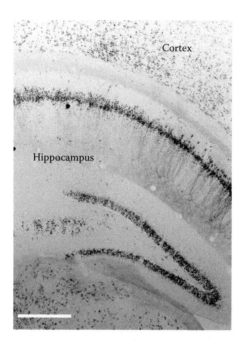

FIGURE 7.2 (see color insert following page 166) The ZW transgene is expressed throughout the CNS. This section is stained for expression of the floxed lacZ transgene (blue). Although we anticipated ubiquitous expression of the transgene, because we used the strong, chicken beta-actin promoter, random insertion of the ZW transgene nevertheless still resulted in a mosaic expression in the CNS. Scale bar 100 μm.

7.5 GENETIC TRACING OF ASCENDING "PAIN" PATHWAYS: SELECTIVE INDUCTION OF A TRANSNEURONAL TRACER IN PRIMARY AFFERENT NOCICEPTORS

In our first studies of pain transmission circuits, we targeted the WGA to DRG neurons that express the Nav1.8 subtype of voltage-gated sodium channel, which is concentrated in nociceptors (Amaya et al. 2000). To this end, we crossed the ZW mouse with a Nav1.8-Cre mouse (Stirling et al. 2005). In double transgenic, Nav1.8xZW (N-ZW) animals, the WGA is synthesized exclusively in Nav1.8-expressing DRG neurons. Consistent with the known expression pattern of Nav1.8, we observed that most WGA-positive (red, Figure 7.3) neurons were small to medium and expressed the neurofilament peripherin (~82% Per+), a marker of neurons with unmyelinated axons. Double labeling with N52, a marker of neurons with myelinated axons, confirmed that the great majority of WGA-positive neurons were unmyelinated (~10% N52+). Surprisingly however, we found that within the subset of small WGA-positive DRG neurons, ~65% bind isolectin B4 (IB4), and at most ~8% express the neuropeptide substance P (SP). This indicated that in N-ZW animals, the mosaic expression of the WGA resulted in preferential expression in nonpeptidergic, unmyelinated sensory neurons that express Nav1.8.

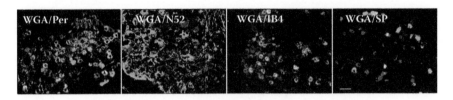

FIGURE 7.3 (see color insert following page 166) Preferential expression of WGA in dorsal root ganglion cells at the origin of unmyelinated (C) fibers of the nonpeptidergic class. We crossed the ZW mice with others that express Cre in nociceptors that express the Nav1.8 subtype of voltage-gated Na channel (Stirling et al. 2005). In double transgenic (N-ZW) animals, most WGA-positive DRG neurons co-stain for the neurofilament peripherin (Per) and bind the isolectin IB4 (yellow denotes double-labeled neurons). In contrast, very few DRG neurons express N52 (a marker of neurons with myelinated axons) or immunostain with antibodies against the peptide, substance P (SP). Scale bar 100 µm.

When we followed the anterograde transport and transneuronal transfer of WGA in the spinal cord, we observed labeling throughout the dorsal horn. In addition to dense terminal labeling in lamina IIi (consistent with the termination pattern of IB4-positive neurons), we also observed large numbers of WGA-positive cell bodies, extending from laminae II to V (Figure 7.4, sagittal section). Rarely did we find labeled neurons in lamina I. Importantly, the labeling pattern differed considerably in newborn animals. At P0, the WGA terminals were restricted to lamina II (and to a lesser extent lamina I), suggesting that the adult pattern of labeling in deeper laminae resulted from transneuronal transfer of the WGA from second-order interneurons of lamina II. This observation is consistent with the transneuronal transport of WGA being time-dependent.

A further advantage of the transneuronal labeling procedure is that it is possible to double label for markers that delineate the neurochemical signature of the postsynaptic neurons. For example, we never found labeled neurons that express the substance P/neurokinin 1 (NK1) receptor. Nor did we find double labeling of

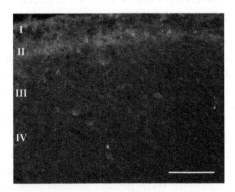

FIGURE 7.4 (see color insert following page 166) Anterograde and transneuronal transfer of WGA to the dorsal horn of the spinal cord in the adult N-ZW mouse. In double transgenic N-ZW animals, in addition to a dense band of labeled terminals in lamina IIi, we detected the WGA tracer (red) in large numbers of postsynaptic neurons located in laminae III–V of the dorsal horn. Scale bar 100 µm.

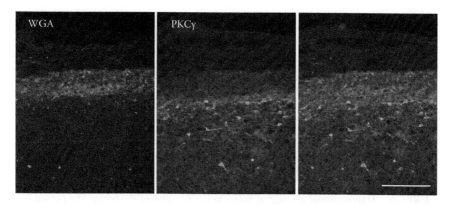

FIGURE 7.5   (see color insert following page 166) PKCγ-expressing interneurons of inner lamina II do not receive inputs from nonpeptidergic primary sensory neurons. In double-transgenic N-ZW animals, we found that most WGA+ afferents terminate (red) dorsal to the band of PKCγ interneurons (green), overlapping the termination zone of the IB4 positive afferents. Scale bar 100 μm.

the PKCγ-expressing interneurons of the most inner part of lamina II. In fact, the WGA-positive afferents that derive from the IB4 population terminate in a very limited band in the superficial dorsal horn, ventral to lamina I and dorsal to the band of PKCγ interneurons (Figure 7.5). This result was unexpected and indicates that these two major neuronal populations of dorsal horn neuron do not receive direct (or even indirect) inputs from the Nav1.8 subset of nonpeptidergic class of nociceptors. Our results highlight the stratification that exists in the inner part of lamina II, with the dorsal part receiving IB4-positive afferent terminals and the most ventral part containing the PKCγ interneuron population. This view is supported by recent studies from our and other laboratories that showed that in the mouse, the IB4-binding unmyelinated afferents terminate immediately dorsal to the PKCγ layer (Zylka et al. 2005; Neumann et al. 2008).

We next followed the transneuronal transport of WGA to the brain. In general, we observed a pattern of labeling that corresponds to regions previously recognized as major targets of spinal cord nociresponsive neurons (Cliffer et al. 1991; Newman et al. 1996; Gauriau and Bernard 2004). The transneuronal labeling predominated in limbic regions of the brain, including the hypothalamus, amygdala, and bed nucleus of the stria terminalis (BNST, Figure 7.6). Very unexpectedly, we also found extensive transneuronal transport to neurons of the lateral aspect of the globus pallidus (GP, Figure 7.6), a region not generally associated with the processing of pain messages. However, Chudler and colleagues showed, in fact, that a large proportion of neurons within the globus pallidus receive both thermal and mechanical nociceptive information (Chudler et al. 1993; Chudler 1998). Perhaps the most surprising observation was that we never observed transneuronal labeling of neurons in the parabrachial nuclei or in the thalamus, even though the spinoparabrachial and spinothalamic pathways are undoubtedly among the main ascending pathways for the transmission of nociceptive information from the spinal cord. Furthermore, the lack of transport to the parabrachial nuclei suggests that the connection to the limbic loci is direct from the spinal cord.

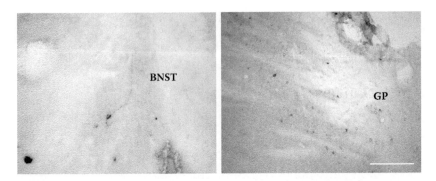

FIGURE 7.6 Transneuronal transfer of WGA to limbic areas of the brain. The presence of WGA in neurons of the bed nucleus of the stria terminalis (BNST) and the globus pallidus (GP) indicates that these neurons are part of a circuit that arises from the nonpeptidergic population of DRG that express the Nav1.8 channel. Immunocytochemistry with DAB, scale bar 100 μm.

7.5.1 Transneuronal Transport versus Local Expression

A critical issue in the analysis of transneuronal transport is to distinguish the cells in which the WGA is synthesized from those that took up the lectin after its transneuronal transport. This can readily be achieved by double-labeling for WGA and β-gal. In fact, because the expression of lacZ and WGA is mutually exclusive, a cell that co-stains for both can only have taken up the WGA after its transneuronal transport. Indeed in the brain of N-ZW animals, we found that at least half of the WGA-labeled neurons also labeled for β-gal, confirming that there was transneuronal transfer of WGA to these regions of the brain.

7.5.2 Parallel and Largely Independent Circuits Are Engaged by the Two Major Primary Afferent Nociceptor Populations

The fortuitous WGA expression pattern in the N-ZW animals proved to be very useful for dissecting the CNS circuits engaged by the nonpeptide class of nociceptors. In fact, prior to our analysis, almost all the anatomical information known about the IB4 nociceptors was that these neurons terminate in the inner part of lamina II of the superficial dorsal horn. Using the transneuronal tracing approach, we now have a much more detailed picture of the circuits engaged by the Nav1.8 subset of IB4 neurons. Our results indicate that the nonpeptidergic population of nociceptors provides a major afferent input to ascending pathways that terminate in limbic areas of the brain. We suggest that the information carried by IB4+ nociceptors is rapidly transmitted to limbic forebrain regions, via direct spino-limbic projections, where the input likely contributes more to the affective component of the pain experience than to its sensory discriminative component. We suggest that the peptide population of nociceptors likely provides the major input to lamina I projection neurons. Taken together, we conclude that the peptidergic and nonpeptidergic neurons engage parallel but independent ascending pathways that arise from laminae I and V dorsal horn neurons, respectively.

7.6   NERVE INJURY-INDUCED EXPRESSION OF WGA: A WINDOW INTO THE STUDY OF CIRCUIT PLASTICITY

In addition to studying normal circuit organization, several variations of the ZW mouse make it possible to study circuit reorganization after injury. Of particular interest for the present discussion is that there is now considerable evidence that tissue and nerve injury-induced persistent pain result, in significant part, from a reorganization of spinal cord circuits, including significant anatomical rearrangement. Among the changes observed are phenotypic changes in the neurochemistry of primary afferent neurons after tissue (Neumann et al. 1996) or nerve injury (Miki et al. 1998), enhanced activation of descending pain facilitatory systems (Carlson et al. 2007; Neubert et al. 2004), and perhaps most dramatically, maladaptive sprouting of primary sensory neurons in the dorsal horn (Jancsó 1992; Fitzgerald et al. 1990; Koerber and Brown 1995; Wilson and Kiltchener 1996; Doubell et al. 1997). Some recent studies have questioned the magnitude of the dorsal horn sprouting that occurs after peripheral nerve injury (Shehab et al., 2003), but there is little doubt that rather dramatic changes occur after injury during development, which suggests that better methods of detection might reveal the adult correlate of this profound developmental plasticity (Fitzgerald 1985; Ruda et al. 2000; Walker et al. 2003). For example, inflammation in the neonatal rat dramatically alters the central projections of small-diameter afferents (transganglionically labeled with WGA-HRP), and this reorganization is associated with functional changes in the properties of dorsal horn nociresponsive neurons (Shortland et al. 1990; Ruda et al. 2000; Peng et al. 2003; Torsney and Fitzgerald 2003).

Although these anatomical studies, which follow upon the pioneering observations of Liu and Chambers (1958), demonstrate profound anatomical rearrangement of primary afferent terminals, they cannot provide information about the circuits that are altered following injury. For example, sprouting can manifest as movement of terminals into a novel terminal field, such as sprouting of A-beta afferents dorsally into the superficial dorsal horn (Woolf et al. 1995; Wilson and Kitchener 1996; Doubell et al. 1997; Nakamura and Myers 1999), but it may also represent increased arborization of axons within their normal zone of termination (as occurs for small-diameter afferents in the setting of tissue injury). In either case, the functional consequences of injury-induced sprouting depend not only and perhaps not so much on the number of new terminals, but also on the extent to which new connections are made by the sprouting afferents. Unfortunately, what is rarely assessed in these studies is the identity of the postsynaptic neuron with which new contacts are made. Some electrophysiological studies have recorded from neurons with which new sprouts have made contact (Baba et al. 1999; Nakatsuka et al. 1999; Yoshimura et al. 2004) but the numbers of cells examined in these studies is extremely limited. Similarly, although electron microscopy can unequivocally establish that new synaptic connections are made, this approach is particularly time-consuming and cannot provide information about the Gestalt of the novel connections that are made.

7.6.1 NERVE INJURY-INDUCED EXPRESSION OF A TRANSNEURONAL TRACER

With a view to adapting the ZW mouse for studies of the reorganization of afferent projections after peripheral nerve injury, we designed a simple strategy that uses peripheral nerve injury to induce expression of the WGA in a neurochemically defined subpopulation of primary sensory neurons. Our studies took advantage of the fact that neuropeptide Y (NPY), which is not normally expressed in adult sensory neurons, is strongly induced in myelinated DRG neurons after peripheral nerve injury (Wakizaka et al. 1991; Noguchi et al. 1993). In these experiments, we crossed the ZW mouse line with a BAC transgenic mouse that expresses Cre recombinase under the influence of the NPY promoter (DeFalco et al. 2001). In double-transgenic NPY-ZW mice, Cre-mediated excision of the floxed lacZ cDNA only occurs in NPY-expressing neurons. Inasmuch as NPY is not expressed in uninjured DRG neurons (Figure 7.7A: NPY, contralateral uncut side), in the absence of injury there is no baseline WGA expression in the DRG (WGA lower left, contralateral uncut side, Figure 7.7C). However, after sciatic nerve transection, we found NPY immunoreactivity in large numbers of DRG neurons, most of which were of medium-to-large diameter (Figure 7.7B). As expected, NPY upregulation occurred concurrently with an induction of the WGA (Figure 7.7D), in a majority of myelinated neurons (90% N52-positive, a marker of neurons with myelinated axons, inset lower right).

We also observed WGA labeling in the spinal cord (Figure 7.8A) and in the nucleus gracilis (DCN) (Figure 7.8B), ipsilateral to the side of the transection. Consistent with the fact that the WGA synthesis was initiated by sciatic nerve transection, we found that the WGA immunoreactivity was restricted to lumbar segments (L2–L6), being concentrated in the medial regions of the deep laminae of the dorsal horn (III–V) as well as in the ventral horn. Only sparse terminals were recorded in superficial laminae (I–II). We presume that this derives from high-threshold A-delta afferents in which the NPY was induced. Surprisingly, however, even with long survival times, we did not see convincing evidence of an expansion of the WGA terminal field into the superficial dorsal horn, which would be expected if the large-diameter afferents had sprouted dorsally following the peripheral nerve injury. In this respect, our results are more consistent with the lack of sprouting reported by Hughes and colleagues (2003).

In addition to the dense terminal pattern in the neck of the dorsal horn, we detected the tracer in large numbers of postsynaptic neurons, secondary to transneuronal transfer of the WGA from the NPY-positive primary afferents. These observations illustrate the great power of this tracing method over existing approaches. For example, although it is possible to study the central projections of myelinated afferents using transganglionic transport of the B fragment of cholera toxin, the latter approach does not reveal the postsynaptic neurons that are contacted by the myelinated afferents.

The ability to identify the postsynaptic targets of the afferents, of course, makes it possible to classify the populations of neurons that are contacted. For example,

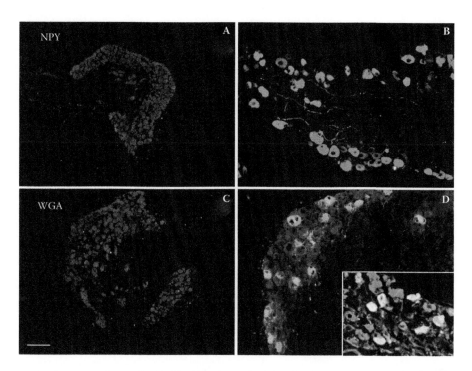

FIGURE 7.7   (see color insert following page 166) Peripheral nerve injury–induced expression of the transneuronal tracer in DRG neurons. We crossed the ZW mice with others that express the Cre recombinase in neurons that express neuropeptide Y (NPY-Cre; DeFalco et al. 2001). Under normal conditions, there is no expression of NPY (A) or WGA (C) in DRG of double transgenic NPY-ZW mice. However, a sciatic nerve transection induces expression of NPY (and thus Cre) (B), which in turn induces expression of WGA (D) in myelinated (N52+; yellow) sensory neurons (inset in D). Scale bar 100 μm.

in double-labeling experiments we showed that a large number of PKCγ interneurons in lamina IIi contained the tracer, indicating that these PKCγ interneurons receive direct inputs from myelinated primary afferent neurons (Figure 7.9). This result supports a recent study from our laboratory showing that the lamina IIi band of PKCγ interneurons overlaps with the central projection of a marker of myelinated primary afferent terminals (namely the VGLUT1 subtype of vesicular glutamate transporter). In fact, we demonstrated that synaptic contacts exist between VGLUT1-positive DRG neurons and PKCγ interneurons (Neumann et al. 2008). Because VGLUT1 marks a population of large-diameter afferents, these latter observations provide a possible mechanism by which the PKCγ interneurons come into play in the generation of injury-induced mechanical allodynia, a condition in which non-noxious stimuli transmitted by myelinated (A-beta) afferents can provoke pain. Finally, despite the rather limited WGA projections in lamina I, we did observe postsynaptic labeling in PKCγ neurons. Whether the contribution of the lamina I PKCγ neurons differs from that of the more abundant population remains to be determined.

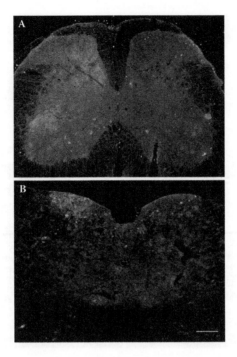

FIGURE 7.8 (see color insert following page 166) Anterograde and transneuronal transfer of WGA to postsynaptic neurons located in deep laminae of the spinal cord (A) and in the dorsal column nuclei (B) in NPY-ZW mice. In NPY-ZW animals, most WGA+ afferents terminate in the medial part of deep laminae of the spinal cord and in the nucleus gracilis ipsilateral to the sciatic nerve section, which is consistent with the projection pattern of myelinated sensory neurons. Even after long survival times, we did not see an expansion of the WGA terminal field into the superficial dorsal horn, indicating that there is limited if any sprouting of these afferents after nerve injury. Scale bar 100 μm in A, 50 μm in B.

7.7 GENETIC TRACING OF DESCENDING "MODULATORY" CIRCUITS: SELECTIVE INDUCTION OF WGA IN SEROTONERGIC NETWORKS

Stimulation of the rostroventral medulla (RVM) or the periaqueductal gray (PAG) produces antinociception that is associated with spinal cord release of serotonin (5HT; Hammond et al. 1985). This analgesic effect can be reduced by administration of 5HT antagonists (Jensen and Yaksh 1984; Barbaro et al. 1985), indicating that 5HT neurons exert an inhibitory effect on the transmission of nociceptive messages at the level of the spinal cord. This apparently straightforward characterization is likely far more complex, as both facilitating and inhibitory controls exist (Sufka et al. 1992; Green et al. 2000; Zeitz et al. 2002; Porreca et al. 2002; Suzuki et al. 2004; Sasaki et al. 2006; Zhao et al. 2007). In fact, some studies even argue against a significant contribution of medullary 5HT neurons to pain modulation. For example, electrophysiological studies indicate that the firing of 5HT neurons is state-dependent, modulated differentially across sleep/wake cycles (Mason 2001) and only weakly, and in some

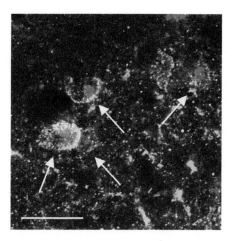

FIGURE 7.9   (see color insert following page 166) PKCγ-expressing interneurons of inner lamina II receive inputs from myelinated sensory neurons (transneuronal transport in the NPY-ZW mouse). The presence of WGA (red) in PKCγ-expressing interneurons (green) indicates that these interneurons have taken up the tracer after its release from myelinated primary afferents. Yellow fluorescence illustrates double-labeled neurons. Scale bar 30 μm.

cases not at all, driven by noxious stimulation (Gao and Mason 2001; Zhang et al. 2006). If anything, this indicates that medullary 5HT neurons are not essential for the regulation of pain control (Jacobs and Azmitia 1992).

In fact, because of the strong evidence for a contribution of descending 5HT systems to pain control at the level of the spinal cord, it is surprising that electrophysiological characterization of neurons of the RVM that either facilitate or inhibit spinal cord nociceptive processing seem not to include the 5HT population. Thus, two major classes of neurons in the RVM have been implicated in the descending control of pain transmission: *on* cells and *off* cells. Whereas activation of *on* cells facilitates the transmission of pain message by spinal cord neurons, *off* cell activation is largely inhibitory (for reviews see Fields et al. 1991; Fields 2004). The fact that opioids activate *off* cells (Barbaro et al. 1986; Cheng et al. 1986) is consistent with an opposing effect of *on* and *off* cells on the processing of nociceptive messages.

One possibility, of course, is that the 5HT contribution occurs at the level of the medulla, through circuits made with the spinally projecting *on* and *off* cell populations. In fact, although anatomical studies have shown that *on*, *off*, and 5HT cells all project to the spinal cord, little is known about the extent to which these different classes of cells interact with one another. Electrophysiological studies have shown that 5HT exerts inhibitory effects on RVM cells (Hentall et al. 1993; Pan et al. 1993), but whether or not this 5HT control is independent of the circuits through which *on* and *off* cells regulate pain transmission neurons is unclear. To better characterize the pain-relevant neuronal networks engaged by serotonergic neurons, we turned again to the ZW mouse. In these studies, we targeted expression of the WGA tracer to 5HT neurons of the raphe and followed transneuronal transport of the WGA within the brainstem and to the spinal cord.

In these studies, we crossed the ZW mice with mice that express Cre driven off of the Pet-1 promoter (Scott et al. 2005), a transcription factor exclusively expressed in serotonergic neurons. Figure 7.10 illustrates the expression of WGA in the nucleus raphe magnus (Figure 7.10A) and the dorsal raphe (Figure 7.10B) nuclei in double-transgenic Pet-ZW mice in which Cre recombination occurred. In these mice, we detected the WGA tracer in all raphe nuclei (medullary and midbrain cell groups), as well as in fibers throughout the CNS (for example, in the parabrachial nuclei, Figure 7.10C). Double-labeling experiments established that both 5HT (green, Figure 7.10D) and non-5HT neurons contained the tracer. The latter correspond to second-order, postsynaptic neurons that clearly took up the tracer after its transneuronal transfer from first-order 5HT neurons. We also observed large numbers of WGA-positive non-5HT neurons in the reticular formation (Figure 7.11A), including the medullary nucleus reticularis gigantocellularis (RGc) and magnocellularis (RMc), as well as in the spinal cord (Figure 7.11B). Fluorogold injections into the

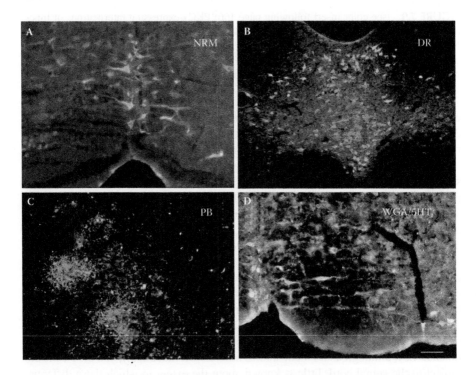

FIGURE 7.10 (see color insert following page 166) Triggering WGA expression in bulbospinal serotonergic neurons. We crossed the ZW mice with others that express the Cre recombinase exclusively in 5HT neurons of the brainstem. This was achieved using the ePet1-Cre mice (Scott et al. 2005). In double-transgenic mice (ePet-ZW), the WGA (red) is expressed in all raphe nuclei, including the nucleus raphe magnus (NRM, A) and the dorsal raphe (DR, B). Anterograde transport of WGA was also detected in axonal terminals in various brainstem nuclei, including the parabrachial nuclei (PB, C). Both 5HT (green) and non-5HT neurons within the raphe magnus and pallidus (D) contain the WGA tracer. Scale bar 100 μm in A, B, and D; 50 μm in C.

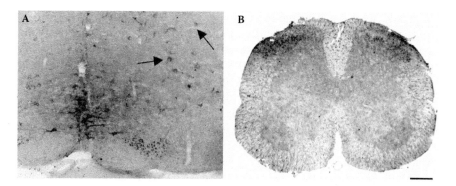

FIGURE 7.11   Anterograde transport and transneuronal transfer of the WGA from brainstem serotonergic neurons. Selective expression of WGA in 5HT neurons of the medullary raphe resulted in transneuronal transfer of the tracer to non-5HT neurons of the reticularis magnocellularis (A) and to the dorsal horn of the spinal cord (B). Scale bar 100 μm.

spinal cord further established that 5HT neurons of the medullary (but not midbrain) raphe nuclei as well as non-5HT neurons of the RGc and RMc that receive a 5HT input are at the origin of descending, bulbospinal pathways. By contrast, we found no evidence that neurons of the midbrain PAG that project to the RVM are postsynaptic to midbrain or medullary 5HT neurons. In other words, the major descending projection systems that originate in the medullary reticular formation are under local 5HT control, from the NRM, but not from the DR. Finally, we found very few examples of WGA-positive noradrenergic (NE) neurons, which indicates that there is considerable independence of the 5HT and NE bulbospinal pathways.

From these studies, we conclude that the major 5HT projection from the RVM is to interneurons and to spinally projecting, largely non-5HT neurons within the NRM and adjacent medullary reticular formation. Based on these observations we suggest that 5HT neurons are critical integrators of outputs that arise from the RVM, and that these neurons influence "pain" processing at the spinal cord level both directly and indirectly via feed-forward connections with multiple non-5HT descending control pathways. Given the large number of 5HT puncta that appose *off* cells (Potrebic et al. 1995), it is likely that some WGA-labeled neurons of the RVM correspond to *off* cells. Whether this 5HT input has a facilitatory or inhibitory net effect on spinal cord processing of nociceptive messages remains to be determined. This net effect, of course, depends on the subtype of 5HT receptor expressed by the postsynaptic neuron.

7.9   CONCLUSIONS

The results obtained in double-transgenic animals illustrate how synthesis of a transneuronal tracer can be triggered in the CNS, provided an appropriate Cre-mediated recombination event can be initiated in a critical subpopulation of neurons. Importantly, because transfer of the tracer to higher-order neurons can be observed, this technique can be adapted to multineuronal circuit analysis in diverse regions of the brain and spinal cord. The sensitivity of the technique is, however, variable.

For example, when we triggered expression of WGA in Nav1.8-expressing sensory neurons, we were able to detect fourth-order neurons (in limbic areas of the brain). By contrast, when we induced WGA expression in 5HT neurons, we revealed at best third-order neurons. We presume that the extent to which we can detect transneuronal transport across multiple synapses depends on the amount of WGA that is produced. We also believe that transport is enhanced if there is considerable convergence of axons that transport the lectin. Future studies that use a transneuronal tracer that is less susceptible to enzymatic degradation should increase our ability to detect neurons further downstream in the circuit that arises from primary afferent neurons. In addition, as more mice are generated in which Cre is driven off of neuronal specific promoters, the value of the ZW mice will continue to grow. Clearly, a better understanding of the circuits engaged by the remarkably diverse populations of primary nociceptors is key to the development of more selective modes of regulating pain processing. This new approach will hopefully contribute to progress in that area.

ACKNOWLEDGMENTS

The work presented in this chapter was supported by NIH grants NS14627 and 48499.

REFERENCES

Abbadie C, Taylor BK, Peterson MA, Basbaum AI. 1997. Differential contribution of the two phases of the formalin test to the pattern of c-fos expression in the rat spinal cord: studies with remifentanil and lidocaine. *Pain.* 69:101–110.

Alvarez FJ, Villalba RM, Zerda R, Schneider SP. 2004. Vesicular glutamate transporters in the spinal cord, with special reference to sensory primary afferent synapses. *J Comp Neurol.* 472:257–280.

Amaya F, Decosterd I, Samad TA, Plumpton C, Tate S, Mannion RJ, Costigan M, Woolf CJ. 2000. Diversity of expression of the sensory neuron-specific TTX-resistant voltage-gated sodium ion channels SNS and SNS2. *Mol Cell Neurosci.* 15:331–342.

Baba H, Doubell TP, Woolf CJ. 1999. Peripheral inflammation facilitates A-beta fiber-mediated synaptic input to the substantia gelatinosa of the adult rat spinal cord. *J Neurosci.* 19:859–867.

Babic N, Mettenleiter TC, Ugolini G, Flamand A, Coulon P. 1994. Propagation of pseudorabies virus in the nervous system of the mouse after intranasal inoculation. *Virology.* 204:616–625.

Barbaro NM, Hammond DL, Fields HL. 1985. Effects of intrathecally administered methysergide and yohimbine on microstimulation-produced antinociception in the rat. *Brain Res.* 343:223–229.

Barbaro NM, Heinricher MM, Fields HL. 1986. Putative pain modulating neurons in the rostral ventral medulla: reflex-related activity predicts effects of morphine. *Brain Res.* 366:203–210.

Bernard JF, Besson JM. 1990. The spino(trigemino)pontoamygdaloid pathway: electrophysiological evidence for an involvement in pain processes. *J Neurophysiol.* 63:473–490.

Bernard JF, Peschanski M, Besson JM. 1989. A possible spino (trigemino)-ponto-amygdaloid pathway for pain. *Neurosci Lett.* 100:83–88.

Borges LF, Sidman RL. 1982. Axonal transport of lectins in the peripheral nervous system. *J Neurosci.* 2:647–653.

Braz JM, Rico B, Basbaum AI. 2002. Transneuronal tracing of diverse CNS circuits by Cre-mediated induction of wheat germ agglutinin in transgenic mice. *Proc. Natl. Acad. Sci. (USA).* 99:15148–15153.

Braz JM, Nassar MA, Wood JN Basbaum AI. 2005. Parallel "pain" pathways arise from sub-populations of primary afferent nociceptor. *Neuron.* 47:787–793.

Braz JM, Basbaum AI. 2008. A genetically expressed transneuronal tracer reveals direct and indirect serotonergic descending control circuits. *J. Comp. Neurol.* 507:1990–2003.

Braz JM and Basbaum AI. 2009. Triggering genetically-expressed transneuronal tracers by peripheral axotomy reveals convergent and segregated sensory neuron-spinal cord connectivity. *Neuroscience.* In press.

Braz JM, Enquist LW, Basbaum AI. 2009. Inputs to serotonergic neurons revealed by conditional transneuronal viral tracing. *J. Comp. Neurol.* In press.

Burstein R, Cliffer KD, Giesler GJ Jr. 1987. Direct somatosensory projections from the spinal cord to the hypothalamus and telencephalon. *J Neurosci.* 7:4159–4164.

Cabot JB, Mennone A, Bogan N, Carroll J, Evinger C, Erichsen JT. 1991. Retrograde, trans-synaptic and transneuronal transport of fragment C of tetanus toxin by sympathetic preganglionic neurons. *Neuroscience.* 40:805–823.

Card JP. 1998. Exploring brain circuitry with neurotropic viruses: new horizons in neuro-anatomy. *Anat. Rec.* 253:176–185.

Card JP, Enquist LW. 2001. Transneuronal circuit analysis with pseudorabies viruses. *Curr. Protoc. Neurosci.* Chapter 1: Unit 1.5. John Wiley, NY.

Carlson JD, Maire JJ, Martenson ME, Heinricher MM. 2007. Sensitization of pain-modulating neurons in the rostral ventromedial medulla after peripheral nerve injury. *J. Neurosci.* 27:13222–13231.

Caterina MJ, Julius D. 1999. Sense and specificity: a molecular identity for nociceptors. *Curr. Opin. Neurobiol.* 9:525–530.

Cheng ZF, Fields HL, Heinricher MM. 1986. Morphine microinjected into the periaqueductal gray has differential effects on 3 classes of medullary neurons. *Brain Res.* 375:57–65.

Chudler EH. 1998. Response properties of neurons in the caudate-putamen and globus pallidus to noxious and non-noxious thermal stimulation in anesthetized rats. *Brain Res.* 812:283–288.

Chudler EH, Sugiyama K, Dong WK. 1993. Nociceptive responses in the neostriatum and globus pallidus of the anesthetized rat. *J Neurophysiol.* 69:1890–1903.

Cliffer KD, Burstein R, Giesler GJ Jr. 1991. Distributions of spinothalamic, spinohypothalamic, and spinotelencephalic fibers revealed by anterograde transport of PHA-L in rats. *J Neurosci.* 11:852–868.

Craig AD. 2004. Lamina I, but not lamina V, spinothalamic neurons exhibit responses that correspond with burning pain. *J. Neurophysiol.* 92:2604–2609.

Cummins TR, Sheets PL, Waxman SG. 2007. The roles of sodium channels in nociception: implications for mechanisms of pain. *Pain.* 131:243–257.

DeFalco J, Tomishima M, Liu H, Zhao C, Cai X, Marth JD, Enquist L, Friedman JM. 2001. Virus-assisted mapping of neural inputs to a feeding center in the hypothalamus. *Science.* 291:2608–2613.

Doubell TP, Mannion RJ, Woolf CJ. 1997. Intact sciatic myelinated primary afferent terminals collaterally sprout in the adult rat dorsal horn following section of a neighbouring peripheral nerve. *J Comp Neurol.* 380:95–104.

Fields H. 2004. State-dependent opioid control of pain. *Nat. Rev. Neurosci.* 5:565–575.

Fields HL, Heinricher MM, Mason P. 1991. Neurotransmitters in nociceptive modulatory circuits. *Annu. Rev. Neurosci.* 14:219–245.

Fitzgerald M. 1985. The sprouting of saphenous nerve terminals in the spinal cord following early postnatal sciatic nerve section in the rat. *J. Comp. Neurol.* 240:407–413.

Fitzgerald M, Woolf CJ, Shortland P. 1990. Collateral sprouting of the central terminals of cutaneous primary afferent neurons in the rat spinal cord: pattern, morphology, and influence of targets. *J. Comp. Neurol.* 300:370–385.

Gao K, Mason P. 2001. Physiological and anatomic evidence for functional subclasses of serotonergic raphe magnus cells. *J. Comp. Neurol.* 439:426–439.

Gauriau C, Bernard JF. 2004. A comparative reappraisal of projections from the superficial laminae of the dorsal horn in the rat: the forebrain. *J Comp Neurol* 468:24–56.

Giesler GJ Jr, Katter JT, Dado RJ. 1994. Direct spinal pathways to the limbic system for nociceptive information. *Trends Neurosci.* 17:244–250.

Green GM, Scarth J, Dickenson A. 2000. An excitatory role for 5-HT in spinal inflammatory nociceptive transmission; state-dependent actions via dorsal horn 5-HT(3) receptors in the anaesthetized rat. *Pain.* 89:81–88.

Hammond DL, Tyce GM, Yaksh TL. 1985. Efflux of 5-hydroxytryptamine and noradrenaline into spinal cord superfusates during stimulation of the rat medulla. *J. Physiol.* 359:151–162.

Hentall ID, Andresen MJ, Taguchi K. 1993. Serotonergic, cholinergic and nociceptive inhibition or excitation of raphe magnus neurons in barbiturate-anesthetized rats. *Neuroscience.* 5:303–310.

Hughes DI, Scott DT, Todd AJ, Riddell JS. 2003. Lack of evidence for sprouting of A-beta afferents into the superficial laminas of the spinal cord dorsal horn after nerve section. *J Neurosci.* 23:9491–9499.

Hwang SJ, Burette A, Rustioni A, Valtschanoff JG. 2004. Vanilloid receptor VR1-positive primary afferents are glutamatergic and contact spinal neurons that co-express neurokinin receptor NK1 and glutamate receptors. *J. Neurocytol.* 33:321–329.

Jacobs BL, Azmitia EC. 1992. Structure and function of the brain serotonin system. *Physiol. Rev.* 72:165–229.

Jancsó G. 1992. Pathobiological reactions of C-fibre primary sensory neurones to peripheral nerve injury. *Exp. Physiol.* 77:405–431.

Jasmin L, Burkey AR, Card JP, Basbaum AI. 1997. Transneuronal labeling of a nociceptive pathway, the spino-(trigemino-)parabrachio-amygdaloid, in the rat. *J. Neurosci.* 17:3751–3765.

Jensen TS, Yaksh TL. 1984. Spinal monoamine and opiate systems partly mediate the antinociceptive effects produced by glutamate at brainstem sites. *Brain Res.* 321:287–297.

Kissa K, Mordelet E, Soudais C, Kremer EJ, Demeneix BA, Brulet P, Coen L. 2002. In vivo neuronal tracing with GFP-TTC gene delivery. *Mol. Cell. Neurosci.* 20:627–637.

Klop EM, Mouton LJ, Holstege G. 2005. Segmental and laminar organization of the spinothalamic neurons in cat: evidence for at least five separate clusters. *J. Comp. Neurol.* 493:580–595.

Kobbert C, Apps R, Bechmann I, Lanciego JL, Mey J, and Thanos S. 2000. Current concepts in neuroanatomical tracing. *Prog. Neurobiol.* 62:327–351.

Koerber HR, Brown PB. 1995. Quantitative analysis of dorsal horn cell receptive fields following limited deafferentation. *J. Neurophysiol.* 74:2065–2076.

Kristensson K, Olsson Y. 1973. Diffusion pathways and retrograde axonal transport of protein tracers in peripheral nerves. *Prog. Neurobiol.* 1:87–109.

Kuypers HGJM, Huisman AM. 1984. Fluorescent neuronal tracers. *Adv. Cell. Neurobiol.* 5:307–340

Liu CN, Chambers WW. 1958. Intraspinal sprouting of dorsal root axons; development of new collaterals and preterminals following partial denervation of the spinal cord in the cat. *AMA Arch. Neurol. Psychiatry.* 79:46–61.

Loewy AD. 1998. Viruses as transneuronal tracers for defining neural circuits. *Neurosci. Behav. Rev.* 22:679–684.

Malmberg AB, Chen C, Tonegawa S, Basbaum AI. 1997. Preserved acute pain and reduced neuropathic pain in mice lacking PKCgamma. *Science.* 278:279–283.

Martin WJ, Cao Y, Basbaum AI. 2004. Characterization of wide dynamic range neurons in the deep dorsal horn of the spinal cord in preprotachykinin-a null mice in vivo. *J Neurophysiol.* 91:1945–1954.

Martin X, Dolivo M. 1983. Neuronal and transneuronal tracing in the trigeminal system of the rat using the herpes virus suis. *Brain Res.* 273:253–276.

Maskos U, Kissa K, St Cloment C, Brulet P. 2002. Retrograde trans-synaptic transfer of green fluorescent protein allows the genetic mapping of neuronal circuits in transgenic mice. *Proc. Natl. Acad. Sci. (USA).* 99:10120–10125.

Mason P. 2001. Contributions of the medullary raphe and ventromedial reticular region to pain modulation and other homeostatic functions. *Annu. Rev. Neurosci.* 24:737–777.

Mazarío J, Basbaum AI. 2007. Contribution of substance P and neurokinin A to the differential injury-induced thermal and mechanical responsiveness of lamina I and V neurons. *J. Neurosci.* 27:762–770.

McCleskey EW, Gold MS. 1999. Ion channels of nociception. *Annu. Rev. Physiol.* 61:835–856.

Menétrey D, Gannon A, Levine JD, Basbaum AI. 1989. Expression of c-fos protein in interneurons and projection neurons of the rat spinal cord in response to noxious somatic, articular, and visceral stimulation. *J. Comp. Neurol.* 285:177–195.

Miki K, Fukuoka T, Tokunaga A, Noguchi K. 1998. Calcitonin gene-related peptide increase in the rat spinal dorsal horn and dorsal column nucleus following peripheral nerve injury: up-regulation in a subpopulation of primary afferent sensory neurons. *Neuroscience.* 82:1243–1252.

Morris R, Cheunsuang O, Stewart A, Maxwell D. 2004. Spinal dorsal horn neurone targets for nociceptive primary afferents: do single neurone morphological characteristics suggest how nociceptive information is processed at the spinal level. *Brain Res Brain Res Rev.* 46:173–190.

Nakamura S, Myers RR. 1999. Myelinated afferents sprout into lamina II of L3–5 dorsal horn following chronic constriction nerve injury in rats. *Brain Res.* 818:285–290.

Nakatsuka T, Park JS, Kumamoto E, Tamaki T, Yoshimura M. 1999. Plastic changes in sensory inputs to rat substantia gelatinosa neurons following peripheral inflammation. *Pain.* 82:39–47.

Neubert MJ, Kincaid W, Heinricher MM. 2004. Nociceptive facilitating neurons in the rostral ventromedial medulla. *Pain.* 110:158–165.

Neumann S, Doubell TP, Leslie T, Woolf CJ. 1996. Inflammatory pain hypersensitivity mediated by phenotypic switch in myelinated primary sensory neurons. *Nature.* 384:360–364.

Neumann S, Braz JM, Skinner K, Llewellyn-Smith I, and Basbaum AI. 2008. Innocuous, not noxious input, activates PKCγ interneurons of the spinal dorsal horn via myelinated afferent fibers. *J. Neurosci.* 28:7936–7944.

Newman HM, Stevens RT, Apkarian AV. 1996. Direct spinal projections to limbic and striatal areas: anterograde transport studies from the upper cervical spinal cord and the cervical enlargement in squirrel monkey and rat. *J. Comp. Neurol.* 365:640–658.

Noguchi K, De Leon M, Nahin RL, Senba E, Ruda MA. 1993. Quantification of axotomy-induced alteration of neuropeptide mRNAs in dorsal root ganglion neurons with special reference to neuropeptide Y mRNA and the effects of neonatal capsaicin treatment. *J. Neurosci. Res.* 35:54–66.

Norgren RB Jr, McLean JH, Bubel HC, Wander A, Bernstein DI, Lehman MN. 1992. Anterograde transport of HSV-1 and HSV-2 in the visual system. *Brain Res. Bull.* 28:393–399.

Pan ZZ, Wessendorf MW, Williams JT. 1993. Modulation by serotonin of the neurons in rat nucleus raphe magnus in vitro. *Neuroscience.* 54:421–429.

Peng YB, Ling QD, Ruda MA, Kenshalo DR. 2003. Electrophysiological changes in adult rat dorsal horn neurons after neonatal peripheral inflammation. *J. Neurophysiol.* 90:73–80.

Porreca F, Ossipov MH, Gebhart GF. 2002. Chronic pain and medullary descending facilitation. *Trends Neurosci.* 25:319–325.

Potrebic SB, Mason P, Fields HL. 1995. The density and distribution of serotonergic appositions onto identified neurons in the rat rostral ventromedial medulla. *J. Neurosci.* 15:3273–3283.

Price DD, Greenspan JD, Dubner R. 2003. Neurons involved in the exteroceptive function of pain. *Pain.* 106:215–9.

Rotto-Percelay DM, Wheeler JG, Osorio FA, Platt KB, Loewy AD. 1992. Transneuronal labeling of spinal interneurons and sympathetic preganglionic neurons after pseudorabies virus injections in the rat medial gastrocnemius muscle. *Brain Res.* 574:291–306.

Ruda MA, Ling QD, Hohmann AG, Peng YB, Tachibana T. 2000. Altered nociceptive neuronal circuits after neonatal peripheral inflammation. *Science.* 289:628–631.

Sasaki M, Obata H, Kawahara K, Saito S, Goto F. 2006. Peripheral 5-HT2A receptor antagonism attenuates primary thermal hyperalgesia and secondary mechanical allodynia after thermal injury in rats. *Pain.* 122:130–136.

Sawchenko PE, Gerfen CR. 1985. Plant lectins and bacterial toxins as tools for tracing neuronal connections. *Trends Neurosci.* 8:378–384

Scott MM, Wylie CJ, Lerch JK, Murphy R, Lobur K, Herlitze S, Jiang W, Conlon RA, Strowbridge BW, Deneris ES. 2005. A genetic approach to access serotonin neurons for in vivo and in vitro studies. *Proc. Natl. Acad. Sci. (USA).* 102:16472–16477.

Shehab SA, Spike RC, Todd AJ. 2003. Evidence against cholera toxin B subunit as a reliable tracer for sprouting of primary afferents following peripheral nerve injury. *Brain Res.* 964:218–227.

Shields SD, Mazario J, Skinner K, Basbaum AI. 2007. Anatomical and functional analysis of aquaporin 1, a water channel in primary afferent neurons. *Pain.* 131:8–20.

Shortland P, Molander C, Woolf CJ, Fitzgerald M. 1990. Neonatal capsaicin treatment induces invasion of the substantia gelatinosa by the terminal arborizations of hair follicle afferents in the rat dorsal horn. *J. Comp. Neurol.* 296:23–31.

Sík A, Coté A, Boldogkõi Z. 2006. Selective spread of neurotropic herpesviruses in the rat hippocampus. *J. Comp. Neurol.* 496:229–243.

Snider WD, McMahon SB. 1998. Tackling pain at the source: new ideas about nociceptors. *Neuron.* 20:629–32.

Stirling LC, Forlani G, Baker MD, Wood JN, Matthews EA, Dickenson AH, Nassar MA. 2005. Nociceptor-specific gene deletion using heterozygous NaV1.8-Cre recombinase mice. *Pain.* 113:27–36.

Su HC, Polak JM. 1987. Combined axonal transport tracing and immunocytochemistry for mapping pathways of peptide-containing nerves in the peripheral nervous system. *Experientia.* 43:761–767.

Sufka KJ, Schomburg FM, Giordano J. 1992. Receptor mediation of 5-HT-induced inflammation and nociception in rats. *Pharmacol. Biochem. Behav.* 41:53–56.

Suzuki R, Rygh LJ, Dickenson AH. 2004. Bad news from the brain: descending 5-HT pathways that control spinal pain processing. *Trends Pharmacol. Sci.* 25:613–617.

Talavera K, Nilius B, Voets T. 2008. Neuronal TRP channels: thermometers, pathfinders and life-savers. *Trends Neurosci.* 31:287–295.

Todd AJ. 2002. Anatomy of primary afferents and projection neurones in the rat spinal dorsal horn with particular emphasis on substance P and the neurokinin 1 receptor. *Exp. Physiol.* 87:245–249.

Torsney C, Fitzgerald M. 2003. Spinal dorsal horn cell receptive field size is increased in adult rats following neonatal hindpaw skin injury. *J. Physiol.* 550:255–261.

Ugolini G. 2008. Use of rabies virus as a transneuronal tracer of neuronal connections: implications for the understanding of rabies pathogenesis. *Dev. Biol. (Basel).* 131:493–506.

Vercelli A, Repici M, Garbossa D, and Grimaldi A. 2000. Recent techniques for tracing pathways in the central nervous system of developing and adult mammals. *Brain Res. Bull.* 51:11–28.

Wakisaka S, Kajander KC, Bennett GJ. 1991. Increased neuropeptide Y (NPY)-like immunoreactivity in rat sensory neurons following peripheral axotomy. *Neurosci. Lett.* 124:200–203.

Walker SM, Meredith-Middleton J, Cooke-Yarborough C, Fitzgerald M. 2003. Neonatal inflammation and primary afferent terminal plasticity in the rat dorsal horn. *Pain.* 105:185–195.

Westlund KN, Carlton SM, Zhang D, Willis WD. 1992. Glutamate-immunoreactive terminals synapse on primate spinothalamic tract cells. *J. Comp. Neurol.* 322:519–527.

Wickersham IR, Finke S, Conzelmann KK, Callaway EM. 2007a. Retrograde neuronal tracing with a deletion-mutant rabies virus. *Nat. Methods.* 4:47–49.

Wickersham IR, Lyon DC, Barnard RJ, Mori T, Finke S, Conzelmann KK, Young JA, Callaway EM. 2007b. Monosynaptic restriction of transsynaptic tracing from single, genetically targeted neurons. *Neuron.* 53:639–647.

Wilson P, Kitchener PD. 1996. Plasticity of cutaneous primary afferent projections to the spinal dorsal horn. *Prog. Neurobiol.* 48:105–129.

Woolf CJ, Shortland P, Reynolds M, Ridings J, Doubell T, Coggeshall RE. 1995. Reorganization of central terminals of myelinated primary afferents in the rat dorsal horn following peripheral axotomy. *J. Comp. Neurol.* 360:121–134.

Yoshihara Y, Mizuno T, Nakahira M, Kawasaki M, Watanabe Y, Kagamiyama H, Jishage K, et al. 1999. A genetic approach to visualization of multisynaptic neural pathways using plant lectin transgene. *Neuron.* 22:33–41.

Yoshimura M, Furue H, Nakatsuka T, Matayoshi T, Katafuchi T. 2004. Functional reorganization of the spinal pain pathways in developmental and pathological conditions. *Novartis Found. Symp.* 261:116–24; discussion 124–31, 149–154.

Zeitz KP, Guy N, Malmberg AB, Dirajlal S, Martin WJ, Sun L, Bonhaus DW, Stucky CL, Julius D, Basbaum AI. 2002. The 5-HT3 subtype of serotonin receptor contributes to nociceptive processing via a novel subset of myelinated and unmyelinated nociceptors. *J. Neurosci.* 22:1010–1019.

Zhang L, Sykes KT, Buhler AV, Hammond DL. 2006. Electrophysiological heterogeneity of spinally projecting serotonergic and nonserotonergic neurons in the rostral ventromedial medulla. *J. Neurophysiol.* 95:1853–1856.

Zhao ZQ, Chiechio S, Sun YG, Zhang KH, Zhao CS, Scott M, Johnson RL, et al. 2007. Mice lacking central serotonergic neurons show enhanced inflammatory pain and an impaired analgesic response to antidepressant drugs. *J. Neurosci.* 27:6045–6053.

Zou Z, Horowitz LF, Montmayeur JP, Snapper S, Buck LB. 2001. Genetic tracing reveals a stereotyped sensory map in the olfactory cortex. *Nature.* 414:173–179.

Zylka MJ, Rice FL, Anderson DJ. 2005. Topographically distinct epidermal nociceptive circuits revealed by axonal tracers targeted to Mrgprd. *Neuron.* 45:17–25.

8 Cytokines in Pain

*Veronica I. Shubayev, Kinshi Kato,
and Robert R. Myers*

CONTENTS

8.1 INTRODUCTION

The development of insights into the role of cytokine modulation of pain has been the prototypical story of translational research from mouse to man. Pain has been understood in the context of inflammation associated with tissue injury: nociception in a milieu of an inflammatory soup bathing small nerve fibers following tissue injury. Translational research with mice during the past decade has now revealed that the pathogenesis of neuropathic pain, the devastating pain condition associated with direct injury or disease to the somatosensory nervous system, is also a consequence of inflammation of a type described as cytokine-mediated neuroinflammation regulated by glia and neurons.

In this chapter we will review the current understanding of cytokines in the pathogenesis of pain, and especially neuropathic pain, based on basic science research with rodent models of peripheral nerve injury. This knowledge has expanded understanding of the role of cytokines in neural dysfunction and provided an additional rationale for therapeutic use of anti-cytokine agents in human painful degenerative diseases. Herein, we review recent data to support the concept that the proinflammatory cytokine-driven processes of degeneration are at the basis of the neuropathic pain condition, and that anti-cytokine therapy represents a promising approach to treating human neuropathic pain states. Additionally, we suggest that the relationship between cytokines and matrix metalloproteinases can be therapeutically exploited.

8.2   CYTOKINES

The term *cytokine* originates from the Greek *cyto* (cell) and *kinos* (movement). Cytokines are small regulatory proteins that are released in a wide variety of cells to modulate cell–cell interaction and other functions especially important for inflammation and immune responses. Cytokines have pleiotropic activities that can trigger several cellular responses depending on cell type, timing, and molecular environment. They act through a respective receptor (as reviewed below) on the cell of their production in an autocrine mode, a neighboring cell in a paracrine mode, or through direct juxtacrine cell–cell interaction.

Cytokines are clustered into several classes: interleukins (IL), tumor necrosis factors (TNF), interferons (IFN), colony-stimulating factors, transforming growth factors, and chemokines. They are also categorized according to the structural homology of their receptors as class I or class II cytokines (Boulay et al. 2003; Langer et al. 2004). Most ILs, colony-stimulating factors, and IFNs belong to one of these two classes of cytokines and mediate their effects through the Janus kinase-signal transducers and activators of transcription (JAK-STAT) pathway. Three other major cytokine families encompass the IL-1 and TNF family members that activate the nuclear factor-κB (NF-κB) and mitogen-activated protein (MAP) kinase signaling pathways, and TGF-β superfamily members that activate signaling proteins of the Smad family. In addition, cytokines are often classified according to their functional ability to contribute to inflammation into proinflammatory cytokines, such as TNF-α, IL-1β, IL-6, IL-12, IL-18, and IFNγ or anti-inflammatory cytokines, such as IL-4, IL-10, IL-13, and TGF-β.

8.3   NOCICEPTION AND NEUROPATHIC PAIN

The Kyoto protocol of the International Association for the Study of Pain (IASP) Basic Pain Terminology (Loeser and Treede 2008) now clearly distinguishes between the terms *pain* and *nociception*. *Pain* is "an unpleasant sensory and emotional experience associated with actual or potential tissue damage or described in terms of such damage," and *nociception* is now redefined as "the neural processes of encoding and processing noxious stimuli." Previously the terms had been used less discriminately. It is now noted that pain is a subjective experience, and nociception is a physiological sensory process. The definition of neuropathic pain remains basically the same: "pain arising as a direct consequence of a lesion or disease affecting the somatosensory system." We will explore how nociceptive and neuropathic pain states are both influenced by expression of proinflammatory cytokines.

For the purposes of this discussion discriminating the role of cytokines in nociceptive and neuropathic pain states, we characterize nociceptive pain as having the following qualities:

- Pain caused by tissue injury
- Pain that is stimulus evoked

- Pain mediated by normally quiescent C polymodal nociceptors through DRG and spinal neurons
- Pain associated with increased neuronal activity through wide dynamic-range (WDR) neurons in the spinal cord
- Pain that can be managed by opioids

whereas neuropathic pain has the following qualities:

- Pain caused by injury or disease of the nervous system, but typically by injury to the peripheral nervous system
- Pain that can be either spontaneous in origin or evoked by minor mechanical stimuli
- Pain that typically develops in days or months after injury
- Pain mediated by abnormal firing of C polymodal nociceptors and in many cases by A-beta fibers through sensitized WDR neurons
- Pain that is often refractory to traditional therapies

Apart from the cognitive and emotional aspects of the pain definitions, the nociceptively mediated pain state is typically associated with tissue injury that produces an acute electrophysiologic discharge in nerve terminals and a subacute response in C-polymodal nerve fibers associated with local inflammation of the damaged tissue. The sensitivity of cutaneous afferent nerve endings to an inflammatory environment was made clear by Kessler et al. (Kessler et al. 1992) when they produced an experimental "inflammatory soup" from a broad mixture of inflammatory mediators consisting of bradykinin, serotonin, prostaglandin E2 (PGE2), and histamine, all at pH 7, to investigate the responsiveness of substance P–conditioned primary afferents. This experimental approach has been widely used in many important studies, such that the role of inflammation has become the hallmark of pain and the single best understood aspect of nociception, the processes of mechanical sensitization of acutely injured nerve fibers (Michaelis et al. 1998) and the enhanced excitability of sensory neurons (Ma et al. 2006) (Figure 8.1). These findings are, in fact, at the heart of the definition of nociception.

What has been less well elaborated is the role of cytokines in the production of the inflammatory soup. It is now known that local expression and upregulation of proinflammatory cytokines such as IL-1β induce cPLA2 and COX-2 mRNA and protein expression and subsequent PGE2 release (Moolwaney and Igwe 2005). These experiments further indicated that p38 MAPK cytokine signaling mechanisms (see below) play a role in IL-1β–mediated PGE2 release. Considering the complex inter-relationship of cytokines, other proinflammatory cytokines can certainly be presumed to be involved in the local generation of inflammatory soup. The interleukins, such as IL-1β and IL-6, in association with TNF-α are complementary and synergistic, and they control many other key inflammatory factors such as inducible nitric oxide synthase mRNA that also plays an important role in nociception (Covey et al. 2000, Myers et al. 2006). However, we believe that TNF-α plays a dominant role, in that it stimulates expression of IL-1β and other cytokines, directly causes nociception, and regulates the process of neural remodeling.

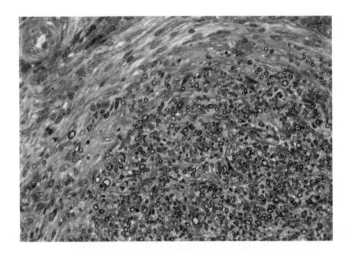

FIGURE 8.1   (see color insert following page 166) Sciatic nerve following transection injury. Plastic-embedded section of rat sciatic nerve stained with Methylene Blue Azure II. Note the lack of a defining perineurial sheath surrounding the nerve fibers and the disorganized interstitial structure. This abnormal environment is a consequence of the tissue injury, which has compromised the integrity of the endoneurial compartment and in so doing has altered the chemistry of the fluids bathing the nerve fibers. Tissue injury produces an inflammatory environment that can be modeled as an "inflammatory soup" of chemicals that are in part regulated by cytokines, and that cause spontaneous neurophysiological activity that activates nociceptive pathways.

In 1850, Augustus Waller (Stoll et al. 2002) described a pathologic process following nerve transection that included an initial reaction at the site of injury and then degeneration and phagocytosis of myelin and axons distal to the injury. Figure 8.1 depicts a nerve bundle after transection injury. What we now know as Wallerian degeneration is fundamental to neuropathology because it occurs after axonal injury of any type, including crush and severe ischemia. Its mechanisms differ importantly from other diseases with axonal degeneration, primarily the dying-back neuropathies, and with pruning of axons during development; these differences are of intense current interest (Coleman 2005; Koike et al. 2008). We focus on the role of cytokines in Wallerian degeneration because they control the invasion and activity of nonresident, hematogenous macrophages that drive the pathology of Wallerian degeneration and that are so closely related to the development of neuropathic pain and the peak periods of hyperalgesia (Myers et al. 1993). The pathology includes gradual axoplasmic disintegration during which the axolemma fragments and its contents undergo granular dissolution (Sommer et al. 1995). The Schwann cell is activated initially and expresses proinflammatory cytokines while the myelin sheath of the Schwann cell forms lamellar ovoids surrounded by Schwann cell cytoplasm. Schwann cells may phagocytose myelin debris, but hematogenous macrophages reinforce and then dominate the process of degeneration and phagocytosis, and are effectively required for Wallerian degeneration (George et al. 1995).

The mechanisms by which macrophages potentiate Wallerian degeneration are not completely understood, although cytokine signaling and secretion of proteases, such as the matrix metalloproteinases (MMPs), play a central role. Following upregulation of proinflammatory cytokines stimulating expression of endothelial adhesion molecules (Beuche and Friede 1984; Stoll et al. 2002), macrophages migrate through the endothelium following additional cytokine and other chemotactic gradients present in injured nerve. Activated macrophages secrete components of the complement cascade, coagulation factors, proteases, hydrolases, interferons, TNF-α, and other cytokines, which facilitate the degeneration and phagocytosis of the nerve fiber. The activation of macrophages may be related to cytokines expressed by activated Schwann cells, mast cells, and endothelial cells following local MMP-induced changes in TNF-α form. Liefner et al. (2000) have used TNF-α gene knockout mice to demonstrate that the main function of TNF-α during Wallerian degeneration is the induction of macrophage recruitment from circulation.

Our previous work shows a progression in neuropathological change related to TNF-α dose. At low doses of TNF-α injected in rat sciatic nerve, there is substantial endoneurial edema accumulating in the subperineurial, perivascular, and endoneurial spaces of the nerve bundle (Wagner and Myers 1996a). At doses in the range of 10 microliters of a 2.5 pg/ml solution of murine recombinant TNF-α, extensive demyelination was observed along the injection tract. Studies of tissue injected with a higher dose of TNF-α showed extensive splitting of myelin lamellae, which formed large vacuoles prior to demyelination. Schwann cells were activated and contained lipid debris consistent with their phagocytic role and macrophages invaded the tissue after 3 days to reinforce the phagocytic process. Activated fibroblasts were present in the endoneurium, and there were reactive changes in endothelial cells. Occasional axons were undergoing Wallerian-like degeneration.

Some normal Schwann cells constitutively express TNF-α and IL-1β *in vivo*, and there is a significant increase in immunoreactivity during Wallerian degeneration (Wagner and Myers 1996). Documented by immunohistochemistry staining for TNF-α protein and *in situ* hybridization for TNF-α mRNA, there is an increase in the number and density of Schwann cell cytoplasmic staining for both TNF-α protein and mRNA following nerve injury. Other endoneurial cells immunopositive for TNF-α and IL-1β that are activated during Wallerian degeneration include endothelial cells, fibroblasts, mast cells, and macrophages (Figure 8.2). The initial increase in TNF-α immunoreactivity in endoneurial cells also positive for IL-1β present within hours after nerve injury and is doubled by 7 days. It is thought that the initial increase in Schwann cell proinflammatory cytokine immunoreactivity following nerve injury serves several important functions, including recruitment and activation of macrophages to the injury site and facilitation of the phagocytic role of Schwann cells. The phagocytic process is then extended and amplified by recruited macrophages, which may, in turn, attack Schwann cells. Thus, it is of particular relevance that TNF-α is an active component of nucleus pulposus in herniated lumbar disc tissue (Olmarker and Larsson 1998) and that the related clinical problem of sciatica might be treated by anti-cytokine therapy (see below).

In fact, it is now clearly understood that interfering with the time-course and magnitude of Wallerian degeneration by altering macrophage activity and/or TNF-α

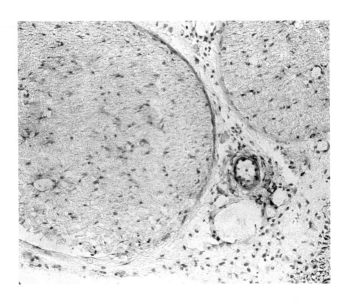

FIGURE 8.2 TNF-α immunoreactivity in damaged nerve. Light micrograph of frozen section of rat sciatic nerve shortly after chronic constriction injury. Note cellular staining by TNF-α antibody in both epineurial and endoneurial compartments. Dense staining includes macrophages extravasated from epineurial vessels and Schwann cells within the endoneurial compartment. Other cells also up-regulate TNF-α at this time including endothelial cells, fibroblasts, and perineurial cells.

expression can modulate the pain experience (Myers et al. 1996; Sommer et al. 2001). This was first demonstrated in experiments with a mouse model of neuropathic pain. Prior to 1996, experimental models of neuropathic pain following nerve injury were conducted primarily in rats using several quantitative measures of pain behavior, such as the Hargreaves test of thermal hyperalgesia (Hargreaves et al. 1988). We adapted these techniques to a mouse model of chronic constriction nerve injury (CCI) using the Wld[s] (slow Wallerian degeneration) strain of mice to test the hypothesis that a delay in macrophage recruitment would be reflected in a delayed onset and reduced magnitude of the thermal hyperalgesia behavior that is characteristic of the CCI model of neuropathic pain. The hypothesis was initially supported by quantitative measures of macrophage numbers and the rate of Wallerian degeneration, as is inferred by the pathological analysis of the histology in Figure 8.3. Subsequent molecular studies and cytokine analyses have reinforced the hypothesis (Shubayev et al. 2006).

The use of the Wld[s] mouse for these studies and other fundamental studies of nerve degeneration typifies the importance of translational research. The Wld[s] mouse has provided insights into neurological degeneration caused by Alzheimer's disease, Parkinson's disease, Creutzfeld-Jakob disease, HIV dementia, and multiple sclerosis (Coleman 2005). The Wld[s] mouse is a spontaneous mutant on chromosome 4 (Lyon et al. 1993) identified as an 85-kb tandem triplication (Coleman et al. 1998) producing a neuroprotective fusion protein of ubiquitin assembly factor E4B (Ube4b) and mononucleotide adenylyltransferase (Nmnat) (Conforti et al. 2000). This genetic

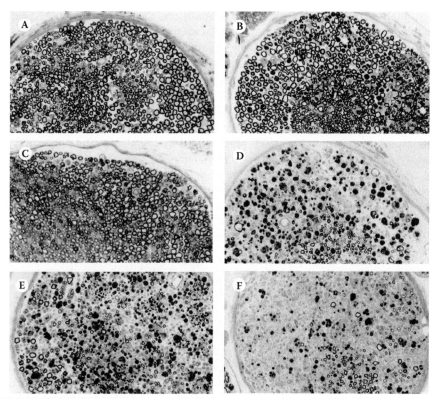

FIGURE 8.3 Delayed Wallerian degeneration of Wld[s] mice. Histological sections (plastic embedded) of peripheral nerve were compared from wild-type (WT) and Wld[s] animals at days 3, 7, and 28 post-injury to verify that Wallerian degeneration had been delayed in the Wld[s] group (A,C,E: Wld[s] days 3, 7, and 28, respectively; B,D,F: WT comparison). At 3 days post-injury, there are very few degenerating axons in tissue from Wld[s] mice (A), whereas WT mice already show a considerable amount of axon and myelin breakdown (B). At 7 days, many large myelinated fibers are still well preserved in the Wld[s] mice (C). In the WT mice, most large myelinated fibers are undergoing Wallerian degeneration (D). Twenty-eight days after injury, there are still more large myelinated nerve fibers present in the Wld[s] mice (E) than in the controls (F). (Reprinted with permission from Myers, R. R., Heckman, H. M., and Rodriguez, M., 1996. Reduced hyperalgesia in nerve-injured WLD mice: relationship to nerve fiber phagocytosis, axonal degeneration, and regeneration in normal mice. *Exp Neurol* 141, 94–101.)

defect is intrinsic to neurons causing delay in Wallerian degeneration and, as a result, delay in macrophage migration (Lunn et al. 1989; Perry et al. 1990a, b).

8.4 CYTOKINES IN CENTRAL NEUROINFLAMMATION AND GLIAL ACTIVATION

Persistent and recurring exposure of injured sensory axons to proinflammatory cytokines by activated stimulation of glia and immune cells promotes recurrent ectopic depolarization and leads to spinal sensitization and enhanced painful behavior. This mechanism occurs not only at the site of injury but in the segmental dorsal root ganglia (DRG) and spinal cord (Milligan and Watkins 2009; Scholz and Woolf 2007).

In the DRG, resident macrophages are normally present, alongside satellite "glia," and though remote from the lesion site, they react to nerve injury by releasing cytokines that foster the influx of infiltrating macrophages (Scholz and Woolf 2007). A week after nerve transection, macrophages are diffused throughout the DRG, where they remain elevated for at least 3 months surrounding the cell bodies of reactive sensory neurons (Hu and McLachlan 2003; Tandrup et al. 2000). The mechanisms of remote neuroinflammation in DRG well after the inflammatory reaction at the nerve lesion site subsides are still not clear, but continuous apoptosis of sensory DRG neurons, resulting in more than 50% neuronal loss by one month after mouse nerve transection, is believed to replenish the proinflammatory cytokine milieu in DRG and contribute to the sustained influx of immune cells (Scholz and Woolf 2007). Proinflammatory cytokines then directly elicit ectopic action potential discharges (see below) and alter the phenotype of sensory neurons, altering the efficacy of their synaptic input into the spinal cord and promoting neuropathic activity (Scholz and Woolf 2007).

Activation of spinal cord glia is both necessary and sometimes even sufficient to the development of persistent pain states associated with various etiologies, including diabetic neuropathy, chemotherapy-induced neuropathy, peripheral nerve inflammation and trauma, and spinal cord inflammation (DeLeo and Winkelstein 2002; Watkins et al. 2007; Wieseler-Frank et al. 2004). Both spinal astrocytes and microglia activate mitogen-activated protein kinases (MAPKs) to induce the synthesis and release of proinflammatory cytokines, such as IL-1β, IL-6, TNF-α, PGE$_2$, and nitric oxide (NO) (Ji and Suter 2007; Zhuang et al. 2005). Resident microglia act as a first line of defense to proinflammatory stimuli, which rapidly proliferate to produce inflammatory and anti-inflammatory cytokines and other substances, and to activate nearby astrocytes, microglia, and neurons (Romero-Sandoval et al. 2008). During chronic neuropathic conditions, sustained activation of astrocytes is believed to explain the ongoing activity of neuropathic cascades (Ji and Suter 2007). In addition, astrocytes encapsulate synapses and remain in close contact with neuronal soma where they can directly alter neuronal communication via expression of neurotransmitter receptors, such as ionotropic non-N-methyl-D-aspartic acid (NMDA) and NMDA receptors, metabotropic glutamate, purinergic and substance P receptors (Haydon 2001; Porter and McCarthy 1997). Differential action of MMPs on continuous IL-1β release in microglia and astrocytes is believed to contribute to the development and the maintenance of neuropathic pain (Kawasaki et al. 2008; Myers et al. 2006), as discussed below.

8.5   AXONAL TRANSPORT OF CYTOKINES

The development of pain states is characterized by acute stimulation of the injured sensory axis through electrophysiological stimuli, synaptic release of neurotransmitters, and kinase activity, followed by longer term modulation through gene expression at the injury site and neuronal soma of DRG, in part as a function of retrograde axonal flow (Myers et al. 2006). The importance of injury-induced factors in stimulation of ectopic hyperexcitability of neuronal soma has been demonstrated through the injection of injured axoplasm into the cell bodies of uninjured sensory neurons

(Ambron et al. 1995). This neuronal hyperexcitability was diminished with exposure of injured nerves to colchicine, a blocker of axonal transport (Gunstream et al. 1995). Axonal transport of small proteins to the cell body (retrograde) influences cell function and from the cell body (anterograde) influences axon viability (Kristensson 1984; Redshaw and Bisby 1984, 1987). Disruption of bi-directional axonal transport of trophic factors, referred to as "trophic currencies" (Altar and DiStefano 1998; von Bartheld 2004), is considered to be an early and perhaps causative event in many neurodegenerative pathologies (De Vos et al. 2008).

We hypothesized that retrograde axonal transport of proinflammatory cytokines contributes to the pathogenesis of neuropathic pain and have studied this mechanism in the model of peripheral neuropathic pain of rat sciatic nerve chronic constriction injury (CCI) (Schafers et al. 2002; Shubayev and Myers 2001, 2002a). Endogenous TNF-α transport was activated within one day of nerve damage and returned to basal levels within a week of CCI (Shubayev and Myers 2001), a typical temporal pattern for axonal transport of trophic factors after nerve injury (Curtis et al. 1998; Leitner et al. 1999; Tonra et al. 1998). To characterize the speed, direction, and other features of axonal TNF-α transport, we used a biotinylated TNF-α tracer, which was injected at the nerve injury site or directly into neuronal soma of DRG and traced along the sensory nerve tract. TNF-α transported axonally at about 300 mm/day, a speed characteristic of small protein retrograde transport. To compare the specific features of biotin-mediated and biotinylated TNF-α-mediated transport, we used Neurobiotin™, an amino derivative of biotin (N-(2-aminoethyl)biotinamide) used for intracellular labeling and neuronal tracing studies (Kita and Armstrong 1991). Through a series of studies, we identified the following unique characteristics of TNF-α transport (Shubayev et al. 2005; Shubayev and Myers 2001, 2002a): (1) intraaxonal TNF-α uptake and transport was selective to a subset of fibers, whereas neurobiotin was internalized by all visible fibers; (2) TNF-α tracer localized intraaxonally and in association with the myelin sheath and/or Schwann cell cytoplasm, whereas neurobiotin was found exclusively intraaxonally; (3) upon its retrograde transport along sensory afferents, TNF-α uptake by DRG soma was only observed in uninjured nerve. In injured afferents, TNF-α tracer accumulated in the axons adjacent to neuronal soma of DRG; in contrast, neurobiotin labeled comparably neuronal soma of injured or normal nerve; (4) TNF-α tracer injected into DRG transported anterogradely back to peripheral nerve and to the spinal cord after peripheral nerve injury and in uninjured neuroaxis, similarly to neurobiotin. However, only TNF-α tracer, and only after injection into injured peripheral nerve, reached the segmental spinal cord, implicating the importance of injury-induced factors in TNF-α transport; (5) endogenously expressed TNF-α receptors I and II were found to co-localize with the TNF-α tracer, suggesting that TNF-α transport is receptor mediated; (6) TNF-α tracer co-localized with both calcitonin gene-related peptide (CRPR)-positive and NF200-positive fibers of the ipsilateral spinal cord, and induced activation of spinal glia consistent with its role in the development of neuropathic pain.

Long axonal processes of neurons necessitate efficient means of axonal transport of small proteins, achieved via different cargo systems (De Vos et al. 2008). Cytokine-related signaling molecules, including MAPKs and JNK (see below), undergo axonal transport (Cavalli et al. 2005); however, future studies need to identify whether and how generation of cytokine signaling endosome occurs (Cosker et al. 2008). Overall, axonal cytokine transport represents a valid mechanism that explains central glial activation and remote neuroinflammation and transcriptional activation of cytokine-dependent factors in the generation of persistent pain states (DeLeo and Winkelstein 2002; Watkins and Maier 2002), including phantom and mirror-image pain.

8.6   CYTOKINES AND PAIN SIGNALING

8.6.1   Cytokine-Dependent Intracellular Signaling

Each cytokine has one or more cell-surface receptors that ensure their autocrine, paracrine, or endocrine signaling to phosphorylation of mitogen-activated protein kinases (MAPKs) and regulation of various cellular activities in the neuronal system, such as gene expression, mitosis, differentiation, and cell survival/apoptosis. Studies with knockout mice for proinflammatory cytokine receptor genes have been useful in exploring their specific signaling roles in the development of pain states. For instance, using TNFRI-/- and TNFRII-/- mice have shown that thermal hyperalgesia requires TNFRI, while mechanical and cold allodynia depend on TNFRI or TNFRII (George et al. 2005; Vogel et al. 2006).

Activation of TNF-α receptors and recruitment of TNF-α receptor-associated factors (TRAFs), an important group of intracellular adaptor proteins, leads to phosphorylation of p38 MAPK, extracellular signal-regulated kinases (ERK), and Jun N-terminal kinase (JNK), potentially activating the NF-κB transcription pathways. The complexity of cytokine cross-talk and cytokine signaling can be illustrated, as shown in Figure 8.4. Peripheral nerve injury increases phosphorylation of p38, ERK, and JNK in neurons and non-neuronal cells (e.g., satellite cells) in the DRG (Jin et al. 2003; Kenney and Kocsis 1998; Obata et al. 2003). In addition, nerve injury–induced spinal microglial activation is characterized by phosphorylation of p38 MAP kinase, ERK isoforms 1 and 2 (Jin et al. 2003; Svensson et al. 2005; Zhuang et al. 2005).

In 1994, p38 was first identified as a MAPK targeted by endotoxin and hyperosmolarity in mammalian cells (Han et al. 1994). At the same time, the cloned target for an anti-inflammatory drug (SB203580) was found to be identical to p38 (Lee et al. 1994). It is now known that the regulation of cytokine biosynthesis in many different cell types is mediated through p38 activation. The p38 pathway phosphorylates and enhances the activity of many transcription factors, such as ELK-1, NF-κB, heat shock transcription factor-1, and SAP-1. Activation of p38 in spinal microglia peaks 3 days after peripheral nerve injury, followed by a slow decline over several weeks (Jin et al. 2003). Intrathecal administration of p38 MAPK inhibitors has been demonstrated to inhibit the development of allodynia and hyperalgesia in several models of pain, presumably by blocking p38 phosphorylation in spinal microglia (Ji et al. 2002; Jin et al. 2003; Milligan et al. 2003; Svensson et al. 2003). Although effective in preventing enhanced pain responses, most p38 inhibition has failed to reverse established pain states (Svensson et al. 2003). However, Cytokine Pharmascience's p38 inhibitor (CNI-1493), administered intrathecally, has been reported to prevent and reverse allodynia associated with extraneural inflammation (Milligan et al. 2003).

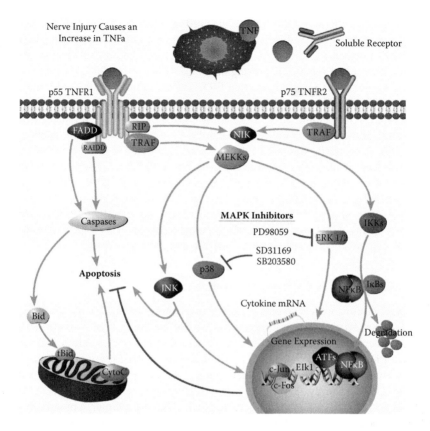

FIGURE 8.4 Mechanisms of TNF-α signaling in injured nerve. By occupying either of the two TNF-α receptors, the soluble TNF-α protein can produce different effects ranging from apoptosis to up-regulation of itself, other proinflammatory cytokines, or anti-inflammatory cytokines. While the full details of these effects are not known, it is clear that stimulation of the p38 mitogen-activated protein kinase (MAPK) pathway is a critical signaling mechanism. Activation of the JNK pathway can produce radically different effects, depending in part on activity in the pathway. Studies with TNF-/- and TNFRI-/- and RII-/- mice are helpful in defining the inter-relationships of cytokines and their receptors.

ERK was originally identified as a primary effector of growth factor receptor signaling, a cascade that involves sequential activation of Ras, Raf, MEK, and ERK (Ji et al. 2008). Activation of ERK in spinal cord dorsal horn neurons is associated with the activation of nociceptive-specific sensory fibers and promotes intracellular events that contribute to central sensitization, which can manifest at both behavioral and cellular levels (Woolf and Salter 2000). Several studies have shown sequential activation of ERK in dorsal horn neurons, then microglia, and finally astrocytes in a neuropathic pain model (Cheng et al. 2003; Ma and Quirion 2002; Zhuang et al. 2005). Intrathecal MEK inhibitor PD98059 has been demonstrated to attenuate spinal nerve ligation-induced mechanical allodynia (Zhuang et al. 2005). ERK activation is also induced in spinal microglia after streptozotocin (STZ)-induced diabetes, and intrathecal U0126 can suppress STZ-induced neuropathic pain (Tsuda et al. 2008).

In addition to peripheral nerve injury, spinal cord injury also activates ERK in spinal microglia, and PD98059 reduces neuropathic pain after spinal cord injury (Zhao et al. 2007). Further, PD98059 inhibits the induction of cyclooxygenase-2 in spinal microglia and spinal release of prostaglandin E2 (Zhao et al. 2007). Collectively, the studies described above indicate that ERK activation in the spinal cord plays an important role in the development and maintenance of neuropathic pain.

Compared with p38 and ERK, little is known about how JNK regulates neuropathic pain. Unlike the activation patterns of ERK and p38, increased phosphorylation of JNK is primarily observed in spinal cord astrocytes at later stages of nerve injury (Zhuang et al. 2005). Intrathecal infusion of the JNK inhibitor SP600125 attenuates neuropathic pain in the SNL model (Obata et al. 2004; Zhuang et al. 2006) and diabetic neuropathy model (Daulhac et al. 2006). Although TNF-α only induces a transient activation of JNK in astrocytes (Zhang et al. 1996), the growth factor FGF-2 (fibroblast growth factor-2) can induce persistent JNK activation in the spinal cord *in vivo* and in astrocytes *in vitro* (Ji et al. 2006).

8.6.2 EXTRACELLULAR CYTOKINE-PROTEASE NETWORK

Cytokines exist in a variety of biologically active isoforms that allow for a well-coordinated, complex, and self-sustained network. For example, TNF-α is synthesized as a monomeric transmembrane 26 kDa protein that is inserted into a cell membrane as a homotrimer (Wajant et al. 2003) and proteolytically cleaved into a 51 kDa homotrimer by an extracellular metalloprotease of a disintegrin and metalloproteinase (ADAMs) family, TNF-α trimer converting enzyme (TACE or ADAM-17) (Black et al. 1997). Extracellular processing of TNF-α into a dimer, a monomer, and smaller peptides is performed by several members of the extracellular matrix metalloproteinase (MMP) family (Gearing et al. 1994). But the relationship of TNF-α with MMPs is multi-fold and far more complex than merely release of an activated TNF-α ligand. Because proinflammatory cytokines, including TNF-α, are potent inducers of MMP gene expression in various cells (Nagase 1997; Saren et al. 1996), including peripheral glia (Chattopadhyay et al. 2007; Shubayev et al. 2006), TNF-α increases the expression of MMPs to facilitate degradation of neurovascular barriers and infiltration of inflammatory cells (Kieseier et al. 2006; Rosenberg 2002; Shubayev et al. 2006). Therefore, direct injection of exogenous recombinant MMP-9 into the injured nerves of macrophage-deficient TNF-α knockout mice resumed their deficient ability to recruit hematogenous macrophages (Shubayev et al. 2006). Through the use of TNF-α receptors knockout mice, we found that TNFRI and TNFRII are partially responsible for MMP-9 induction in injured sciatic nerve and that other proinflammatory cytokines contribute (Chattopadhyay et al. 2007). At the same time, multiple MMPs regulate TNF-α signaling through processing of TNF-α ligand, leading to activation of TNF-α-dependent signaling and its surface receptors (leading to inactivation), TNFRI and TNFRII (Williams et al. 1996), generating a soluble receptor that can act as a direct TNF-α antagonist, as reviewed further.

MMPs represent a large family of highly potent extracellular proteases that comprise collagenases, stromelysins, gelatinases, and membrane-type (MT)-MMPs (Page-McCaw et al. 2007). In injured peripheral nerve, MMPs control the

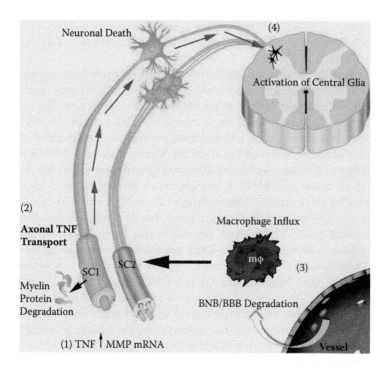

FIGURE 8.5   A proposed diagram of cytokine-metalloprotease network in pain processing. (1) Proinflammatory cytokines induce the expression of the early-gene MMPs, such as MMP-9, in myelinated Schwann cell (Sc) within 1 day after injury. (2) MMPs mediate TNF-α-induced myelin protein degradation, contributing to primary demyelination. Myelin degradation exposes axonal plasma membrane of mechanosensory myelinated Aβ fibers for ectopic hyperexcitability to Sc-derived nociceptive stimuli. Injury-released TNF-α undergoes axonal transport to DRG and spinal cord, where it contributes to ectopic hyperexcitability of neuronal soma (3). MMPs are the only proteases able to degrade blood–nerve barrier, mediating cytokine-induced macrophage infiltration, local neuroinflammation, and neuropathic pain. (4). Activated spinal microglia drives the initiation, whereas astrocytes contribute to the maintenance of the persistent pain states. It is stimulated by axonally transported TNF-α, in part responsible for remote activation of MMP expression, neuroinflammation, and central glial activation. Thus, broad-spectrum MMP inhibition is effective in attenuation of acute and persistent mechanical nociception. Note: MMP, matrix metalloproteinase; TNF, tumor necrosis factor alpha; SC1, myelinating Schwann cell; SC2, non-myelinating Schwann cell; mø, macrophage; BNB, blood–nerve barrier; BBB, blood–brain barrier.

integrity of blood–brain barrier, myelin protein turnover, survival, and phenotypic remodeling of glia and neurons as summarized in Figure 8.5 (Chattopadhyay et al. 2007; Kobayashi et al. 2008; Shubayev et al. 2006). The specific role of MMPs in generation of pain states is believed to relate to their control of cytokine release (Kawasaki et al. 2008; Myers et al. 2006) as well as myelin protein processing on mechanosensory Aβ fibers (Kobayashi et al. 2008). As an early-gene MMP family member, MMP-9 participates in the initiation of the neuropathic pain state through the immediate changes at the injury site and activation of spinal microglia, whereas a late-gene family member, MMP-2, facilitates the development and the maintenance of neuropathic pain via sustaining activation of spinal astrocytes (Kawasaki et al. 2008; Kobayashi et al. 2008).

Pharmacologic inhibition of MMPs using synthetic hydroxamate-based MMP inhibitors (MMPi) has been considered an important therapeutic approach to cytokine-related neuropathic pain states (Myers et al. 2006). Administration of a broad-spectrum MMP inhibitor or small interfering RNA (siRNA) for MMP-9 and MMP-2 produced an immediate and sustained attenuation of mechanical allodynia (Kawasaki et al. 2008; Kobayashi et al. 2008). Mutant mice with MT5-MMP gene deletion failed to develop mechanical allodynia after partial sciatic nerve ligation (Komori et al. 2004) and MMP-9 knockout mice showed reduced spontaneous pain scores after nerve crush (Chattopadhyay et al. 2007). Little is known about the mechanisms of TACE action in damaged nerve, but its expression and distribution patterns parallel that of TNF-α in CCI-injured sciatic nerve (Shubayev and Myers 2002). Synthetic inhibitor of TACE, TAPI, attenuated mechanical allodynia, and thermal hyperalgesia of CCI (Sommer et al. 1997). It is important to note that hyperalgesia caused by tissue injury resulting from intraplantar endotoxin also responds to MMPi pre-treatment (Talhouk et al. 2000), implicating MMPs in the development of inflammatory pain states. The anti-nociceptive action of minocycline, a broad-spectrum antimicrobial tetracycline that has shown promise in treating persistent pain, is believed to relate to selective inhibition of microglial activation and reduced expression of proinflammatory cytokines and MMPs (Ledeboer et al. 2005; Raghavendra et al. 2003).

8.6.3 CYTOKINES IN DIRECT NEURONAL EXCITABILITY

There is an increasing appreciation for the ability of cytokines to directly impact excitatory neuronal function, resulting in spontaneous (ectopic) activity and pain facilitation (Schafers and Sorkin 2008; Scholz and Woolf 2007). For example, application of TNF-α perineurially rapidly evokes ongoing activity in C fibers by reducing mechanical activation thresholds in C nociceptors of sciatic nerve (Junger and Sorkin 2000; Sorkin et al. 1997), and its application to DRG neurons elicits neuronal discharges in A- and C-fibers (Ozaktay et al. 2006; Schafers et al. 2003; Zhang et al. 2002). Dorsal root exposure to TNF-α evoked spontaneous discharges in dorsal horn wide dynamic–range neurons (Onda et al. 2002), and its intrathecal administration enhances responses to C fiber stimulation and increases wind-up and the post-discharge response of deep dorsal horn neurons (Reeve et al. 2000). Intraplantar, DRG, or dorsal root exposure to IL-1β induces spontaneous acute discharges and hyperalgesia, potentiates heat-activated inward currents, or increases mechanosensitivity of the peripheral receptive fields (Fukuoka et al. 1994; Ozaktay et al. 2002). Understanding of the mechanism of cytokine action on neuronal excitability specifically in pain processing is only emerging. Overall, cytokine action on neuronal excitability is believed to relate to their action on ion channels and be (1) acute (within seconds or minutes), modulating post-translational modifications and immediate activity of ion channels subunits, or (2) sustained, modulating gene expression of ion channel proteins.

Ion channels are integral gated pore-forming cell membrane proteins that permit ions to flow down their electrochemical gradients into or out of a cell, referred to as *gating*. Gating involves a conformational change of one or more subunits and varies depending on the stimuli and the type of a channel. Voltage-gated sodium, calcium, potassium, and transient receptor potential (TRP) channels respond to a difference in electrical membrane potential, as ligand-gated ionotropic channels, such as nicotinic acetylcholine receptor, glutamate-gated N-methyl-D-aspartic acid (NMDA) and α-amino-3-hydroxyl-5-methyl-4-isoxazole-propionate (APMA) receptors, the anion-permeable α-aminobutyric acid (GABA)$_A$-gated receptor or purinergic ATP-gated P2X receptors, are all stimulated by respective ligands.

Cytokine-stimulated activation of MAPKs promoted phosphorylation of channel subunits, leading to a conformational change affecting their stabilization. In DRG neurons, tetrodotoxin-resistant Na^+ currents were controlled by rapid TNF-α/TRNRI-induced activation of p38 MAPK (Jin and Gereau 2006). TNF-α increased the cell surface levels of AMPA receptor via TNFR1-mediated PI3K activation (Stellwagen et al. 2005). AMPA receptor subunit GluR1 was most responsive to TNF-α, resulting in rapid insertion of Ca^{2+}-permeable AMPA channels (Beattie et al. 2002; Ogoshi et al. 2005; Yang et al. 2005). IL-1β can reduce the frequency of AMPA-dependent spontaneous excitatory postsynaptic currents (Yang et al. 2005) and enhance NMDAR-mediated current (Viviani et al. 2003; Yang et al. 2005). Brief IL-1β application sensitized TRPV1 currents (Obreja et al. 2002) and inhibited Na^+ currents via IL-1β receptor, whereas its extended exposure increased total Na^+ current via the PKC and G-protein-coupled signaling pathway (Liu et al. 2006). TNF-α and IL-1β can acutely reduce the outward K^+ and/or inward Na^+ current in neurons (Diem et al. 2001; Houzen et al. 1997; Sawada et al. 1991). However, IL-1β and not other tested cytokines rapidly inhibited neuronal activity of voltage-dependent calcium channels by reducing the expression of Ca^{2+} channel protein (Plata-Salaman and Ffrench-Mullen 1992, 1994; Zhou et al. 2006). These changes impact a variety of mechanisms, including neurotransmitter release (Murray et al. 1997; Rada et al. 1991), long-term potentiation (Cunningham et al. 1996; Katsuki et al. 1990; Schneider et al. 1998), and synaptic transmission (D'Arcangelo et al. 1997). TNF-α can reduce the expression of glutamate EAAT2/GLT-1 transporter (Sitcheran et al. 2005) and IL-1β decreases the expression of Na^+ channel subunits in epithelial cells (Choi et al. 2007; Roux et al. 2005). It is not only neuronal, but glia-released cytokines also contribute to neuronal excitability. For example, TNF-α can regulate synaptic scaling via the changes in AMPA receptor subtype density (Stellwagen and Malenka 2006) and can reduce PKC-dependent K^+ conductance, affecting glial capacity to buffer extracellular K^+ released via neuronal firing (Koller et al. 1998).

Overall, our understanding of the mechanisms of cytokine action on ectopic neuronal excitability in generating pain states is only emerging. But from other models of neuronal excitability, we have learned that cytokines regulate the acute and long-term changes in voltage-gated and ligand-gated ion channels through modulation of their expression levels, conformational changes, and compartmentalization, such as endocytosis of GABA receptor and glutamate uptake by glial transporters.

8.7 CYTOKINE-RELATED THERAPEUTICS

A strong case can be made for the use of anti-TNF-α therapy in painful inflammatory diseases. Indeed, the FDA has approved several anti-TNF-α drugs for use in arthritis, and there is considerable interest in expanding the indications for anti-TNF-α therapy. In particular, the use of anti-TNF-α drugs to treat the pain and neuropathological consequences of herniated discs and sciatica is already undergoing formal clinical study in Europe, and off-label use of anti-TNF-α drugs for sciatica has begun in the United States. The purpose of this section is to review the indications for the use of anti-TNF-α therapy to treat painful neuropathic conditions.

The commercial development of specific soluble TNF-α receptors has greatly advanced the field (Schafers and Sommer 2007). These biologics provide an excess of soluble TNF-α receptors that can occupy the biologically active TNF-α protein before it encounters the receptors on the cell surface. By doing this they effectively neutralize the TNF-α protein in tissue, and therefore reduce its biologic effect. These biologics typically consist of recombinant human tumor necrosis factor p75 TNF-α receptor-Fc fusion protein. Effectively, this provides a therapeutic counter-effect to TNF-α itself. The products are made by encoding the DNA of the soluble portion of human tumor necrosis factor receptor p75 and combining it with the DNA encoded Fc portion of immunoglobulin IgG, translating it to act as an immunoglobulin-like dimer that competitively inhibits TNF-α and prevents its binding to cell surface receptors. Therefore, by binding to the soluble receptor, it reduces the effect of circulating TNF-α. The therapeutic use of the soluble receptor is restricted because the half-life of the molecule is short, and thus combinations of the soluble receptor with immunoglobulin or other molecules has been attempted to try and increase the half life to enable therapeutic applications. Several drugs are now commercially available including the following:

- etanercept (Enbrel), which has been widely used in rheumatoid arthritis but is also approved for ankylosing spondylitis, psoriasis, and psoriatic arthritis. It is being used off-label for low back pain and other neuroinflammatory diseases (Tobinick 2009).
- infliximab (Remicade), which is a chimeric monoclonal TNF-α antibody approved for rheumatoid arthritis with methotrexate and for Crohn's disease. It is currently undergoing formal double-blind, randomized clinical trials for sciatica and has been studied extensively in laboratory animals with experimental spinal disc herniation.
- adalimumab (Humira) is the newest TNF-α biological response modifier to be approved in the United States and is a fully human recombinant immunoglobulin G1 anti-TNF-α monoclonal antibody.

Several other similar anti-TNF-α biologics are under development.

The rationale for the use of these compounds in human disease comes directly from experimental knowledge of the role of TNF-α in painful radicular neuropathy and from case reports and small clinical trials, particularly with respect to the prevalent problem of lumbar and cervical radiculopathy. Lifetime incidence of lumbar radiculopathy (sciatica) is 20%–40% (Nachemson 2004). Radicular pain is intense, unremitting, and often unresponsive to analgesics. When physical therapy and non-steroidal anti-inflammatory drugs (NSAIDs) provide inadequate relief, treatment progresses to opioid analgesics and to more invasive, expensive measures such as epidural injections of steroids or of local anesthetics. Even these measures are often inadequate in the degree and/or durability of pain relief provided. For patients with MRI-confirmed disc herniation and persistent radicular pain of 6 to 8 weeks' duration, evaluation for lumbar disc surgery is usually recommended. Natural history studies and randomized controlled trials confirm that more than 50% of first-episode radiculopathy patients recover without surgery if they can obtain adequate pain relief during the initial 2- to 3-month recovery period. Therefore, the decision of whether to elect surgery is often difficult. Indeed, 56% of patients offered surgery declined after viewing a balanced summary of the risks, benefits, and alternatives (Weinstein et al. 2006). Many patients strongly prefer to postpone or avoid lumbar disc surgery, in order to allow spontaneous disc resorption and recovery from disability and pain. Surgeons and patients are actively seeking more effective approaches to relief of radicular pain that are less expensive, less invasive, and more reversible than lumbar disc surgery. Anti-cytokine therapy is therefore actively being explored.

The potential of TNF-α as a target for neuropathic pain therapy is based upon its critical role in mediating neuroinflammation. Of particular importance is the relationship of upregulated TNF-α to spontaneous electrophysiological activity in pain fibers and the process of Wallerian degeneration discussed above, since axonal injury underlies the pathophysiology of radicular pain.

Data from multiple pre-clinical models support the potential of TNF-α inhibition to treat the pain and disability of disc injury by addressing its pathophysiology (Cooper and Freemont 2004; Myers et al. 2006; Olmarker et al. 2004). In the pig nucleus pulposus puncture model, selective TNF-α inhibition protects against nerve root injury, preventing thrombus formation, intraneural edema, and reduced nerve conduction velocity (reviewed by Olmarker et al. 2004, 2005). Similarly, TNF-α inhibition protects from peripheral neuropathy and pain (Schafers et al. 2001). For example, etanercept reduces hyperalgesia in experimental painful neuropathy (Sommer et al. 2001).

TNF-α inhibitors have shown to be of potential efficacy in small academic trials in radiculopathy patients (Karppinen et al. 2003; Korhonen et al. 2004, 2005, 2006; Shin et al. 2006; Tobinick and Davoodifar 2004). A blinded randomized, controlled trial of epidural etanercept has been initiated in patients with MRI-confirmed lumbar disc herniation and persistent lumbar radiculopathy (Cohen and Griffith 2006). Initial data are encouraging.

Additionally, anti-IL-1β therapy is available with the drug anakinra (Kineret, Amben), which is a recombinant, nonglycosylated form of the human interleukin-1 receptor antagonist (IL-1Ra). Kineret blocks the biologic activity of IL-1β by competitively inhibiting IL-1β binding to the interleukin-1 type 1 receptor (IL-1R1). It is indicated for the reduction in signs and symptoms and slowing the progression of structural damage in active rheumatoid arthritis in patients 18 years of age or older who have failed one or more disease-modifying antirheumatic drugs. Kineret should not be used with TNF-α blocking agents. It is given subcutaneously in a dose of 100 mg on a daily basis.

Before the commercial development of monoclonal antibodies to TNF-α or soluble receptors, we tested the ability of thalidomide to reduce the painful and pathological consequences of CCI (Sommer and Schafers 1998). Typically, CCI produces a thermal hyperalgesia that peaks within the first week of injury and Wallerian degeneration and then slowly resolves during the phase of nerve regeneration. Thalidomide reduces the production of TNF-α by activated macrophages and significantly reduces the degree of thermal hyperalgesia during the peak of macrophage influx. When treatment was stopped, hyperalgesia increased.

Gene transfer therapy approaches with subcutaneous inoculation of soluble TNFR1-expressing herpes simplex virus vector inducing a local release of soluble TNFR1 to the DRG successfully decreased pain-related behavior after SNL by reducing membrane-bound TNF-α and concomitant reductions in IL-1β and phosphorylated p38 (Hao et al. 2007). Recently, an alternative strategy using active anti-TNF-α vaccination successfully reduced TNF-α-driven chronic and acute inflammation in mice (Le Buanec et al. 2006).

8.7.1 Anti-Inflammatory Cytokines

Upregulation of TNF-α also leads to compensatory increases in anti-inflammatory cytokines, such as IL-4 and IL-10, that down-regulate TNF-α production. We injected IL-10 directly into rat sciatic nerve following CCI to determine if this form of anti-TNF-α therapy reduced CCI hyperalgesia (Wagner et al. 1998). Hyperalgesia was significantly diminished throughout the experimental period. Thus, exogenous IL-10 therapy could be a means through which the detrimental effects of TNF-α can be regulated, a strategy now being explored in clinical trials for the management of sepsis and congestive heart failure (Kumar et al. 2005). More recently, intrathecal IL-10 gene delivery has been shown to produce a prolonged reversal of neuropathic pain (Milligan et al. 2006a, b). Anti-inflammatory cytokines may play a crucial role because of tight interactions with endogenous analgesia, for instance the opioidergic system (Kraus et al. 2001). The anti-inflammatory cytokine IL-4 also has analgesic effects in animal models of neuropathic pain (Hao et al. 2006; Vale et al. 2003).

In summary, cytokines are potent modulators of pain cascades through local reorganization in damaged tissue and nerve, activating acute pain cascades through direct neuronal excitability; intracellular MAPK; and other signaling cascades in local glial, endothelial, and mast cells, resulting in macrophage recruitment, sustained activation of local glia, and localized neuroinflammation. Through axonal transport and sustained neuronal excitability, cytokines can activate central glia in segmental spinal cord and cephalad pathways, promoting central neuroinflammation and persistent pain. Anti-nociceptive strategies include inhibition of proinflammatory cytokines and use of anti-inflammatory cytokines.

REFERENCES

Altar, C. A., and DiStefano, P. S., 1998. Neurotrophin trafficking by anterograde transport. *Trends Neurosci* 21, 433–437.

Ambron, R. T., Dulin, M. F., Zhang, X. P., Schmied, R., and Walters, E. T., 1995. Axoplasm enriched in a protein mobilized by nerve injury induces memory-like alterations in Aplysia neurons. *J Neurosci* 15, 3440–3446.

Beattie, E. C., Stellwagen, D., Morishita, W., Bresnahan, J. C., Ha, B. K., Von Zastrow, M., Beattie, M. S., and Malenka, R. C., 2002. Control of synaptic strength by glial TNF-alpha. *Science* 295, 2282–2285.

Beuche, W., and Friede, R. L., 1984. The role of non-resident cells in Wallerian degeneration. *J Neurocytol* 13, 767–796.

Black, R. A., Rauch, C. T., Kozlosky, C. J., Peschon, J. J., Slack, J. L., Wolfson, M. F., Castner, B. J., et al. 1997. A metalloproteinase disintegrin that releases tumour-necrosis factor-alpha from cells. *Nature* 385, 729–733.

Boulay, J. L., O'Shea, J. J., and Paul, W. E., 2003. Molecular phylogeny within type I cytokines and their cognate receptors. *Immunity* 19, 159–163.

Cavalli, V., Kujala, P., Klumperman, J., and Goldstein, L. S., 2005. Sunday Driver links axonal transport to damage signaling. *J Cell Biol* 168, 775–787.

Chattopadhyay, S., Myers, R. R., Janes, J., and Shubayev, V., 2007. Cytokine regulation of MMP-9 in peripheral glia: implications for pathological processes and pain in injured nerve. *Brain Behav Immun* 21, 561–568.

Cheng, X. P., Wang, B. R., Liu, H. L., You, S. W., Huang, W. J., Jiao, X. Y., and Ju, G., 2003. Phosphorylation of extracellular signal-regulated kinases 1/2 is predominantly enhanced in the microglia of the rat spinal cord following dorsal root transection. *Neuroscience* 119, 701–712.

Choi, J. Y., Choi, Y. S., Kim, S. J., Son, E. J., Choi, H. S., and Yoon, J. H., 2007. Interleukin-1 beta suppresses epithelial sodium channel beta-subunit expression and ENaC-dependent fluid absorption in human middle ear epithelial cells. *Eur J Pharmacol* 567, 19–25.

Cohen, S., and Griffith, S., 2006. Efficacy of epidural etanercept in the treatment of radicular low back pain (LBP). Clinical Trials.gov NCT00374572.

Coleman, M., 2005. Axon degeneration mechanisms: commonality amid diversity. *Nat Rev Neurosci* 6, 889–898.

Coleman, M. P., Conforti, L., Buckmaster, E. A., Tarlton, A., Ewing, R. M., Brown, M. C., Lyon, M. F., and Perry, V. H., 1998. An 85-kb tandem triplication in the slow Wallerian degeneration (Wlds) mouse. *Proc Natl Acad Sci U S A* 95, 9985–9990.

Conforti, L., Tarlton, A., Mack, T. G., Mi, W., Buckmaster, E. A., Wagner, D., Perry, V. H., and Coleman, M. P., 2000. A Ufd2/D4Cole1e chimeric protein and overexpression of Rbp7 in the slow Wallerian degeneration (WldS) mouse. *Proc Natl Acad Sci U S A* 97, 11377–11382.

Cooper, R. G., and Freemont, A. J., 2004. TNF-alpha blockade for herniated intervertebral disc-induced sciatica: a way forward at last? *Rheumatology* (Oxford) 43, 119–121.

Cosker, K. E., Courchesne, S. L., and Segal, R. A., 2008. Action in the axon: generation and transport of signaling endosomes. *Curr Opin Neurobiol* 18, 270–275.

Covey, W. C., Ignatowski, T. A., Knight, P. R., and Spengler, R. N., 2000. Brain-derived TNF-alpha: involvement in neuroplastic changes implicated in the conscious perception of persistent pain. *Brain Res* 859, 113–122.

Cunningham, A. J., Murray, C. A., O'Neill, L. A., Lynch, M. A., and O'Connor, J. J., 1996. Interleukin-1 beta (IL-1 beta) and tumour necrosis factor (TNF) inhibit long-term potentiation in the rat dentate gyrus in vitro. *Neurosci Lett* 203, 17–20.

Curtis, R., Tonra, J. R., Stark, J. L., Adryan, K. M., Park, J. S., Cliffer, K. D., Lindsay, R. M., and DiStefano, P. S., 1998. Neuronal injury increases retrograde axonal transport of the neurotrophins to spinal sensory neurons and motor neurons via multiple receptor mechanisms. *Mol Cell Neurosci* 12, 105–118.

D'Arcangelo, G., Dodt, H. U., and Zieglgansberger, W., 1997. Reduction of excitation by interleukin-1 beta in rat neocortical slices visualized using infrared-darkfield videomicroscopy. *Neuroreport* 8, 2079–2083.

Daulhac, L., Mallet, C., Courteix, C., Etienne, M., Duroux, E., Privat, A. M., Eschalier, A., and Fialip, J., 2006. Diabetes-induced mechanical hyperalgesia involves spinal mitogen-activated protein kinase activation in neurons and microglia via N-methyl-D-aspartate-dependent mechanisms. *Mol Pharmacol* 70, 1246–1254.

De Vos, K. J., Grierson, A. J., Ackerley, S., and Miller, C. C., 2008. Role of axonal transport in neurodegenerative diseases. *Annu Rev Neurosci* 31, 151–173.

DeLeo, J. A., and Winkelstein, B. A., 2002. Physiology of chronic spinal pain syndromes: from animal models to biomechanics. *Spine* 27, 2526–2537.

Diem, R., Meyer, R., Weishaupt, J. H., and Bahr, M., 2001. Reduction of potassium currents and phosphatidylinositol 3-kinase-dependent AKT phosphorylation by tumor necrosis factor-(alpha) rescues axotomized retinal ganglion cells from retrograde cell death in vivo. *J Neurosci* 21, 2058–2066.

Fukuoka, H., Kawatani, M., Hisamitsu, T., and Takeshige, C., 1994. Cutaneous hyperalgesia induced by peripheral injection of interleukin-1 beta in the rat. *Brain Res* 657, 133–140.

Gearing, A. J., Beckett, P., Christodoulou, M., Churchill, M., Clements, J., Davidson, A. H., Drummond, A. H., et al. 1994. Processing of tumour necrosis factor-alpha precursor by metalloproteinases. *Nature* 370, 555–557.

George, A., Buehl, A., and Sommer, C., 2005. Tumor necrosis factor receptor 1 and 2 proteins are differentially regulated during Wallerian degeneration of mouse sciatic nerve. *Exp Neurol* 192, 163–166.

George, E. B., Glass, J. D., and Griffin, J. W., 1995. Axotomy-induced axonal degeneration is mediated by calcium influx through ion-specific channels. *J Neurosci* 15, 6445–6452.

Gunstream, J. D., Castro, G. A., and Walters, E. T., 1995. Retrograde transport of plasticity signals in Aplysia sensory neurons following axonal injury. *J Neurosci* 15, 439–448.

Han, J., Lee, J. D., Bibbs, L., and Ulevitch, R. J., 1994. A MAP kinase targeted by endotoxin and hyperosmolarity in mammalian cells. *Science* 265, 808–811.

Hao, S., Mata, M., Glorioso, J. C., and Fink, D. J., 2006. HSV-mediated expression of interleukin-4 in dorsal root ganglion neurons reduces neuropathic pain. *Mol Pain* 2, 6.

Hao, S., Mata, M., Glorioso, J. C., and Fink, D. J., 2007. Gene transfer to interfere with TNF-alpha signaling in neuropathic pain. *Gene Ther* 14, 1010–1016.

Hargreaves, K., Dubner, R., Brown, F., Flores, C., and Joris, J., 1988. A new and sensitive method for measuring thermal nociception in cutaneous hyperalgesia. *Pain* 32, 77–88.

Haydon, P. G., 2001. GLIA: listening and talking to the synapse. *Nat Rev Neurosci* 2, 185–193.

Houzen, H., Kikuchi, S., Kanno, M., Shinpo, K., and Tashiro, K., 1997. Tumor necrosis factor enhancement of transient outward potassium currents in cultured rat cortical neurons. *J Neurosci Res* 50, 990–999.

Hu, P., and McLachlan, E. M., 2003. Distinct functional types of macrophage in dorsal root ganglia and spinal nerves proximal to sciatic and spinal nerve transections in the rat. *Exp Neurol* 184, 590–605.

Ji, R. R., Gereau, R. W. t., Malcangio, M., and Strichartz, G. R., 2008. MAP kinase and pain. *Brain Res Rev.*

Ji, R. R., Kawasaki, Y., Zhuang, Z. Y., Wen, Y. R., and Decosterd, I., 2006. Possible role of spinal astrocytes in maintaining chronic pain sensitization: review of current evidence with focus on bFGF/JNK pathway. *Neuron Glia Biol* 2, 259–269.

Ji, R. R., Samad, T. A., Jin, S. X., Schmoll, R., and Woolf, C. J., 2002. p38 MAPK activation by NGF in primary sensory neurons after inflammation increases TRPV1 levels and maintains heat hyperalgesia. *Neuron* 36, 57–68.

Ji, R. R., and Suter, M. R., 2007. p38 MAPK, microglial signaling, and neuropathic pain. *Mol Pain* 3, 33.

Jin, S. X., Zhuang, Z. Y., Woolf, C. J., and Ji, R. R., 2003. p38 mitogen-activated protein kinase is activated after a spinal nerve ligation in spinal cord microglia and dorsal root ganglion neurons and contributes to the generation of neuropathic pain. *J Neurosci* 23, 4017–4022.

Jin, X., and Gereau, R. W. t., 2006. Acute p38-mediated modulation of tetrodotoxin-resistant sodium channels in mouse sensory neurons by tumor necrosis factor-alpha. *J Neurosci* 26, 246–255.

Junger, H., and Sorkin, L. S., 2000. Nociceptive and inflammatory effects of subcutaneous TNF-alpha. *Pain* 85, 145–151.

Karppinen, J., Korhonen, T., Malmivaara, A., Paimela, L., Kyllonen, E., Lindgren, K. A., Rantanen, P., et al. 2003. Tumor necrosis factor-alpha monoclonal antibody, infliximab, used to manage severe sciatica. *Spine* 28, 750–753; discussion 753–754.

Katsuki, H., Nakai, S., Hirai, Y., Akaji, K., Kiso, Y., and Satoh, M., 1990. Interleukin-1 beta inhibits long-term potentiation in the CA3 region of mouse hippocampal slices. *Eur J Pharmacol* 181, 323–326.

Kawasaki, Y., Xu, Z. Z., Wang, X., Park, J. Y., Zhuang, Z. Y., Tan, P. H., Gao, Y. J., et al. 2008. Distinct roles of matrix metalloproteases in the early- and late-phase development of neuropathic pain. *Nat Med* 14, 331–336.

Kenney, A. M., and Kocsis, J. D., 1998. Peripheral axotomy induces long-term c-Jun amino-terminal kinase-1 activation and activator protein-1 binding activity by c-Jun and junD in adult rat dorsal root ganglia In vivo. *J Neurosci* 18, 1318–1328.

Kessler, W., Kirchhoff, C., Reeh, P. W., and Handwerker, H. O., 1992. Excitation of cutaneous afferent nerve endings in vitro by a combination of inflammatory mediators and conditioning effect of substance P. *Exp Brain Res* 91, 467–476.

Kieseier, B. C., Hartung, H. P., and Wiendl, H., 2006. Immune circuitry in the peripheral nervous system. *Curr Opin Neurol* 19, 437–445.

Kita, H., and Armstrong, W., 1991. A biotin-containing compound N-(2-aminoethyl)biotin-amide for intracellular labeling and neuronal tracing studies: comparison with biocytin. *J Neurosci Methods* 37, 141–150.

Kobayashi, H., Chattopadhyay, S., Kato, K., Dolkas, J., Kikuchi, S., Myers, R. R., and Shubayev, V. I., 2008. MMPs initiate Schwann cell-mediated MBP degradation and mechanical nociception after nerve damage. *Mol Cell Neurosci* 39, 619–627.

Koike, T., Yang, Y., Suzuki, K., and Zheng, X., 2008. Axon & dendrite degeneration: its mechanisms and protective experimental paradigms. *Neurochem Int* 52, 751–760.

Koller, H., Allert, N., Oel, D., Stoll, G., and Siebler, M., 1998. TNF-alpha induces a protein kinase C-dependent reduction in astroglial K+ conductance. *Neuroreport* 9, 1375–1378.

Komori, K., Nonaka, T., Okada, A., Kinoh, H., Hayashita-Kinoh, H., Yoshida, N., Yana, I., and Seiki, M., 2004. Absence of mechanical allodynia and A-beta-fiber sprouting after sciatic nerve injury in mice lacking membrane-type 5 matrix metalloproteinase. *FEBS Lett* 557, 125–128.

Korhonen, T., Karppinen, J., Malmivaara, A., Autio, R., Niinimaki, J., Paimela, L., Kyllonen, E., et al. 2004. Efficacy of infliximab for disc herniation-induced sciatica: one-year follow-up. *Spine* 29, 2115–2119.

Korhonen, T., Karppinen, J., Paimela, L., Malmivaara, A., Lindgren, K. A., Bowman, C., Hammond, A., et al. 2006. The treatment of disc-herniation-induced sciatica with infliximab: one-year follow-up results of FIRST II, a randomized controlled trial. *Spine* 31, 2759–2766.

Korhonen, T., Karppinen, J., Paimela, L., Malmivaara, A., Lindgren, K. A., Jarvinen, S., Niinimaki, J., Veeger, N., Seitsalo, S., and Hurri, H., 2005. The treatment of disc herniation-induced sciatica with infliximab: results of a randomized, controlled, 3-month follow-up study. *Spine* 30, 2724–2728.

Kraus, J., Borner, C., Giannini, E., Hickfang, K., Braun, H., Mayer, P., Hoehe, M. R., Ambrosch, A., Konig, W., and Hollt, V., 2001. Regulation of mu-opioid receptor gene transcription by interleukin-4 and influence of an allelic variation within a STAT6 transcription factor binding site. *J Biol Chem* 276, 43901–43908.

Kristensson, K., 1984. Retrograde signaling after nerve injury. In: Elam, J. S., and Cancalon, P., (Eds.), *Axonal Transport in Growth and Regeneration.* Plenum, New York, pp. 31–43.

Kumar, A., Zanotti, S., Bunnell, G., Habet, K., Anel, R., Neumann, A., Cheang, M., Dinarello, C. A., Cutler, D., and Parrillo, J. E., 2005. Interleukin-10 blunts the human inflammatory response to lipopolysaccharide without affecting the cardiovascular response. *Crit Care Med* 33, 331–340.

Langer, J. A., Cutrone, E. C., and Kotenko, S., 2004. The Class II cytokine receptor (CRF2) family: overview and patterns of receptor-ligand interactions. *Cytokine Growth Factor Rev* 15, 33–48.

Le Buanec, H., Delavallee, L., Bessis, N., Paturance, S., Bizzini, B., Gallo, R., Zagury, D., and Boissier, M. C., 2006. TNF-alpha kinoid vaccination-induced neutralizing antibodies to TNF-alpha protect mice from autologous TNF-alpha-driven chronic and acute inflammation. *Proc Natl Acad Sci U S A* 103, 19442–19447.

Ledeboer, A., Sloane, E. M., Milligan, E. D., Frank, M. G., Mahony, J. H., Maier, S. F., and Watkins, L. R., 2005. Minocycline attenuates mechanical allodynia and proinflammatory cytokine expression in rat models of pain facilitation. *Pain* 115, 71–83.

Lee, J. C., Laydon, J. T., McDonnell, P. C., Gallagher, T. F., Kumar, S., Green, D., McNulty, D., et al. 1994. A protein kinase involved in the regulation of inflammatory cytokine biosynthesis. *Nature* 372, 739–746.

Leitner, M. L., Molliver, D. C., Osborne, P. A., Vejsada, R., Golden, J. P., Lampe, P. A., Kato, A. C., Milbrandt, J., and Johnson, E. M., Jr., 1999. Analysis of the retrograde transport of glial cell line-derived neurotrophic factor (GDNF), neurturin, and persephin suggests that in vivo signaling for the GDNF family is GFRalpha coreceptor-specific. *J Neurosci* 19, 9322–9331.

Liefner, M., Siebert, H., Sachse, T., Michel, U., Kollias, G., and Bruck, W., 2000. The role of TNF-alpha during Wallerian degeneration. *J Neuroimmunol* 108, 147–152.

Liu, L., Yang, T. M., Liedtke, W., and Simon, S. A., 2006. Chronic IL-1-beta signaling potentiates voltage-dependent sodium currents in trigeminal nociceptive neurons. *J Neurophysiol* 95, 1478–1490.

Loeser, J. D., and Treede, R. D., 2008. The Kyoto protocol of IASP basic pain terminology. *Pain* 137, 473–477.

Lunn, E. R., Perry, V. H., Brown, M. C., Rosen, H., and Gordon, S., 1989. Absence of Wallerian degeneration does not hinder regeneration in peripheral nerve. *Eur J Neurosci* 1, 27–33.

Lyon, M. F., Ogunkolade, B. W., Brown, M. C., Atherton, D. J., and Perry, V. H., 1993. A gene affecting Wallerian nerve degeneration maps distally on mouse chromosome 4. *Proc Natl Acad Sci U S A* 90, 9717–9720.

Ma, C., Greenquist, K. W., and Lamotte, R. H., 2006. Inflammatory mediators enhance the excitability of chronically compressed dorsal root ganglion neurons. *J Neurophysiol* 95, 2098–2107.

Ma, W., and Quirion, R., 2002. Partial sciatic nerve ligation induces increase in the phosphorylation of extracellular signal-regulated kinase (ERK) and c-Jun N-terminal kinase (JNK) in astrocytes in the lumbar spinal dorsal horn and the gracile nucleus. *Pain* 99, 175–184.

Michaelis, M., Vogel, C., Blenk, K. H., Arnarson, A., and Janig, W., 1998. Inflammatory mediators sensitize acutely axotomized nerve fibers to mechanical stimulation in the rat. *J Neurosci* 18, 7581–7587.

Milligan, E. D., Sloane, E. M., Langer, S. J., Hughes, T. S., Jekich, B. M., Frank, M. G., Mahoney, J. H., et al. 2006a. Repeated intrathecal injections of plasmid DNA encoding interleukin-10 produce prolonged reversal of neuropathic pain. *Pain* 126, 294–308.

Milligan, E. D., Soderquist, R. G., Malone, S. M., Mahoney, J. H., Hughes, T. S., Langer, S. J., Sloane, E. M., et al. 2006b. Intrathecal polymer-based interleukin-10 gene delivery for neuropathic pain. *Neuron Glia Biol* 2, 293–308.

Milligan, E. D., Twining, C., Chacur, M., Biedenkapp, J., O'Connor, K., Poole, S., Tracey, K., Martin, D., Maier, S. F., and Watkins, L. R., 2003. Spinal glia and proinflammatory cytokines mediate mirror-image neuropathic pain in rats. *J Neurosci* 23, 1026–1040.

Milligan, E. D., and Watkins, L. R., 2009. Pathological and protective roles of glia in chronic pain. *Nat Rev Neurosci* 10, 23–36.

Moolwaney, A. S., and Igwe, O. J., 2005. Regulation of the cyclooxygenase-2 system by interleukin-1 beta through mitogen-activated protein kinase signaling pathways: a comparative study of human neuroglioma and neuroblastoma cells. *Brain Res Mol Brain Res* 137, 202–212.

Murray, C. A., McGahon, B., McBennett, S., and Lynch, M. A., 1997. Interleukin-1 beta inhibits glutamate release in hippocampus of young, but not aged, rats. *Neurobiol Aging* 18, 343–348.

Myers, R. R., Campana, W. M., and Shubayev, V. I., 2006. The role of neuroinflammation in neuropathic pain: mechanisms and therapeutic targets. *Drug Discov Today* 11, 8–20.

Myers, R. R., Heckman, H. M., and Rodriguez, M., 1996. Reduced hyperalgesia in nerve-injured WLD mice: relationship to nerve fiber phagocytosis, axonal degeneration, and regeneration in normal mice. *Exp Neurol* 141, 94–101.

Myers, R. R., Yamamoto, T., Yaksh, T. L., and Powell, H. C., 1993. The role of focal nerve ischemia and Wallerian degeneration in peripheral nerve injury producing hyperesthesia. *Anesthesiology* 78, 308–316.

Nachemson, A., 2004. Epidemiology and the economics of low back pain. In: Herkowitz, H., Nordin, M., Dvorak, J., Bell, G., and Grob, D., (Eds.), *The Lumbar Spine*. Lippincott Williams & Wilkins.

Nagase, H., 1997. Activation mechanisms of matrix metalloproteinases. *Biol Chem* 378, 151–160.

Obata, K., Yamanaka, H., Dai, Y., Mizushima, T., Fukuoka, T., Tokunaga, A., and Noguchi, K., 2004. Activation of extracellular signal-regulated protein kinase in the dorsal root ganglion following inflammation near the nerve cell body. *Neuroscience* 126, 1011–1021.

Obata, K., Yamanaka, H., Dai, Y., Tachibana, T., Fukuoka, T., Tokunaga, A., Yoshikawa, H., and Noguchi, K., 2003. Differential activation of extracellular signal-regulated protein kinase in primary afferent neurons regulates brain-derived neurotrophic factor expression after peripheral inflammation and nerve injury. *J Neurosci* 23, 4117–4126.

Obreja, O., Rathee, P. K., Lips, K. S., Distler, C., and Kress, M., 2002. IL-1 beta potentiates heat-activated currents in rat sensory neurons: involvement of IL-1RI, tyrosine kinase, and protein kinase C. *Faseb J* 16, 1497–1503.

Ogoshi, F., Yin, H. Z., Kuppumbatti, Y., Song, B., Amindari, S., and Weiss, J. H., 2005. Tumor necrosis-factor-alpha (TNF-alpha) induces rapid insertion of Ca2+-permeable alpha-amino-3-hydroxyl-5-methyl-4-isoxazole-propionate (AMPA)/kainate (Ca-A/K) channels in a subset of hippocampal pyramidal neurons. *Exp Neurol* 193, 384–393.

Olmarker, K., and Larsson, K., 1998. Tumor necrosis factor alpha and nucleus-pulposus-induced nerve root injury. *Spine* 23, 2538–2544.

Olmarker, K., Myers, R. R., Kikuchi, S., and Rydevik, B., 2004. Pathophysiology of nerve root pain in disc herniation and spinal stenosis. In: Herkowitz, H., Nordin, M., Dvorak, J., Bell, G., and Grob, D., (Eds.), *The Lumbar Spine*. Lippincott Williams & Wilkins.

Olmarker, K., Rydevik, B., Kikuchi, S., and Myers, R. R., 2005. Sciatic and nerve root pain in disc herniation and spinal stenosis. A basic science review and clinical perspective. In: Herkowitz, H., Balderston, R. A., Garfin, S. R., Eismont, F. J., and Bell, G. R., (Eds.), *Rothman-Simeone: The Spine*. WB Saunders.

Onda, A., Hamba, M., Yabuki, S., and Kikuchi, S., 2002. Exogenous tumor necrosis factor-alpha induces abnormal discharges in rat dorsal horn neurons. *Spine* 27, 1618–1624; discussion 1624.

Ozaktay, A. C., Cavanaugh, J. M., Asik, I., DeLeo, J. A., and Weinstein, J. N., 2002. Dorsal root sensitivity to interleukin-1 beta, interleukin-6 and tumor necrosis factor in rats. *Eur Spine J* 11, 467–475.

Ozaktay, A. C., Kallakuri, S., Takebayashi, T., Cavanaugh, J. M., Asik, I., DeLeo, J. A., and Weinstein, J. N., 2006. Effects of interleukin-1 beta, interleukin-6, and tumor necrosis factor on sensitivity of dorsal root ganglion and peripheral receptive fields in rats. *Eur Spine J* 15, 1529–1537.

Page-McCaw, A., Ewald, A. J., and Werb, Z., 2007. Matrix metalloproteinases and the regulation of tissue remodelling. *Nat Rev Mol Cell Biol* 8, 221–233.

Perry, V. H., Brown, M. C., Lunn, E. R., Tree, P., and Gordon, S., 1990a. Evidence that very slow Wallerian degeneration in c57bl/ola mice is an intrinsic property of the peripheral nerve. *Eur J Neurosci* 2, 802–808.

Perry, V. H., Lunn, E. R., Brown, M. C., Cahusac, S., and Gordon, S., 1990b. Evidence that the rate of Wallerian degeneration is controlled by a single autosomal dominant gene. *Eur J Neurosci* 2, 408–413.

Plata-Salaman, C. R., and Ffrench-Mullen, J. M., 1992. Interleukin-1 beta depresses calcium currents in CA1 hippocampal neurons at pathophysiological concentrations. *Brain Res Bull* 29, 221–223.

Plata-Salaman, C. R., and Ffrench-Mullen, J. M., 1994. Interleukin-1 beta inhibits Ca2+ channel currents in hippocampal neurons through protein kinase C. *Eur J Pharmacol* 266, 1–10.

Porter, J. T., and McCarthy, K. D., 1997. Astrocytic neurotransmitter receptors in situ and in vivo. *Prog Neurobiol* 51, 439–455.

Rada, P., Mark, G. P., Vitek, M. P., Mangano, R. M., Blume, A. J., Beer, B., and Hoebel, B. G., 1991. Interleukin-1 beta decreases acetylcholine measured by microdialysis in the hippocampus of freely moving rats. *Brain Res* 550, 287–290.

Raghavendra, V., Tanga, F., and DeLeo, J. A., 2003. Inhibition of microglial activation attenuates the development but not existing hypersensitivity in a rat model of neuropathy. *J Pharmacol Exp Ther* 306, 624–630.

Redshaw, J. D., and Bisby, M. A., 1984. Fast axonal transport in central nervous system and peripheral nervous system axons following axotomy. *J Neurobiol* 15, 109–117.

Redshaw, J. D., and Bisby, M. A., 1987. Proteins of fast axonal transport in regenerating rat sciatic sensory axons: a conditioning lesion does not amplify the characteristic response to axotomy. *Exp Neurol* 98, 212–221.

Reeve, A. J., Patel, S., Fox, A., Walker, K., and Urban, L., 2000. Intrathecally administered endotoxin or cytokines produce allodynia, hyperalgesia and changes in spinal cord neuronal responses to nociceptive stimuli in the rat. *Eur J Pain* 4, 247–257.

Romero-Sandoval, E. A., Horvath, R. J., and DeLeo, J. A., 2008. Neuroimmune interactions and pain: focus on glial-modulating targets. *Curr Opin Investig Drugs* 9, 726–734.

Rosenberg, G. A., 2002. Matrix metalloproteinases in neuroinflammation. *Glia* 39, 279–291.

Roux, J., Kawakatsu, H., Gartland, B., Pespeni, M., Sheppard, D., Matthay, M. A., Canessa, C. M., and Pittet, J. F., 2005. Interleukin-1-beta decreases expression of the epithelial sodium channel alpha-subunit in alveolar epithelial cells via a p38 MAPK-dependent signaling pathway. *J Biol Chem* 280, 18579–18589.

Saren, P., Welgus, H. G., and Kovanen, P. T., 1996. TNF-alpha and IL-1-beta selectively induce expression of 92-kDa gelatinase by human macrophages. *J Immunol* 157, 4159–4165.

Sawada, M., Hara, N., and Maeno, T., 1991. Tumor necrosis factor reduces the ACh-induced outward current in identified Aplysia neurons. *Neurosci Lett* 131, 217–220.

Schafers, M., Brinkhoff, J., Neukirchen, S., Marziniak, M., and Sommer, C., 2001. Combined epineurial therapy with neutralizing antibodies to tumor necrosis factor-alpha and interleukin-1 receptor has an additive effect in reducing neuropathic pain in mice. *Neurosci Lett* 310, 113–116.

Schafers, M., Geis, C., Brors, D., Yaksh, T. L., and Sommer, C., 2002. Anterograde transport of tumor necrosis factor-alpha in the intact and injured rat sciatic nerve. *J Neurosci* 22, 536–545.

Schafers, M., Lee, D. H., Brors, D., Yaksh, T. L., and Sorkin, L. S., 2003. Increased sensitivity of injured and adjacent uninjured rat primary sensory neurons to exogenous tumor necrosis factor-alpha after spinal nerve ligation. *J Neurosci* 23, 3028–3038.

Schafers, M., and Sommer, C., 2007. Anticytokine therapy in neuropathic pain management. *Expert Rev Neurother* 7, 1613–1627.

Schafers, M., and Sorkin, L., 2008. Effect of cytokines on neuronal excitability. *Neurosci Lett* 437, 188–193.

Schneider, H., Pitossi, F., Balschun, D., Wagner, A., del Rey, A., and Besedovsky, H. O., 1998. A neuromodulatory role of interleukin-1-beta in the hippocampus. *Proc Natl Acad Sci U S A* 95, 7778–7783.

Scholz, J., and Woolf, C. J., 2007. The neuropathic pain triad: neurons, immune cells and glia. *Nat Neurosci* 10, 1361–1368.

Shin, K. C., Lee, S., Moon, S., Min, H., Park, Y., Cho, J., and An, H., 2006. A prospective controlled trial of TNF-alpha inhibitor for symptomatic patients with cervical disc herniation. *Spine J* 5, 45S.

Shubayev, V. I., Angert, M., Dolkas, J., Campana, W. M., Palenscar, K., and Myers, R. R., 2006. TNF-alpha-induced MMP-9 promotes macrophage recruitment into injured peripheral nerve. *Mol Cell Neurosci* 31, 407–415.

Shubayev, V. I., Dolkas, J., Angert, M., and Myers, R. R., 2005. TNF-alpha retrograde axonal transport activates ipsilateral and contralateral spinal glia and DRG gene expression after peripheral nerve injury. *J Peripher Nerv Syst* 10, 86.

Shubayev, V. I., and Myers, R. R., 2000. Upregulation and interaction of TNF-alpha and gelatinases A and B in painful peripheral nerve injury. *Brain Res* 855, 83–89.

Shubayev, V. I., and Myers, R. R., 2001. Axonal transport of TNF-alpha in painful neuropathy: distribution of ligand tracer and TNF receptors. *J Neuroimmunol* 114, 48–56.

Shubayev, V. I., and Myers, R. R., 2002a. Anterograde TNF-alpha transport from rat dorsal root ganglion to spinal cord and injured sciatic nerve. *Neurosci Lett* 320, 99–101.

Shubayev, V. I., and Myers, R. R., 2002b. Endoneurial remodeling by TNF-alpha- and TNF-alpha-releasing proteases. A spatial and temporal co-localization study in painful neuropathy. *J Peripher Nerv Syst* 7, 28–36.

Sitcheran, R., Gupta, P., Fisher, P. B., and Baldwin, A. S., 2005. Positive and negative regulation of EAAT2 by NF-kappaB: a role for N-myc in TNF-alpha-controlled repression. *Embo J* 24, 510–520.

Sommer, C., Lalonde, A., Heckman, H. M., Rodriguez, M., and Myers, R. R., 1995. Quantitative neuropathology of a focal nerve injury causing hyperalgesia. *J Neuropathol Exp Neurol* 54, 635–643.

Sommer, C., and Schafers, M., 1998. Painful mononeuropathy in C57BL/Wld mice with delayed Wallerian degeneration: differential effects of cytokine production and nerve regeneration on thermal and mechanical hypersensitivity. *Brain Res* 784, 154–162.

Sommer, C., Schafers, M., Marziniak, M., and Toyka, K. V., 2001. Etanercept reduces hyperalgesia in experimental painful neuropathy. *J Peripher Nerv Syst* 6, 67–72.

Sommer, C., Schmidt, C., George, A., and Toyka, K. V., 1997. A metalloprotease-inhibitor reduces pain associated behavior in mice with experimental neuropathy. *Neurosci Lett* 237, 45–48.

Sorkin, L. S., Xiao, W. H., Wagner, R., and Myers, R. R., 1997. Tumour necrosis factor-alpha induces ectopic activity in nociceptive primary afferent fibres. *Neuroscience* 81, 255–262.

Stellwagen, D., Beattie, E. C., Seo, J. Y., and Malenka, R. C., 2005. Differential regulation of AMPA receptor and GABA receptor trafficking by tumor necrosis factor-alpha. *J Neurosci* 25, 3219–3228.

Stellwagen, D., and Malenka, R. C., 2006. Synaptic scaling mediated by glial TNF-alpha. *Nature* 440, 1054–1059.

Stoll, G., Jander, S., and Myers, R. R., 2002. Degeneration and regeneration of the peripheral nervous system: from Augustus Waller's observations to neuroinflammation. *J Peripher Nerv Syst* 7, 13–27.

Svensson, C. I., Fitzsimmons, B., Azizi, S., Powell, H. C., Hua, X. Y., and Yaksh, T. L., 2005. Spinal p38-beta isoform mediates tissue injury-induced hyperalgesia and spinal sensitization. *J Neurochem* 92, 1508–1520.

Svensson, C. I., Marsala, M., Westerlund, A., Calcutt, N. A., Campana, W. M., Freshwater, J. D., Catalano, R., et al. 2003. Activation of p38 mitogen-activated protein kinase in spinal microglia is a critical link in inflammation-induced spinal pain processing. *J Neurochem* 86, 1534–1544.

Talhouk, R. S., Hajjar, L., Abou-Gergi, R., Simaa'n, C. J., Mouneimne, G., Saade, N. E., and Safieh-Garabedian, B., 2000. Functional interplay between gelatinases and hyperalgesia in endotoxin-induced localized inflammatory pain. *Pain* 84, 397–405.

Tandrup, T., Woolf, C. J., and Coggeshall, R. E., 2000. Delayed loss of small dorsal root ganglion cells after transection of the rat sciatic nerve. *J Comp Neurol* 422, 172–180.

Tobinick, E., 2009. Perispinal etanercept for neuroinflammatory disorders. *Drug Discov Today* 14, 168–177.

Tobinick, E., and Davoodifar, S., 2004. Efficacy of etanercept delivered by perispinal administration for chronic back and/or neck disc-related pain: a study of clinical observations in 143 patients. *Curr Med Res Opin* 20, 1075–1085.

Tonra, J. R., Curtis, R., Wong, V., Cliffer, K. D., Park, J. S., Timmes, A., Nguyen, T., Lindsay, R. M., Acheson, A., and DiStefano, P. S., 1998. Axotomy upregulates the anterograde transport and expression of brain-derived neurotrophic factor by sensory neurons. *J Neurosci* 18, 4374–4383.

Tsuda, M., Ueno, H., Kataoka, A., Tozaki-Saitoh, H., and Inoue, K., 2008. Activation of dorsal horn microglia contributes to diabetes-induced tactile allodynia via extracellular signal-regulated protein kinase signaling. *Glia* 56, 378–386.

Vale, M. L., Marques, J. B., Moreira, C. A., Rocha, F. A., Ferreira, S. H., Poole, S., Cunha, F. Q., and Ribeiro, R. A., 2003. Antinociceptive effects of interleukin-4, -10, and -13 on the writhing response in mice and zymosan-induced knee joint incapacitation in rats. *J Pharmacol Exp Ther* 304, 102–108.

Viviani, B., Bartesaghi, S., Gardoni, F., Vezzani, A., Behrens, M. M., Bartfai, T., Binaglia, M., et al. 2003. Interleukin-1-beta enhances NMDA receptor-mediated intracellular calcium increase through activation of the Src family of kinases. *J Neurosci* 23, 8692–8700.

Vogel, C., Stallforth, S., and Sommer, C., 2006. Altered pain behavior and regeneration after nerve injury in TNF receptor deficient mice. *J Peripher Nerv Syst* 11, 294–303.

von Bartheld, C. S., 2004. Axonal transport and neuronal transcytosis of trophic factors, tracers, and pathogens. *J Neurobiol* 58, 295–314.

Wagner, R., Janjigian, M., and Myers, R. R., 1998. Anti-inflammatory interleukin-10 therapy in CCI neuropathy decreases thermal hyperalgesia, macrophage recruitment, and endoneurial TNF-alpha expression. *Pain* 74, 35–42.

Wagner, R., and Myers, R. R., 1996. Endoneurial injection of TNF-alpha produces neuropathic pain behaviors. *Neuroreport* 7, 2897–2901.

Wagner, R., and Myers, R. R., 1996. Schwann cells produce tumor necrosis factor alpha: expression in injured and non-injured nerves. *Neuroscience* 73, 625–629.

Wajant, H., Pfizenmaier, K., and Scheurich, P., 2003. Tumor necrosis factor signaling. *Cell Death Differ* 10, 45–65.

Watkins, L. R., Hutchinson, M. R., Milligan, E. D., and Maier, S. F., 2007. "Listening" and "talking" to neurons: implications of immune activation for pain control and increasing the efficacy of opioids. *Brain Res Rev* 56, 148–169.

Watkins, L. R., and Maier, S. F., 2002. Beyond neurons: evidence that immune and glial cells contribute to pathological pain states. *Physiol Rev* 82, 981–1011.

Weinstein, J. N., Tosteson, T. D., Lurie, J. D., Tosteson, A. N., Hanscom, B., Skinner, J. S., Abdu, W. A., Hilibrand, A. S., Boden, S. D., and Deyo, R. A., 2006. Surgical vs nonoperative treatment for lumbar disk herniation: the Spine Patient Outcomes Research Trial (SPORT): a randomized trial. *JAMA* 296, 2441–2450.

Wieseler-Frank, J., Maier, S. F., and Watkins, L. R., 2004. Glial activation and pathological pain. *Neurochem Int* 45, 389–395.

Williams, L. M., Gibbons, D. L., Gearing, A., Maini, R. N., Feldmann, M., and Brennan, F. M., 1996. Paradoxical effects of a synthetic metalloproteinase inhibitor that blocks both p55 and p75 TNF receptor shedding and TNF-alpha processing in RA synovial membrane cell cultures. *J Clin Invest* 97, 2833–2841.

Woolf, C. J., and Salter, M. W., 2000. Neuronal plasticity: increasing the gain in pain. *Science* 288, 1765–1769.

Yang, S., Liu, Z. W., Wen, L., Qiao, H. F., Zhou, W. X., and Zhang, Y. X., 2005. Interleukin-1-beta enhances NMDA receptor-mediated current but inhibits excitatory synaptic transmission. *Brain Res* 1034, 172–179.

Zhang, J. M., Li, H., Liu, B., and Brull, S. J., 2002. Acute topical application of tumor necrosis factor alpha evokes protein kinase A-dependent responses in rat sensory neurons. *J Neurophysiol* 88, 1387–1392.

Zhang, P., Miller, B. S., Rosenzweig, S. A., and Bhat, N. R., 1996. Activation of C-jun N-terminal kinase/stress-activated protein kinase in primary glial cultures. *J Neurosci Res* 46, 114–121.

Zhao, P., Waxman, S. G., and Hains, B. C., 2007. Extracellular signal-regulated kinase-regulated microglia-neuron signaling by prostaglandin E2 contributes to pain after spinal cord injury. *J Neurosci* 27, 2357–2368.

Zhou, C., Ye, H. H., Wang, S. Q., and Chai, Z., 2006. Interleukin-1-beta regulation of N-type Ca2+ channels in cortical neurons. *Neurosci Lett* 403, 181–185.

Zhuang, Z. Y., Gerner, P., Woolf, C. J., and Ji, R. R., 2005. ERK is sequentially activated in neurons, microglia, and astrocytes by spinal nerve ligation and contributes to mechanical allodynia in this neuropathic pain model. *Pain* 114, 149–159.

Zhuang, Z. Y., Wen, Y. R., Zhang, D. R., Borsello, T., Bonny, C., Strichartz, G. R., Decosterd, I., and Ji, R. R., 2006. A peptide c-Jun N-terminal kinase (JNK) inhibitor blocks mechanical allodynia after spinal nerve ligation: respective roles of JNK activation in primary sensory neurons and spinal astrocytes for neuropathic pain development and maintenance. *J Neurosci* 26, 3551–3560.

9 Glial Modulation in Pain States

Translation into Humans

Ryan J. Horvath, Edgar Alfonso Romero-Sandoval, and Joyce A. De Leo

CONTENTS

Chronic pain is a common and debilitating ailment that affects over 80 million Americans each year at an estimated cost of estimated $61.2 billion per year in health expenditures and lost productivity (Stewart et al. 2003). In this chapter we will discuss the inadequacy of current pain therapeutics and present several classes of novel analgesics that modify glial cell function. In each of the following sections, we will present the preclinical and basic science research behind each class of therapeutic and the potential reality of their clinical use. We begin with a discussion of the failure of opioid therapy to adequately treat chronic pain populations and highlight serious side effects that only recently have come under scrutiny: for example, hormonal imbalances, hyperalgesia, heightened abuse, and diversion. Our laboratory's focus on glial biology has begun to elucidate the role of these cells in both the mechanisms of some of these opioid side effects as well as primary glial effects related to their analgesic actions. Next, we will discuss the use of cannabinoids as analgesics and preclinical research that may increase their clinical efficacy by uncovering novel central nervous system (CNS) glial/immune mechanisms. Finally, we will end with a discussion of a new class of analgesics that has come directly out of basic science research, that is, more specific glial modulators.

9.1 OPIOIDS: THE ROLE OF GLIA IN INCREASED USE AND ABUSE

9.1.1 CLINICAL USE OF OPIOIDS: ISSUES OF LONG-TERM EFFICACY AND ABUSE

Opioids are among the most potent analgesic agents available for acute and post-surgical analgesia. Despite a lack of efficacy in many pain syndromes, their use in the treatment of chronic pain has increased dramatically. Long-term analgesia for those patients for whom opioids are initially effective is limited by side effects. Short-term side effects include respiratory depression, constipation, nausea and vomiting, urinary retention, pruritus, and myoclonus. These side effects range from mild, as with constipation, which can lead to poor patient compliance but can be mitigated pharmacologically, to severe, as with respiratory depression, which is the leading cause of death in opioid toxicity. Long-term side effects include analgesic tolerance, hyperalgesia (enhanced pain perception), hormonal imbalance, and immune modulation (Ballantyne and Mao 2003). It is now the standard of care to prescribe testosterone replacement for male patients and monitor all patients for immune suppression throughout long-term opioid therapy. Currently there are no effective strategies to reduce tolerance and opioid-induced hyperalgesia, which are thought to be the cause of failed long-term analgesia (Ballantyne 2007).

The clinical use of opioids has increased dramatically over the past 20 years with increased recognition of the under-treatment of chronic pain and better marketing by pharmaceutical companies (Cicero, Inciardi, and Munoz 2005; Joranson et al. 2000). A salient example of this trend was the increase in the off-label use of the non-opioid gabapentin (Neurontin) for chronic pain treatment, which was driven by misinformation and false claims (Landefeld and Steinman 2009). Portenoy and Foley (1986) were the first to report on the efficacy of opioid therapy for chronic, non-malignant pain. Their retrospective case study of 38 patients found that low-dose (<20 mg per day for most patients) opioids provided adequate analgesia in a majority of cases. However, opioid treatment did not improve employment or social function. Based largely upon this landmark study, the medical use of prescription opioids increased dramatically. From 1990 to 1996, the medical use of opioids including morphine, fentanyl, and oxycodone rose by an astounding 59%, 1168%, and 23%, respectively (Joranson et al. 2000).

It was not until 1997, however, that national guidelines for the expanded use of opioid analgesics to treat chronic pain were published by the American Society of Anesthesiologists and the American Pain Society (Anesthesiology 1997; Pain 1997). Following the publication of these guidelines, from 1997 to 2001 the therapeutic use of morphine, fentanyl, and oxycodone increased still further by 48.8%, 151.2%, and 347.9%, respectively. The diversion, misuse, and abuse of opioids have mirrored the increase in opioid prescriptions. From 1997 to 2001 there was an increase in the abuse of morphine-containing compounds by 161.8%, fentanyl-containing compounds by 249.8%, and oxycodone-containing compounds by 267.3% (Novak, Nemeth, and Lawson 2004). Much of this increase in the abuse of opioids is thought to be due to the increased availability of prescription opioids and the popular misconception that prescription drugs are less dangerous than non-regulated, illegal "street" drugs. The diversion and misuse of opioids has become a critical issue in their therapeutic use (Ballantyne and LaForge 2007).

The alarming increase in the amount of opioids being prescribed and the ever-increasing misuse and abuse of opioids has led to the generation of new guidelines for their use (Pergolizzi et al. 2008; Trescot et al. 2008). These updated guidelines highlight many recent studies that have shown that opioids prescribed for chronic pain may be effective for short-term analgesia (Portenoy and Foley 1986) but are of indeterminate or highly variable effectiveness for long-term (>6 months) treatment. Much of this variability is due to the differences in the pain populations studied, the wide dosage ranges of opioids currently being prescribed for chronic pain (from 20 to >100 mg morphine equivalents per day), and non-uniformity in the outcomes measured (Ballantyne and Shin 2008). Limited opioid efficacy, increasing diversion and misuse, and side effects are key problems in the treatment of chronic pain. These clinical realities and insights from preclinical investigation highlight the need for the development of new analgesic agents that target non-opioid based mechanisms that include other cell types and interactions beyond neuronal actions.

9.1.2 THE ROLE OF GLIA IN OPIOID-INDUCED SIDE EFFECTS

Much of the early work in determining the mechanisms of opioid side effects including tolerance and hyperalgesia had focused on neuronal mechanisms of adaptation and sensitization. Early studies showed that antagonism of the μ-opioid receptor with naloxone limited the development of tolerance (Beitner-Johnson, Guitart, and Nestler 1993). This and other studies led to the development of synthetic μ-opioid agonists, mixed μ-opioid agonist and κ-opioid antagonists, and partial μ-opioid agonists, which have all failed to significantly improve upon the efficacy or side-effect profile of morphine. Much attention and research was also directed toward findings that opioid receptor desensitization involved the N-methyl-D-aspartate (NMDA) receptor cascade. It was hoped that by antagonizing NMDA receptors, opioids would maintain analgesic efficacy while reducing tolerance formation. Although preclinical animal studies (Mao et al. 1996; Mao, Price, and Mayer 1995) and preliminary clinical studies (Caruso 2000; Katz 2000) supported this hypothesis, the results of large-scale clinical trials using NMDA antagonists in conjunction with opioids to limit tolerance formation have been disappointing (Galer et al. 2005).

Glia (astrocytes and microglia) have historically received little attention for their role in chronic pain and opioid side effects (Ronnback and Hansson 1988), despite constituting approximately 70% of the total cell population in the central nervous system (CNS) (Kettenmann 2005). Our lab, among others, has proposed that glia play a critical role in the development of neuropathic pain (DeLeo et al. 1996; DeLeo and Yezierski 2001; Watkins, Maier, and Goehler 1995) and morphine tolerance (Raghavendra, Rutkowski, and DeLeo 2002; Tawfik et al. 2005). Song and Zhao were the first to provide evidence for a role of glia in morphine tolerance (Song and Zhao 2001). They showed that 9 days of intraperitoneal morphine administration increased astrocytic glial fibrillary acid protein (GFAP) expression across various regions of the CNS including the spinal cord, posterior cingulate cortex, and hippocampus. This enhanced GFAP expression followed the loss of the analgesic effect of morphine (tolerance). They were also able to attenuate morphine tolerance through intrathecal injection of the metabolic inhibitor fluorocitrate. Subsequently our lab has shown that chronic morphine caused glial hypertrophy and enhanced spinal expression of the astrocytic marker GFAP and of the microglia marker CD11b (Raghavendra, Rutkowski, and DeLeo 2002; Tawfik et al. 2005) (Figure 9.1). We have also shown, using an L5 nerve transection model, that chronic pain reduced acute morphine analgesia, enhanced the formation of analgesic tolerance, and increased morphine withdrawal-induced hyperalgesia (Raghavendra, Rutkowski, and DeLeo 2002).

It is now recognized that neuropathic pain and morphine tolerance share many common mechanisms including a central role for glia (Mayer et al. 1999; Mika 2008; Romero-Sandoval, Horvath, and DeLeo 2008). Nerve injury and chronic morphine exposure enhance the spinal expression of IL-1β, IL-6, and TNF-α; and inhibition of these pro-inflammatory cytokines attenuates both chronic pain and morphine tolerance (DeLeo and Yezierski 2001; Raghavendra and DeLeo 2006; Raghavendra, Rutkowski, and DeLeo 2002; Raghavendra, Tanga, and DeLeo 2004). The glial modulator drugs propentofylline (Raghavendra et al. 2003; Tawfik et al. 2007) and

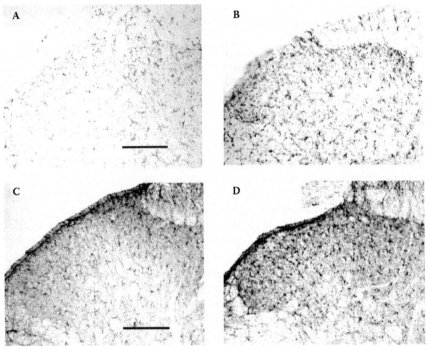

FIGURE 9.1 Increased microglia and astrocyte activation by chronic morphine treatment in sham-operated rats. Morphine was injected subcutaneously (10mg/kg) twice daily for five days. Enhanced CD11b immunostaining in the dorsal horn of the L5 lumbar spinal cord was observed in chronic-morphine-treated rats (B) compared with saline-treated controls (A). Enhanced GFAP immunostaining was also observed in chronic-morphine-treated rats (D) compared with saline-treated controls (C). Scale bar, 150 mm (n=4 per group). (Modified with permission from Raghavendra, V., M. D. Rutkowski, and J. A. DeLeo. 2002. The role of spinal neuroimmune activation in morphine tolerance/hyperalgesia in neuropathic and sham-operated rats. *J Neurosci* 22 (22):9980–9989.)

minocycline (Ledeboer et al. 2005; Raghavendra, Tanga, and DeLeo 2003) have been shown to attenuate chronic pain states by reducing glial reactivity. Similarly, propentofylline (Raghavendra, Tanga, and DeLeo 2004) and minocycline (Cui et al. 2008) have been shown to attenuate the development of morphine tolerance and morphine withdrawal-induced hyperalgesia (Figure 9.2).

These similarities lead one to speculate that morphine-induced side effects including tolerance and hyperalgesia might be mediated through a direct primary effect on CNS glia. We and others have now begun to look at specific effects of morphine on microglia and astrocytes. Astrocytes have been directly implicated in the rewarding

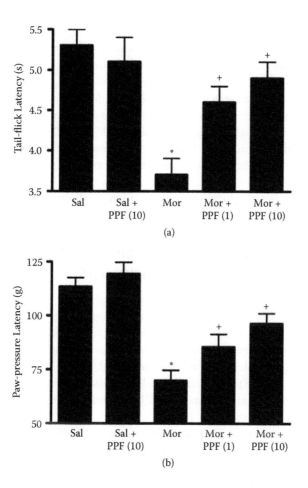

FIGURE 9.2 Propentofylline attenuates morphine-withdrawal-induced hyperalgesia. Morphine (Mor) was injected subcutaneously (10mg/kg) twice daily for 5 days. Propentofylline (PPF 1mg and 10mg) was administered once daily during the induction of morphine tolerance. Withdrawal-induced hyperalgesia as measured by tail-flick and paw-pressure latencies was recorded 16hr after the last injection of morphine or saline (Sal). Data are expressed as mean ± SEM (n=8). *p<0.05 vs. saline-treated group, +p<0.05 vs. morphine-treated group. (Reprinted with permission from *Neuropsychopharmacology* (2004) 29:327–334.)

effects of morphine (Narita et al. 2006). These studies showed that propentofylline reduced morphine-induced astrocytic reactivity and suppressed the drug's rewarding effects as measured through conditioned place preference testing. These effects of morphine administration are thought to be mediated by astrocytic derived proinflammatory cytokine and chemokines (Narita et al. 2006, 2008).

Our lab has recently completed a series of experiments assessing the direct effects of morphine on cultured microglia to build upon our prior *in vivo* findings. We have shown that morphine directly enhances the reactivity of microglia as shown by enhanced ionized calcium adaptor binding protein (Iba1) and $P2X_4$ receptor expression. Morphine also enhances the migration of microglia via modulation of $P2X_4$ signaling through the pAkt/PI3K pathway (Horvath et al. 2009). In previous studies, we have shown that both propentofylline and minocycline attenuate microglial migration (Horvath et al. 2008). These glial modulators are known to attenuate morphine tolerance (Cui et al. 2008; Raghavendra et al. 2003), suggesting a possible role for microglial migration in morphine tolerance. Microglial $P2X_4$ receptors have previously been shown to gate the formation of pain after nerve injury, purportedly through enhanced microglial migration (Tsuda et al. 2003). Our most recent results show that inhibition of microglial $P2X_4$ receptor expression through intrathecal antisense oligonucleotide administration inhibits the formation of morphine tolerance (Horvath et al., unpublished results). These results suggest that non-opioid-based analgesics that target the function and activation of glia may prove a significant improvement over current analgesics.

9.2 CANNABINOIDS/ENDOCANNABINOIDS FOR ANALGESIA: PROBLEMS AND PROMISES

Δ^9-Tetrahydrocannabinol, the major active ingredient of the marijuana plant (*Cannabis sativa*), is a psychotropic agent that is thought to exert most of its effects by binding to G protein-coupled cannabinoid receptor (CBR) type 1 and 2. CBR1 mainly exist in neural structures and are expressed in brain, spinal cord, and peripheral nerves (Bridges et al. 2003; Farquhar-Smith et al. 2000; Lim et al. 2003; Pettit et al. 1998; Tsou et al. 1998). Additionally, CBR1 are also expressed in microglial cells and may act as immune modulators in the CNS (Cabral, Harmon, and Carlisle 2001; Sinha et al. 1998; Waksman et al. 1999).

CBR2 are expressed in immune cells (Klein et al. 2003) and keratinocytes (Ibrahim et al. 2005). Even though they were originally believed to be present solely in the periphery, recent studies have demonstrated that CBR2 exist in peripheral sensory neurons (Anand et al. 2008) and in the CNS (Van Sickle et al. 2005). In accordance with the role of CBR2 in immune cells, they are expressed in microglia and perivascular cells in normal human and rat brain (Ashton et al. 2006; Nunez et al. 2004) and in microglia and astrocytes, especially during inflammation (Ramirez et al. 2005; Sheng et al. 2005). These findings suggest that the cannabinoid system may have immune modulatory functions also in the CNS via CBR2.

Cannabinoids or/and endocannabinoids also seem to act on other non-CBR1/CBR2, such as the "abnormal-cannabidiol" receptors (Jarai et al. 1999) and the orphan G protein–coupled receptor 55 (GPR55) and 119 (GPR119) (Brown 2007; Lauckner et al. 2008; Overton et al. 2006). Recently, it has been shown that GPR55 activation induces analgesia in inflammatory and neuropathic pain (Staton et al. 2008). The clear biological implications and the pharmacodynamic profile of these receptors are yet to be determined. More work is needed to clearly identify the relevance of these receptors in the endocannabinoid system.

9.2.1 EXOGENOUS CANNABINOIDS AND PAIN MODULATION

Acute activation of CBR1 using exogenous cannabinoids produces analgesia in several pain models (Iversen and Chapman 2002; Scott, Wright, and Angus 2004; Yoon and Choi 2003). Even though peripheral neuronal CBR1 activation is sufficient to induce analgesia (Agarwal et al. 2007), central CBR1 activation also reduces pain behaviors (Romero-Sandoval and Eisenach 2007). However, CBR1 located in the CNS are responsible for the classic cannabinoid-induced neurological side effects: reduced reflexes, vocalization, catalepsy, hypothermia, muscle rigidity, and reduced motor activity (Ledent et al. 1999; Romero-Sandoval and Eisenach 2007; Zimmer et al. 1999). Additionally, preclinical studies have shown that sustained spinal activation of CBR1 in neuropathic pain conditions induces hypersensitivity and antinociceptive tolerance formation (Gardell et al. 2002; Smith et al. 2007). CBR1- or cannabis-based drugs have shown modest efficacy in patients suffering from neuropathic and multiple sclerosis-related pain (Campbell et al. 2001; Iskedjian et al. 2007; Perez and Ribera 2008). The potential acute neurological side effects, their lack of clear advantages compared with existing analgesic drugs and route of administration (Scully 2007), cannabis-induced schizophrenia (Rathbone, Variend, and Mehta 2008), physical dependence, withdrawal syndrome (Iversen and Chapman 2002), tolerance (D'Souza et al. 2008), and cannabis-induced hyperalgesia (Clark et al. 1981) are true limitations for the CBR1-based or nonselective CBR analgesic clinical use. In spite of this information, CBR1 have been extensively studied, but the function of spinal cord CBR2 has not been fully described in the processing of nociceptive information (Walker and Hohmann 2005).

9.2.2 CBR2 ACTIVATION AND GLIAL FUNCTION: POTENTIAL TARGETS TO TREAT PAIN

Peripheral CBR2 activation induces antinociception by inducing the release of beta-endorphin from keratinocytes, which acts at local neuronal mu-opioid receptors (Ibrahim et al. 2005). Peripheral nerve injury induces CBR2 expression in peripheral fibers and spinal cord (Anand et al. 2008; Beltramo et al. 2006; Wotherspoon et al. 2005; Zhang et al. 2003) and increases CBR2 functionality in the thalamus (Jhaveri et al. 2008). In normal rat brain CBR2 exist in microglia and perivascular cells (Ashton et al. 2006). CBR2 mRNA is presumably increased in microglia following peripheral nerve injury (Zhang et al. 2003). Recently, we have shown that peripheral nerve injury induces the expression of CBR2 protein in spinal cord microglia and perivascular microglia (but not in neurons or astrocytes, Figure 9.3) four days after the injury in rats (Romero-Sandoval, Nutile-McMenemy, and DeLeo 2008). These findings are interesting because, at this time, microglial cells have been shown to play a key role in the development of pain following peripheral nerve injury (Ledeboer et al. 2005; Raghavendra, Tanga, and DeLeo 2003). These findings suggest that the cannabinoid system may have immune modulatory functions also in the CNS. In accordance with the potential immune modulatory effects of CBR2 activation in the CNS, microglial CBR2 activation induces neuroprotection by modulating the microglial extracellular signal-regulated kinase (ERK) pathway and reducing nitric oxide production (Eljaschewitsch et al. 2006), suppresses CD40 expression induced by interferon γ and attenuates tumor necrosis factor-α (Ehrhart et al. 2005), regulates microglial migration (Miller and Stella 2008; Walter et al. 2003), and induces the production of the anti-inflammatory factor interleukin 1-receptor antagonist from neurons and glia (Molina-Holgado et al. 2003).

Supporting the role of glial cell function in pain modulation and glial modulation by CBR2 activation, we have shown that *in vivo* spinal microglial CBR2 activation induces antinociception while producing a concomitant reduction of spinal Iba-1 and CD11b (microglial marker) and glial fibrillary acidic protein (GFAP) (astrocytic marker) expression in postoperative and neuropathic pain models (Romero-Sandoval et al. 2008; Romero-Sandoval and Eisenach 2007), and an increase in the anti-inflammatory factor ED2/CD163 in spinal perivascular microglia following peripheral nerve injury (Romero-Sandoval, Nutile-McMenemy, and DeLeo 2008). Interestingly, spinal CBR2 activation does not induce any cannabinoid-related neurological side effects. Furthermore, the sub-chronic activation of spinal CBR2 does not induce antinociceptive tolerance (Romero-Sandoval et al. 2008). In humans CBR2

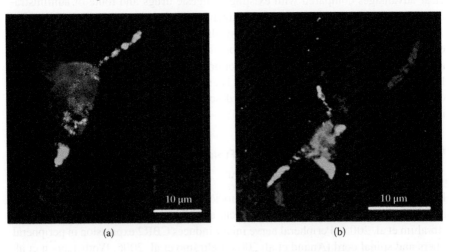

(a)   (b)

FIGURE 9.3 (see color insert following page 166) CBR2 are expressed in perivascular microglia (A, ED2/CD163) and microglia (B, Iba-1). Representative confocal images of CBR2 (red), ED2/CD163 (green in A), Iba-1 (green in B), and DAPI (blue) staining of the dorsal horn L5 spinal cord ipsilateral to L5 nerve transection (4 days after surgery).

are expressed in microglia and perivascular cells in normal brain (Ashton et al. 2006; Nunez et al. 2004) and in microglia and astrocytes, especially during inflammation (Benito et al. 2007; Ramirez et al. 2005; Sheng et al. 2005; Yiangou et al. 2006). Microglial CBR2 activation has been shown to be effective in preventing the neurodegenerative process that occurs in Alzheimer's disease (Ramirez et al. 2005).

Together, these findings strongly suggest that drugs directed to selectively activate glial CBR2 possess a strong translational potential. Among the strengths of CBR2 agonists are, first, their lack of neurological side effects; second, their characteristic of not inducing tolerance after sub-chronic administration; third, their glial modulatory effects, which make them attractive drugs to treat several neurological diseases; and fourth, their substantial penetration of the CNS after systemic administration (Valenzano et al. 2005), which allows these drugs to act not only in the periphery, but also in the CNS. This is important because better efficacy could be reached in conditions where a central effect is desirable, such as in chronic pain states.

There are some concerns regarding the existing CBR2 agonists that limit their therapeutic potential: first, their selectivity over CBR1; second, their short-lasting effects or half-life; and third, their route of administration or oral bioavailability. These limitations are being overcome by the development of more selective CBR2 compounds (Yao et al. 2009), with improved oral bioavailability with a longer half-life (Cheng et al. 2008; Kai et al. 2007; Ohta et al. 2007), and with antinociceptive effects after oral administration (Ohta et al. 2008).

9.3 POTENTIAL FOR DIRECT TRANSLATION: TREATMENT AND DIAGNOSIS

There is obviously a need for the development of new agents for the treatment of chronic pain that target alternative mechanisms from the currently approved drugs. The primary concentration of target discovery in the past has been to decrease neuronal excitability and/or local inflammation. The focus of this chapter is the role of glial cells in the etiology of persistent pain, tolerance, hyperalgesia, and/or analgesia. Unfortunately, the number of agents that selectively modulate glial function is very limited. The term *glial modulator* is routinely used for these agents without a clear understanding of how modulation is defined. Drugs that alter specific glial function including migration, proliferation, algesic mediator release, and neurotransmitter expression may provide either beneficial or deleterious effects. Similarly, the term *glial activity* is also non-descriptive or useful since cannabinoids that produce analgesia increase the expression of a microglial/perivascular cell marker, ED2/CD163, while spinal ED2/CD163 is decreased following nerve injury (Romero-Sandoval, Nutile-McMenemy, and DeLeo 2008). These findings suggest a far more complex series of events that lead to CNS glia responding to perturbations than a simple enhanced expression of surface or cytosolic glial proteins.

Currently available glial modulators include a broad spectrum of chemical structures and potential mechanisms; for example, fluorocitrate, a non-selective metabolic inhibitor (Meller et al. 1994); minocycline, a tetracycline derivative that inhibits microglia *in vitro* (Raghavendra, Tanga, and DeLeo 2003); ibudilast, a non-selective phosphodiesterase (PDE) inhibitor (Ledeboer, Hutchinson, et al. 2007; Ledeboer, Liu, et al. 2007); methionine sulfoximine (MSO), an astrocytic glutamine synthetase inhibitor (Chiang et al. 2007); and propentofylline, an atypical methylxanthine (DeLeo, Schubert, and Kreutzberg 1988; Sweitzer, Schubert, and DeLeo 2001). For the last two decades, we have investigated the role of propentofylline in animal models of stroke, pain and opioid tolerance, and hyperalgesia. Using an unanesthetized gerbil model of cerebral ischemia, we initially demonstrated in 1987 that propentofylline prevented hippocampal calcium accumulation, neuronal cell death, and astrocytic hypertrophy following transient global ischemia (DeLeo et al. 1987). This was the first demonstration that an agent that modified glial GFAP expression was also neuroprotective, supporting the hypothesis that glial reactivity/ hypertrophy was associated with CNS damage.

Based on our growing interest in the role of glia in the genesis and maintenance of persistent pain states, we tested propentofylline in a rat model of neuropathic pain. Systemic or intrathecal administration of propentofylline attenuated nerve injury-induced allodynia as well as decreased spinal GFAP, CD11b, IL-1β, IL-6, and TNF-α expression *in vivo* (Raghavendra, Tanga, and DeLeo 2003; Sweitzer, Schubert, and DeLeo 2001). In a separate series of studies, we also showed that propentofylline restored the analgesic activity in both opioid tolerance and neuropathic pain models (Raghavendra et al. 2003; Raghavendra, Tanga, and DeLeo 2004). These behavioral results were accompanied by a decrease in spinal glial reactivity and proinflammatory cytokine expression following propentofylline administration. In addition to being neuroprotective and anti-inflammatory, propentofylline has been shown to reduce glial proliferation and migration (Horvath et al. 2008; Schubert et al. 1997; Si et al. 1996).

The exact mechanism of its anti-allodynic effects remains unknown; however, several specific actions have been proposed. Propentofylline inhibits adenosine reuptake as well as the cyclic-adenosine-5',3'-monophosphate (cAMP)-specific phosphodiesterase (PDE IV) (Meskini et al. 1994; Nagata et al. 1985; Parkinson and Fredholm 1991). In support of this mechanism, we have demonstrated that propentofylline increases cAMP levels (Tawfik et al. 2006). In order to probe a direct synaptic action in modulating central sensitization, we investigated whether propentofylline altered astrocytic glutamate transporter expression. It has been previously shown that peripheral nerve injury induces a transient increase in GLT-1 followed by a sustained decrease in the dorsal horn of the lumbar spinal cord (Sung, Lim, and Mao 2003). Astrocytes play a key role in maintaining homeostatic levels of synaptic glutamate by acting as a sink to recycle excessive glutamate, preventing post-synaptic neurotoxicity. We demonstrated that propentofylline suppresses an activated astrocytic phenotype, enhances a specific glutamate transporter, GLT-1 at the mRNA and protein levels, and increases glutamate uptake both *in vivo* and *in vitro* (Tawfik et al. 2006; Tawfik et al. 2008) (Figure 9.4). This may translate into decreased synaptic glutamate, preventing glutamatergic neuronal sensitization. Interestingly, in a recent study, systemic propentofylline was effective on existing mechanical allodynia following L5 spinal nerve transection. At the end of drug administration, evoked behavioral hypersensitivity remained similar to the sham surgery group two weeks after the last dose, suggesting a disease-modifying effect

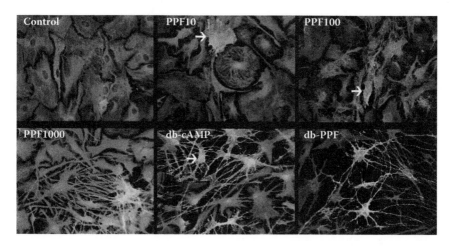

FIGURE 9.4   (see color insert following page 166) Seven-day treatment with propentofylline (PPF) enhances GLT-1 expression in cultured astrocytes. Double-label immunocytochemical localization of GLT-1 (green) and GFAP (red). Note the increasing areas of co-localization (yellow) with PPF treatment that mimics that seen with db-cAMP. Arrows identify astrocytes that are low GFAP, high GLT-1 expressing. PPF10, 100, 1000: PPF-treated 10, 100, or 1000 μM; db-cAMP: 250 μM db-cAMP; db-PPF: 250 μM db-cAMP plus 1000 μM PPF. (Reprinted with permission from *Glia* (2006) 54(3):193–203.)

by modulating glial function (Tawfik et al. 2007) (Figure 9.5). This finding has obvious clinical implications for both the treatment and even the prevention of numerous chronic pain syndromes.

It is important to consider that some of the CNS glial modulating properties of these agents may be downstream from their primary mechanism of action. Thus, the design of specific glial modulators without other neuronal actions may be difficult to achieve. Two agents, ibudilast and propentofylline, with known clinical safety data have progressed to human proof-of-concept pain trials. Ibudilast has been shown to enhance morphine analgesia while decreasing morphine tolerance and morphine withdrawal symptoms and thus may have clinical utility as an opioid adjuvant to decrease dose escalation and adverse opioid side effects. The clinical outcome for propentofylline is not yet available, but the promise remains intact for a novel class of agents for the treatment of chronic pain.

Although there have been copious publications documenting CNS glial changes in animal models that produce behavioral hypersensitivity that may mimic human pain conditions, the evidence for human glial changes have been absent. Recently, however, it was reported that increased expression of spinal GFAP and CD68-positive microglia was observed in autopsy tissue from a patient diagnosed with complex regional pain syndrome (Del Valle, Schwartzman, and Alexander 2009). There were limitations to this study in addition to the sample size of one, but it is compelling that spinal glial reactivity observed in animal models was seen in a patient with a chronic pain syndrome as compared with no obvious glial reactivity in control patients without chronic pain as defined by enhanced GFAP and CD68 expression.

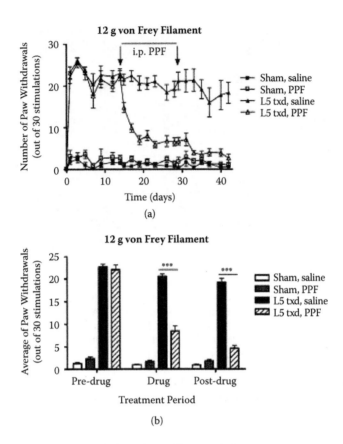

FIGURE 9.5 Propentofylline reverses existing allodynia to a 12 g von Frey filament in an L5 spinal nerve transection model of neuropathic pain. Rats received L5 spinal nerve transection (L5 txd) or sham surgery on day 0, and the development of mechanical allodynia to a 12 g von Frey filament was assessed. On days 14–27 post-surgery, animals received either 10 mg/kg propentofylline (PPF) or saline by intraperitoneal injection, and behavioral responses were tested 15 h later. Administration of propentofylline to L5 spinal nerve-transected rats (L5 txd, PPF) resulted in a significant decrease in mechanical allodynia to a 12 g von Frey filament compared with L5 spinal nerve-transected saline (L5 txd, sal) controls (***$P < .001$ for days 15–42, "L5 txd, sal" vs. "L5 txd, PPF"). (A) Mechanical allodynia is reported as the average number of paw withdrawals out of 30 ± SEM. ($n = 6$–8/treatment). (B) Average number of paw withdrawals for the predrug (days 0–14), drug (days 15–28), and post-drug (days 29–42) testing days revealed a significant effect of propentofylline on mechanical allodynia during treatment and post-drug/washout phases (***$P < .001$, "L5 txd, sal" vs. "L5 txd, PPF"). Day 0 represents preinjury baseline responses. (Reprinted with permission from *Brain, Behavior and Immunity* (2007) 21(2):238–246.)

One of the challenges in the translation of a plethora of compelling preclinical data to clinical trials has been the absence of a biological marker to predict efficacy in the more complicated human subject. There are limitations and concerns with our currently used animal models of acute and chronic pain. For example, it is not evident whether the measurements of spontaneous or evoked nocifensive behaviors relate to human pain. Although the success of preclinical data to the translation of clinically approved agents is high, there still have been many instances of animal models being totally wrong in predicting clinical efficacy (Boyce and Hill 2000). Therefore, the identification of biomarkers to facilitate drug development is much needed (Chizh et al. 2008).

One such potential biomarker is the use of non-invasive neuroimaging both to study pain processing and as a tool to assess drug efficacy (Stephenson and Arneric 2008). *In vivo* imaging of microglia is possible using selective positron emission tomography (PET) radiotracers such as PK-11195. For example, PK-11195 labels the peripheral benzodiazepine receptor that is selectively increased in reactive microglia (Banati 2002). There have been numerous reports both in animal models and in humans of the successful use of this application (Ji et al. 2008; Stephenson et al. 1995; Yasuno et al. 2008). It is a provocative and intriguing concept to consider the use of *in vivo* microglial imaging as a diagnostic tool to predict efficacy with novel glial modulating agents. As is similar to other health fields, trial and error are used as a means of "informed" drug treatment, and therefore we cannot predict drug efficacy for chronic pain prior to administration. It has been postulated that genetic polymorphisms, for example, in receptors or metabolic enzymes as described by pharmacogenomics, underlie the variability in patient drug response. We do not know the time course, extent, or even presence of glial spinal and/or supraspinal changes in patients that experience acute or chronic pain. By using neuroimaging and PET radiotracers such as the peripheral benzodiazepine receptor, it may be possible to target drug therapy specifically to the patient and syndrome presentation and, therefore, improve clinical efficacy.

9.4 CONCLUSIONS: LOOK TO THE FUTURE

Historically, the focus of development for new analgesics has been to incrementally improve upon established drugs. This has led to a glut of "me too" therapeutics that have not fundamentally improved analgesia for the chronic pain population. Here we showed that opioids, the current standard of care, are inadequate analgesics due their insufficient efficacy, their side effects, and the potential misuse and abuse. We proposed two exciting new classes of analgesics, cannabinoids and glial modulators. Cannabinoids are in the beginning stages of clinical use and may prove potent analgesics with improvements based upon preclinical basic science research. Glial modulators are at the initial stages of proving efficacy in clinical trials and are exciting examples of novel compounds specifically designed against targets identified in basic science research. We look forward to the generation of many new analgesics developed against a wide range of preclinically identified targets. This will give clinicians a much broader and more capable arsenal of analgesics to treat chronic pain patients.

REFERENCES

Agarwal, N., P. Pacher, I. Tegeder, F. Amaya, C. E. Constantin, G. J. Brenner, T. Rubino, et al. 2007. Cannabinoids mediate analgesia largely via peripheral type 1 cannabinoid receptors in nociceptors. *Nat Neurosci* 10 (7):870–9.

Anand, U., W. R. Otto, D. Sanchez-Herrera, P. Facer, Y. Yiangou, Y. Korchev, R. Birch, C. Benham, C. Bountra, I. P. Chessell, and P. Anand. 2008. Cannabinoid receptor CB2 localisation and agonist-mediated inhibition of capsaicin responses in human sensory neurons. *Pain* 138 (3):667–80.

Anesthesiology. 1997. Practice guidelines for chronic pain management. A report by the American Society of Anesthesiologists Task Force on Pain Management, Chronic Pain Section. *Anesthesiology* 86 (4):995–1004.

Ashton, J. C., D. Friberg, C. L. Darlington, and P. F. Smith. 2006. Expression of the cannabinoid CB2 receptor in the rat cerebellum: an immunohistochemical study. *Neurosci Lett* 396 (2):113–6.

Ballantyne, J. C. 2007. Opioid analgesia: perspectives on right use and utility. *Pain Physician* 10 (3):479–91.

Ballantyne, J. C., and K. S. LaForge. 2007. Opioid dependence and addiction during opioid treatment of chronic pain. *Pain* 129 (3):235–55.

Ballantyne, J. C., and J. Mao. 2003. Opioid therapy for chronic pain. *N Engl J Med* 349 (20):1943–53.

Ballantyne, J. C., and N. S. Shin. 2008. Efficacy of opioids for chronic pain: a review of the evidence. *Clin J Pain* 24 (6):469–78.

Banati, R. B. 2002. Visualising microglial activation in vivo. *Glia* 40 (2):206–17.

Beitner-Johnson, D., X. Guitart, and E. J. Nestler. 1993. Glial fibrillary acidic protein and the mesolimbic dopamine system: regulation by chronic morphine and Lewis-Fischer strain differences in the rat ventral tegmental area. *J Neurochem* 61 (5):1766–73.

Beltramo, M., N. Bernardini, R. Bertorelli, M. Campanella, E. Nicolussi, S. Fredduzzi, and A. Reggiani. 2006. CB2 receptor-mediated antihyperalgesia: possible direct involvement of neural mechanisms. *Eur J Neurosci* 23 (6):1530–8.

Benito, C., J. P. Romero, R. M. Tolon, D. Clemente, F. Docagne, C. J. Hillard, C. Guaza, and J. Romero. 2007. Cannabinoid CB1 and CB2 receptors and fatty acid amide hydrolase are specific markers of plaque cell subtypes in human multiple sclerosis. *J Neurosci* 27 (9):2396–402.

Boyce, S., and R. G. Hill, eds. 2000. *Discrepant results from preclinical and clinical studies on the potential of substance P-receptor antagonist compounds as analgesics.* Edited by M. R. M. Devor, Wiesenfeld-Halin Z. Seattle: IASP Press.

Bridges, D., A. S. Rice, M. Egertova, M. R. Elphick, J. Winter, and G. J. Michael. 2003. Localisation of cannabinoid receptor 1 in rat dorsal root ganglion using in situ hybridisation and immunohistochemistry. *Neuroscience* 119 (3):803–12.

Brown, A. J. 2007. Novel cannabinoid receptors. *Br J Pharmacol* 152 (5):567–75.

Cabral, G. A., K. N. Harmon, and S. J. Carlisle. 2001. Cannabinoid-mediated inhibition of inducible nitric oxide production by rat microglial cells: evidence for CB1 receptor participation. *Adv Exp Med Biol* 493:207–14.

Campbell, F. A., M. R. Tramer, D. Carroll, D. J. Reynolds, R. A. Moore, and H. J. McQuay. 2001. Are cannabinoids an effective and safe treatment option in the management of pain? A qualitative systematic review. *BMJ* 323 (7303):13–6.

Caruso, F. S. 2000. MorphiDex pharmacokinetic studies and single-dose analgesic efficacy studies in patients with postoperative pain. *J Pain Symptom Manage* 19 (1 Suppl):S31–6.

Cheng, Y., B. K. Albrecht, J. Brown, J. L. Buchanan, W. H. Buckner, E. F. DiMauro, R. Emkey, et al. 2008. Discovery and optimization of a novel series of N-arylamide oxadiazoles as potent, highly selective and orally bioavailable cannabinoid receptor 2 (CB2) agonists. *J Med Chem* 51 (16):5019–34.

Chiang, C. Y., J. Wang, Y. F. Xie, Y. Zhang, J. W. Hu, J. O. Dostrovsky, and B. J. Sessle. 2007. Astroglial glutamate-glutamine shuttle is involved in central sensitization of nociceptive neurons in rat medullary dorsal horn. *J Neurosci* 27 (34):9068–76.

Chizh, B. A., J. D. Greenspan, K. L. Casey, M. I. Nemenov, and R. D. Treede. 2008. Identifying biological markers of activity in human nociceptive pathways to facilitate analgesic drug development. *Pain* 140 (2):249–53.

Cicero, T. J., J. A. Inciardi, and A. Munoz. 2005. Trends in abuse of OxyContin and other opioid analgesics in the United States: 2002–2004. *J Pain* 6 (10):662–72.

Clark, W. C., M. N. Janal, P. Zeidenberg, and G. G. Nahas. 1981. Effects of moderate and high doses of marihuana on thermal pain: a sensory decision theory analysis. *J Clin Pharmacol* 21 (8–9 Suppl):299S–310S.

Cui, Y., X. X. Liao, W. Liu, R. X. Guo, Z. Z. Wu, C. M. Zhao, P. X. Chen, and J. Q. Feng. 2008. A novel role of minocycline: attenuating morphine antinociceptive tolerance by inhibition of p38 MAPK in the activated spinal microglia. *Brain Behav Immun* 22 (1):114–23.

D'Souza, D. C., M. Ranganathan, G. Braley, R. Gueorguieva, Z. Zimolo, T. Cooper, E. Perry, and J. Krystal. 2008. Blunted psychotomimetic and amnestic effects of delta-9-tetrahydrocannabinol in frequent users of cannabis. *Neuropsychopharmacology* 33 (10):2505–16.

Del Valle, L., RJ. Schwartzman, and G. Alexander. 2009. Spinal cord histopathological alterations in a patient with long-standing complex regional pain syndrome. *Brain Behav Immun* 23 (1):85–91.

DeLeo, J. A., R. W. Colburn, M. Nichols, and A. Malhotra. 1996. Interleukin-6-mediated hyperalgesia/allodynia and increased spinal IL-6 expression in a rat mononeuropathy model. *J Interferon Cytokine Res* 16 (9):695–700.

DeLeo, J. A., and R. P. Yezierski. 2001. The role of neuroinflammation and neuroimmune activation in persistent pain. *Pain* 90 (1–2):1–6.

DeLeo, J., P. Schubert, and G. W. Kreutzberg. 1988. Propentofylline (HWA 285) protects hippocampal neurons of Mongolian gerbils against ischemic damage in the presence of an adenosine antagonist. *Neurosci Lett* 84 (3):307–11.

DeLeo, J., L. Toth, P. Schubert, K. Rudolphi, and G. W. Kreutzberg. 1987. Ischemia-induced neuronal cell death, calcium accumulation, and glial response in the hippocampus of the Mongolian gerbil and protection by propentofylline (HWA 285). *J Cereb Blood Flow Metab* 7 (6):745–51.

Ehrhart, J., D. Obregon, T. Mori, H. Hou, N. Sun, Y. Bai, T. Klein, F. Fernandez, J. Tan, and R. D. Shytle. 2005. Stimulation of cannabinoid receptor 2 (CB2) suppresses microglial activation. *J Neuroinflammation* 2:29.

Eljaschewitsch, E., A. Witting, C. Mawrin, T. Lee, P. M. Schmidt, S. Wolf, H. Hoertnagl, et al. 2006. The endocannabinoid anandamide protects neurons during CNS inflammation by induction of MKP-1 in microglial cells. *Neuron* 49 (1):67–79.

Farquhar-Smith, W. P., M. Egertova, E. J. Bradbury, S. B. McMahon, A. S. Rice, and M. R. Elphick. 2000. Cannabinoid CB(1) receptor expression in rat spinal cord. *Mol Cell Neurosci* 15 (6):510–21.

Galer, B. S., D. Lee, T. Ma, B. Nagle, and T. G. Schlagheck. 2005. MorphiDex (morphine sulfate/dextromethorphan hydrobromide combination) in the treatment of chronic pain: three multicenter, randomized, double-blind, controlled clinical trials fail to demonstrate enhanced opioid analgesia or reduction in tolerance. *Pain* 115 (3):284–95.

Gardell, L. R., S. E. Burgess, A. Dogrul, M. H. Ossipov, T. P. Malan, J. Lai, and F. Porreca. 2002. Pronociceptive effects of spinal dynorphin promote cannabinoid-induced pain and antinociceptive tolerance. *Pain* 98 (1–2):79–88.

Horvath, R. J. and DeLeo, J. A. 2009. Morphine enhances microglial migration through modulation of P2X4 receptor signaling. *J Neurosci* 29 (4):998–1005.

Horvath, R. J., N. Nutile-McMenemy, M. S. Alkaitis, and J. A. DeLeo. 2008. Differential migration, LPS-induced cytokine, chemokine, and NO expression in immortalized BV-2 and HAPI cell lines and primary microglial cultures. *J Neurochem* 107 (2):557–69.

Ibrahim, M. M., F. Porreca, J. Lai, P. J. Albrecht, F. L. Rice, A. Khodorova, G. Davar, et al. 2005. CB2 cannabinoid receptor activation produces antinociception by stimulating peripheral release of endogenous opioids. *Proc Natl Acad Sci U S A* 102 (8):3093–8.

Iskedjian, M., B. Bereza, A. Gordon, C. Piwko, and T. R. Einarson. 2007. Meta-analysis of cannabis based treatments for neuropathic and multiple sclerosis-related pain. *Curr Med Res Opin* 23 (1):17–24.

Iversen, L., and V. Chapman. 2002. Cannabinoids: a real prospect for pain relief? *Curr Opin Pharmacol* 2 (1):50–5.

Jarai, Z., J. A. Wagner, K. Varga, K. D. Lake, D. R. Compton, B. R. Martin, A. M. Zimmer, et al. 1999. Cannabinoid-induced mesenteric vasodilation through an endothelial site distinct from CB1 or CB2 receptors. *Proc Natl Acad Sci U S A* 96 (24):14136–41.

Jhaveri, M. D., S. J. Elmes, D. Richardson, D. A. Barrett, D. A. Kendall, R. Mason, and V. Chapman. 2008. Evidence for a novel functional role of cannabinoid CB(2) receptors in the thalamus of neuropathic rats. *Eur J Neurosci* 27 (7):1722–30.

Ji, B., J. Maeda, M. Sawada, M. Ono, T. Okauchi, M. Inaji, M. R. Zhang, et al. 2008. Imaging of peripheral benzodiazepine receptor expression as biomarkers of detrimental versus beneficial glial responses in mouse models of Alzheimer's and other CNS pathologies. *J Neurosci* 28 (47):12255–67.

Joranson, D. E., K. M. Ryan, A. M. Gilson, and J. L. Dahl. 2000. Trends in medical use and abuse of opioid analgesics. *JAMA* 283 (13):1710–4.

Kai, H., Y. Morioka, M. Tomida, T. Takahashi, M. Hattori, K. Hanasaki, K. Koike, et al. 2007. 2-Arylimino-5,6-dihydro-4H-1,3-thiazines as a new class of cannabinoid receptor agonists. Part 2: Orally bioavailable compounds. *Bioorg Med Chem Lett* 17 (14):3925–9.

Katz, N. P. 2000. MorphiDex (MS:DM) double-blind, multiple-dose studies in chronic pain patients. *J Pain Symptom Manage* 19 (1 Suppl):S37–41.

Kettenmann, H., Ransom, B. R. 2005. *Neuroglia: Second Edition.* New York: Oxford University Press.

Klein, T. W., C. Newton, K. Larsen, L. Lu, I. Perkins, L. Nong, and H. Friedman. 2003. The cannabinoid system and immune modulation. *J Leukoc Biol* 74 (4):486–96.

Landefeld, C. S., and M. A. Steinman. 2009. The Neurontin legacy—marketing through misinformation and manipulation. *N Engl J Med* 360 (2):103–6.

Lauckner, J. E., J. B. Jensen, H. Y. Chen, H. C. Lu, B. Hille, and K. Mackie. 2008. GPR55 is a cannabinoid receptor that increases intracellular calcium and inhibits M current. *Proc Natl Acad Sci U S A* 105 (7):2699–704.

Ledeboer, A., M. R. Hutchinson, L. R. Watkins, and K. W. Johnson. 2007. Ibudilast (AV-411). A new class therapeutic candidate for neuropathic pain and opioid withdrawal syndromes. *Expert Opin Investig Drugs* 16 (7):935–50.

Ledeboer, A., T. Liu, J. A. Shumilla, J. H. Mahoney, S. Vijay, M. I. Gross, J. A. Vargas, et al. 2007. The glial modulatory drug AV411 attenuates mechanical allodynia in rat models of neuropathic pain. *Neuron Glia Biol* 2 (4):279–291.

Ledeboer, A., E. M. Sloane, E. D. Milligan, M. G. Frank, J. H. Mahony, S. F. Maier, and L. R. Watkins. 2005. Minocycline attenuates mechanical allodynia and proinflammatory cytokine expression in rat models of pain facilitation. *Pain* 115 (1–2):71–83.

Ledent, C., O. Valverde, G. Cossu, F. Petitet, J. F. Aubert, F. Beslot, G. A. Bohme, et al. 1999. Unresponsiveness to cannabinoids and reduced addictive effects of opiates in CB1 receptor knockout mice. *Science* 283 (5400):401–4.

Lim, G., B. Sung, R. R. Ji, and J. Mao. 2003. Upregulation of spinal cannabinoid-1-receptors following nerve injury enhances the effects of Win 55,212-2 on neuropathic pain behaviors in rats. *Pain* 105 (1–2):275–83.

Mao, J., D. D. Price, F. S. Caruso, and D. J. Mayer. 1996. Oral administration of dextromethorphan prevents the development of morphine tolerance and dependence in rats. *Pain* 67 (2–3):361–8.

Mao, J., D. D. Price, and D. J. Mayer. 1995. Mechanisms of hyperalgesia and morphine toler-ance: a current view of their possible interactions. *Pain* 62 (3):259–74.

Mayer, D. J., J. Mao, J. Holt, and D. D. Price. 1999. Cellular mechanisms of neuropathic pain, morphine tolerance, and their interactions. *Proc Natl Acad Sci U S A* 96 (14):7731–6.

Meller, S. T., C. Dykstra, D. Grzybycki, S. Murphy, and G. F. Gebhart. 1994. The possible role of glia in nociceptive processing and hyperalgesia in the spinal cord of the rat. *Neuropharmacology* 33 (11):1471–8.

Meskini, N., G. Nemoz, I. Okyayuz-Baklouti, M. Lagarde, and A. F. Prigent. 1994. Phosphodiesterase inhibitory profile of some related xanthine derivatives phar-macologically active on the peripheral microcirculation. *Biochem Pharmacol* 47 (5):781–8.

Mika, J. 2008. Modulation of microglia can attenuate neuropathic pain symptoms and enhance morphine effectiveness. *Pharmacol Rep* 60 (3):297–307.

Miller, A. M., and N. Stella. 2008. CB2 receptor-mediated migration of immune cells: it can go either way. *Br J Pharmacol* 153 (2):299–308.

Molina-Holgado, F., E. Pinteaux, J. D. Moore, E. Molina-Holgado, C. Guaza, R. M. Gibson, and N. J. Rothwell. 2003. Endogenous interleukin-1 receptor antagonist mediates anti-inflammatory and neuroprotective actions of cannabinoids in neurons and glia. *J Neurosci* 23 (16):6470–4.

Nagata, K., T. Ogawa, M. Omosu, K. Fujimoto, and S. Hayashi. 1985. In vitro and in vivo inhibitory effects of propentofylline on cyclic AMP phosphodiesterase activity. *Arzneimittelforschung* 35 (7):1034–6.

Narita, M., M. Miyatake, M. Shibasaki, K. Shindo, A. Nakamura, N. Kuzumaki, Y. Nagumo, and T. Suzuki. 2006. Direct evidence of astrocytic modulation in the development of reward-ing effects induced by drugs of abuse. *Neuropsychopharmacology* 31 (11):2476–88.

Narita, M., M. Suzuki, N. Kuzumaki, M. Miyatake, and T. Suzuki. 2008. Implication of acti-vated astrocytes in the development of drug dependence: differences between metham-phetamine and morphine. *Ann N Y Acad Sci* 1141:96–104.

Novak, S., W. C. Nemeth, and K. A. Lawson. 2004. Trends in medical use and abuse of sus-tained-release opioid analgesics: a revisit. *Pain Med* 5 (1):59–65.

Nunez, E., C. Benito, M. R. Pazos, A. Barbachano, O. Fajardo, S. Gonzalez, R. M. Tolon, and J. Romero. 2004. Cannabinoid CB2 receptors are expressed by perivascular microglial cells in the human brain: an immunohistochemical study. *Synapse* 53 (4):208–13.

Ohta, H., T. Ishizaka, M. Tatsuzuki, M. Yoshinaga, I. Iida, T. Yamaguchi, Y. Tomishima, N. Futaki, Y. Toda, and S. Saito. 2008. Imine derivatives as new potent and selective CB2 cannabinoid receptor agonists with an analgesic action. *Bioorg Med Chem* 16 (3):1111–24.

Ohta, H., T. Ishizaka, M. Yoshinaga, A. Morita, Y. Tomishima, Y. Toda, and S. Saito. 2007. Sulfonamide derivatives as new potent and selective CB2 cannabinoid receptor agonists. *Bioorg Med Chem Lett* 17 (18):5133–5.

Overton, H. A., A. J. Babbs, S. M. Doel, M. C. Fyfe, L. S. Gardner, G. Griffin, H. C. Jackson, et al. 2006. Deorphanization of a G protein-coupled receptor for oleoylethanolamide and its use in the discovery of small-molecule hypophagic agents. *Cell Metab* 3 (3):167–75.

Pain, Clin J. 1997. The use of opioids for the treatment of chronic pain. A consensus statement from the American Academy of Pain Medicine and the American Pain Society. *Clin J Pain* 13 (1):6–8.

Parkinson, F. E., and B. B. Fredholm. 1991. Effects of propentofylline on adenosine A1 and A2 receptors and nitrobenzylthioinosine-sensitive nucleoside transporters: quantitative autoradiographic analysis. *Eur J Pharmacol* 202 (3):361–6.

Perez, J., and M. V. Ribera. 2008. Managing neuropathic pain with Sativex: a review of its pros and cons. *Expert Opin Pharmacother* 9 (7):1189–95.

Pergolizzi, J., R. H. Boger, K. Budd, A. Dahan, S. Erdine, G. Hans, H. G. Kress, et al. 2008. Opioids and the management of chronic severe pain in the elderly: consensus statement of an International Expert Panel with focus on the six clinically most often used World Health Organization step III opioids (buprenorphine, fentanyl, hydromorphone, methadone, morphine, oxycodone). *Pain Pract* 8 (4):287–313.

Pettit, D. A., M. P. Harrison, J. M. Olson, R. F. Spencer, and G. A. Cabral. 1998. Immunohistochemical localization of the neural cannabinoid receptor in rat brain. *J Neurosci Res* 51 (3):391–402.

Portenoy, R. K., and K. M. Foley. 1986. Chronic use of opioid analgesics in non-malignant pain: report of 38 cases. *Pain* 25 (2):171–86.

Raghavendra, V., and J. A. DeLeo. 2006. Cytokine modulation of opioid action. In: Schmidt, R.F., Willis, W.D. (Eds.) *Encyclopedia of Pain* (227–235). New York: Springer.

Raghavendra, V., M. D. Rutkowski, and J. A. DeLeo. 2002. The role of spinal neuroimmune activation in morphine tolerance/hyperalgesia in neuropathic and sham-operated rats. *J Neurosci* 22 (22):9980–9.

Raghavendra, V., F. Tanga, and J. A. DeLeo. 2003. Inhibition of microglial activation attenuates the development but not existing hypersensitivity in a rat model of neuropathy. *J Pharmacol Exp Ther* 306 (2):624–30.

Raghavendra, V., F. Tanga, M. D. Rutkowski, and J. A. DeLeo. 2003. Anti-hyperalgesic and morphine-sparing actions of propentofylline following peripheral nerve injury in rats: mechanistic implications of spinal glia and proinflammatory cytokines. *Pain* 104 (3):655–64.

Raghavendra, V., F. Y. Tanga, and J. A. DeLeo. 2004. Attenuation of morphine tolerance, withdrawal-induced hyperalgesia, and associated spinal inflammatory immune responses by propentofylline in rats. *Neuropsychopharmacology* 29 (2):327–34.

Ramirez, B. G., C. Blazquez, T. Gomez del Pulgar, M. Guzman, and M. L. de Ceballos. 2005. Prevention of Alzheimer's disease pathology by cannabinoids: neuroprotection mediated by blockade of microglial activation. *J Neurosci* 25 (8):1904–13.

Rathbone, J., H. Variend, and H. Mehta. 2008. Cannabis and schizophrenia. *Cochrane Database Syst Rev* (3):CD004837.

Romero-Sandoval, A., N. Chai, N. Nutile-McMenemy, and J. A. DeLeo. 2008. A comparison of spinal Iba1 and GFAP expression in rodent models of acute and chronic pain. *Brain Res* 1219:116–26.

Romero-Sandoval, A., and J. C. Eisenach. 2007. Spinal cannabinoid receptor type 2 activation reduces hypersensitivity and spinal cord glial activation after paw incision. *Anesthesiology* 106 (4):787–94.

Romero-Sandoval, A., N. Nutile-McMenemy, and J. A. DeLeo. 2008. Spinal microglial and perivascular cell cannabinoid receptor type 2 activation reduces behavioral hypersensitivity without tolerance after peripheral nerve injury. *Anesthesiology* 108 (4):722–34.

Romero-Sandoval, E. A., R. J. Horvath, and J. A. DeLeo. 2008. Neuroimmune interactions and pain: focus on glial-modulating targets. *Curr Opin Investig Drugs* 9 (7):726–34.

Ronnback, L., and E. Hansson. 1988. Are astroglial cells involved in morphine tolerance? *Neurochem Res* 13 (2):87–103.

Schubert, P., T. Ogata, K. Rudolphi, C. Marchini, A. McRae, and S. Ferroni. 1997. Support of homeostatic glial cell signaling: a novel therapeutic approach by propentofylline. *Ann N Y Acad Sci* 826:337–47.

Scott, D. A., C. E. Wright, and J. A. Angus. 2004. Evidence that CB-1 and CB-2 cannabinoid receptors mediate antinociception in neuropathic pain in the rat. *Pain* 109 (1–2):124–31.

Scully, C. 2007. Cannabis: adverse effects from an oromucosal spray. *Br Dent J* 203 (6):E12; discussion 336–7.

Sheng, W. S., S. Hu, X. Min, G. A. Cabral, J. R. Lokensgard, and P. K. Peterson. 2005. Synthetic cannabinoid WIN55,212-2 inhibits generation of inflammatory mediators by IL-1 beta-stimulated human astrocytes. *Glia* 49 (2):211–9.

Si, Q. S., Y. Nakamura, P. Schubert, K. Rudolphi, and K. Kataoka. 1996. Adenosine and propentofylline inhibit the proliferation of cultured microglial cells. *Exp Neurol* 137 (2):345–9.

Sinha, D., T. I. Bonner, N. R. Bhat, and L. A. Matsuda. 1998. Expression of the CB1 cannabinoid receptor in macrophage-like cells from brain tissue: immunochemical characterization by fusion protein antibodies. *J Neuroimmunol* 82 (1):13–21.

Smith, P. A., D. E. Selley, L. J. Sim-Selley, and S. P. Welch. 2007. Low dose combination of morphine and delta9-tetrahydrocannabinol circumvents antinociceptive tolerance and apparent desensitization of receptors. *Eur J Pharmacol* 571 (2–3):129–37.

Song, P., and Z. Q. Zhao. 2001. The involvement of glial cells in the development of morphine tolerance. *Neurosci Res* 39 (3):281–6.

Staton, P. C., J. P. Hatcher, D. J. Walker, A. D. Morrison, E. M. Shapland, J. P. Hughes, E. Chong, et al. 2008. The putative cannabinoid receptor GPR55 plays a role in mechanical hyperalgesia associated with inflammatory and neuropathic pain. *Pain* 139 (1):225–36.

Stephenson, D. T., and S. P. Arneric. 2008. Neuroimaging of pain: advances and future prospects. *J Pain* 9 (7):567–79.

Stephenson, D. T., D. A. Schober, E. B. Smalstig, R. E. Mincy, D. R. Gehlert, and J. A. Clemens. 1995. Peripheral benzodiazepine receptors are colocalized with activated microglia following transient global forebrain ischemia in the rat. *J Neurosci* 15 (7 Pt 2):5263–74.

Stewart, W. F., J. A. Ricci, E. Chee, D. Morganstein, and R. Lipton. 2003. Lost productive time and cost due to common pain conditions in the US workforce. *JAMA* 290 (18):2443–54.

Sung, B., G. Lim, and J. Mao. 2003. Altered expression and uptake activity of spinal glutamate transporters after nerve injury contribute to the pathogenesis of neuropathic pain in rats. *J Neurosci* 23 (7):2899–910.

Sweitzer, S. M., P. Schubert, and J. A. DeLeo. 2001. Propentofylline, a glial modulating agent, exhibits antiallodynic properties in a rat model of neuropathic pain. *J Pharmacol Exp Ther* 297 (3):1210–7.

Tawfik, V. L., M. L. Lacroix-Fralish, K. K. Bercury, N. Nutile-McMenemy, B. T. Harris, and J. A. Deleo. 2006. Induction of astrocyte differentiation by propentofylline increases glutamate transporter expression in vitro: heterogeneity of the quiescent phenotype. *Glia* 54 (3):193–203.

Tawfik, V. L., M. L. LaCroix-Fralish, N. Nutile-McMenemy, and J. A. DeLeo. 2005. Transcriptional and translational regulation of glial activation by morphine in a rodent model of neuropathic pain. *J Pharmacol Exp Ther* 313 (3):1239–47.

Tawfik, V. L., N. Nutile-McMenemy, M. L. Lacroix-Fralish, and J. A. DeLeo. 2007. Efficacy of propentofylline, a glial modulating agent, on existing mechanical allodynia following peripheral nerve injury. *Brain Behav Immun* 21 (2):238–46.

Tawfik, V. L., M. R. Regan, C. Haenggeli, M. L. Lacroix-Fralish, N. Nutile-McMenemy, N. Perez, J. D. Rothstein, and J. A. DeLeo. 2008. Propentofylline-induced astrocyte modulation leads to alterations in glial glutamate promoter activation following spinal nerve transection. *Neuroscience* 152 (4):1086–92.

Trescot, A. M., S. Helm, H. Hansen, R. Benyamin, S. E. Glaser, R. Adlaka, S. Patel, and L. Manchikanti. 2008. Opioids in the management of chronic non-cancer pain: an update of American Society of the Interventional Pain Physicians' (ASIPP) Guidelines. *Pain Physician* 11 (2 Suppl):S5–S62.

Tsou, K., S. Brown, M. C. Sanudo-Pena, K. Mackie, and J. M. Walker. 1998. Immunohistochemical distribution of cannabinoid CB1 receptors in the rat central nervous system. *Neuroscience* 83 (2):393–411.

Tsuda, M., Y. Shigemoto-Mogami, S. Koizumi, A. Mizokoshi, S. Kohsaka, M. W. Salter, and K. Inoue. 2003. P2X4 receptors induced in spinal microglia gate tactile allodynia after nerve injury. *Nature* 424 (6950):778–83.

Valenzano, K. J., L. Tafesse, G. Lee, J. E. Harrison, J. M. Boulet, S. L. Gottshall, L. Mark, et al. 2005. Pharmacological and pharmacokinetic characterization of the cannabinoid receptor 2 agonist, GW405833, utilizing rodent models of acute and chronic pain, anxiety, ataxia and catalepsy. *Neuropharmacology* 48 (5):658–72.

Van Sickle, M. D., M. Duncan, P. J. Kingsley, A. Mouihate, P. Urbani, K. Mackie, N. Stella, et al. 2005. Identification and functional characterization of brainstem cannabinoid CB2 receptors. *Science* 310 (5746):329–32.

Waksman, Y., J. M. Olson, S. J. Carlisle, and G. A. Cabral. 1999. The central cannabinoid receptor (CB1) mediates inhibition of nitric oxide production by rat microglial cells. *J Pharmacol Exp Ther* 288 (3):1357–66.

Walker, J. M., and A. G. Hohmann. 2005. Cannabinoid mechanisms of pain suppression. *Handb Exp Pharmacol* (168):509–54.

Walter, L., A. Franklin, A. Witting, C. Wade, Y. Xie, G. Kunos, K. Mackie, and N. Stella. 2003. Nonpsychotropic cannabinoid receptors regulate microglial cell migration. *J Neurosci* 23 (4):1398–405.

Watkins, L. R., S. F. Maier, and L. E. Goehler. 1995. Immune activation: the role of pro-inflammatory cytokines in inflammation, illness responses and pathological pain states. *Pain* 63 (3):289–302.

Wotherspoon, G., A. Fox, P. McIntyre, S. Colley, S. Bevan, and J. Winter. 2005. Peripheral nerve injury induces cannabinoid receptor 2 protein expression in rat sensory neurons. *Neuroscience* 135 (1):235–45.

Yao, B. B., G. Hsieh, A. V. Daza, Y. Fan, G. K. Grayson, T. R. Garrison, O. El Kouhen, et al. 2009. Characterization of a cannabinoid CB2 receptor-selective agonist, A-836339 [2,2,3,3-tetramethyl-cyclopropanecarboxylic acid [3-(2-methoxy-ethyl)-4,5-dimethyl-3H-thiazol-(2Z)-ylidene]-amide], using in vitro pharmacological assays, in vivo pain models, and pharmacological magnetic resonance imaging. *J Pharmacol Exp Ther* 328 (1):141–51.

Yasuno, F., M. Ota, J. Kosaka, H. Ito, M. Higuchi, T. K. Doronbekov, S. Nozaki, et al. 2008. Increased binding of peripheral benzodiazepine receptor in Alzheimer's disease measured by positron emission tomography with [11C]DAA1106. *Biol Psychiatry* 64 (10):835–41.

Yiangou, Y., P. Facer, P. Durrenberger, I. P. Chessell, A. Naylor, C. Bountra, R. R. Banati, and P. Anand. 2006. COX-2, CB2 and P2X7-immunoreactivities are increased in activated microglial cells/macrophages of multiple sclerosis and amyotrophic lateral sclerosis spinal cord. *BMC Neurol* 6:12.

Yoon, M. H., and J. I. Choi. 2003. Pharmacologic interaction between cannabinoid and either clonidine or neostigmine in the rat formalin test. *Anesthesiology* 99 (3):701–7.

Zhang, J., C. Hoffert, H. K. Vu, T. Groblewski, S. Ahmad, and D. O'Donnell. 2003. Induction of CB2 receptor expression in the rat spinal cord of neuropathic but not inflammatory chronic pain models. *Eur J Neurosci* 17 (12):2750–4.

Zimmer, A., A. M. Zimmer, A. G. Hohmann, M. Herkenham, and T. I. Bonner. 1999. Increased mortality, hypoactivity, and hypoalgesia in cannabinoid CB1 receptor knockout mice. *Proc Natl Acad Sci U S A* 96 (10):5780–5.

10 On the Role of ATP-Gated P2X Receptors in Acute, Inflammatory and Neuropathic Pain

Estelle Toulme, Makoto Tsuda,
Baljit S. Khakh, and Kazuhide Inoue

CONTENTS

P2X receptors are a family of ATP-gated cation channels with distinct functional properties and distributions in the body. In humans, P2X receptors comprise seven gene products that can form at least 14 known homo- and heterotrimeric receptors. In this chapter we focus on three P2X subunits (P2X3, P2X4, and P2X7) that each form homomeric receptors, and their roles in acute and chronic pain.

10.1 INTRODUCTION

The somatosensory system is the part of the nervous system that has evolved to integrate sensory input signals from the body including touch, heat and pain sensations. The primary afferent neurons, located on the dorsal side of the spinal cord, dorsal root ganglion (DRG) neurons are pseudomonopolar neurons that conduct sensory information encoded by the frequency of action potentials along their processes from the peripheral sites (e.g. skin) to the dorsal horn of the spinal cord. Broadly speaking, there are two different types of pain: acute and chronic. Acute pain, also called nociceptive pain, is a sensation experienced in response to injury or tissue damage (Millan 1999). The relatively short duration of such pain acts as a signal to warn the subject about an imminent danger (e.g. stepping on a sharp object like a pin). This type of pain is conducted through nociceptor myelinated A-delta and unmyelinated C-fiber neurons classified depending on their myelination status (Snider and McMahon 1998). In contrast, chronic pain has no "warning" function for the subject and is usually the manifestation of an underlying injury, disorder or disease this type of pain can last for several years. Chronic pain is subclassified into neuropathic pain and inflammatory pain. Neuropathic pain is due to damaged or misfiring peripheral nerves, whereas inflammatory pain is derived from a non-specific immune response that alters nerve function.

Understanding the cellular and molecular basis of pain has been a stated goal of biological research for several decades accounting for increased recent interest in extracellular ATP signaling. Adenosine, 5′-triphosphate (ATP) is a ubiquitous molecule found in every cell in the millimolar concentration range and released into the extracellular milieu after tissue injury or visceral distension once released activates ATP receptor molecules on nearby sensory nerves. Extracellular ATP is an endogenous agonist at P2 purinoceptors, which comprise metabotropic G protein-coupled P2Y receptors and ionotropic cation-permeable P2X receptors (Ralevic and Burnstock 1998). To date, seven P2X subunits have been cloned. They share a common topology displaying two transmembrane domains linked by a large extracellular loop and intracellular N- and C-terminal tails (North 2002). Their activation by ATP induces the opening of a pore permeable to Na^+, K^+ and Ca^{2+}, inducing an overall depolarization of the cell (Khakh and North 2006). From the perspective of human medicine, a role for ATP signaling in pain has its roots in pioneering studies carried out several decades ago. In 1966, Collier and colleagues, followed 10 years later by Bleehen and Keele, showed that local application of ATP onto human skin blisters induced a persistent sensation of pain (Bleehen and Keele 1977; Collier et al. 1966). These papers instigated subsequent studies where a role for ATP as an extracellular chemical messenger that transmits sensory information has been investigated in humans and animals. A boost to the field has derived from the use of genetic methods such as knockout mice and antisense oligonucleotides, as well as the availability of P2X receptor specific antagonists. Using such approaches, the involvement of ATP-activated receptors has been related to different types of pain (Table 10.1). The latest data suggest that whereas acute pain seems to be linked to the activation of P2X3 receptors expressed in sensory neurons, neuropathic pain more

TABLE 10.1

Summary for P2X Receptor Involvement in Acute and Chronic Pain

	Cellular Distribution	Selective Agonist	Antagonists	Pain Involvement	Key References
P2X3	sensory neurons	ATP, αβmeATP	TNP-ATP A-317491	Nociception: pain initiation and pain transmission, inflammatory and neuropathic pain	Cockayne et al. 2000 Souslova et al. 2000 Jarvis et al. 2002
P2X4	microglia	ATP	TNP-ATP	Chronic pain: neuropathic pain	Tsuda et al. 2003 Ulmann et al. 2008
P2X7	immune cells	ATP BzATP	A-438079 AZ-11645373	Chronic pain: inflammatory and neuropathic pain	Chessell et al. 2005 Honore et al. 2006a

likely involves P2X4 receptors on the surfaces of glial cells, and inflammatory pain involves P2X7 receptors on immune cells.

10.2   P2X3 RECEPTORS AND ACUTE PAIN

ATP released from different cell types is implicated in the initiation of pain by activating P2 receptors on sensory nerve terminals (Burnstock 2001). Known P2X subtypes with a role in nociception include P2X3 and P2X2/3, which are considered potential therapeutic targets for the management of pathological conditions.

10.2.1   P2X3 Receptor Distribution

The expression profile of P2X subunits has been investigated at the RNA level in sensory neurons, revealing that with the exception of P2X7, all other P2X subunits are detected in DRG (Chen et al. 1995; Lewis et al. 1995). Interestingly, P2X3 receptors were first successfully cloned from, and functionally characterized, in DRG neurons (Chen et al. 1995). Further experiments on P2X3 subunit distribution led to the recognition that it was almost exclusively expressed in primary sensory neurons, located on both their peripheral and central terminals (Dunn et al. 2001; North 2004; Vulchanova et al. 1997). Immunocytochemistry for P2X3 subunits revealed that these receptors are expressed by small- and medium-sized neurons but absent from large diameter neurons (Novakovic et al. 1999; Vulchanova et al. 1997, 1998). However, P2X3 immunoreactivity has also been detected in two other structures indirectly involved in the transmission of the nociceptive message: the nucleus tractus solitarius and the trigeminal spinal nucleus (Llewellyn-Smith and Burnstock 1998). One other P2X subunit, P2X2, has been identified by immunocytochemistry in DRG neurons. P2X2 subunits were located in medium-diameter neurons where they co-localized with P2X3 subunits in some neurons (Vulchanova et al. 1997).

At their central termini primary afferent fibers form synapses with dorsal horn neurons in the spinal cord—the first sensory synapses in the somatosensory pathway, utilizing the fast excitatory neurotransmitter, glutamate. Immunohistochemistry experiments performed on spinal cord slices using an antibody directed against P2X3 revealed specific localization of this subunit in the inner lamina II (IIi) of the dorsal horn. At the presynaptic level a co-culture model mimicking the first somatosensory synapse of DRG and dorsal horn neurons isolated from the spinal cord, showed that ATP application could induce glutamate release that activated postsynaptic glutamatergic receptors expressed in the dorsal horn neurons (Gu and MacDermott 1997). These experiments provided the first evidence for the existence of presynaptic P2X receptors in DRG neuronal processes, and for a role of P2X receptors in modulating fast synaptic transmission in the central nervous system. Furthermore, inhibiting P2X receptors by application of the P2 antagonist pyridoxal-phosphate-6-azophenyl-2',4'-disulfonate (PPADS) has been shown to abolish the glutamate-dependent postsynaptic currents of dorsal horn neurons (Li et al. 1998). The facilitation of sensory transmission through presynaptic P2X receptors has been reproduced *in vitro* by the exogenous application of ATP and by endogenous release of ATP after stimulation of primary afferent fibers (Nakatsuka and Gu 2001).

P2X receptors can also act postsynaptically when they are expressed in a subpopulation of dorsal horn neurons of the spinal cord (Bardoni et al. 1997; Jo and Schlichter 1999). In these neurons, ATP acts as a fast neurotransmitter and is thought to be released in combination with GABA in order to activate suramin and PPADS sensitive receptors, suggesting that GABAergic interneurons are a physiologically relevant source of ATP (Jo and Schlichter 1999). Consistent with these experiments, applications of ATP elicit inward currents in cultured (Hugel and Schlichter 2000; Jahr and Jessell 1983) or acutely dissociated (Bardoni et al. 1997; Rhee et al. 2000) dorsal horn neurons, providing strong evidence for the existence of postsynaptic P2X receptors in these cells. Elegant *in vitro* experiments also show that P2X receptors are found in the nerve endings of sensory neurons near their cutaneous targets. In this model damaged skin cells release ATP, which activates P2X receptors on sensory nerve endings (Cook and McCleskey 2002).

10.2.2 Functional Properties of Sensory P2X3 Receptors

Significant progress has been made in the functional characterization of P2X receptors in the somatosensory system based on electrophysiological recordings from the somatic compartments of sensory neurons. A detailed characterization of homomeric P2X3 receptors expressed in HEK cells (Lewis et al. 1995, and subsequently by many other labs) has revealed that P2X3 channels displayed some distinguishing characteristics. Indeed, ATP application induced an inward current with fast activation rise time, fast desensitization, and a slow recovery time from desensitization. These currents were mimicked by α,β-methylene-ATP ($\alpha\beta$meATP), a non-hydrolysable ATP analog, and inhibited by 2', 3'-O-(2,4,6- trinitrophenyl) adenosine 5'-triphosphate (TNP-ATP) (Lewis et al. 1995). P2X3 subunits are also able to co-assemble with P2X2 subunits and form functional heteromeric receptors sharing some characteristics from both homomeric P2X2 and P2X3 receptors, but resulting in a novel phenotype. Heteromeric P2X2/3 receptors have a fast activation phase but slow desensitization kinetics and like homomeric P2X3 receptors they are sensitive to αβmeATP (Chen et al. 1995). Functional studies of P2X receptors in sensory neurons are based on the assumption that receptors expressed on cell bodies are also expressed on nerve endings. Whole-cell patch clamp recordings performed on acutely dissociated DRG neurons have generally revealed three different types of responses to ATP (10 μM) that differ in terms of desensitization kinetics: a transient current, a sustained current and a mixed current with both transient and sustained components. Fast currents were usually seen in small-diameter DRG neurons. Due to its fast desensitization (typically in less than ~100 ms) this ATP-activated current has been assigned to homomeric P2X3 receptors since it shows remarkably similar electrophysiological properties to recombinant P2X3 receptors studied in heterologous expression systems. The pharmacological profile of this receptor includes activation by ATP and αβmeATP and inhibition by TNP-ATP (Cook and McCleskey 1997; Rae et al. 1998; Robertson et al. 1996). Its identity as P2X3 receptor population has been further supported by the generation of P2X3 subunit knockout mice that did not display any rapidly desensitizing ATP-induced currents in response to an application of ATP (Cockayne et al. 2000; Souslova et al. 2000). Slow currents and mixed currents are characterized by a plateau phase with weak desensitization in the range of seconds in the presence of the agonists like ATP. They are believed to be mediated by a mixed population of homomeric P2X2 receptors and heteromeric P2X2/3 receptors. They have been further divided into two subpopulations based on their characteristic pharmacological profiles (Lewis et al. 1995), particularly their αβmeATP sensitivity. The specific distribution of P2X3 receptors on terminal nerve endings and cell bodies of sensory neurons along with their ability to induce membrane potential changes suggested the involvement of this receptor subtype in processing pain states.

10.2.3   PAIN PHENOTYPES REVEALED BY GENETIC AND PHARMACOLOGICAL METHODS

Exogenously applied P2X receptor agonists induce acute pain both in humans and animals. For example, a subplantar injection of ATP or αβmeATP in rodents resulted in pain behavior characterized by spontaneous licking, biting, and lifting of the injected hindpaw (Bland-Ward and Humphrey 1997; Hamilton et al. 1999). The generation of P2X3 subunit knockout mice by two different laboratories shed significant light on the involvement of P2X3 receptors in pain-mediated processes. The most obvious phenotype displayed by P2X3 knockout mice was the loss of their ability to sense bladder filling. This effect has been reported to be a consequence of the absence of P2X3 receptors in nerve terminals of sensory neurons that innervate the bladder urothelium. Moreover, they showed that injection of ATP in the hindpaw of P2X3-null mice evoked a significantly decreased nociceptive behavior compared to wild-type mice (Cockayne et al. 2000). However, the response of the mice to nociceptive thermal and mechanical stimuli was normal, although there was an alteration of the responses to warm thermal stimuli (Souslova et al. 2000). These data indicated a major role for P2X3 receptors in mediating acute nociceptive pain and suggested that targeting the receptor with selective antagonists could modulate this pain state. This hypothesis was subsequently addressed by use of the highly selective P2X3 and P2X2/3 receptor antagonist, A-317491. Intraplantar and intrathecal injection of A-317491 into rodents produced antinociceptive effects in complete Freund adjuvant (CFA)-induced chronic hyperalgesia and nerve injury-induced tactile allodynia (Jarvis et al. 2002). Finally, P2X3-subunit-containing receptors may play a role in neuropathic pain states following peripheral nerve ligation in rats. Intrathecal administration of P2X3 antisense oligonucleotides attenuated both mechanically triggered allodynia (Honore et al. 2002) and thermal- and mechanically triggered hyperalgesia (Barclay et al. 2002; Honore et al. 2002), which were previously induced by spinal nerve ligation in the rat.

10.3   P2X4 RECEPTORS AND NEUROPATHIC PAIN

10.3.1   P2X4 RECEPTORS IN SPINAL MICROGLIA ARE NECESSARY FOR NEUROPATHIC PAIN

Several findings indicate the activation of spinal microglia in neuropathic pain states, but until recently it remained an open question as to whether spinal microglia play a causal role in neuropathic pain behaviors. Tsuda et al. (2003) directly implicated activated microglia in the pathogenesis of neuropathic pain by determining the role of the P2X4 purinoceptor. A clue to identifying P2X4 receptors in the spinal cord as being required for neuropathic pain first came from a pharmacological investigation of pain behavior after nerve injury using the P2X receptor antagonists TNP-ATP and PPADS (Tsuda et al. 2003). The marked tactile allodynia after injury of a spinal nerve was found to be reversed by acutely administering intrathecally TNP-ATP but was unaffected by administering PPADS. From the pharmacological profiles of TNP-ATP and PPADS, it was inferred that the tactile allodynia depended upon P2X4 receptors in the spinal cord (Tsuda et al. 2003). The expression of P2X4 receptor protein progressively increased in the days following nerve injury, the time-course of which was parallel to that of the development of tactile allodynia. In the immunohistochemical analysis, it was found that many small cells in the dorsal horn of the nerve-injured side were positive for P2X4 receptor protein and these cells were identified as microglia, rather than neurons or astrocytes (Tsuda et al. 2003). The cells expressing P2X4 receptors in the nerve-injured side of the dorsal horn showed high levels of OX-42 labeling and morphological hypertrophy, all of which are characteristic markers of activated microglia. Moreover, it was found that reducing the up-regulation of P2X4 receptor protein in spinal microglia by P2X4 antisense oligodeoxynucleotides prevented the development of the nerve injury–induced tactile allodynia. Together, this evidence implied that P2X4 receptor activation in spinal microglia is necessary for neuropathic pain. The sufficiency of P2X4 receptor activation in microglia for the development of allodynia was demonstrated by intrathecal administration of activated, cultured microglia in which these receptors had been stimulated *in vitro* by ATP (Tsuda et al. 2003). In otherwise naïve animals, allodynia develops progressively over a period of 3–5 h following the administration of P2X4 receptor-stimulated microglia. Moreover, in rats in which tactile allodynia was caused by the ATP-stimulated microglia, this allodynia was reversed by administering TNP-ATP. Thus, the allodynia caused by ATP-stimulated microglia is pharmacologically similar to that caused by peripheral nerve injury. These aforementioned data have been extended with the generation of P2X4 knockout mice that show a remarkable reduction of tactile allodynia after spinal nerve injury in comparison with wild-type animals (Ulmann et al. 2008). They also demonstrate a complete absence of mechanical hypersensitivity after peripheral nerve injury. The development of genetically modified mice has been an important step to directly assess the role of microglia P2X4 receptors. Together these findings indicate that P2X4 receptor stimulation of microglia is not only necessary for tactile allodynia but is also sufficient to cause allodynia.

10.3.2 PLAUSIBLE MECHANISMS OF UP-REGULATION OF MICROGLIAL P2X4 RECEPTORS IN NEUROPATHIC PAIN

The up-regulation of P2X4 receptor expression in microglia is evidently a key process in neuropathic pain. How peripheral nerve injury increases expression of P2X4 in microglia is not clearly understood, but recent studies have shown that the extracellular matrix protein, fibronectin, could be involved. It was found that microglia cultured on fibronectin-coated dishes showed a marked increase in P2X4 expression both at the mRNA and protein levels (Nasu-Tada et al. 2006). The up-regulated P2X4 protein on microglia might be functional since the P2X4-mediated Ca^{2+} response was enhanced in fibronectin-treated microglia. Furthermore, intrathecal delivery of ATP-stimulated microglia to the rat lumbar spinal cord revealed that microglia treated with fibronectin more effectively induced allodynia than control microglia. In the dorsal horn of a model of neuropathic pain, the level of fibronectin protein was elevated 3 to 7 days after nerve injury (Nasu-Tada et al. 2006), the time when P2X4 protein levels start to increase (Tsuda et al. 2003). Blockade of the fibronectin receptor attenuated nerve injury-induced P2X4 up-regulation and allodynia (Tsuda et al. 2008a). Intrathecal delivery of fibronectin increased P2X4 expression and produced allodynia, a behavior that was not evoked in P2X4-deficient mice. Moreover, we have recently shown that fibronectin fails to induce up-regulation of P2X4 expression in microglial cells lacking Lyn tyrosine kinase, a member of the Src family kinases (Tsuda et al. 2008b). Importantly, Lyn-deficient mice show reduction of nerve injury-induced P2X4 up-regulation and neuropathic pain. Thus Lyn may be a key kinase in the molecular machinery mediating the up-regulation of P2X4 in microglia (Figure 10.1).

It was also reported that activating both toll-like receptors and NOD2 (another pattern-recognition receptor) in cultured microglia increased the expression of P2X4 at the mRNA level (Guo et al. 2006), thus suggesting the involvement of these receptors in the regulation of P2X4. However, the functional relevance of these receptors *in vivo* is still unknown.

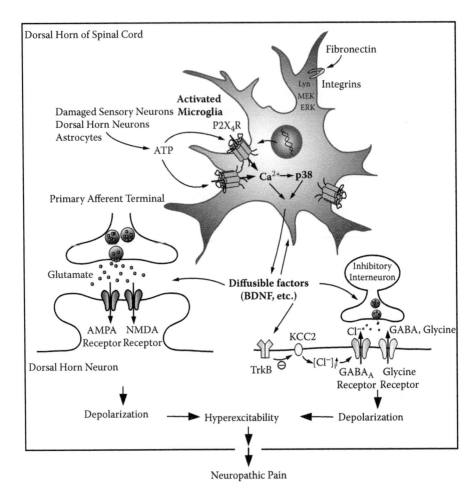

FIGURE 10.1 Schematic illustration of potential mechanisms by which P2X4 receptors up-regulated in microglia may modulate neuropathic pain signaling in the dorsal horn after peripheral nerve injury.

There is evidence that P2X4 receptors in microglial cells are located predominantly within lysosomal compartments and are targeted there by N- and C-terminal motifs (Qureshi et al. 2007). Notably, P2X4 receptors remain stable within the lysosome and resist degradation by virtue of their N-linked glycans. It is of particular interest that stimulating lysosome exocytosis, for example, using a Ca^{2+} ionophore, results in the trafficking of P2X4 to the plasma membrane (Qureshi et al. 2007). Given that P2X4 receptor activity on the plasma membrane of microglia is critical for microglial function, it is conceivable that mechanisms underlying P2X4 recycling between the lysosome and the plasma membrane are important for neuropathic pain. However, this is not yet demonstrated *in vitro* or *in vivo* and warrants further experimental investigation.

10.3.3 MECHANISM OF NEUROPATHIC PAIN EVOKED BY MICROGLIAL P2X4 RECEPTOR STIMULATION

Enhanced excitability and firing discharge of neurons in the dorsal horn pain-processing network following nerve injury is crucial for neuropathic pain syndromes (Scholz and Woolf 2002; Woolf and Salter 2000). The first evidence that microglia may participate in the hyperexcitability in dorsal horn neurons was revealed by Coull et al. (2005). They used spinal cord slices taken from rats that had displayed allodynia by intrathecal administration of P2X4 receptor-stimulated microglia and found that ATP-stimulated microglia positively shifted the anion reversal potential (E_{anion}) in spinal lamina I neurons and rendered GABA effects depolarizing, rather than hyperpolarizing, in these neurons (Coull et al. 2005). Moreover, TNP-ATP, which can reverse nerve injury-induced allodynia (Tsuda et al. 2003), acutely reverses the depolarizing shift in E_{anion} in lamina I neurons after peripheral nerve injury. Together with the findings that the depolarizing shift in E_{anion} and the excitatory response to GABA are key events in dorsal horn neurons in neuropathic pain (Coull et al. 2003), these results imply that spinal microglia stimulated by P2X4 receptors causes neuropathic pain through a rise in intracellular Cl$^-$ in spinal lamina I neurons.

What is the mechanism(s) by which microglia affect the anion balance in lamina I neurons in the dorsal horn? Coull et al. (2005) determined the role of brain-derived neurotrophic factor (BDNF) as a signaling factor between microglia and dorsal horn lamina I neurons. It was found that intrathecal application of BDNF mimicked tactile allodynia and the depolarizing shift in E_{anion} in lamina I neurons by peripheral nerve injury or by intrathecal administering P2X4 receptor stimulated microglia. Furthermore, interfering with signaling between BDNF and its receptor (TrkB) either by a function-blocking TrkB antibody or by a BDNF-sequestering fusion protein (TrkB-Fc) prevented tactile allodynia caused by peripheral nerve injury or by intrathecal administering P2X4-stimulated microglia. Applying ATP to microglia caused the release of BDNF, which was prevented by TNP-ATP. These results indicate that P2X-stimulated microglia release BDNF as a crucial factor to signal to lamina I neurons, causing a collapse of their transmembrane anion gradient and the subsequent neuronal hyperexcitability (Figure 10.1). A direct demonstration that P2X4 receptors expressed on activated microglia control the release of BDNF comes from studies in P2X4-/- mice. Indeed, BDNF immunoreactivity in the dorsal horn of the spinal cord was higher in P2X4-/- mice than in wild-type animals after peripheral nerve injury (Ulmann et al. 2008). Moreover, primary cultures of dorsal horn microglia revealed a reduction in BDNF staining after ATP stimulation in wild-type cultures, whereas in cultures from P2X4-/- mice, an application of ATP failed to induce a significant change (Ulmann et al. 2008). These results, along with the previous study by Coull et al., strongly argue for the involvement of P2X4 receptors in controlling BDNF release from activated microglia. The cellular mechanisms by which microglial P2X4 receptor activation could trigger the release of BDNF have been investigated recently showing that stimulation of P2X4 receptors causes SNARE-mediated synthesis and release of BDNF that was dependent on the Ca^{2+} flux through P2X4 receptors and p38-MAPK activation (Trang et al. 2009). The GABA$_A$R-mediated depolarization might also produce an excitation through voltage-sensitive Ca^{2+} channels and NMDA receptors. There is evidence that several proinflammatory cytokines that are known to be released from microglia (Inoue 2002, 2006) modulate excitatory synaptic transmission. Interleukin-1β was reported to enhance NMDA receptor-mediated Ca^{2+} response (Viviani et al. 2003). Long-term treatment with interferon-γ produced an increase in neuronal excitability in dorsal horn neurons (Vikman et al. 2003). Thus, the net enhanced transmission in the dorsal horn pain network by these factors might be responsible for the nerve injury-induced neuropathic pain.

10.4 P2X7 RECEPTORS AND INFLAMMATORY PAIN

10.4.1 P2X7 RECEPTOR DISTRIBUTION

Among the P2X receptor superfamily, P2X7 receptors are considered the most unusual with respect to their functional and molecular characteristics. First, although it is thought to share a common transmembrane topology with other P2X receptor subunits, its C-terminal domain is 200 amino acids longer than the other six members of the family. This tail is thought to confer atypical properties on the P2X7 receptor. In addition, homomeric P2X7 receptors require 10-fold higher concentrations of ATP (>100 μM) to be activated. Prolonged agonist exposure induces the formation of large cytolytic pores in the cell membrane (Surprenant et al. 1996). P2X7 subunits were initially cloned from rat (Surprenant et al. 1996) and human macrophages (Rassendren et al. 1997). They are predominantly found on cells of the immune system including lymphocytes, and peripheral macrophages (Collo et al. 1997). In rat central nervous system, functional P2X7 receptors have been described on microglia and astrocytes, but their existence on central neurons is still controversial. The specific distribution of P2X7 receptors has led to a large interest in this receptor in immune cell biology and in drug development research programs where it is targeted to treat inflammatory diseases.

10.4.2 FUNCTIONAL PROPERTIES OF P2X7 RECEPTORS

The characterization of P2X7 receptors in HEK293 cells has revealed that unlike other P2X receptors, high concentrations of ATP are needed to activate them. It has also been shown that 2′,3′-O-(4-benzoylbenzoyl)-ATP (BzATP) is a more potent agonist at homomeric P2X7 receptors than ATP (EC_{50}= 20 μM for BzATP versus >100 μM for ATP) (Jacobson et al. 2002) even though it is not selective and is also an agonist at P2X1 and P2X3 receptors. P2X7 receptors are inhibited by PPADS and suramin, which are both non-selective P2X antagonists. They are also blocked by a more potent and selective antagonist known as Brilliant Blue G (BBG). As with the other P2X receptors, upon brief ATP application, the activation of the P2X7 protein opens a channel that will induce the passage of Ca^{2+}, Na^+, and K^+. A physiologically relevant characteristic of this state is that changes in the intracellular concentration of potassium induce the release of a signaling molecule, interleukin-1β (IL-1β) (Colomar et al. 2003). The active form of IL-1β is a potent pro-inflammatory cytokine that will induce a cascade of events including the production of superoxide products (Parvathenani et al. 2003) and tumor necrosis factor α (TNF-α; [Woolf et al. 1997]), all of which have roles in generation or maintenance of pain. On the other hand, when ATP is applied for longer periods of time, the receptor undergoes a change in ionic permeability whereby its pore is thought to become larger such that molecules of molecular mass up to 800 Da (Surprenant et al. 1996). This process is characterized by the uptake of high-molecular weight fluorescent dyes such as YO-PRO or Lucifer yellow and is followed by cytoskeletal rearrangement such as membrane blebbing, leading eventually to cell death in immune cells. Considering the overall distribution of P2X7 receptors on proinflammatory cells and the functional properties of P2X7 receptors, it is not surprising that many studies have been carried out to determine the role of this receptor type in inflammation.

10.4.3 P2X7 Receptor Pain Phenotypes Revealed by Genetic and Pharmacological Methods

The *in vivo* investigation of the role of P2X7 receptors in inflammation was aided immensely by the development of P2X7 receptor knockout (P2X7-/-) mice and has allowed investigation of its roles in inflammation. Labasi et al. (2002) examined the response of the P2X7-/- mice in a monoclonal antibody-induced arthritis model. They showed that whereas this procedure induced a severe arthritic phenotype in wildtype mice, in P2X7-/- mice arthritis severity was significantly attenuated (Labasi et al. 2002). Another group investigated the phenotype of the mice P2X7-/- further. Indeed, in a more detailed study, Chessell et al. showed that P2X7-deficient animals did not develop measurable symptoms of pain following a standard induction of inflammatory status (Chessell et al. 2005).

In light of the role of P2X7 receptors in mediating inflammatory pain, a search for selective antagonists has been initiated by different pharmaceutical companies. Systematic compound screening has enabled the discovery of several selective P2X7 antagonists. Thus, AZ-11645373, a highly potent antagonist at human P2X7 receptors, has been shown to be effective at inhibiting ATP- and Bz-ATP-elicited currents (Stokes et al. 2006). In parallel, another group has synthesized two compounds that have been tested in inflammatory pain models with some success. Indeed, systemic administration of the selective P2X7 antagonists A-438079 and A-740003 reduced thermal hyperalgesia in two models of inflammatory pain: intraplantar administration of complete Freund's adjuvant or carrageenan (Honore et al. 2006a, 2006b).

There is some evidence to suggest that P2X7 receptors may also play roles in neuropathic pain conditions. P2X7 receptor deficient mice exhibit a reduction of thermal and mechanical hypersensitivities after partial sciatic nerve ligation (Chessell et al. 2005). Systemic (i.p.) administration of selective P2X7 antagonists A-740003 and A-438079 reduces tactile allodynia in three different models of neuropathic pain in rats (Honore et al. 2006a). A-438079 (i.v.) reduced innocuous stimuli-evoked activity of *in vivo* dorsal horn neurons in neuropathic rats, but not in sham-operated rats (McGaraughty et al. 2001). The effect of intrathecal administration of these antagonists has not yet been examined; thus the mechanisms underlying the role of spinal P2X7 receptors (presumably expressed in microglia) in neuropathic pain remains unclear. Collectively, these data reinforce the strong role of P2X7 receptor activation in inflammatory and neuropathic pain states. Several P2X7 receptor antagonists are in clinical trials in order to treat inflammatory pain such as osteoarthritis.

10.5   SUMMARY

We have presented an overview of the multiple lines of evidence that implicate ATP signaling via P2X receptors to pain phenotypes. The progress made in the last few years strongly suggests that receptor-specific P2X antagonists may be useful analgesics in humans. Such drugs could be used to treat a variety of painful conditions that are currently nonresponsive to conventional analgesics such as nonsteroidal antiinflammatory drugs and opioid analgesics. The availability of P2X receptor-specific antagonists also holds the promise of revealing the cellular and molecular neurobiology underlying pain states.

REFERENCES

Barclay J, Patel S, Dorn G, Wotherspoon G, Moffatt S, Eunson L, Abdel'al S, et al. (2002) Functional downregulation of P2X3 receptor subunit in rat sensory neurons reveals a significant role in chronic neuropathic and inflammatory pain. *J Neurosci* 22(18): 8139–8147.

Bardoni R, Goldstein PA, Lee CJ, Gu JG, MacDermott AB (1997) ATP P2X receptors mediate fast synaptic transmission in the dorsal horn of the rat spinal cord. *J Neurosci* 17(14): 5297–5304.

Bland-Ward PA, Humphrey PP (1997) Acute nociception mediated by hindpaw P2X receptor activation in the rat. *Br J Pharmacol* 122(2): 365–371.

Bleehen T, Keele CA (1977) Observations on the algogenic actions of adenosine compounds on the human blister base preparation. *Pain* 3(4): 367–377.

Burnstock G (2001) Purine-mediated signalling in pain and visceral perception. *Trends Pharmacol Sci* 22(4): 182–188.

Chen CC, Akopian AN, Sivilotti L, Colquhoun D, Burnstock G, Wood JN (1995) A P2X purinoceptor expressed by a subset of sensory neurons. *Nature* 377(6548): 428–431.

Chessell IP, Hatcher JP, Bountra C, Michel AD, Hughes JP, Green P, Egerton J, et al. (2005) Disruption of the P2X7 purinoceptor gene abolishes chronic inflammatory and neuropathic pain. *Pain* 114(3): 386–396.

Cockayne DA, Hamilton SG, Zhu QM, Dunn PM, Zhong Y, Novakovic S, Malmberg AB, et al. (2000) Urinary bladder hyporeflexia and reduced pain-related behaviour in P2X3-deficient mice. *Nature* 407(6807): 1011–1015.

Collier HO, James GW, Schneider C (1966) Antagonism by aspirin and fenamates of bronchoconstriction and nociception induced by adenosine-5′-triphosphate. *Nature* 212(5060): 411–412.

Collo G, Neidhart S, Kawashima E, Kosco-Vilbois M, North RA, Buell G (1997) Tissue distribution of the P2X7 receptor. *Neuropharmacology* 36(9): 1277–1283.

Colomar A, Marty V, Medina C, Combe C, Parnet P, Amedee T (2003) Maturation and release of interleukin-1 beta by lipopolysaccharide-primed mouse Schwann cells require the stimulation of P2X7 receptors. *J Biol Chem* 278(33): 30732–30740.

Cook SP, McCleskey EW (1997) Desensitization, recovery and Ca(2+)-dependent modulation of ATP-gated P2X receptors in nociceptors. *Neuropharmacology* 36(9): 1303–1308.

Cook SP, McCleskey EW (2002) Cell damage excites nociceptors through release of cytosolic ATP. *Pain* 95(1–2): 41–47.

Coull JA, Beggs S, Boudreau D, Boivin D, Tsuda M, Inoue K, Gravel C, Salter MW, De Koninck Y (2005) BDNF from microglia causes the shift in neuronal anion gradient underlying neuropathic pain. *Nature* 438(7070): 1017–1021.

Coull JA, Boudreau D, Bachand K, Prescott SA, Nault F, Sik A, De Koninck P, De Koninck Y (2003) Trans-synaptic shift in anion gradient in spinal lamina I neurons as a mechanism of neuropathic pain. *Nature* 424(6951): 938–942.

Dunn PM, Zhong Y, Burnstock G (2001) P2X receptors in peripheral neurons. *Prog Neurobiol* 65(2): 107–134.

Gu JG, MacDermott AB (1997) Activation of ATP P2X receptors elicits glutamate release from sensory neuron synapses. *Nature* 389(6652): 749–753.

Guo LH, Guo KT, Wendel HP, Schluesener HJ (2006) Combinations of TLR and NOD2 ligands stimulate rat microglial P2X4R expression. *Biochem Biophys Res Commun* 349(3): 1156–1162.

Hamilton SG, Wade A, McMahon SB (1999) The effects of inflammation and inflammatory mediators on nociceptive behaviour induced by ATP analogues in the rat. *Br J Pharmacol* 126(1): 326–332.

Honore P, Donnelly-Roberts D, Namovic MT, Hsieh G, Zhu CZ, Mikusa JP, Hernandez G, et al. (2006a) A-740003 [N-(1-{[(cyanoimino)(5-quinolinylamino) methyl]amino}-2,2-dimethylpropyl)-2-(3,4-dimethoxyphenyl)acetamide], a novel and selective P2X7 receptor antagonist, dose-dependently reduces neuropathic pain in the rat. *J Pharmacol Exp Ther* 319(3): 1376–1385.

Honore P, Kage K, Mikusa J, Watt AT, Johnston JF, Wyatt JR, Faltynek CR, Jarvis MF, Lynch K (2002) Analgesic profile of intrathecal P2X(3) antisense oligonucleotide treatment in chronic inflammatory and neuropathic pain states in rats. *Pain* 99(1–2): 11–19.

Honore P, Wade CL, Zhong C, Harris RR, Wu C, Ghayur T, Iwakura Y, et al. (2006b) Interleukin-1 alpha beta gene-deficient mice show reduced nociceptive sensitivity in models of inflammatory and neuropathic pain but not post-operative pain. *Behav Brain Res* 167(2): 355–364.

Hugel S, Schlichter R (2000) Presynaptic P2X receptors facilitate inhibitory GABAergic transmission between cultured rat spinal cord dorsal horn neurons. *J Neurosci* 20(6): 2121–2130.

Inoue K (2002) Microglial activation by purines and pyrimidines. *Glia* 40(2): 156–163.

Inoue K (2006) The function of microglia through purinergic receptors: neuropathic pain and cytokine release. *Pharmacol Ther* 109(1–2): 210–226.

Jacobson KA, Jarvis MF, Williams M (2002) Purine and pyrimidine (P2) receptors as drug targets. *J Med Chem* 45(19): 4057–4093.

Jahr CE, Jessell TM (1983) ATP excites a subpopulation of rat dorsal horn neurones. *Nature* 304(5928): 730–733.

Jarvis MF, Burgard EC, McGaraughty S, Honore P, Lynch K, Brennan TJ, Subieta A, et al. (2002) A-317491, a novel potent and selective non-nucleotide antagonist of P2X3 and P2X2/3 receptors, reduces chronic inflammatory and neuropathic pain in the rat. *Proc Natl Acad Sci U S A* 99(26): 17179–17184.

Jo YH, Schlichter R (1999) Synaptic corelease of ATP and GABA in cultured spinal neurons. *Nat Neurosci* 2(3): 241–245.

Khakh BS, North RA (2006) P2X receptors as cell-surface ATP sensors in health and disease. *Nature* 442(7102): 527–532.

Labasi JM, Petrushova N, Donovan C, McCurdy S, Lira P, Payette MM, Brissette W, Wicks JR, Audoly L, Gabel CA (2002) Absence of the P2X7 receptor alters leukocyte function and attenuates an inflammatory response. *J Immunol* 168(12): 6436–6445.

Lewis C, Neidhart S, Holy C, North RA, Buell G, Surprenant A (1995) Coexpression of P2X2 and P2X3 receptor subunits can account for ATP-gated currents in sensory neurons. *Nature* 377(6548): 432–435.

Li P, Calejesan AA, Zhuo M (1998) ATP P2x receptors and sensory synaptic transmission between primary afferent fibers and spinal dorsal horn neurons in rats. *J Neurophysiol* 80(6): 3356–3360.

Llewellyn-Smith IJ, Burnstock G (1998) Ultrastructural localization of P2X3 receptors in rat sensory neurons. *Neuroreport* 9(11): 2545–2550.

McGaraughty S, Chu KL, Wismer CT, Mikusa J, Zhu CZ, Cowart M, Kowaluk EA, Jarvis MF (2001) Effects of A-134974, a novel adenosine kinase inhibitor, on carrageenan-induced inflammatory hyperalgesia and locomotor activity in rats: evaluation of the sites of action. *J Pharmacol Exp Ther* 296(2): 501–509.

Millan MJ (1999) The induction of pain: an integrative review. *Prog Neurobiol* 57(1): 1–164.

Nakatsuka T, Gu JG (2001) ATP P2X receptor-mediated enhancement of glutamate release and evoked EPSCs in dorsal horn neurons of the rat spinal cord. *J Neurosci* 21(17): 6522–6531.

Nasu-Tada K, Koizumi S, Tsuda M, Kunifusa E, Inoue K (2006) Possible involvement of increase in spinal fibronectin following peripheral nerve injury in upregulation of microglial P2X4, a key molecule for mechanical allodynia. *Glia* 53(7): 769–775.

North RA (2002) Molecular physiology of P2X receptors. *Physiol Rev* 82(4): 1013–1067.

North RA (2004) P2X3 receptors and peripheral pain mechanisms. *J Physiol* 554(Pt 2): 301–308.

Novakovic SD, Kassotakis LC, Oglesby IB, Smith JA, Eglen RM, Ford AP, Hunter JC (1999) Immunocytochemical localization of P2X3 purinoceptors in sensory neurons in naive rats and following neuropathic injury. *Pain* 80(1–2): 273–282.

Parvathenani LK, Tertyshnikova S, Greco CR, Roberts SB, Robertson B, Posmantur R (2003) P2X7 mediates superoxide production in primary microglia and is up-regulated in a transgenic mouse model of Alzheimer's disease. *J Biol Chem* 278(15): 13309–13317.

Qureshi OS, Paramasivam A, Yu JC, Murrell-Lagnado RD (2007) Regulation of P2X4 receptors by lysosomal targeting, glycan protection and exocytosis. *J Cell Sci* 120(Pt 21): 3838–3849.

Rae MG, Rowan EG, Kennedy C (1998) Pharmacological properties of P2X3-receptors present in neurones of the rat dorsal root ganglia. *Br J Pharmacol* 124(1): 176–180.

Ralevic V, Burnstock G (1998) Receptors for purines and pyrimidines. *Pharmacol Rev* 50(3): 413–492.

Rassendren F, Buell GN, Virginio C, Collo G, North RA, Surprenant A (1997) The permeabilizing ATP receptor, P2X7. Cloning and expression of a human cDNA. *J Biol Chem* 272(9): 5482–5486.

Rhee JS, Wang ZM, Nabekura J, Inoue K, Akaike N (2000) ATP facilitates spontaneous glycinergic IPSC frequency at dissociated rat dorsal horn interneuron synapses. *J Physiol* 524 Pt 2: 471–483.

Robertson SJ, Rae MG, Rowan EG, Kennedy C (1996) Characterization of a P2X-purinoceptor in cultured neurones of the rat dorsal root ganglia. *Br J Pharmacol* 118(4): 951–956.

Scholz J, Woolf CJ (2002) Can we conquer pain? *Nat Neurosci* 5 Suppl: 1062–1067.

Snider WD, McMahon SB (1998) Tackling pain at the source: new ideas about nociceptors. *Neuron* 20(4): 629–632.

Souslova V, Cesare P, Ding Y, Akopian AN, Stanfa L, Suzuki R, Carpenter K, et al. (2000) Warm-coding deficits and aberrant inflammatory pain in mice lacking P2X3 receptors. *Nature* 407(6807): 1015–1017.

Stokes L, Jiang LH, Alcaraz L, Bent J, Bowers K, Fagura M, Furber M, et al. (2006) Characterization of a selective and potent antagonist of human P2X(7) receptors, AZ11645373. *Br J Pharmacol* 149(7): 880–887.

Surprenant A, Rassendren F, Kawashima E, North RA, Buell G (1996) The cytolytic P2Z receptor for extracellular ATP identified as a P2X receptor (P2X7). *Science* 272(5262): 735–738.

Trang T, Beggs S, Wan X, Salter MW (2009) P2X4-receptor-mediated synthesis and release of brain-derived neurotrophic factor in microglia is dependent on calcium and p38-mitogen-activated protein kinase activation. *J Neurosci* 29(11): 3518–3528.

Tsuda M, Shigemoto-Mogami Y, Koizumi S, Mizokoshi A, Kohsaka S, Salter MW, Inoue K (2003) P2X4 receptors induced in spinal microglia gate tactile allodynia after nerve injury. *Nature* 424(6950): 778–783.

Tsuda M, Toyomitsu E, Komatsu T, Masuda T, Kunifusa E, Nasu-Tada K, Koizumi S, Yamamoto K, Ando J, Inoue K (2008a) Fibronectin/integrin system is involved in P2X(4) receptor upregulation in the spinal cord and neuropathic pain after nerve injury. *Glia* 56(5): 579–585.

Tsuda M, Tozaki-Saitoh H, Masuda T, Toyomitsu E, Tezuka T, Yamamoto T, Inoue K (2008b) Lyn tyrosine kinase is required for P2X(4) receptor upregulation and neuropathic pain after peripheral nerve injury. *Glia* 56(1): 50–58.

Ulmann L, Hatcher JP, Hughes JP, Chaumont S, Green PJ, Conquet F, Buell GN, Reeve AJ, Chessell IP, Rassendren F (2008) Up-regulation of P2X4 receptors in spinal microglia after peripheral nerve injury mediates BDNF release and neuropathic pain. *J Neurosci* 28(44): 11263–11268.

Vikman KS, Hill RH, Backstrom E, Robertson B, Kristensson K (2003) Interferon-gamma induces characteristics of central sensitization in spinal dorsal horn neurons in vitro. *Pain* 106(3): 241–251.

Viviani B, Bartesaghi S, Gardoni F, Vezzani A, Behrens MM, Bartfai T, Binaglia M, et al. (2003) Interleukin-1 beta enhances NMDA receptor-mediated intracellular calcium increase through activation of the Src family of kinases. *J Neurosci* 23(25): 8692–8700.

Vulchanova L, Riedl MS, Shuster SJ, Buell G, Surprenant A, North RA, Elde R (1997) Immunohistochemical study of the P2X2 and P2X3 receptor subunits in rat and monkey sensory neurons and their central terminals. *Neuropharmacology* 36(9): 1229–1242.

Vulchanova L, Riedl MS, Shuster SJ, Stone LS, Hargreaves KM, Buell G, Surprenant A, North RA, Elde R (1998) P2X3 is expressed by DRG neurons that terminate in inner lamina II. *Eur J Neurosci* 10(11): 3470–3478.

Woolf CJ, Allchorne A, Safieh-Garabedian B, Poole S (1997) Cytokines, nerve growth factor and inflammatory hyperalgesia: the contribution of tumour necrosis factor alpha. *Br J Pharmacol* 121(3): 417–424.

Woolf CJ, Salter MW (2000) Neuronal plasticity: increasing the gain in pain. *Science* 288(5472): 1765–1769.

11 Myalgia and Fatigue
Translation from Mouse Sensory Neurons to Fibromyalgia and Chronic Fatigue Syndromes

Alan R. Light, Charles J. Vierck, and Kathleen C. Light

CONTENTS

11.1   INTRODUCTION

Muscle fatigue and pain are among the most common complaints at emergency rooms and clinics across the country. Fatigue and pain are often acute, remitting spontaneously or appearing to be attenuated by a variety of drugs and treatment modalities. In spite of these remissions, popular magazines (e.g., *Time*) estimate that each year Americans spend over $30 billion on herbal remedies and $50 billion on alternative therapies to treat symptoms that include muscle pain and fatigue. These statistics indicate that even acute muscle pain and fatigue are serious health problems that are not adequately addressed by current medical practice.

Occasionally, muscle pain and fatigue take on a chronic nature, leading to syndromes including chronic fatigue syndrome (CFS) and fibromyalgia syndrome (FMS)—devastating conditions characterized by continuing, debilitating fatigue, which is made worse by even mild exercise in the case of CFS and by chronic widespread pain (CWP) with a particular emphasis in the muscles, which can prevent most or all activities in the case of FMS. Both of these conditions are frequently associated with each other and with a variety of other illnesses, such as temporomandibular disorder (TMD), irritable bowel syndrome (IBS), and multiple chemical sensitivity. These syndromes destroy lives, respond poorly to current treatment strategies, and can lead to exhaustion of the financial resources of afflicted patients. Together, these disorders affect 7 to 20 million people in the United States each year, as reported by various authorities (Reeves et al. 2007). Clearly, patients with these syndromes deserve a concerted research effort to understand, treat, and eventually cure these illnesses.

In contrast to cutaneous pain, which has been thoroughly studied and is comparatively well understood, the molecular mechanisms for muscle pain are still unknown. Even more enigmatic is the symptom of debilitating fatigue. Mosso, in his compendious volume on the subject a century ago, remarked that all cultures seem to have just one word for *fatigue* (Mosso 1904). Yet *fatigue* describes many conditions, including failure of muscle fibers to shorten normally, deficient motor command signals, feelings of tiredness, heaviness, pressure, and weakness from muscles, and a feeling of mental fatigue that impedes concentration and performance of conceptual tasks.

The subject of most physiological investigations of fatigue has been *voluntary* muscle contraction. Decreased function causing failure of voluntary muscle contraction can occur at all levels of the neuromuscular system, including the motor cortex, signaling to motoneurons, motoneuron signals to the muscle, excitation–contraction coupling in the muscle, and actin–myosin filament interactions. However, the most common failure is a decrease in the motor command signal from the motor cortex (see recent reports and reviews by Bellinger et al. 2008;

Gibson et al. 2003; Noakes et al. 2005; St Clair and Noakes 2004). A recent review suggests that failures in voluntary muscle contraction are most often caused by a central comparator that integrates homeostatic inputs from many physiological systems and shuts down motor commands when energy resources are threatened (Noakes 2007). One of the homeostatic inputs is suggested to "originate from a difference between subconscious representations of baseline physiological homeo-static state and the state of physiological activity induced by physical activity, which creates a second order representation which is perceived by consciousness-producing structures as the sensation of fatigue" (Gibson et al. 2003, page 174). We suggest that there is a simpler sensation of fatigue that is triggered by inputs from specific receptors that are sensitive to metabolites produced by muscle contraction. We further propose that this elementary sensation is transduced, conducted, and perceived within a unique sensory system with properties analogous to other sensory modalities such as pain. We call it the "sensation of muscle fatigue."

11.1.1 Fatigue and Pain as Separate Specific Sensory Systems

What defines a specific sensory system? While Müller (1840) and Bell are credited with the "doctrine of specific nerve energies," the concept of a specific sensory system has since been modified to require the following three basic elements, at a minimum:

1. A qualitatively unique perception consistently associated with a specific form of energy
2. Specific receptors for the energy or stimulus modality
3. Specific transmission systems and brain regions specialized for transmitting and integrating information from the receptors

11.1.2 Does Pain Satisfy These Requirements?

Pain has been shown to satisfy these three requirements and more (Perl 1971, 1998, 2007). (1) Pain is a unique qualitative experience that is normally not confused with any other sensory experience. While there is not a specific form of energy associated with pain, all peripheral stimuli that induce pain, damage tissue, or threaten tissue damage if the stimulus is sustained constitute a consistent *effective* form of energy. Therefore, the form of energy is a "damaging form of energy." (2) The specific receptors for the energy are "nociceptors" that are tuned to energy that can damage tissues. (3) A specific transmission system for pain includes the superficial dorsal horn of the spinal cord, the spinothalamic tract and portions of the thalamus, insular cortex, and cingulate gyri of the brain (Brooks et al. 2005; Craig 2003). Understanding the concept of pain as a specific sensory system has enabled the discovery of molecular receptors that allow sensory endings to be sensitive to the stimuli causing damage and pain. This concept has also led to the discovery of modulating substances, such as opioids that can selectively reduce pain by acting at particular regions of the brain and spinal cord.

11.1.3 DOES THE SENSATION OF MUSCLE FATIGUE SATISFY THE TENETS FOR A SPECIFIC SENSORY SYSTEM?

(1) Although early experiments documented the sensation of "muscle tiredness or pressure" that occurred with non-painful accumulations of muscle contraction-induced metabolites, the uniqueness of this sensation has not been thoroughly elucidated. (2) Specific receptors for the sensation of muscle fatigue have not been identified. (3) While specific spinal cord pathways for sensations of muscle fatigue have some basis (Wilson et al. 2002), more research is necessary. Similarly, some brain areas have been identified as related to the sensation of fatigue (Caseras et al. 2008; Cook et al. 2004, 2007; Williamson et al. 1997), the most common being prefrontal cortex, insular cortex, and the anterior and/or posterior cingulate gyrus. However, much more work is necessary here as well. Thus, for the sensation of muscle fatigue, satisfaction of the three requirements for serving as a specific sensory system is incomplete.

Sensory fatigue shares many similarities with pain and with dyspnea (air hunger). All are protective sensory phenomena. Pain elicits behavior that protects from tissue injury caused by mechanical, chemical, or thermal insults. Dyspnea protects from increases in CO_2 and decreases in O_2 that would lead to cell death from anoxia. Fatigue is the first line of defense that protects from the over-utilization of energy stores that would lead to rigor and death. All three of these sensory phenomena activate protective reflexes. Nociceptive input activates withdrawal reflexes, and pain motivates conscious actions that remove the injured body part from the stimulus, and it activates sympathetic reflexes that enhance escape. Dyspnea increases respiratory processes, and it activates sympathetic reflexes to preserve vital functions. Sensory fatigue reduces motor commands, and it activates sympathetic reflexes to increase energy stores by increasing blood flow in working muscles and decreasing blood flow in non-working muscles. Each of these sensations evokes a unique cognitive percept with potent aversive connotations, allowing humans to recognize which system is being affected and engaging behavior to alter the conditions that are causing the sensory experience. So why are dyspnea and pain considered by most scientists to have the characteristics of unique sensory systems, while sensory muscle fatigue is rarely considered as such?

Part of the problem has been semantic. Mosso (1904) promoted and popularized confusion between the sensory phenomenon of fatigue and a definition of fatigue as the loss of ability to contract muscles. The sensory experience of fatigue and the loss of muscle contraction were so strongly interrelated in his writings that they could not be separated, even though he clearly recognized and remarked on a special sensory component of fatigue. This historic emphasis on the failure of muscle contraction as the definition of fatigue has persisted in most influential publications to this day (Davis and Bailey 1997), although some have begun to consider fatigue separate from motor insufficiency, albeit as more of an emotional response than a sensory system (Gibson et al. 2003). The semantics of pain went through a similar evolution with a persistent confusion of sensory phenomena with associated consequences in cognition, memory, emotions, autonomic activation, and motor behavior.

Another problem for both pain and sensory muscle fatigue is that special receptors that encode the sensory signals lack a special, macroscopic external apparatus like an eye, ear, nose, or tongue for us to easily observe. For pain, the special nature of nociceptors was painstakingly demonstrated through careful physiological recording experiments, anatomical demonstrations, and unique psychophysical experiments (summarized by Perl 2007). The molecular characteristics that determined the properties of nociceptors were discovered by the use of convenient special agents such as capsaicin that could be used to selectively activate them. The spinal cord, brain stem, and cerebral pathways and integration centers for pain have been mapped out using procedures that uniquely activate cutaneous nociceptors.

A third problem is that the nature of muscle makes it very difficult to expose sensory receptors that are uniquely associated with muscle pain or sensory muscle fatigue. Exposing muscle receptors for direct activation by dissecting the overlying structures almost inevitably disrupts the circulation and activates nociceptors, confounding any attempts to selectively activate receptors for pain or sensory muscle fatigue. In spite of these problems, many attempts have been made to determine the nature of the sense organs involved in muscle pain and the sensory signals involved in the regulation of sympathetic reflexes that are known to be activated by alterations in muscle metabolites that are associated with muscle contraction. Most of the earliest experiments were indirect, using sympathetic reflexes as readouts for the signals from the sensory receptors in the muscles.

11.2 DO SKELETAL MUSCLES HAVE TWO UNIQUE TYPES OF SENSORY RECEPTORS THAT DETECT METABOLITES: ONE TYPE NOCICEPTIVE (CAPABLE OF SIGNALING PAIN), THE OTHER ERGORECEPTIVE (CAPABLE OF DETECTING MUSCLE WORK)?

More than 70 years ago Alam and Smirk (1937) clearly indicated that both a pre-pain "tiredness" and muscle pain could be evoked by metabolites that were produced by muscle contraction, with the former activated by lower concentrations of metabolites than the latter. Metabolites produced by muscle contraction caused blood pressure increases and the sensations of muscle fatigue and muscle pain.

The initial blood pressure increases (see Figure 11.1) were accompanied not by pain, but by a pre-pain phenomenon described as (1) "tiredness or heaviness but not causing appreciable discomfort." As metabolites continued to increase the subjects experienced (2) "a period where the sensation in the forearm becomes definitely aching in character and is described as pain which, however, is easily tolerated," and then (3) "a period with severe pain" follows. They demonstrated that the sensations and the blood pressure increases were mediated by afferent nerve fibers innervating the exercising muscle. Thus, the concept of at least two populations of sensory afferents conveying non-painful versus painful information has been extant for more than 70 years. Subsequent experiments confirmed and refined these observations, clearly demonstrating that contraction-produced metabolites detected by sensory afferents in muscle activate a sympathetic exercise pressor reflex (reviewed by Kaufman and

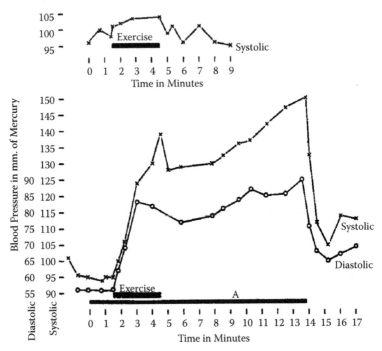

FIGURE 11.1 Top: Exercise without arrest of the circulation. Bottom: Exercise with arrest of the circulation. Exercise increases metabolites in the working muscles, but also activates sympathetic reflexes that enhance blood flow and reduce metabolites. If the circulation is stopped, metabolites continue to increase, causing an increase in sympathetic reflexes until a plateau is reached. (Reprinted with permission from Alam M, Smirk FH. 1937. Observations in man upon a blood pressure raising reflex arising from the voluntary muscles. *J Physiol.* 89: 372–383.)

Hayes 2002 and by McCord and Kaufman in Chapter 12 in this volume). The primary neural pathway for this reflex has been assumed to be afferent neurons from skeletal muscle that project to sympathetic centers in the spinal cord. These centers project to the brain stem and hypothalamus, with efferent pathways projecting to sympathetic nuclei in the spinal cord, which project to blood vessels in skeletal muscles.

An interesting development in this area was the discovery that sensory afferents innervating blood vessels often contain calcitonin-gene-related-peptide (CGRP) (Kruger et al. 1989; Silverman and Kruger 1989). Even more interesting is the hypothesis that CGRP could be released by sensory endings in a "neuroeffector" mechanism, meaning that the sensory endings themselves could directly cause effects on peripheral tissues. Given that CGRP is a potent vasodilator, the activation of muscle sensory afferents could release CGRP that directly evokes the vasodilation part of the exercise pressor reflex that occurs in the working muscle (Micevych and Kruger 1992). A second finding also indicates a very direct effect of sensory afferents on sympathetic output (Papka and McNeill 1993). These mechanisms could greatly enhance or independently control the exercise pressor response if the sensory neurons containing CGRP are responsive to metabolites.

Mense and colleagues demonstrated at least two different populations of muscle receptors in experiments based on mechanical stimulation (see Figure 11.2) (Kniffki et al. 1978, 1981; Mense and Meyer 1985). Most of the non-spindle sensory afferents were group III (corresponding to Aδ afferent fibers in cutaneous nerves) or group IV (non-myelinated fibers corresponding to C-afferent fibers in cutaneous nerves). Some of each group of fibers could be classified as "nociceptors" (responding to tissue-damaging stimuli) or non-nociceptors capable of detecting muscle movements, and therefore possible "ergoreceptors" (receptors capable of detecting muscle work). Recent estimates from Mense's lab suggest that at least 40% of mechanoreceptive group IV afferents could be non-nociceptive (Mense 2009) (see Figure 11.2).

Identification of responses to muscle-produced metabolites has been more problematic than characterizing responses to mechanical stimulation. Adreani et al. (Adreani et al. 1997; Adreani and Kaufman 1998) reported that about half of a small sample of Group III and IV muscle afferents responded to metabolites produced by muscle contraction that were likely to be non-nociceptive, while the other half responded to metabolites that would probably elicit a painful response. Attempts to identify the metabolites responsible for activating large populations of muscle sensory afferents have been ambiguous. While numerous substances have been found to activate muscle afferents, when these substances are applied individually to sensory endings in concentrations produced by non-painful or even painful muscle contraction, very few of the afferent endings are activated.

Mosso had determined that substances produced by tetanic muscle contraction could lead to fatigue. He cites his work from 1890 showing that blood from a fatigued

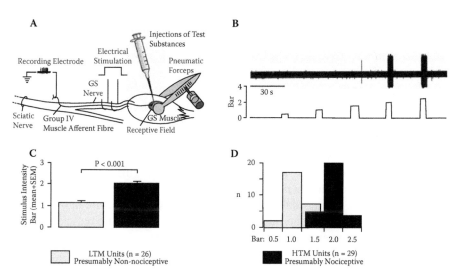

FIGURE 11.2 Experimental setup and results for muscle Group IV afferent recordings. A: Setup. B: Mechanical threshold determination for Group IV unit. C: Pressure thresholds of 26 LTM (grey) and 26 HTM (black) units. D: Distribution of pressure thresholds of LTM and HTM units showing little overlap of thresholds. LTM = Low Threshold Mechanoreceptor. HTM = High Threshold Mechanoreceptor. (Reprinted with permission from Mense S. 2009. Algesic agents exciting muscle nociceptors. *Exp Brain Res.* 196:89–100.)

animal could cause fatigue-like behavior in an unfatigued animal (Mosso 1904). Mosso inferred from available evidence that fatigue was not caused by a lack of oxygen or increased blood CO_2, but it might have been caused by a buildup of lactic acid, a lowering of pH, and a temperature increase. While some of Mosso's conclusions proved incorrect, mostly because of the lack of knowledge of basic neurophysiology (action potentials, membrane potentials, etc., were unknown at this time), his speculation on the substances that might mediate sensory muscle fatigue were clearly prescient.

11.3   MOLECULAR RECEPTORS THAT ARE ACTIVATED BY METABOLITES PRODUCED BY MUSCLE CONTRACTION

Major advances have been made in the last 10 years in identifying metabolites produced by muscle contraction. The most likely ion channels to detect pH changes, acid-sensing ion channels (ASICs), also called amiloride-sensitive ion channels or amiloride-sensitive ion currents, were cloned and sequenced in 1997 (Waldmann et al. 1997a, b). However, initially it appeared that these channels could not play a role in detecting the normal pH values found in humans. They could not gate enough current to activate sensory endings with the small changes in pH observed in conditions known to cause muscle pain in humans, for example, a pH change from 7.4 to 7.0 that can trigger angina in the heart. Immke and McCleskey (2001a) showed that lactate played a role in enhancing the response of ASICs to pH values between pH 7.4 and 7.0. Thus, lactic acid, at concentrations produced during ischemia and extreme active muscle contractions, was more effective in activating muscle sensory afferents than pH reduction alone, as produced by non-ischemic events (e.g., metabolic acidosis).

Later, another critical co-factor was found. ATP in the low concentrations found in muscle interstitium during normal muscle contractions greatly enhanced the sensitivity of ASICs to pH and lactate (Naves and McCleskey 2005; Yagi et al. 2006). Altogether, these combinations enable the ASIC3 receptor to gate sustained current in a range from pH 7.6–6.6 that is capable of activating sensory endings.

Other researchers in McCleskey's laboratory showed that ASIC receptors were more abundant in sensory nerve fibers that innervated skeletal and cardiac muscle than skin. The sensory nerve fibers that contained ASICs were found on the outside of small arterioles and venules in the fascia surrounding the muscles and separating fascicle bundles (Connor et al. 2005; Molliver et al. 2005). This location is optimal for detecting levels of muscle produced metabolites as they are being removed by the vascular system. Interestingly, the fascia seems to be the structure most heavily innervated by nociceptors, and most likely involved with eliciting muscle pain in human muscle (Gibson et al. 2009; Takahashi et al. 2005).

11.4   THE NEED FOR TRANSLATIONAL BRIDGES TO FIBROMYALGIA AND CHRONIC FATIGUE SYNDROME

Our group recognized that McCleskey's data could help answer a number of longstanding questions about how muscle sensory afferents detect muscle work and also signal metabolite-induced muscle pain. First, the results showing that combinations of metabolites were more effective in activating DRG neurons than individual metabolites (Immke and McCleskey 2001b, 2003; Naves and McCleskey, 2005; Yagi et al. 2006) suggested to us a way to understand activation of muscle metaboreceptor sensory neurons. By presenting muscle sensory neurons with various combinations (pairs, triplets, etc.) of the metabolites that are produced by muscle contraction, the metabolites necessary and sufficient to activate the sensory neurons in the fashion they are activated by muscle contraction could be determined. Secondly, once an effective combination of metabolites was determined, the actual molecular receptors utilized by muscle sensory neurons to detect this combination could be confirmed by using selective antagonists to the likely molecular receptors (mostly suggested by Immke and McCleskey 2001b, 2003; Naves and McCleskey, 2005; Yagi et al. 2006). In addition, the identity of the molecular receptors could be verified using PCR and real-time, quantitative PCR to determine if sufficient quantities of mRNA for these specific receptors exist in muscle sensory neurons. Using this one could determine the essential molecular difference between muscle sensory neurons that signal sensory fatigue and those that signal muscle pain.

This knowledge became apparent at the same time that we (A.L. and K.L.) became associated with the Pain Research Center at the University of Utah, headed by Dr. C. Richard Chapman (see Chapter 1). A primary research focus of this center has been the pathophysiology and treatment of fibromyalgia syndrome (FMS), chronic widespread pain (CWP), and chronic fatigue syndrome (CFS)—three syndromes that share chronic pain and fatigue as common symptoms. We recognized that a better understanding of sensory muscle fatigue and muscle pain, as well as sympathetic reflexes activated by muscle afferents, might enhance the understanding of FMS and CFS. Further, a better understanding of FMS could provide insight into the basic nature of sensory muscle fatigue and muscle pain. Because it was apparent that multidisciplinary research would be of great benefit, we formed a larger working group to attempt to determine causes, mechanisms, and treatments for sensory fatigue and muscle pain. The interactions of this group instigated the laboratory animal and translational clinical experiments described below.

To determine if combinations of metabolites might solve the problem of how the exercise pressor response is normally triggered, and possibly also how metabolites trigger the sense of muscle fatigue, we employed calcium imaging of dorsal root ganglion neurons, enabling observation of large numbers of randomly selected sensory neurons. To compare neurons innervating all structures with those innervating muscle, we retrogradely labeled dorsal root ganglion (DRG) neurons from hindlimb muscles by injecting them with the tracer DiI 7–10 days before we harvested the cells. We chose to do these experiments in C57 Bl/6 mice because of availability of mutant mice with deleted genes that putatively contribute to detection of metabolites. (These experiments are described in detail in Light et al. 2008.)

We applied metabolites individually to dissociated DRG neurons cultured for 24 hours as illustrated in Figure 11.3. As shown in this figure even pH 6 (very low physiologically) activated very few DRG neurons, although pH 4, which is definitely out of the physiological range, did activate ~26% of the DRG neurons, as noted by others (Leffler et al. 2006). We also found, as had previous studies, that individual metabolites were relatively ineffective, except for ATP, which could excite up to 60%

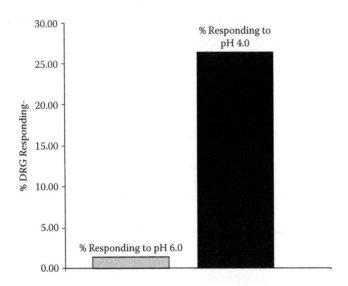

FIGURE 11.3 Percent of all DRG neurons responding to direct application of pH 6.0 or pH 4.0 (from 159 DRG neurons). Acid pH activates few DRG neurons at very low pH (6.0). Non-physiological pH (4.0) is r.

of DRG neurons when applied at 10 μM (Connor et al. 2005; Reinohl et al. 2003). Interestingly, in our experiments, if the normal resting level of ATP outside muscle cells was used to pre-adapt the DRG neurons, very few responses to ATP alone were found when ATP was increased in a graded fashion (see Figure 11.4). This implies that ATP receptors (such as P2X5) sensitive to low ATP concentrations on DRG neurons are normally in an adapted state and incapable of generating activating currents. However, receptors that are activated only by higher concentrations of ATP (e.g., P2X4) can be directly activated by metabolite levels produced by painful levels of metabolites (2 μM, or higher).

When two metabolites putatively involved in metaboreception were applied at physiologic levels, only the combination of ATP and protons produced responses that would elicit the sympathetic reflexes observed in humans with increased metabolites (compare Figure 11.4 with Figure 11.1). However, when ATP, protons, and lactate were applied in a graded manner, similar to the way in which they would increase with exercise in a circulation occluded muscle, the responses of DRG neurons, and particularly of muscle-innervating DRG neurons (determined by retrograde labeling), mimicked very closely the sympathetic responses used by others as a read-out of the metaboreceptive responses (Figure 11.4).

Some DRG neurons responded to low levels of metabolites, while others responded best to high levels of metabolites (see Figure 11.5). When we graphed the average amplitude of the calcium responses, the results revealed response profiles of the low-metabolite-responding neurons that were consistent with signaling that elicits the exercise pressor response. These profiles also fit best with perceptions of muscle tiredness that accompany increases in muscle metabolites.

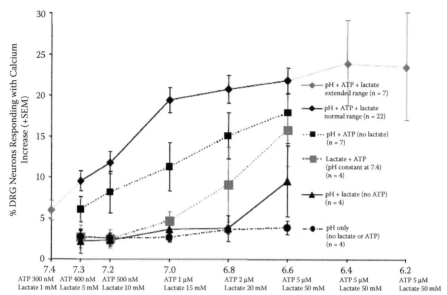

FIGURE 11.4 Percent of DRG neurons responding to various combinations of metabolites. Calcium responses of mouse DRG neurons to various combinations of metabolites at indicated concentrations. Note that neurons are most responsive to the combination of three metabolites in the range most commonly found at the interface between muscle and blood vessels (top trace). (Reprinted with permission from Light AR, Hughen RW, Zhang J, Rainier J, Liu Z, Lee J. 2008. Dorsal root ganglion neurons innervating skeletal muscle respond to physiological combinations of protons, ATP, and lactate mediated by ASIC, P2X, and TRPV1. *J Neurophysiol.* 100: 1184–1201.)

The responses of the high metabolite responders, on the other hand, fit best with perceptions of pain that are caused by greater increases in muscle metabolites (see Figure 11.6).

To determine if the molecular receptors mediating calcium responses were those suggested by McCleskey's group, we determined the metabolite-induced responses of DRG neurons innervating muscle when selective antagonists were applied. These experiments demonstrated that the most likely molecular receptors mediating responses to metabolites are an ASIC receptor (likely ASIC3), a P2X receptor (likely both P2X5, and P2X4), and surprisingly, TRPV1 (see Figure 11.7).

The inclusion of TRPV1 (half of the metabolite responsiveness was abolished by TRPV1 antagonists) was surprising, because pH responses at 6.6 are unlikely to be mediated by activation of this receptor. TRPV1 has been shown to be inactive until lower pH levels are reached (>pH 6.0), but we had previously noted that most of the pH-responsive neurons labeled from muscle were responsive to capsaicin, indicating they probably express functional TRPV1 receptors. But what is the functional role of TRPV1 receptors on DRG neurons if not the detection of activation by protons? One possibility is detection of endogenous vanilloids, like anandamide (Cristino et al. 2008; Toth et al. 2005). They also might detect muscle temperature, which like the metabolites is a by-product of muscle contraction. If activated by temperature, the TRPV1 receptors may act like P2X5 receptors in McCleskey's experiments and may not contribute directly to excitation, but rather enhance the responses of DRG

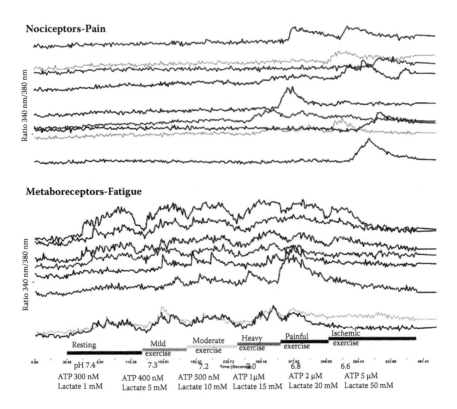

FIGURE 11.5 Responses of DRG neurons to application of indicated levels of metabolites.

neurons when muscle temperature rises above normal body temperature. Sugiura et al. (2003) demonstrated a similar phenomenon in DRG neurons—that pH 6.0-evoked, sustained currents were enhanced by increasing temperature. Interestingly, we found that low-metabolite-detecting DRG neurons responded to capsaicin just as frequently as high-metabolite-detecting neurons. In addition, the responses to low metabolites were partially blocked by the TRPV1 antagonist LJO 328, implicating TRPV1 receptors in these responses. Thus, temperature could be a factor in activation of muscle "fatigue receptors" as well as nociceptors. This could explain the increased sensory fatigue caused by increased muscle and core temperatures (Drust et al. 2005). Increased blood flow generated by sympathetic reflexes activated by fatigue receptor stimulation would help moderate muscle temperature increases and would decrease the other metabolites generated by muscle contraction.

Are other metabolites and receptors involved in signaling sensory muscle fatigue and pain? McCleskey (personal communication) found that the precursor molecule pyruvate could substitute for lactate but was only ~50% as effective in enhancing metabolite responsiveness. Lactate does not act alone and enhances responsiveness in the most common range of metabolites, but the fact that pyruvate can act as a substitute might explain recent experiments showing that sympathetic reflexes evoked by contracting muscle were blocked in patients with McCardle's disease (genetic loss

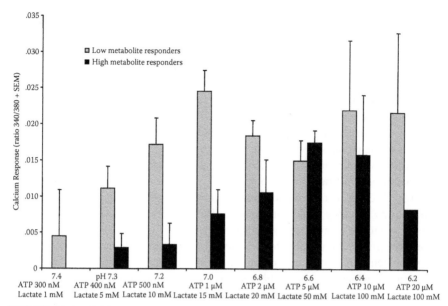

FIGURE 11.6 Calcium response of muscle-labeled DRG neurons responding to changes in metabolites indicated at bottom of graph. (Reprinted with permission from Light AR, Hughen RW, Zhang J, Rainier J, Liu Z, Lee J. 2008. Dorsal root ganglion neurons innervating skeletal muscle respond to physiological combinations of protons, ATP, and lactate mediated by ASIC, P2X, and TRPV1. *J Neurophysiol.* 100: 1184–1201.)

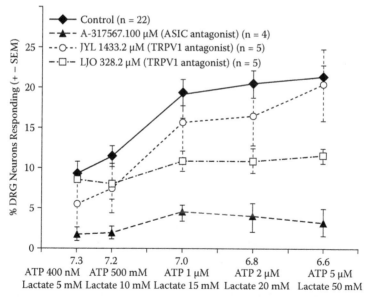

FIGURE 11.7 Percent of DRG neurons responding to metabolites at values indicated at bottom of graphs when ASIC and TRPV1 antagonists were co-applied. (Reprinted with permission from Light AR, Hughen RW, Zhang J, Rainier J, Liu Z, Lee J. 2008. Dorsal root ganglion neurons innervating skeletal muscle respond to physiological combinations of protons, ATP, and lactate mediated by ASIC, P2X, and TRPV1. *J Neurophysiol.* 100: 1184–1201.)

of ability to express lactate in muscle while preserving pyruvate) under some conditions, but present under other conditions (see review by Kaufman 2003).

Several other metabolites associated with muscle contraction have been suggested to be important in signaling that elicits the exercise pressor reflex and/or muscle pain. These include ammonia, mono or di-protonated phosphate, CO_2, ADP, adenosine, prostaglandins, and potassium. We tested ammonia and ammonium (the species most abundant at low pH values), CO_2, and ADP and found them to be ineffective or inhibitory in directly activating muscle-innervating DRG neurons or in enhancing responses to the metabolites shown in the figures above. We assume that the effects of potassium would be non-specific, as increases in extracellular potassium increase the excitability of all neurons, as noted by (Rybicki et al. 1985). We have not yet tested the effects of increased phosphate, adenosine, or prostaglandins, leaving open the possibility that they could contribute to the normal sensations of muscle fatigue and pain and to the exercise pressor response.

Retrospectively, the metabolites we found to synergistically activate a combination of molecular receptors suggests a unique detection system for the rapid use of energy stores, which can signal the body to protection against metabolic and traumatic injury. The combination of protons, lactate, ATP, and temperature increases seems especially relevant to muscle contraction and also applies to brain function. A false signal of sensory muscle fatigue or muscle pain would not occur with metabolic acidosis or respiratory acidosis, as only protons would be increased in these conditions, which would be ineffective in activating muscle afferent neurons unless extreme. On the other hand, traumatic muscle injury, which would cause release of high concentrations of ATP, would activate muscle nociceptors and fatigue receptors, leading to protective behavioral effects (decreased muscle contraction and removal of the injured muscle from the source of trauma).

11.5   ARE THESE MOLECULAR RECEPTORS RESPONSIBLE FOR SENSORY MUSCLE FATIGUE AND MUSCLE PAIN?

As part of a larger investigative group interested in muscle pain and sensations related to muscle fatigue, we recognized that these are greatly increased in people following exhausting exercise. Muscle pain and sensory muscle fatigue ameliorate quickly once exercise is ended, but increase again in untrained muscles 12–24 hours later and can last 48 hours or longer before diminishing. The pain has been studied extensively and is often called *delayed-onset muscle soreness* (DOMS), and most of us (having experienced it personally) know that a sense of muscle fatigue is also increased during DOMS. The explanation for DOMS has been that exercise of untrained muscle causes micro-tears that result in inflammation; however, this explanation is controversial (Cheung et al. 2003). Whether or not micro trauma is involved, the sensory receptors encoding muscle pain and fatigue must be sensitized in order for these enhanced sensations to be apparent because the metabolite levels are not increased at rest in DOMS. As increases in ASIC3 have been shown in some models of long-term muscle pain (Voilley et al. 2001), we hypothesized that one or more of the molecular receptors involved in metabolite signaling of muscle pain

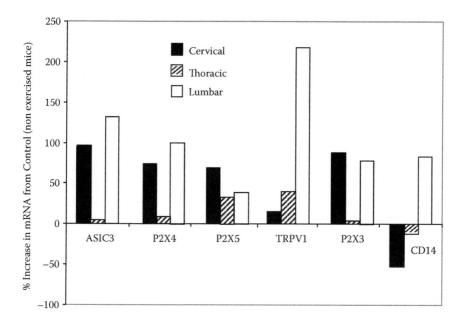

FIGURE 11.8 Mouse DRG mRNA increases caused by exhausting exercise, measured 16 hours after a bout of exercise (n = 6).

would reveal increased mRNA (possibly leading to increased number of receptors, and therefore, increased signaling) following exercise that produced DOMS.

We exercised mice by running them in hamster balls until they would not run any longer (~2 hours), then sacrificed the mice 16 hours later and collected the DRGs from cervical, thoracic, and lumbar regions of the spinal cord. We hypothesized that increases in mRNA for ASIC3, P2X5, P2X4, and/or TRPV1 would be observed in cervical and lumbar DRG neurons innervating the forelimb and hindlimb muscle groups exercised extensively by running.

As Figure 11.8 shows, mRNA for ASIC3, P2X4, and P2X5 were increased in those lumbar and cervical DRG neurons innervating the exercised limbs. Increases were much smaller in the thoracic region, which contains few neurons innervating the limbs. We interpret this to mean that ASIC3, P2X4, and P2X5 contribute to the enhanced pain and sensory fatigue experienced during DOMS, and possibly during other events that cause long-term increases in muscle pain and sensory muscle fatigue.

Several other mediators are also likely to be involved in enhancement of pain and fatigue caused by inflammatory conditions. These include products of the arachidonic acid pathways (Hayes et al. 2006), cytokines (Hoheisel et al. 2005), and neuropeptides such as CGRP (Ambalavanar et al. 2006, 2007; Kruger et al. 1989; Micevych and Kruger 1992). Further work is needed to determine all the contributors to this process.

11.6 TRANSLATION FROM MOUSE TO HUMAN

The next question was, do the molecular receptors we identified in mouse play a role in the human sensory phenomena of muscle pain and fatigue? This presented a problem because sensory muscle fatigue and muscle pain induce the same behavior in mice: They quit running. Thus, discriminating sensory muscle fatigue from muscle pain in the rodent model is difficult. In human patients with CFS, fatigue is the defining symptom, while muscle pain is a common secondary symptom. In patients with FMS or CWP, widespread muscle pain is the defining symptom, and fatigue is the most common secondary symptom. These seemed like appropriate human models to test the idea that ASIC, P2X, and TRPV1 receptors were involved in muscle pain and muscle tiredness. The question, however, was how to access the sensory afferents.

We knew that leukocytes in human subjects show changes and accumulate in the exercised muscle in some cases following exhausting exercise (Kruger et al. 2008). In order to do so, these cells must detect a signal from muscle, and we hypothesized that this signal might be the same metabolites detected by afferent neurons innervating muscle. These leukocytes might utilize the same molecular receptors to detect the metabolites.

Analysis of mRNA from mouse blood indicated that all the molecular receptors we had found to be important for detecting increases in metabolites in muscle sensory neurons were present in leukocytes, and a similar mRNA analysis of human blood also showed that human leukocytes had substantial quantities of these receptor mRNAs. Was the mRNA in leukocytes altered by exercise? This proved to be difficult to determine in mice, because the amount of blood we could obtain was insufficient to measure the mRNA of these genes accurately. On the other hand, sufficient quantities of blood could easily be obtained at multiple times from human subjects. We evaluated mRNA increases in blood from human subjects after exercise and measured the homologous mRNA of ion channels that had been altered in DRG neurons innervating mouse muscle and also mRNA for $\alpha 2A$, $\beta 1$, and $\beta 2$ adrenergic receptors, catechol-o-methyl transferase (COMT) and several pro-inflammatory and anti-inflammatory cytokines and cytokine receptors that might be altered by exercise.

Increases in mRNA obtained from leukocytes from a healthy subject who performed maximal, exhausting strength and aerobic whole-body exercise for 45 minutes (Figure 11.9) were found for the same molecular receptors important for detection of metabolites in mouse DRG neurons that were increased by exhausting exercise in mouse DRGs. Clearly, mRNA increases were strongly related to the timing of DOMS and sensory muscle fatigue that appeared in this subject following exercise.

11.7 MOLECULAR RECEPTORS INVOLVED IN CHRONIC FATIGUE SYNDROME

The preceding observations encouraged us to determine if the mRNA of sensory muscle fatigue and pain receptors was increased in CFS patients following exercise that exacerbates their symptoms. A major hallmark of CFS is that even mild exercise greatly increases sensory muscle fatigue as well as muscle pain (Holmes et al. 1988). With guidance from Andrea White, our exercise scientist, we developed a 25-minute, moderate, whole-body exercise protocol in which each subject performed continuously at 70% of his/her age-predicted maximal heart rate. We confirmed that in

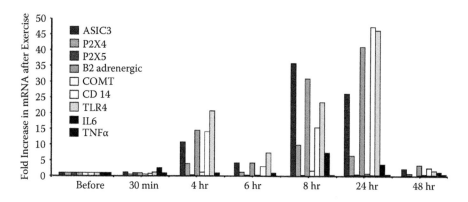

FIGURE 11.9 (see color insert following page 166) mRNA increases in human leukocytes at times indicated after 45 minutes of exhausting exercise. All data are from one female subject. Exercise consisted of aerobic and strength exercises including indoor rock climbing, treadmill, and weight machines.

normal, healthy women and men, this moderate exercise produced minimal muscle pain or sensory muscle fatigue during or immediately after completion, and DOMS did not develop. We compared healthy controls versus patients with CFS and FMS, collecting leukocytes and extracting their mRNA at five intervals: at pre-exercise baseline, and at 30 minutes, 8, 24, and 48 hours after the end of exercise to examine ASIC3, P2X4, P2X5, TRPV1, adrenergic α2A, adrenergic β1 and β2, COMT, IL6, IL10, TNF-α, TLR4, and CD14.

The results (Figure 11.10; Light et al., 2009a) show that following moderate exercise, control subjects revealed no significant increase in mRNA for any of the genes tested. CFS patients (70% with chronic widespread pain meeting diagnostic criteria for FMS) exhibited significant increases in mRNA expression of nearly all of the genes studied, except for IL6 and IL10. The surprise was that in CFS patients, increases in metabolite-detecting receptor and adrenergic receptor gene expression were obvious at 30 minutes after onset of exercise. A second surprise was the magnitude of the effects that greatly exceeded any increases we had seen even with maximum exercise in control subjects. In addition to the mean differences, we found strong correlations between increases in mRNA of ASIC3, P2X4, P2X5, TRPV1, adrenergic α2A, adrenergic β1, β2, COMT, IL10, TLR4, and CD14 in the CFS patients and their reports of increased mental fatigue during the 2-day time period following exercise. We found similarly strong correlations between ASIC3, TRPV1, adrenergic α2A and β2, and IL10 and the patients' reports of pain during the 2 days following the exercise challenge.

These data indicate that a common transcription factor is likely altered in CFS, causing strong upregulation of mRNA production in leukocyte genes that detect metabolite increases, and presumably affecting leukocyte production of cytokines. Whether similar action takes place on muscle sensory neurons in CFS patients is unknown, but seems likely, given information suggesting that peripheral signals for pain can be enhanced in at least some of these patients (Staud et al. 2003).

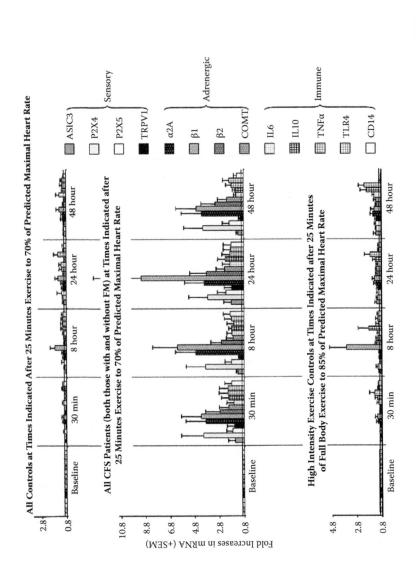

FIGURE 11.10 (see color insert following page 166) mRNA increases in CFS patients after 25 minutes of moderate exercise (70% of age adjusted maximum heart rate) (middle) but not in control subjects (top). Much more intense exercise (85% of age-adjusted maximum heart rate) in control subjects (bottom) increases mRNA for only a few of the genes. Faint horizontal line near the bottom of each panel indicates the baseline levels.

11.8 DYSREGULATED MRNAS MAY ALSO EXPLAIN COMMON CO-MORBIDITIES OF CFS

Many of the genes we found to be dysregulated in CFS patients after exercise may also be implicated in related disorders. In our CFS patients, 70% met criteria for FMS, which is characterized by widespread pain dominated by muscle pain. The muscle nociceptors found in our mouse studies utilized most of the molecular receptors upregulated in CFS patients, including ASIC, P2X4, and TRPV1. Colon sensitivity to inflammation has been shown to be mediated by ASIC receptors and TRPV1 receptors (Christianson et al. 2006; Jones III et al. 2005; Page et al. 2005; Sugiura et al. 2007), suggesting that dysregulation of these receptors could increase the likelihood of irritable bowel syndrome (IBS), another condition often co-morbid with CFS. Multiple chemical sensitivity may be mediated by a TRP channel, possibly TRPV1 or TRPA1 (Pall and Anderson 2004).

Several sensory modalities (but not all) become more sensitive in some cases of CFS. Hyperacusis in CFS may be related to increased ASIC receptors (Geisser et al. 2008; Hildebrand et al. 2004; Peng et al. 2004), as ASIC receptors appear to enhance hearing. Clinical depression may share some of the genes that were upregulated in CFS patients; for example, TRPV1 has recently been implicated in depression (Di et al. 2008), and ASICs may play a role in depression and anxiety (Coryell et al. 2007; Wemmie et al. 2003, 2004). In addition, polymorphisms in COMT may be risk factors for depression, although this issue is not without controversy (Baekken et al. 2008; Funke et al. 2005; Ohara et al. 1998; Tsai et al. 2009). Perhaps genes contributing directly to the sensory experience of fatigue share upstream regulation with the genes that are modified in these co-morbid conditions that share sensory muscle fatigue, hyperalgesia, and chronic pain as central symptoms, and some of the same regulated genes.

A number of other conditions that exhibit unexplained sensory muscle fatigue may also share dysregulation of some of these same sensory, adrenergic, and cytokine genes. We have preliminary evidence that unexplained fatigue in multiple sclerosis, for example, is associated with baseline changes in TRPV1 and adrenergic β2 receptors as well as alterations in several of the ion channel genes and adrenergic receptors following moderate exercise. A similar dysregulation might also explain the greatly enhanced sensory muscle fatigue often observed in heart failure, even when the vascular dynamics appear to be adequate for perfusion (Falk et al. 2006; Khan and Sinoway 2000). If the cardiac sensory receptors, for example, were dysregulated by primary insult to the heart, a profound sensation of fatigue might ensue, unattributable to any particular muscle.

11.9 CAN ACUTE FATIGUE LEAD TO CHRONIC FATIGUE?

A key question about CFS is whether it is similar to chronic pain in the sense that acute pain might constitute a cause of chronic pain (Vierck, Jr., 2006). A clue in this regard would be whether receptors that detect metabolites associated with sensory muscle fatigue appear to be under positive feedback control. Activation of ASICs, P2X, and TRPV1 increases the mRNA of these receptors in dorsal root ganglia of mice following exhausting exercise, and indeed they are upregulated following acute activation of these receptors by metabolites, possibly resulting in more robust activation by future episodes of metabolite production. If the mRNA increases we have observed in normal subjects and particularly in CFS patients are translated into functional receptors, further production of metabolites will produce more receptors, causing an increase in the signal for muscle sensory fatigue if compensating mechanisms are absent (for example, rest, increased sympathetic effects on vascular smooth muscle to eliminate metabolites, and increased capillary capacity—all of which will reduce metabolites). It is also possible that sensory muscle fatigue is like pain in that persistent sensory input may cause central "wind-up" and other plastic changes (such as glial activation; Chacur et al. 2008) that can enhance the afferent signals at many levels in the central nervous system. Thus, sensory experiences for both fatigue and muscle pain could be enhanced over time, and the sympathetic signals caused by sensory muscle fatigue and muscle pain could increase centrally in the spinal cord and brain. There is considerable evidence that signals for fatigue and muscle pain are enhanced in CFS and FMS patients (Caseras et al. 2008; Cook et al. 2007; Geisser et al. 2008; Gracely et al. 2002; Schmaling et al. 2003), although the relative contribution of central components versus peripheral components remains undetermined.

11.10 COMMON CAUSES OF ACUTE FATIGUE MIGHT CAUSE UP-REGULATION OF THE mRNA FOR THE MOLECULAR RECEPTORS SUGGESTED TO MEDIATE SENSORY MUSCLE FATIGUE AND PAIN

At present, we do not know all of the factors that cause upregulation of the molecular receptors responsible for detecting contraction-produced metabolites in muscle sensory neurons. Mamet et al. (2003) provided evidence that nerve growth factor (NGF) can maintain and upregulate ASIC receptors and can mediate inflammatory effects on ASIC receptor upregulation. Thus, any event that increases NGF could increase sensory muscle fatigue and pain.

Cytokine receptors have also been demonstrated on sensory neurons, (e.g., Li et al. 2004). As cytokines have been implicated in the symptoms of a number of viral infections, including Epstein-Barr (Wright-Browne et al. 1998), it is quite likely that viral infections induce muscle fatigue and muscle pain by sensitizing the sensory muscle afferents that signal muscle fatigue and pain. Whether these effects include upregulation of the molecular receptors that encode metabolites is unknown but also seems likely. Thus, acute viral infection could cause myalgia and fatigue through cytokine pathways.

Reductions in mitochondrial efficiency (viral or non-viral causes) could result in decreased aerobic capacity. These decreases could cause increased glycolysis, which would lead to increases in the metabolites that activate sensory muscle afferents. Any such increases could also lead to increased resting and exercise sensations of muscle fatigue and pain. Such increases could also lead to increases in the molecular receptors that encode metabolites, causing further increases in muscle pain and fatigue, and eventually leading to sympathetic dysregulation and possibly CFS and

FMS. Clearly, further investigation of the mechanisms that can cause increases in metabolite-detecting receptors is necessary and could clarify a host of conditions in which muscle fatigue and pain are major symptoms.

11.11   MENTAL FATIGUE

Although beyond the scope of this report, similar processes could lead to dysregulation of receptors encoding mental fatigue. Of course, given that the metabolic demands of the brain are different from those of muscle, the specific molecular receptors encoding brain metabolites and the way these receptors regulate cerebral blood flow may differ. However, recent reports suggest there is dysregulation of global cerebral blood flow in CFS patients (Mathew et al. 2008; Yoshiuchi et al. 2006) that could lead to metabolite increases similar to those identified in skeletal muscle of CFS patients.

11.12   SYMPATHETIC DYSREGULATION CONTRIBUTES
TO ENHANCED SENSORY FATIGUE

The muscle sensory afferents that may encode sensory muscle fatigue have been linked to sympathetic responses caused by muscle contraction for at least 70 years (Alam and Smirk 1937; see also discussion in Light et al. 2008). Thus, sensory fatigue afferents (and possibly muscle pain afferents) may be the sensory arm(s) of sympathetic reflexes associated with muscle contraction. These reflexes normally serve to increase the blood flow to working muscles partly by diverting blood flow from non-working muscles. This increased blood flow serves to reduce metabolites, terminating the signals underlying sensory muscle fatigue. The sympathetic output signal is mediated by adrenergic $\beta2$ receptors on vascular smooth muscle in the working muscle and by adrenergic α receptors in non-working muscles, and by adrenergic $\beta1$ receptors in heart muscle.

If the sympathetic signal is not correctly regulated, increased metabolites will persist, causing a constant signal from "fatigue" sensory afferents. Potentially, a constant signal of fatigue from sensory afferents could lead to sympathetic dysregulation by sending a constant sympathetic signal to adrenergic β receptors on vascular smooth muscle in skeletal muscles leading to down-regulation of these adrenergic β receptors. Such downregulation could lead to enhanced metabolites in working muscles, causing an enhanced signal for sensory muscle fatigue from virtually all muscles in the body. This enhanced signal could lead to further dysregulation of sympathetic reflexes. Similarly, if the sympathetic signal is not correctly terminated, for example if COMT is dysregulated either by environmental or genetic causes (Goertzel et al. 2006) leading to increased norepinephrine and epinephrine action on vascular adrenergic β receptors, down regulation can occur, and the metabolites may not be regulated appropriately. Similarly, alterations in the effectiveness of adrenergic receptors (as is the case in some polymorphisms) could lead to a mismatch between the sensory signal sent to the sympathetic nervous system, and the amount of vasodilation and/or vasoconstriction that results leading to dysregulation of muscle metabolites (Diatchenko et al. 2006). This dysregulation of muscle metabolites could lead to enhanced sensory muscle fatigue both at rest and following exercise.

11.13   ADRENERGIC RECEPTORS ON SENSORY NEURONS?

A dilemma concerning the hypothesized dysregulation of sympathetic reflexes in CFS and FMS patients derives from the view that dysregulation results from an overactive sympathetic nervous system that "burns out" because of desensitization or downregulation of peripheral receptors in vascular smooth muscle resulting from constant bombardment with norepinephrine. In a double-blind, placebo-controlled crossover experiment designed to observe this dysregulation, we (Light et al. 2009b) intravenously administered a low dose of the non-selective β adrenergic antagonist propranolol or placebo to patients with FMS or temporomandibular disorder (TMD). Surprisingly, the sympathetic reflexes (assessed by plasma catecholamine and cardiovascular responses to various challenges) that were dysregulated in FMS patients became relatively normalized by the administration of propranolol. The explanation for this could be a complex reorganization in the central nervous system and/or possible effects on the immune system via adrenergic receptors on leukocytes, or a simple action on sensory neurons that control sympathetic reflexes and muscle pain. The complex CNS and immune explanations seem improbable because most TMD and FMS patients reported a reduction in clinical pain by propranolol (and not by placebo; see Figure 11.11) that was almost immediate (within minutes), without time for reorganization of central sympathetic systems or alterations in immune function.

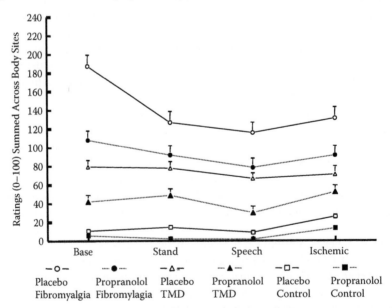

FIGURE 11.11   Propranolol reduces total body pain. (Reprinted with permission from Light KC, Bragdon EE, Grewen KM, Brownley KA. 2009. Adrenergic dysregulation and pain with and without acute beta-blockade in women with fibromyalgia and temporomandibular disorder. *J Pain.* In press.)

The simple explanation was that there are β adrenergic receptors not only on vascular smooth muscle in skeletal muscles, but also on the muscle sensory neurons whose activation signals muscle pain and triggers sympathetic reflexes. To fully explain these phenomena, putative adrenergic receptors on muscle sensory neurons would have to be considerably upregulated in TMD and FMS. A recent case report showing that CFS symptoms could be greatly reduced with propranolol supports this possibility (Wyller et al. 2007). However, there has been no previous demonstration of β adrenergic receptors on muscle sensory neurons.

11.14   BACK TO THE MOUSE TO FIGURE OUT HOW ADRENERGIC RECEPTORS COULD INCREASE MUSCLE PAIN AND FATIGUE

Although paradoxical, alpha-adrenergic receptors have been localized on sensory afferent neurons. Alpha-adrenergic receptors (both α1 and α2) have been shown to upregulate on nociceptive sensory neurons in animal models of neuropathic pain and inflammation, resulting in enhanced nociceptive effects of cutaneous stimulation (Birder and Perl 1999; Lee et al. 1999; O'Halloran and Perl 1997; Sato and Perl 1991). However, alpha-adrenergic receptors on afferent nerves in the spinal cord appear to inhibit pain. In fact, intrathecal treatment with clonidine, an alpha-adrenergic agonist, is often used to reduce various types of pain in patients (Boyd 2001; Brown et al. 2004; Sites et al. 2003).

Only a few reports suggest that beta-adrenergic receptors play a role in enhanced pain following injury (Khasar et al. 1999), and these receptors have not been found on muscle sensory neurons. Given the compelling results from the human experiments, we examined mouse DRG using PCR to determine if beta-adrenergic receptors were present. We also induced gastrocnemius muscle inflammation and used quantitative real-time PCR to determine if beta-adrenergic receptors were upregulated. Finding adrenergic β1 and β2 receptor mRNA present on mouse lumbar DRG neurons, we further observed β2 receptor mRNA is upregulated 24 hours after carrageenan injection into the gastrocnemius muscle (see Figure 11.12), and this upregulation lasted at least 8 days.

To determine if β2 receptor mRNA was upregulated more in neurons innervating the carrageenan-injected muscle, we harvested DRG neurons labeled with DiI from the carrageenan-injected muscle, and compared them with DiI-labeled neurons from the contralateral side, also injected with DiI but not carrageenan-inflamed (Figure 11.13). This procedure revealed that β2 adrenergic mRNA was much higher in DiI-labeled neurons from the carrageenan-inflamed muscle than in muscle-innervating DRG neurons from the contralateral side.

Jon Levine (University of California, San Francisco) was sufficiently intrigued by our animal and human observations to evaluate possible behavioral effects of β-adrenergic upregulation in rats with the gastrocnemius muscles inflamed by injection of carrageenan, as we had done in mice. He tested sensitivity of the inflamed muscle 1 hour, 24 hours, and 7 days after inflammation. He also tested the muscle sensitivity after injection of propranolol at doses comparable to those used in the

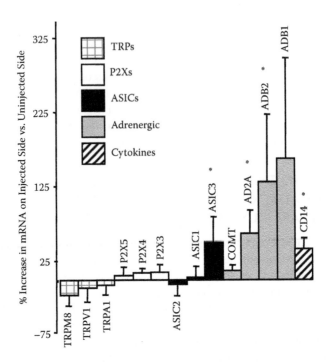

FIGURE 11.12 Twenty-four hour percent increase in mRNA caused by carrageenan-induced muscle inflammation. Note significant increases in adrenergic receptors (n = 8 mice).

human studies described above, and showed (Figure 11.14) that propranolol had no effect 1 hour after induction of inflammation. However, the threshold for a withdrawal response was greatly increased (indicating pain reduction) by propranolol 24 hours after induction of inflammation—when mRNA for beta-adrenergic receptors was increased in DRG neurons.

Propranolol increased withdrawal threshold an even greater percentage 7 days after inflammation—a time at which the only detectable receptors significantly increased in mouse DRGs were β adrenergic receptors. Dr. Levine has also shown that stress via the hypothalamic-pituitary-adrenal axis (HPA) in combination with sympathetic activation can cause a switch in the adrenergic-signaling pathway causing hyperalgesia in rats. Thus, it appears that β adrenergic receptors on DRG sensory neurons contribute to enhanced sensitivity of inflamed muscle. Whether or not beta-adrenergic receptors play a role in the signal for enhanced sensations of muscle fatigue is unknown, but this seems likely. The same situation may apply to enhanced muscle pain and sensory muscle fatigue in patients with a variety of conditions, including FMS and CFS. Understanding the regulation of beta-adrenergic receptors in muscle sensory neurons, as well as how this regulation interacts with the activity of ASICs, P2X, and TRPV1 receptors in signaling muscle fatigue and pain, may greatly enhance our ability to manage a number of muscle pain and sensory muscle fatigue conditions.

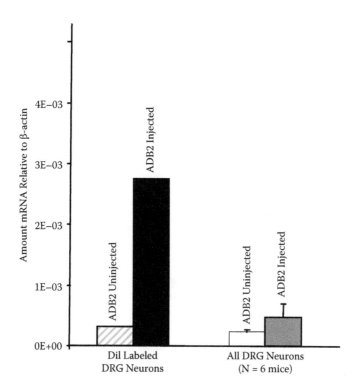

FIGURE 11.13 Quantitative mRNA from DiI-labeled DRG neurons from gastrocnemius muscle injected with 3% carrageenan 24 hours before harvesting neurons indicates that most of the β2 adrenergic receptor increases were in DRG neurons in the inflamed muscle. 100 DRGs—No amplification. In the "All DRG Neurons" group, muscle was not injected with DiI.

11.15 FUTURE RESEARCH DIRECTIONS

Both CFS and FMS are difficult to diagnose and treat due to a deficiency of objective biomarkers for diagnosis or for quantifying effects of treatment on the primary symptoms in these syndromes: muscle pain and sensations of muscle fatigue. The basis for a medically and legally acceptable objective test might be developed from the leukocyte gene expression changes after exercise that we observed in our patient groups. With these assays, post-exercise blood tests could provide objective validation of a prolonged and severe sensory muscle fatigue and muscle pain state. Combined with assays for other genes modified by other fatiguing and muscle pain-causing diseases, mechanisms of CFS and FMS might be distinguished from other sources of fatigue and pain. Observations on how gene modulation evoked by exercise is altered by treatment may also provide further understanding of CFS and FMS and help guide more effective treatments of these syndromes.

Using these mRNA tests in a prospective study on a population of subjects, some of whom develop CFS and/or FMS, may provide a test for those susceptible to

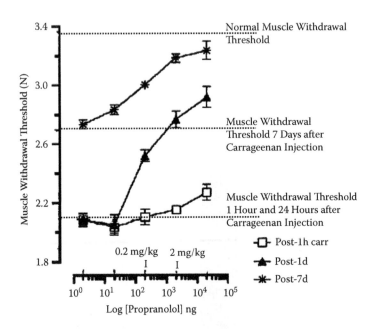

FIGURE 11.14 Effects of propranolol on inflammation-induced muscle pain in male rats. Carr (1% i/m) versus propranolol DRC (n = 6 rats). (Courtesy of Dr. Olayinka Dina and Jon Levine, UCSF.)

developing these conditions, and could provide information necessary for prevention of CFS and FMS.

The experiments described here also suggest alternative research directions that could increase understanding of several functional disorders. Some of the genes dysregulated in CFS and FMS have been implicated in irritable bowel syndrome (IBS), specifically, ASICs, TRPV1, adrenergic β2, IL10, and TLR4 receptors. The concept that some of these may form receptor complexes that interact to create functional units may be helpful in understanding this syndrome. Similarly, multiple chemical sensitivity might benefit from understanding the receptor complexes that can detect a variety of different chemical signals.

The concept of *receptor complexes* might also apply to the elusive molecular nature of mechanical nociceptors activated by noxious *mechanical* stimuli. Identification of these critical molecules could greatly enhance treatment of a variety of pain conditions and syndromes. One particular consideration is that understanding the interaction between the mechanoreceptors and metaboreceptors of Group III and IV muscle sensory neurons might assist description and treatment of restless leg syndrome (RLS). This enigmatic syndrome is characterized by "unpleasant sensations in the legs and an uncontrollable urge to move when at rest in an effort to relieve these feelings" (NINDS 2001). While considered a central neurological problem at present (much like CFS and FMS), the symptoms (the sensation of mechanical stimulation inside the legs) and their alleviation by movement (increasing blood flow to eliminate metabolites) can be explained by activation of muscle Group III, non-nociceptive mechanoreceptors that are also affected by contraction-produced muscle metabolites. Further study of Group III and IV muscle afferent involvement in RLS deserves serious investigation.

Finally, does fatigue constitute a specific sensory system? While our initial efforts indicate that specific receptors may exist, much work needs to be done on characterizing the receptors for fatigue in skeletal muscle, even more on the very important receptors for fatigue in cardiac muscle, and still more on the fatigue receptors in the brain and spinal cord, as well as defining the spinal cord, brain stem, thalamic, and cortical regions that process the sensory aspect of fatigue. With a specific stimulus (the "correct" metabolite "soup" to specifically activate these receptors), brain regions integrating this phenomenon can be identified. It is even possible that the motor regions that receive this signal and are responsible for the noncognitive shutdown of motor command that causes failure of voluntary muscle contraction can be identified by using the knowledge gained by studying the afferent system. It is likely that these motor regions hold a clue to the causes of somatoform disorders that can cause paralysis of specific cognitively controlled movements in some patients.

As with any fertile research area there remain many questions, but we have the tools, and some of the concepts, necessary to begin to investigate the most prevalent symptoms that bring patients to the clinics around the world: fatigue and pain.

ACKNOWLEDGMENTS

Research by the authors reported here was supported by NIH grant R21 NS057821 from NINDS and NIAMS, with additional ancillary support from NIH R21 AT0002209 from NCAM, a catalyst grant from the University of Utah Health Sciences Center, a synergy grant from the University of Utah, and a grant from the Deptartment of Anesthesiology, University of Utah.

The authors would also like to acknowledge the support of the University of Utah Health Sciences Center Genomics and Imaging Cores. Special thanks to Dr. Lucinda Bateman for help with patient referrals. Thanks to Drs. Rajan Radhakrishnan and Jesse Zhang for their help on projects described in this chapter. Also, our thanks to undergraduate students supported by UROP and BIOURP programs including Benjamin Jensen, Cody Larson, Tania Michael, Clay Peterson, Nick Daskalas, Sean Gowen, Shane Hawthorne, Sonia Sarfraz, Tuyet Nguyen, and graduate students from the Exercise and Sport Science Department including K.L. Fitschen, Luke Wendt, and Tim Vanhaitsma.

REFERENCES

Adreani CM, Hill JM, Kaufman MP. 1997. Responses of group III and IV muscle afferents to dynamic exercise. *J Appl Physiol.* 82: 1811–1817.

Adreani CM, Kaufman MP. 1998. Effect of arterial occlusion on responses of group III and IV afferents to dynamic exercise. *J Appl Physiol.* 84: 1827–1833.

Alam M, Smirk FH. 1937. Observations in man upon a blood pressure raising reflex arising from the voluntary muscles. *J Physiol.* 89: 372–383.

Ambalavanar R, Moritani M, Moutanni A, Gangula P, Yallampalli C, Dessem D. 2006. Deep tissue inflammation upregulates neuropeptides and evokes nociceptive behaviors which are modulated by a neuropeptide antagonist. *Pain*. 120: 53–68.

Ambalavanar R, Yallampalli C, Yallampalli U, Dessem D. 2007. Injection of adjuvant but not acidic saline into craniofacial muscle evokes nociceptive behaviors and neuropeptide expression. *Neuroscience*. 149: 650–659.

Baekken PM, Skorpen F, Stordal E, Zwart JA, Hagen K. 2008. Depression and anxiety in relation to catechol-O-methyltransferase Val158Met genotype in the general population: the Nord-Trondelag Health Study (HUNT). *BMC Psychiatry*. 8: 48.

Bellinger AM, Mongillo M, Marks AR. 2008. Stressed out: the skeletal muscle ryanodine receptor as a target of stress. *J Clin Invest*. 118: 445–453.

Birder LA, Perl ER. 1999. Expression of alpha2-adrenergic receptors in rat primary afferent neurones after peripheral nerve injury or inflammation. *J Physiol*. 515 (Pt 2): 533–542.

Boyd RE. 2001. Alpha2-adrenergic receptor agonists as analgesics. *Curr Top Med Chem*. 1: 193–197.

Brooks JC, Zambreanu L, Godinez A, Craig AD, Tracey I. 2005. Somatotopic organisation of the human insula to painful heat studied with high resolution functional imaging. *Neuroimage*. 27: 201–209.

Brown DR, Hofer RE, Patterson DE, Fronapfel PJ, Maxson PM, Narr BJ, Eisenach JH, Blute ML, Schroeder DR, Warner DO. 2004. Intrathecal anesthesia and recovery from radical prostatectomy: a prospective, randomized, controlled trial. *Anesthesiology*. 100: 926–934.

Caseras X, Mataix-Cols D, Rimes KA, Giampietro V, Brammer M, Zelaya F, Chalder T, Godfrey E. 2008. The neural correlates of fatigue: an exploratory imaginal fatigue provocation study in chronic fatigue syndrome. *Psychol Med*. 38: 941–951.

Chacur M, Lambertz D, Hoheisel U, Mense S. 2008. Role of spinal microglia in myositis-induced central sensitisation: An immunohistochemical and behavioural study in rats. *Eur J Pain*.

Cheung K, Hume P, Maxwell L. 2003. Delayed onset muscle soreness: treatment strategies and performance factors. *Sports Med*. 33: 145–164.

Christianson JA, McIlwrath SL, Koerber HR, Davis BM. 2006. Transient receptor potential vanilloid 1-immunopositive neurons in the mouse are more prevalent within colon afferents compared to skin and muscle afferents. *Neuroscience*. 140: 247–257.

Connor M, Naves LA, McCleskey EW. 2005. Contrasting phenotypes of putative proprioceptive and nociceptive trigeminal neurons innervating jaw muscle in rat. *Mol Pain*. 1: 31.

Cook DB, Lange G, Ciccone DS, Liu WC, Steffener J, Natelson BH. 2004. Functional imaging of pain in patients with primary fibromyalgia. *J Rheumatol*. 31: 364–378.

Cook DB, O'Connor PJ, Lange G, Steffener J. 2007. Functional neuroimaging correlates of mental fatigue induced by cognition among chronic fatigue syndrome patients and controls. *Neuroimage*. 36: 108–122.

Coryell MW, Ziemann AE, Westmoreland PJ, Haenfler JM, Kurjakovic Z, Zha XM, Price M, Schnizler MK, Wemmie JA. 2007. Targeting ASIC1a reduces innate fear and alters neuronal activity in the fear circuit. *Biol Psychiatry*. 62 (10): 1140–1148.

Craig AD. 2003. Pain mechanisms: labeled lines versus convergence in central processing. *Annu Rev Neurosci*. 26: 1–30.

Cristino L, Starowicz K, De Petrocellis L, Morishita J, Ueda N, Guglielmotti V, Di M, V. 2008. Immunohistochemical localization of anabolic and catabolic enzymes for anandamide and other putative endovanilloids in the hippocampus and cerebellar cortex of the mouse brain. *Neuroscience*. 151: 955–968.

Davis JM, Bailey SP. 1997. Possible mechanisms of central nervous system fatigue during exercise. *Med Sci Sports Exerc*. 29: 45–57.

Di M, V, Gobbi G, Szallasi A. 2008. Brain TRPV1: a depressing TR(i)P down memory lane? *Trends Pharmacol Sci.* 29: 594–600.

Diatchenko L, Anderson AD, Slade GD, Fillingim RB, Shabalina SA, Higgins TJ, Sama S, et al. 2006. Three major haplotypes of the beta 2 adrenergic receptor define psychological profile, blood pressure, and the risk for development of a common musculoskeletal pain disorder. *Am J Med Genet B Neuropsychiatr Genet.* 141: 449–462.

Drust B, Rasmussen P, Mohr M, Nielsen B, Nybo L. 2005. Elevations in core and muscle temperature impairs repeated sprint performance. *Acta Physiol Scand.* 183: 181–190.

Falk K, Swedberg K, Gaston-Johansson F, Ekman I. 2006. Fatigue is a prevalent and severe symptom associated with uncertainty and sense of coherence in patients with chronic heart failure. *Eur J Cardiovasc Nurs.* 6 (2): 99–104.

Funke B, Malhotra AK, Finn CT, Plocik AM, Lake SL, Lencz T, DeRosse P, Kane JM, Kucherlapati R. 2005. COMT genetic variation confers risk for psychotic and affective disorders: a case control study. *Behav Brain Funct.* 1: 19.

Geisser ME, Strader DC, Petzke F, Gracely RH, Clauw DJ, Williams DA. 2008. Comorbid somatic symptoms and functional status in patients with fibromyalgia and chronic fatigue syndrome: sensory amplification as a common mechanism. *Psychosomatics.* 49: 235–242.

Gibson A, Baden D, Lambert MI, Lambert EV, Harley YXR, Hampson D, Russell VA, Noakes TD. 2003. The conscious perception of the sensation of fatigue. *Sports Med.* 33: 167–176.

Gibson W, Arendt-Nielsen L, Taguchi T, Mizumura K, Graven-Nielsen T. 2009. Increased pain from muscle fascia following eccentric exercise: animal and human findings. *Exp Brain Res.*

Goertzel BN, Pennachin C, de Souza CL, Gurbaxani B, Maloney EM, Jones JF. 2006. Combinations of single nucleotide polymorphisms in neuroendocrine effector and receptor genes predict chronic fatigue syndrome. *Pharmacogenomics.* 7: 475–483.

Gracely RH, Petzke F, Wolf JM, Clauw DJ. 2002. Functional magnetic resonance imaging evidence of augmented pain processing in fibromyalgia. *Arthritis Rheum.* 46: 1333–1343.

Hayes SG, Kindig AE, Kaufman MP. 2006. Cyclooxygenase blockade attenuates responses of group III and IV muscle afferents to dynamic exercise in cats. *Am J Physiol Heart Circ Physiol.* 290: H2239–H2246.

Hildebrand MS, de Silva MG, Klockars T, Rose E, Price M, Smith RJ, McGuirt WT, Christopoulos H, Petit C, Dahl HH. 2004. Characterisation of DRASIC in the mouse inner ear. *Hear Res.* 190: 149–160.

Hoheisel U, Unger T, Mense S. 2005. Excitatory and modulatory effects of inflammatory cytokines and neurotrophins on mechanosensitive group IV muscle afferents in the rat. *Pain.* 114: 168–176.

Holmes GP, Kaplan JE, Gantz NM, Komaroff AL, Schonberger LB, Straus SE, Jones JF, Dubois RE, Cunningham-Rundles C, Pahwa S. 1988. Chronic fatigue syndrome: a working case definition. *Ann Intern Med.* 108: 387–389.

Immke DC, McCleskey EW. 2001a. ASIC3: a lactic acid sensor for cardiac pain. *ScientificWorldJournal.* 1: 510–512.

Immke DC, McCleskey EW. 2001b. Lactate enhances the acid-sensing Na+ channel on ischemia-sensing neurons. *Nat Neurosci.* 4: 869–870.

Immke DC, McCleskey EW. 2003. Protons open acid-sensing ion channels by catalyzing relief of Ca2+ blockade. *Neuron.* 37: 75–84.

Jones RC, III, Xu L, Gebhart GF. 2005. The mechanosensitivity of mouse colon afferent fibers and their sensitization by inflammatory mediators require transient receptor potential vanilloid 1 and acid-sensing ion channel 3. *J Neurosci.* 25: 10981–10989.

Kaufman MP. 2003. Has the phoenix risen? *J Physiol.* 548: 666.

Kaufman MP, Hayes SG. 2002. The exercise pressor reflex. *Clin Auton Res.* 12: 429–439.

Khan MH, Sinoway LI. 2000. Muscle reflex control of sympathetic nerve activity in heart failure: the role of exercise conditioning. *Heart Fail Rev.* 5: 87–100.

Khasar SG, McCarter G, Levine JD. 1999. Epinephrine produces a beta-adrenergic receptor-mediated mechanical hyperalgesia and in vitro sensitization of rat nociceptors. *J Neurophysiol.* 81: 1104–1112.

Kniffki KD, Mense S, Schmidt RF. 1978. Responses of group IV afferent units from skeletal muscle to stretch, contraction and chemical stimulation. *Exp Brain Res.* 31: 511–522.

Kniffki KD, Mense S, Schmidt RF. 1981. Muscle receptors with fine afferent fibers which may evoke circulatory reflexes. *Circ Res.* 48: I25–I31.

Kruger K, Lechtermann A, Fobker M, Volker K, Mooren FC. 2008. Exercise-induced redistribution of T lymphocytes is regulated by adrenergic mechanisms. *Brain Behav Immun.* 22: 324–338.

Kruger L, Silverman JD, Mantyh PW, Sternini C, Brecha NC. 1989. Peripheral patterns of calcitonin-gene-related peptide general somatic sensory innervation: cutaneous and deep terminations. *J Comp Neurol.* 280: 291–302.

Lee DH, Liu X, Kim HT, Chung K, Chung JM. 1999. Receptor subtype mediating the adrenergic sensitivity of pain behavior and ectopic discharges in neuropathic Lewis rats. *J Neurophysiol.* 81: 2226–2233.

Leffler A, Monter B, Koltzenburg M. 2006. The role of the capsaicin receptor TRPV1 and acid-sensing ion channels (ASICS) in proton sensitivity of subpopulations of primary nociceptive neurons in rats and mice. *Neuroscience.* 139: 699–709.

Li Y, Ji A, Weihe E, Schafer MK. 2004. Cell-specific expression and lipopolysaccharide-induced regulation of tumor necrosis factor alpha (TNF alpha) and TNF receptors in rat dorsal root ganglion. *J Neurosci.* 24: 9623–9631.

Light AR, Hughen RW, Zhang J, Rainier J, Liu Z, Lee J. 2008. Dorsal root ganglion neurons innervating skeletal muscle respond to physiological combinations of protons, ATP, and lactate mediated by ASIC, P2X, and TRPV1. *J Neurophysiol.* 100: 1184–1201.

Light AR, White AT, Hughen RW, Light KC. Moderate exercise increases expression for sensory, adrenergic, and immune genes in chronic fatigue syndrome patients, but not in normal subjects. *J Pain* epub July 30.

Light KC, Bragdon EE, Grewen KM, Brownley KA. 2009b. Adrenergic dysregulation and pain with and without acute beta-blockade in women with fibromyalgia and temporomandibular disorder. *J Pain.* 10: 452–552.

Mamet J, Lazdunski M, Voilley N. 2003. How nerve growth factor drives physiological and inflammatory expressions of acid-sensing ion channel 3 in sensory neurons. *J Biol Chem.* 278: 48907–48913.

Mathew SJ, Mao X, Keegan KA, Levine SM, Smith EL, Heier LA, Otcheretko V, Coplan JD, Shungu DC. 2009. Ventricular cerebrospinal fluid lactate is increased in chronic fatigue syndrome compared with generalized anxiety disorder: an in vivo 3.0 T (1)H MRS imaging study. *NMR Biomed.* 22 (3): 251–258.

Mense S. 2009. Algesic agents exciting muscle nociceptors. *Exp Brain Res.* 196:89–100.

Mense S, Meyer H. 1985. Different types of slowly conducting afferent units in cat skeletal muscle and tendon. *J Physiol (Lond).* 363: 403–417.

Micevych PE, Kruger L. 1992. The status of calcitonin gene-related peptide as an effector peptide. *Ann N Y Acad Sci.* 657: 379–396.

Molliver DC, Immke DC, Fierro L, Pare M, Rice FL, McCleskey EW. 2005. ASIC3, an acid-sensing ion channel, is expressed in metaboreceptive sensory neurons. *Mol Pain.* 1: 35.

Mosso A. (1904) *Fatigue.*

Müller J. (1840) Handbuch der Physiologie des Menschen. Koblenz, Germany: J. Holscher.

Naves LA, McCleskey EW. 2005. An acid-sensing ion channel that detects ischemic pain. *Braz J Med Biol Res.* 38: 1561–1569.

NINDS. 2001. Restless Legs Syndrome Fact Sheet. *NIH Publication No 01-4847.* www.ninds. nih.gov/disorders/restless_legs/detail_restless_legs.htm

Noakes TD. 2007. The central governor model of exercise regulation applied to the marathon. *Sports Med.* 37: 374–377.

Noakes TD, St Clair GA, Lambert EV. 2005. From catastrophe to complexity: a novel model of integrative central neural regulation of effort and fatigue during exercise in humans: summary and conclusions. *Br J Sports Med.* 39: 120–124.

O'Halloran KD, Perl ER. 1997. Effects of partial nerve injury on the responses of C-fiber polymodal nociceptors to adrenergic agonists. *Brain Res.* 759: 233–240.

Ohara K, Nagai M, Suzuki Y, Ohara K. 1998. Low activity allele of catechol-o-methyltransferase gene and Japanese unipolar depression. *Neuroreport.* 9: 1305–1308.

Page AJ, Brierley SM, Martin CM, Price MP, Symonds E, Butler R, Wemmie JA, Blackshaw LA. 2005. Different contributions of ASIC channels 1a, 2, and 3 in gastrointestinal mechanosensory function. *Gut.* 54: 1408–1415.

Pall ML, Anderson JH. 2004. The vanilloid receptor as a putative target of diverse chemicals in multiple chemical sensitivity. *Arch Environ Health.* 59: 363–375.

Papka RE, McNeill DL. 1993. Light- and electron-microscopic study of synaptic connections in the paracervical ganglion of the female rat: special reference to calcitonin gene-related peptide-, galanin- and tachykinin (substance P and neurokinin A)-immunoreactive nerve fibers and terminals. *Cell Tissue Res.* 271: 417–428.

Peng BG, Ahmad S, Chen S, Chen P, Price MP, Lin X. 2004. Acid-sensing ion channel 2 contributes a major component to acid-evoked excitatory responses in spiral ganglion neurons and plays a role in noise susceptibility of mice. *J Neurosci.* 24: 10167–10175.

Perl ER. 1971. Is pain a specific sensation. *J Psychiatr Res.* 8: 273–287.

Perl ER. 1998. Getting a line on pain: is it mediated by dedicated pathways? *Nat Neurosci.* 1: 177–178.

Perl ER. 2007. Ideas about pain, a historical view. *Nat Rev Neurosci.* 8: 71–80.

Reeves WC, Jones JF, Maloney E, Heim C, Hoaglin DC, Boneva RS, Morrissey M, Devlin R. 2007. Prevalence of chronic fatigue syndrome in metropolitan, urban, and rural Georgia. *Popul Health Metr.* 5: 5.

Reinohl J, Hoheisel U, Unger T, Mense S. 2003. Adenosine triphosphate as a stimulant for nociceptive and non-nociceptive muscle group IV receptors in the rat. *Neurosci Lett.* 338: 25–28.

Rybicki KJ, Waldrop TG, Kaufman MP. 1985. Increasing gracilis muscle interstitial potassium concentrations stimulate group III and IV afferents. *J Appl Physiol.* 58: 936–941.

Sato J, Perl ER. 1991. Adrenergic excitation of cutaneous pain receptors induced by peripheral nerve injury. *Science.* 251: 1608–1610.

Schmaling KB, Lewis DH, Fiedelak JI, Mahurin R, Buchwald DS. 2003. Single-photon emission computerized tomography and neurocognitive function in patients with chronic fatigue syndrome. *Psychosom Med.* 65: 129–136.

Silverman JD, Kruger L. 1989. Calcitonin-gene-related-peptide-immunoreactive innervation of the rat head with emphasis on specialized sensory structures. *J Comp Neurol.* 280: 303–330.

Sites BD, Beach M, Biggs R, Rohan C, Wiley C, Rassias A, Gregory J, Fanciullo G. 2003. Intrathecal clonidine added to a bupivacaine-morphine spinal anesthetic improves postoperative analgesia for total knee arthroplasty. *Anesth Analg.* 96: 1083–8, table.

St Clair GA, Noakes TD. 2004. Evidence for complex system integration and dynamic neural regulation of skeletal muscle recruitment during exercise in humans. *Br J Sports Med.* 38: 797–806.

Staud R, Cannon RC, Mauderli AP, Robinson ME, Price DD, Vierck CJ, Jr. 2003. Temporal summation of pain from mechanical stimulation of muscle tissue in normal controls and subjects with fibromyalgia syndrome. *Pain.* 102: 87–95.

Sugiura T, Bielefeldt K, Gebhart GF. 2007. Mouse colon sensory neurons detect extracellular acidosis via TRPV1. *Am J Physiol Cell Physiol.* 292: C1768–C1774.

Sugiura T, Kasai M, Katsuya H, Mizumura K. 2003. Thermal properties of acid-induced depolarization in cultured rat small primary afferent neurons. *Neurosci Lett.* 350: 109–112.

Takahashi K, Taguchi T, Itoh K, Okada K, Kawakita K, Mizumura K. 2005. Influence of surface anesthesia on the pressure pain threshold measured with different-sized probes. *Somatosens Mot Res.* 22: 299–305.

Toth A, Boczan J, Kedei N, Lizanecz E, Bagi Z, Papp Z, Edes I, Csiba L, Blumberg PM. 2005. Expression and distribution of vanilloid receptor 1 (TRPV1) in the adult rat brain. *Brain Res Mol Brain Res.* 135: 162–168.

Tsai SJ, Gau YT, Hong CJ, Liou YJ, Yu YW, Chen TJ. 2009. Sexually dimorphic effect of catechol-O-methyltransferase val158met polymorphism on clinical response to fluoxetine in major depressive patients. *J Affect Disord.* 113: 183–187.

Vierck CJ, Jr. 2006. Mechanisms underlying development of spatially distributed chronic pain (fibromyalgia). *Pain.* 124: 242–263.

Voilley N, de Weille J, Mamet J, Lazdunski M. 2001. Nonsteroid anti-inflammatory drugs inhibit both the activity and the inflammation-induced expression of acid-sensing ion channels in nociceptors. *J Neurosci.* 21: 8026–8033.

Waldmann R, Bassilana F, de Weille J, Champigny G, Heurteaux C, Lazdunski M. 1997a. Molecular cloning of a non-inactivating proton-gated Na+ channel specific for sensory neurons. *J Biol Chem.* 272: 20975–20978.

Waldmann R, Champigny G, Bassilana F, Heurteaux C, Lazdunski M. 1997b. A proton-gated cation channel involved in acid-sensing. *Nature.* 386: 173–177.

Wemmie JA, Askwith CC, Lamani E, Cassell MD, Freeman JH, Jr., Welsh MJ. 2003. Acid-sensing ion channel 1 is localized in brain regions with high synaptic density and contributes to fear conditioning. *J Neurosci.* 23: 5496–5502.

Wemmie JA, Coryell MW, Askwith CC, Lamani E, Leonard AS, Sigmund CD, Welsh MJ. 2004. Overexpression of acid-sensing ion channel 1a in transgenic mice increases acquired fear-related behavior. *Proc Natl Acad Sci U S A.* 101: 3621–3626.

Williamson JW, Nobrega AC, McColl R, Mathews D, Winchester P, Friberg L, Mitchell JH. 1997. Activation of the insular cortex during dynamic exercise in humans. *J Physiol.* 503 (Pt 2): 277–283.

Wilson LB, Andrew D, Craig AD. 2002. Activation of spinobulbar lamina I neurons by static muscle contraction. *J Neurophysiol.* 87: 1641–1645.

Wright-Browne V, Schnee AM, Jenkins MA, Thall PF, Aggarwal BB, Talpaz M, Estrov Z. 1998. Serum cytokine levels in infectious mononucleosis at diagnosis and convalescence. *Leuk Lymphoma.* 30: 583–589.

Wyller VB, Thaulow E, Amlie JP. 2007. Treatment of chronic fatigue and orthostatic intolerance with propranolol. *J Pediatr.* 150: 654–655.

Yagi J, Wenk HN, Naves LA, McCleskey EW. 2006. Sustained currents through ASIC3 ion channels at the modest pH changes that occur during myocardial ischemia. *Circ Res.* 99: 501–509.

Yoshiuchi K, Farkas J, Natelson BH. 2006. Patients with chronic fatigue syndrome have reduced absolute cortical blood flow. *Clin Physiol Funct Imaging.* 26: 83–86.

12 Reflex Autonomic Responses Evoked by Group III and IV Muscle Afferents

Jennifer L. McCord and Marc P. Kaufman

CONTENTS

12.1 INTRODUCTION

Exercise is well known to increase mean arterial pressure, heart rate, and ventilation, effects caused, in part, by a reflex arising from contracting skeletal muscles. This phenomenon has been named the exercise pressor reflex (Mitchell, Kaufman, and Iwamoto 1983) and is thought to relay information to the central nervous system regarding the metabolic state and the mechanical activity of the exercising muscles (Hayes and Kaufman 2001; Kaufman and Forster 1996). The afferent arm of the exercise pressor reflex arc is composed of thinly myelinated group III and non-myelinated group IV muscle afferents. Group III afferents primarily transmit information about mechanical stimuli arising in the exercising muscles, whereas the group IV afferents primarily transmit information about metabolic stimuli (Hayes and Kaufman 2001; Kaufman and Forster 1996). Group III and IV muscle afferents are also thought to be activated by nociceptive stimuli and are likely the sole source of pain from skeletal and cardiac muscle. The aim of this account is to correlate what is known about the afferent arm of the exercise pressor reflex and to what extent it is evoked by nociceptive versus non-nociceptive stimulation of skeletal muscle.

12.2 MUSCLE AFFERENTS

Before discussing the role played by muscle afferents in evoking the sensation of pain, we need to provide some basic information about the sensory innervation of limb skeletal muscles and the discharge properties of their afferents. Limb skeletal muscle is innervated by five types of afferents. Group I is subdivided into Ia and Ib. These afferents are thickly myelinated and conduct impulses between 72 and 120 m/s in cats and dogs. Group II afferents are also thickly myelinated and conduct impulses between 31 and 71 m/s in cats and dogs. Group Ia and II afferents innervate muscle spindles, and group Ib afferent innervate Golgi tendon organs. Group III afferents, also known as Aδ fibers, are thinly myelinated and conduct impulses between 2.5 and 30.0 m/s in cats and dogs. Group IV afferents, also known as C-fibers, are unmyelinated and conduct impulses at less than 2.5 m/s in cats and dogs. Group III and IV afferents have free nerve endings often within the connective tissue of skeletal muscle, whereas group IV afferents have free nerve endings often within small vessels of muscle (von During and Andres 1990).

Group Ia and Ib as well as group II afferents do not contribute to cardiovascular response to exercise, nor are they responsible for the evoking the sensation of pain. McCloskey and Mitchell (1972) found that the reflex ventilatory and cardiovascular increases evoked by static contraction were caused by stimulation of group III and IV muscle afferents. Specifically, anodal blockade of the dorsal roots prevented impulses from group I and II afferents from reaching the spinal cord, but did not prevent the contraction-induced increases in ventilatory and cardiovascular function. On the other hand, topical application of a local anesthetic (which blocks group III and IV fibers before blocking group I and II fibers) to the dorsal roots did not block impulse conduction in group I and II afferents, but did block the contraction-induced cardiovascular increases and reduced the ventilatory increases. In subsequent studies in dogs, Tibes (1977), using cold blockade, also reported that group III and IV muscle afferents, but not group I and II afferents, were responsible for evoking the exercise pressor reflex.

There is additional evidence that groups III and IV, but not groups I or II, cause the reflex cardiovascular and ventilatory increases evoked by contraction of hindlimb skeletal muscle. For example, electrical stimulation of the central cut end of the medial and lateral gastrocnemius nerves did not increase either ventilation or arterial pressure until the current intensity recruited group III fibers (Mizumura and Kumazawa 1976; Sato, Sato, and Schmidt 1981; Tibes 1977). Also, activation of group Ia and II spindle afferents by longitudinal vibration, a strong stimulus for these afferents, had almost no effect on ventilation, arterial pressure, or heart rate (Hodgson and Matthews 1968; McCloskey, Matthews, and Mitchell 1972). On the other hand, chemical activation of group III and IV afferents has been shown to increase ventilation, arterial pressure, and heart rate (Crayton, Mitchell, and Payne, III 1981; Kaufman and Forster 1996; Tibes 1977), whereas chemical activation of group Ia and II afferents had no effect on these variables (Hodgson and Matthews 1968; Waldrop, Rybicki, and Kaufman 1984).

Group III and IV afferents, not group I or II, are also the afferents that are responsible for evoking the sensation of pain. Close arterial injections of bradykinin into the triceps surae muscle, in concentrations that were considered painful in man and animals, increased only the activity of group III and IV muscle afferents. In fact, group I and II afferents exhibited no change and even decreased their firing rates. On the other hand, the injections of bradykinin increased the firing rate of over two-thirds of the group III afferents tested and over 50% of group IV afferents tested (Mense 1977). In agreement with this animal study, a human study also found that group III and IV muscle afferents were responsible for evoking the sensation of pain within skeletal muscle (Graven-Nielsen, Mense, and rendt-Nielsen 2004). When these authors anesthetized skin as well as differentially nerve-blocked group I and II afferents within skeletal muscle, the subjects could still feel a painful pressure stimulus.

12.3   DISCHARGE PROPERTIES OF GROUP III AND IV AFFERENTS

One of the first to investigate the discharge properties of group III afferents was Paintal, who found that group III endings located in the hindlimb muscles of dogs responded to pressure applied to the muscle (Paintal 1960). Since then several investigators have independently reached similar conclusions about the discharge properties of these thinly myelinated afferents. These are (1) that about half responded to contraction, be it intermittent tetanic or maintained tetanic (static) contraction (Ellaway, Murphy, and Tripathi 1982; Kaufman et al. 1983; Mense and Stahnke 1983; Paintal 1960); (2) that about half responded to intra-arterial injection of bradykinin, a potent algesic agent (Kaufman et al. 1983; Kumazawa and Mizumura 1977; Mense 1977); (3) that many responded to non-noxious punctate pressure applied to their receptive fields (Hayward, Wesselmann, and Rymer 1991; Kaufman et al. 1983, 1984a; Kumazawa and Mizumura 1977; Mense and Stahnke 1983; Paintal 1960); and (4) that their responses to tendon stretch was variable, some investigators finding a frequent effect (Abrahams, Lynn, and Richmond 1984), others finding a moderate effect (Hayward, Wesselmann, and Rymer 1991; Kaufman et al. 1983; Kaufman and Rybicki 1987; Mense and Stahnke 1983), and still another finding an infrequent effect (Paintal 1960).

Group III muscle afferents are believed to possess polymodal discharge properties because they respond to both chemical and mechanical stimuli (Kumazawa and Mizumura 1977). This belief, however, is controversial because it is based on the responses of group III afferents to large and possibly unphysiological doses of bradykinin (Mense and Meyer 1985). Nevertheless, the mechanical sensitivity of group III afferents might be their most relevant discharge property when assessing their contribution to evoking reflex autonomic adjustments to exercise. Group III muscle afferents, for example, respond vigorously at the onset of tetanic contraction, with the first impulse often being discharged within 200 milliseconds of the start of this maneuver (Kaufman et al. 1983). Moreover, group III afferents increase their responses to tetanic contractions as the peak tension developed by the working muscle increases (Hayward, Wesselmann, and Rymer 1991; Kaufman et al. 1983; Mense and Stahnke 1983). In addition, group III afferents usually decrease their discharge rate during a static contraction as the working muscle fatigues (Hayward, Wesselmann, and Rymer 1991; Kaufman et al. 1983). Further, group III afferents are capable of synchronizing their discharge with a constantly oscillating stimulus, be it either 5 Hz twitch contraction (Kaufman et al. 1984b), intermittent tetanic contraction (Mense and Stahnke 1983), or a true form of dynamic exercise evoked by stimulation of the mesencephalic locomotor region (Adreani, Hill, and Kaufman 1997; Pickar, Hill, and Kaufman 1994). Lastly, gadolinium has been shown to block the mechanical sensitivity of group III afferents (Hayes et al. 2009; Hayes and Kaufman 2001) in the triceps surae muscles of cats.

Group III afferents respond to the metabolic stimuli possibly through direct stimulation or sensitization of their endings. For example, they are stimulated by bradykinin (Kaufman et al. 1983; Kumazawa and Mizumura 1977; Mense 1977), potassium (Hník et al. 1986; Kumazawa and Mizumura 1977; Mense 1977; Rybicki, Waldrop, and Kaufman 1985; Thimm and Baum 1987), arachidonic acid (Rotto and Kaufman 1988), and lactic acid (Rotto and Kaufman 1988; Sinoway et al. 1993; Thimm and Baum 1987). In addition, bradykinin and arachidonic acid increased the responsiveness of group III fibers to contraction of the triceps surae muscles (Mense and Meyer 1988; Rotto et al. 1990).

Group IV afferents respond minimally to mechanical stimuli compared with group III afferents and are much more responsive to metabolic stimuli. In fact, these afferents possess different discharge properties than do group III muscle afferents. These differences are (1) group IV afferents usually display obvious responses to contraction with latencies of 5–30 seconds, whereas group III afferents often respond vigorously within 30–200 milliseconds (Kaufman et al. 1983; Kaufman et al. 1984b; Mense and Stahnke 1983); (2) their responses to dynamic exercise are not attenuated by gadolinium, a mechanogated channel blocker, whereas the responses of group III afferents are blocked by this agent (Hayes et al. 2008a); (3) when compared to contraction while the muscle was freely perfused, contraction while the muscle was ischemic increased the responses of almost half the group IV afferents tested but increased the responses of only 12% of group III afferents tested, indicating responsiveness to metabolites in the former, but much less in the later group (Kaufman et al. 1984a); (4) about half responded to static contraction while the muscles were ischemic but did not respond to a static contraction of equal magnitude when the muscles were freely perfused (Kaufman et al. 1984a; Mense and Stahnke 1983); (5) group IV afferents are much less sensitive than are group III afferents to mechanical stimuli (either probing their receptive fields or to tendon stretch). In contrast to group III afferents, group IV afferents most often require noxious pinching of their receptive fields to discharge them, and frequently do not respond to noxious tendon stretch (Kaufman et al. 1984a; Kaufman and Rybicki 1987; Kniffki, Mense, and Schmidt 1978; Mense and Stahnke 1983).

Group IV afferents are stimulated by the same substances that have been shown to stimulate and/or sensitize group III afferents to other stimuli, such as contraction or probing of their receptive fields. Hence, intra-arterial injection of bradykinin (Kaufman et al. 1983; Mense 1977; Mense and Schmidt 1974), potassium (Hník et al. 1969; Kaufman and Rybicki 1987; Mense 1977), lactic acid (Graham et al. 1986; Rotto and Kaufman 1988; Thimm and Baum 1987), and arachidonic acid (Rotto and Kaufman 1988) have been shown to stimulate at least half of the group IV afferents tested. In addition, other substances stimulate group IV afferents, such as prostaglandin E2 (Mense 1981) and ATP (Hanna, Hayes, and Kaufman 2002; Hanna and

Kaufman 2004; Reinöhl et al. 2003). Histamine and serotonin stimulated only a few group IV afferents and the threshold doses appeared to be quite high (Kniffki, Mense, and Schmidt 1978; Mense 1977).

12.4 METABOLITES RELEASED DURING MUSCULAR CONTRACTION AND PAIN

The evidence concerning ATP, working through purinergic 2 (P2) receptors, as a substance that evokes the exercise pressor reflex is strong. Specifically, injection of P2 agonists into the arterial supply or directly into the gastrocnemius muscle stimulated over two-thirds of the group IV afferents tested in cats (Hanna and Kaufman 2004) and rats (Reinöhl et al. 2003). In addition, close arterial injection of P2 agonists into the vascular supply of feline hindlimb muscle has been shown to evoke a reflex pressor response (Hanna, Hayes, and Kaufman 2002). Perhaps even more compelling evidence is that the pressor response to static exercise was attenuated by injection of PPADS, a P2 receptor antagonist (Hanna and Kaufman 2003; Kindig, Hayes, and Kaufman 2007). Further, blockade of P2 receptors by PPADS attenuated the responses of both group III and IV afferents to static contraction, both while the muscles were freely perfused and while they were ischemic. Likewise, PPADS attenuated the responses of group IV afferents to post contraction ischemia (Kindig et al. 2006; Kindig, Hayes, and Kaufman 2007). All these findings suggest that ATP activates P2 receptors on the endings of group III and IV thin fiber afferents during exercise and that these receptors contribute to the exercise pressor reflex by sensitizing group III mechanoreceptors and stimulating group IV metaboreceptors.

Of the P2 receptor subtypes, the P2X receptor on group IV afferents is thought to evoke the muscle chemoreflex (Chen et al. 1995; Cook et al. 1997; Hayes, McCord, and Kaufman 2008; Lewis et al. 1995). In addition, these receptors respond to levels of ATP that are released in muscle in response to stress (Cook and McCleskey 2002). Lastly, injections of ATP and a P2X receptor agonist, alpha, beta-methylene ATP, into the rat hindpaw increased behavioral indexes of pain (i.e., hindpaw licking, lifting, and flinching) (Bland-Ward and Humphrey 1997; Sawynok and Reid 1997), effects that were prevented by P2X receptor blockade (Bland-Ward and Humphrey 1997; Sawynok and Reid 1997). Therefore, it seems likely that ATP plays a role in the pressor response and the sensation of muscle pain.

Arachidonic acid and its cyclooxygenase products appear to play a role in evoking the exercise pressor reflex. Concentrations of arachidonic acid and PGE2 in the freely perfused gastrocnemius muscles of cats increased significantly when these muscles were contracted. When the muscles were contracted under ischemic conditions, arachidonic acid and PGE2 concentrations rose to an even higher level (Rotto et al. 1989; Symons et al. 1991). Most importantly, the pressor, cardioaccelerator, and cardiac contractility responses to static contraction of the hindlimb muscles of anesthetized cats were significantly attenuated by indomethacin as well as by sodium meclofenamate, agents that block the activity of cyclooxygenase. Further, blood flow to the hindlimb muscles during static contraction was found to be almost the same both before and after indomethacin (Stebbins, Maruoka, and Longhurst 1988). These findings led to the speculation that prostaglandins must be acting on the endings of thin fiber afferents instead of through a regional vascular effect. In a subsequent study, the responses of group III and IV afferents to dynamic exercise, performed under both freely perfused and ischemic conditions, were attenuated by blockade of cyclooxygenase. In addition, during a postexercise ischemic period, cyclooxygenase blockade decreased the discharge of group IV afferents (Hayes, Kindig, and Kaufman 2006).

Nonsteroidal anti-inflammatory agents (NSAIDs) may have actions on acid-sensing ion channels (ASICs) that are independent of their cyclooxygenase activity. For example, arachidonic acid has been shown to potentiate currents passed through ASICs in dorsal root ganglion neurons (Smith, Cadiou, and McNaughton 2007). Moreover, the potentiation is not prevented by inhibition of arachidonic acid metabolism (Smith, Cadiou, and McNaughton 2007). Additional studies would need to determine if the activation of group III and IV afferents by arachidonic acid metabolites during exercise work through prostaglandin receptors or through effects on ASICs. Nonetheless, the attenuation of the pressor response (Stebbins, Maruoka, and Longhurst 1988) and discharge rate of group III and IV afferents (Hayes, Kindig, and Kaufman 2006) during exercise was achieved by cycolooxygenase blockade with indomethacin, an agent, unlike other NSAIDs, that did not inhibit current passing through ASICs (Voilley et al. 2001). This evidence suggests that cyclooxygenase products activate group III and IV afferents during exercise, and prostaglandin receptors contribute to the stimulation of mechanoreceptors and metaboreceptors evoking the exercise pressor reflex.

Bradykinin has also been suggested to play a role in the exercise pressor reflex. Muscle contraction increased levels of bradykinin within skeletal muscle (Stebbins et al. 1990). Moreover, bradykinin injected into the arterial supply of muscle reflexly increased mean arterial pressure, a response that was attenuated by a bradykinin 2 receptor antagonist. A bradykinin 1 receptor agonist did not evoke a reflex pressor response (Pan, Stebbins, and Longhurst 1993). Inflammation and trauma can increase the concentrations of cyclooxygenase products of arachidonic acid, such as PGE2, and bradykinin in skeletal muscle. PGE2 not only stimulated group IV afferents but in concentrations less than that needed for stimulation, PGE2 enhanced the excitatory action of bradykinin on thin fiber afferents (Mense 1981). Bradykinin is also released by noxious stimulation and in greater amounts during an ischemic than in a freely perfused contraction (Stebbins et al. 1990).

Several lines of evidence suggest that lactic acid plays a role in the exercise pressor reflex. First, lactic acid injected into the arterial supply of muscle stimulates groups III and IV afferents as well as evokes a reflex pressor response (Rotto and Kaufman 1988; Rotto, Stebbins, and Kaufman 1989; Sinoway et al. 1993). Second, lactic acid concentrations in the muscle interstitium are increased by contraction (MacLean et al. 1998). Third, blunting lactic acid production by either dichloroacetate or glycogen depletion decreased the exercise pressor reflex (Ettinger et al. 1991; Sinoway et al. 1992). Blockade of receptors to lactic acid provided additional evidence for lactic acid's involvement in evoking the exercise pressor reflex. Lactic acid activates two receptors, transient receptor potential vanilloid type 1 channels (TRPV1) and acid sensing ion channels (ASICs). Whereas TRPV1 receptors were found not to play a role in the exercise pressor reflex (Kindig, Heller, and Kaufman 2005), two structurally different ASIC antagonists (amiloride and A-317567) attenuated the reflex (Hayes et al. 2008; Hayes, Kindig, and Kaufman 2007; McCord, Hayes, and Kaufman 2008). All this evidence suggests that lactic acid, working through ASICs, plays a role in the exercise pressor reflex.

A strong role for lactic acid working through ASICs has also been found in the sensation of pain, especially that caused by ischemia. Of the four ASICs, ASIC3 seems to be the most likely channel responding to muscle stress (e.g., exercise) and pain. First, ASIC3 was found to be expressed almost exclusively in dorsal root ganglion neurons (Krishtal, Marchenko, and Pidoplichko 1983; Waldmann et al. 1997, 1999) and in very high levels on metaboreceptors (Benson, Eckert, and McCleskey 1999; Molliver et al. 2005; Sutherland et al. 2001). In addition, ASIC3 knockout mice did not respond to noxious stimuli, whereas their wild type counterparts did (Price et al. 2001). Moreover, ASIC3 channels open when the pH drops from 7.4 to 7.0 (Sutherland et al. 2001) and can pass a sustained current at a pH of 7.0 (Yagi et al. 2006); this pH is consistent with that found during ischemic exercise (Cornett et al. 2000; Sinoway et al. 1989) and ischemic pain (Cobbe and Poole-Wilson 1980; Remme et al. 1986). Further, the responsiveness of this channel was greatly enhanced when lactate and/or ATP were present, two molecules that are released during exercise and pain (Immke and McCleskey 2001).

During pain, lactic acid has been suggested to stimulate TRPV1. However, pH must reach a level of less than 6.0 to activate TRPV1, a value that is below that occurring during exercise or ischemic pain except in extreme circumstances. Not surprisingly, during an infusion of a solution with a pH of 6.0 directly into human skin, the intensity of the pain was not diminished by blockade of TRPV1 receptors (Ugawa et al. 2002), suggesting that another receptor such as ASICs were responsible of the sensation of pain at this pH level. Nonetheless, TRPV1 receptors were found to work synergistically along with ASICs and P2X receptors to respond to the metabolites produced during muscle contraction and pain (Light et al. 2008).

The data presented thus far suggests that blockade of either the receptor or the production of a metabolite reduces the exercise pressor reflex by half. One wonders how all of these substances can be responsible for the reflex when adding the individual magnitudes of the reduction in the reflex far exceeds 100%. An answer to this question comes from two lines of evidence suggesting that the combination of metabolites is the key to activation of group III and IV afferents during exercise and nociceptive stimulation of muscle. The first line of evidence offered by McCleskey and colleagues found that lactate and ATP greatly potentiated the effect of an acidic pH (~7.0–6.8) on the activation of ASIC3. When ASIC3 was activated by a pH of 7.0 the current was 80% greater in the presence of physiological levels of lactate than when lactate was not present (Immke and McCleskey 2001). Further, it was found that not only does lactate act immediately but it must be present in the interstitium for the ASIC current to be augmented because lactate acts by chelating calcium, allowing the ASIC channel react to protons (Immke and McCleskey 2003). In addition, ATP, in concentrations seen in an ischemic contraction, increased current through ASIC3; moreover, the current through the channel was still high for minutes after the ATP was removed (Naves and McCleskey 2005). The enhanced current from ATP required 15s–1 minute after ATP application for peak effect (Naves and McCleskey 2005).

The second line of evidence by Light and colleagues found that combinations of protons, lactate, and ATP were needed to activate cultured dorsal root ganglion cells. When the cells were exposed to just one of the metabolites, only a small stimulatory effect was measured, whereas the combination of all three metabolites had an effect that exceeded the summation of each one individually (Light et al. 2008). This finding also explains why half of the exercise pressor reflex was attenuated by blockade of one receptor. The conclusion from these investigations provides evidence that not just one metabolite or receptors can be the sole contributor to the reflex, but combinations of metabolites act synergistically on two or more receptors for the full expression of the reflex.

12.5   AFFERENT PATHWAY

The prior sections have described the discharge properties of group III and IV afferents and the metabolites that stimulate them during exercise and pain. The findings show that stimulation of the same afferents can evoke two effects, one resulting in cardiovascular reflex adjustments during exercise and one resulting in the sensation of pain. How can the same afferent endings cause both effects? It is possible that group III and IV receptors are separated into subclasses; that is, some afferents respond to non-noxious stimuli while other afferents respond to noxious stimuli (Mense and Meyer 1985). In agreement, Light et al. (2008) found two populations of metaboreceptors in dorsal root ganglion neurons. They found one population of neurons responded best to low metabolite levels, in concentrations that would be considered consistent with non-painful contractions. The second population of neurons responded best to high levels of metabolites that would be consistent with either ischemic contractions or tissue damage. Thus, during ischemic contraction two different populations of metaboreceptors would respond, one population to contraction and the other to ischemia.

Noxious stimuli can increase the number of dorsal horn cells responding to electrical stimulation. Following a noxious lesion to the gastrocnemius muscle in anesthetized rats, only two hours were needed to see an expansion of the spinal input from the inflamed area. The result was that 30% of neurons within the L3 spinal segment, which receives little input from the gastrocnemius muscles of control rats, responded to electrical stimulation of nerves supplying inflamed muscle, whereas no neurons in the L3 region of rats without inflamed muscles responded to this stimulation (Hoheisel, Koch, and Mense 1994).

Group III and IV afferents synapse onto laminae I and V neurons in the dorsal horn of the spinal cord (Mense and Craig 1988; Willis, Jr. et al. 2001), with the majority terminating in lamina 1 (Craig, Heppelmann, and Schaible 1988; Craig and Mense 1983). Lamina I neurons transmit information from group III and IV afferents to the brain through three relevant pathways. First, projections have been found from lamina I neurons to the posterior part of the ventromedial nucleus of the thalamus, a center for pain and temperature sensation (Craig et al. 1994). Second, projections have been found from lamina I neurons to the rostral ventrolateral medulla, an area that contains sympathetic premotor neurons (Craig 1995). Third, lamina I neurons project to the nucleus tractus solitarius, an area that receives baroreceptor information (Craig 1995).

Iwamoto et al. (1985) completed serial sectioning of the neuraxis to determine what neural structures were necessary to evoke the exercise pressor reflex. Their critical finding was that an intact medulla was needed to evoke the exercise pressor reflex. Further, it has been shown that input from skeletal muscle can excite neurons within the nucleus tractus solitarius, rostral ventrolateral medulla, caudal ventrolateral medulla, the lateral tegmental field, and the ventromedial region of the rostral periaqueductal grey (Ciriello and Calaresu 1977; Iwamoto et al. 1982; Iwamoto and Kaufman 1987; Li et al. 1997; Li and Mitchell 2000; Person 1989).

12.6   SYMPATHETIC ACTIVATION

The sympathetic nervous system is a two-neuron chain and is comprised of a preganglionic neuron, whose axon is usually myelinated, and an unmyelinated postganglionic neuron. The terminal ending of the preganglionic axon releases acetylcholine, which in turn activates a nicotinic receptor on the postganglionic dendrites or cell body. The postganglionic terminal usually releases norepinephrine onto the end organ. In the limbs, stimulation of sympathetic postganglionic fibers has vasomotor, pilomotor, and sudomotor functions. In the kidneys, stimulation of sympathetic postganglionic fibers evokes renin secretion, vasoconstriction, and sodium reabsorption. Likewise in the heart, stimulation of sympathetic postganglionic fibers increases cardiac rate and contractility (via $\beta1$-adrenergic receptors) and causes coronary vascular smooth muscle to constrict (via $\alpha1$-adrenergic receptors).

In humans, the exercise pressor reflex has been shown to increase sympathetic nerve activity to non-exercising muscle (Hill, Adreani, and Kaufman 1996; Mark et al. 1985; Saito 1995; Saito, Naito, and Mano 1990). For example, Victor and colleagues found that attempted exercise in humans that were paralyzed partially by curare yielded a trivial increase in muscle sympathetic nerve activity to non-exercising muscles compared with the increase that was evoked by actual handgrip performed before curare (Victor et al. 1989a). In addition, the exercise pressor reflex has been shown to play a key role in causing renal vasoconstriction, a sympathetically mediated effect that was attributed to group III mechanoreceptors (Momen et al. 2003).

The exercise pressor reflex has been shown to increase whole renal nerve activity in both chloralose-anesthetized and decerebrated cats. Static contraction of the hindlimb muscles tripled renal nerve discharge, an increase that was shown to be reflex in origin because it was prevented by section of the dorsal roots innervating the hindlimb (Victor et al. 1989b). The reflex increase in renal nerve discharge was in part due to the stimulation of group III mechanoreceptors. Static contraction reflexly increased renal nerve activity with an onset latency that averaged less than one second (Matsukawa et al. 1990; Victor et al. 1989b). This brief latency appears to be too short to allow for the activation of group IV metaboreceptors in the contracting muscles. In addition, intermittent static contraction synchronized renal nerve discharge so that a burst of activity was evoked by each contraction. This synchronization appears best explained by activation of group III mechanoreceptors. In addition, the increase in renal sympathetic nerve activity during exercise was attenuated by blockade of group III mechanoreceptors with gadolinium (Kim et al. 2007).

While the list of substances that activate and/or sensitize mechanoreceptors is a work in progress, it does include P2 receptors (Kindig, Hayes, and Kaufman 2007) but does not normally include lactic acid working through ASICs (McCord, Hayes, and Kaufman 2008). Even though it is likely that metaboreceptors do not play a role in the first few seconds of the increased renal sympathetic nerve activity at the onset of exercise, metaboreceptors are thought to provide a continuous signal to keep sympathetic activity increased for the duration of exercise (Matsukawa et al. 1990).

The exercise pressor reflex has also been shown to play a role in the increase of sympathetic nerve activity to the heart. Specifically, the onset latency of the cardiac sympathetic nerve activity response to static contraction was found to always be less than one second and as a consequence was attributed to the stimulation of group III mechanoreceptors in contracting skeletal muscle (Matsukawa et al. 1994; Tsuchimochi et al. 2009).

Activation of the sympathetic nervous system by the exercise pressor reflex has important physiological effects in humans. In static exercise, sympathetically induced vasoconstriction in both the viscera (Middlekauff et al. 1997; Momen et al. 2003) and resting skeletal muscles (Victor et al. 1989a) shunts arterial blood to contracting muscles as well as maintains their perfusion pressure. There is ample evidence that muscle performance is directly related to its perfusion pressure in both humans and animals (Eiken 1987; Fitzpatrick, Taylor, and McCloskey 1996; Hobbs and McCloskey 1987). In dynamic exercise, sympathetic restraint of vasodilation in rhythmically contracting muscles is necessary to maintain arterial pressure during large muscle mass dynamic exercise (Pryor et al. 1990). For example, the vasodilation occurring in 20 kilograms of exercising muscle in humans would require a cardiac output of 50 liters per minute for arterial pressure to be maintained. The human heart is not capable of generating this output. Consequently, the maintenance of arterial pressure by sympathetic vasoconstriction assures the brain and the heart of adequate perfusion during dynamic exercise.

During static exercise the reflex sympathetic activation arising from the contraction-induced stimulation of group III and IV afferents is believed to increase arterial blood pressure to counter the mechanical compression of vessels in contracting muscles. Any vasodilation occurring in contracting muscles is believed to be caused by the production of metabolites, which act directly to relax vascular smooth muscle. An alternative view, however, is also possible. This view is that the vasodilation seen in exercising skeletal muscles is caused by vasodilator peptides, including substance P and CGRP, which are released by group IV afferents (Kruger et al. 1989), whose endings are located in small vessels (von During and Andres 1990). These unmyelinated afferents are stimulated by contraction and send impulses in two directions, the first being toward the dorsal horn where they synapse, and second in a retrograde fashion, where they relax vascular smooth muscle. This mechanism, which is often called the *axon reflex*, would only oppose sympathetic restraint of blood flow in exercising muscle. As a consequence, the combination of sympathetic restraint which would occur in all muscles, and the axon reflex, which would occur only in exercising muscles, could function to shunt blood flow to the muscles in need of an increased blood flow.

12.7   TRANSLATION TO HEART FAILURE

Patients with heart failure display augmented sympathetic nerve activity, vascular resistance, and blood pressure in response to exercise when compared to their healthy counterparts (Middlekauff et al. 2000, 2001; Ponikowski et al. 2001; Sterns et al. 1991). This exaggerated response is thought to be due to alterations in the exercise pressor reflex rather than to alterations in central command (Middlekauff et al. 2001). Further, involuntary biceps contraction, a maneuver that selectively stimulates mechanoreceptors, produced an augmented renal vasoconstrictor response in heart failure patients, suggesting that mechanoreceptors were sensitized (Middlekauff et al. 2001). In contrast, studies using post-contraction ischemia determined that metaboreceptors were not sensitized in humans (Middlekauff et al. 2000; Sterns et al. 1991). Therefore, evidence in humans indirectly suggests that the exaggerated sympathoexcitation of heart failure patients is due to sensitization of mechanoreceptors leading to an exaggerated exercise pressor reflex.

Heart failure studies in rats have shown that group III afferents are responsible for the exaggerated exercise pressor reflex. To determine the contribution of the mechanoreflex to the exercise pressor reflex, gadolinium, which blocks mechanogated channels, was injected into healthy control rats, heart failure rats, and healthy rats that had their group IV afferent neurons destroyed (Smith et al. 2005a). Healthy rats that have selective destruction of group IV afferent neurons exhibited the exaggerated exercise pressor reflex observed in heart failure (Smith et al. 2005b). Gadolinium reduced the pressor response to exercise in all three groups, but the magnitude of the reduction was greater in the heart failure and the group IV afferent-destroyed rats than in the healthy control rats (Smith et al. 2005a). These findings suggest that if group IV afferents are not present or desensitized, group III afferents will overcompensate, thus accounting for the augmented exercise pressor in the heart failure population.

REFERENCES

Abrahams, V. C., B. Lynn, and F. J. R. Richmond. 1984. Organization and sensory properties of small myelinated fibres in the dorsal cervical rami of the cat. *J.Physiol.* 347:177–187.

Adreani, C. M., J. M. Hill, and M. P. Kaufman. 1997. Responses of group III and IV muscle afferents to dynamic exercise. *J.Appl.Physiol.* 82:1811–1817.

Benson, C. J., S. P. Eckert, and E. W. McCleskey. 1999. Acid-evoked currents in cardiac sensory neurons: a possible mediator of myocardial ischemic sensation. *Circ.Res.* 84(8):921–928.

Bland-Ward, P. A., and P. P. A. Humphrey. 1997. Acute nociception mediated by hindpaw P2X receptor activation in the rat. *Br.J.Pharmacol.* 122:365–371.

Chen, C. C., A. N. Akopian, L. Sivilotti, D. Colquhoun, G. Burnstock, and J. N. Wood. 1995. A P2X purinoceptor expressed by a subset of sensory neurons. *Nature* 377:428–431.

Ciriello, J., and F. R. Calaresu. 1977. Lateral reticular nucleus: a site of somatic and cardiovascular integration in the cat. *Am.J.Physiol.* 233:R100–R109.

Cobbe, S. M., and P. A. Poole-Wilson. 1980. The time of onset and severity of acidosis in myocardial ischaemia. *J.Mol.Cell Cardiol.* 12(8):745–760.

Cook, S. P., and E. W. McCleskey. 2002. Cell damage excites nociceptors through release of cytosolic ATP. *Pain* 95(1–2):41–47.

Cook, S. P., L. Vulchanova, K. M. Hargreaves, R. Elde, and E. W. McCleskey. 1997. Distinct ATP receptors on pain-sensing and stretch-sensing neurons. *Nature* 387:505–508.

Cornett, J. A., M. D. Herr, K. S. Gray, M. B. Smith, Q. X. Yang, and L. I. Sinoway. 2000. Ischemic exercise and the muscle metaboreflex. *J.Appl.Physiol.* 89(4):1432–1436.

Craig, A. D. 1995. Distribution of brainstem projections from spinal lamina I neurons in the cat and the monkey. *J.Comp.Neurol.* 361:225–248.

Craig, A. D., M. C. Bushnell, E. T. Zhang, and A. Blomqvist. 1994. A thalamic nucleus specific for pain and temperature sensation. *Nature* 372(6508):770–773.

Craig, A. D., B. Heppelmann, and H. G. Schaible. 1988. The projection of the medial and posterior articular nerves of the cat's knee to the spinal cord. *J.Comp.Neurol.* 276(2):279–288.

Craig, A. D., and S. Mense. 1983. The distribution of afferent fibers from the gastrocnemius-soleus muscle in the dorsal horn of the cat as revealed by the transport of horseradish peroxidase. *Neurosci.Lett.* 41:233–238.

Crayton, S. C., J. H. Mitchell, and F. C. Payne, III. 1981. Reflex cardiovascular response during the injection of capsaicin into skeletal muscle. *Am.J.Physiol.* 240:H315–H319.

Eiken, O. 1987. Responses to dynamic leg exercise in man as influenced by changes in muscle perfusion pressure. *Acta.Physiol.Scand.* 131:1–37.

Ellaway, P. H., P. R. Murphy, and A. Tripathi. 1982. Closely coupled excitation of gamma-motoneurons by group III muscle afferents with low mechanical threshold in a cat. *J.Physiol.* 331:481–498.

Ettinger, S., K. Gray, S. Whisler, and L. Sinoway. 1991. Dichloroacetate reduces sympathetic nerve responses to static exercise. *Am.J.Physiol.* 261:H1653–H1658.

Fitzpatrick, R., J. L. Taylor, and D. I. McCloskey. 1996. Effects of arterial perfusion pressure on force production in working human hand muscles. *J.Physiol.* 495(3):885–891.

Graham, R., Y. Jammes, C. Delpierre, C. Grimaud, and Ch. Roussos. 1986. The effects of ischemia, lactic acid and hypertonic sodium chloride on phrenic afferent discharge during spontaneous diaphragmatic contraction. *Neurosci.Lett.* 67:257–262.

Graven-Nielsen, T., S. Mense, and L. rendt-Nielsen. 2004. Painful and non-painful pressure sensations from human skeletal muscle. *Exp.Brain Res.* 159(3):273–283.

Hanna, R. L., S. G. Hayes, and M. P. Kaufman. 2002. α,β–methylene ATP elicits a reflex pressor response arising from muscle in decerebrate cats. *J.Appl.Physiol.* 93:834–841.

Hanna, R. L., and M. P. Kaufman. 2003. Role played by purinergic receptors on muscle afferents in evoking the exercise pressor reflex. *J.Appl.Physiol.* 94:1437–1445.

———. 2004. Activation of thin-fiber muscle afferents by a P2X agonist in cats. *J.Appl. Physiol.* 96(3):1166–1169.

Hayes, S. G., and M. P. Kaufman. 2001. Gadolinium attenuates exercise pressor reflex in cats. *Am.J.Physiol. Heart Circ.Physiol.* 280:H2153–H2161.

Hayes, S. G., A. E. Kindig, and M. P. Kaufman. 2006. Cyclooxygenase blockade attenuates responses of group III and IV muscle afferents to dynamic exercise in cats. *Am.J.Physiol Heart Circ.Physiol.* 290(6):H2239–H2246.

———. 2007. Blockade of acid sensing ion channels attenuates the exercise pressor reflex in cats. *J.Physiol.* 581:1271–2323.

Hayes, S. G., J. L. McCord, and M. P. Kaufman. 2008. Role played by P2X and P2Y receptors in evoking the muscle chemoreflex. *J.Appl.Physiol.* 104(2):538–541.

Hayes, S. G., J. L. McCord, S. Koba, and M. P. Kaufman. 2009. Gadolinium inhibits group III but not group IV muscle afferent responses to dynamic exercise. *J.Physiol.* 587:873–882.

————. 2008a. Gadolinium inhibits group III but not group IV muscle afferent responses to dynamic exercise. *J.Physiol.*

Hayes, S. G., J. L. McCord, J. Rainier, Z. Liu, and M. P. Kaufman. 2008. Role played by acid-sensitive ion channels in evoking the exercise pressor reflex. *Am.J.Physiol. Heart Circ. Physiol.* 295(4):H1720–H1725.

Hayward, L., U. Wesselmann, and W. Z. Rymer. 1991. Effects of muscle fatigue on mechanically sensitive afferents of slow conduction velocity in the cat triceps surae. *J.Neurophysiol.* 65:360–370.

Hill, J. M., C. M. Adreani, and M. P. Kaufman. 1996. Muscle reflex stimulates sympathetic postganglionic efferents innervating triceps surae muscles of cats. *Am.J.Physiol.* 271:H38–H43.

Hník, P., O. Hudlická, J. Kucera, and R. Payne. 1969. Activation of muscle afferents by non-proprioceptive stimuli. *Am.J.Physiol.* 217:1451–1458.

Hník, P., F. Vyskocil, E. Ujec, R. Vejsada, and H. Rehfeldt. 1986. Work-induced potassium loss from skeletal muscles and its physiological implications. In *Biochemistry of Exercise VI*, ed. Saltin, B., 345–364. (Champaign, IL: Human Kinetics).

Hobbs, S. F., and D. I. McCloskey. 1987. Effects of blood pressure on force production in cat and human muscle. *J.Appl.Physiol.* 63:834–839.

Hodgson, H. J. F., and P. B. C. Matthews. 1968. The ineffectiveness of excitation of the primary endings of the muscle spindle by vibration as a respiratory stimulate in the decerebrate cat. *J.Physiol.* 194:555–563.

Hoheisel, U., K. Koch, and S. Mense. 1994. Functional reorganization in the rat dorsal horn during an experimental myositis. *Pain* 59(1):111–118.

Immke, D. C., and E. W. McCleskey. 2001. Lactate enhances the acid-sensing Na^+ channel on ischemia-sensing neurons. *Nat.Neurosci.* 4(9):869–870.

————. 2003. Protons open acid-sensing ion channels by catalyzing relief of Ca^{2+} blockade. *Neuron* 37(1):75–84.

Iwamoto, G. A., and M. P. Kaufman. 1987. Caudal ventrolateral medullary cells responsive to static muscular contraction. *J.Appl.Physiol.* 62:149–157.

Iwamoto, G. A., M. P. Kaufman, B. R. Botterman, and J. H. Mitchell. 1982. Effects of lateral reticular nucleus lesions on the exercise pressor reflex in cats. *Circ.Res.* 51:400–403.

Iwamoto, G. A., T. G. Waldrop, M. P. Kaufman, B. R. Botterman, K. J. Rybicki, and J. H. Mitchell. 1985. Pressor reflex evoked by muscular contraction: contributions by neuraxis levels. *J.Appl.Physiol.* 59:459–467.

Kaufman, M. P., and H. V. Forster. 1996. Reflexes controlling circulatory, ventilatory and airway responses to exercise. In *Handbook of Physiology,* Section 12: Exercise: Regulation and Integration of Multiple Systems. II. Control of Respiratory and Cardiovascular Systems, eds. Rowell, L. B., and J. T. Shepherd, 381–447. (New York,: Oxford University Press).

Kaufman, M. P., J. C. Longhurst, K. J. Rybicki, J. H. Wallach, and J. H. Mitchell. 1983. Effects of static muscular contraction on impulse activity of groups III and IV afferents in cats. *J.Appl.Physiol.* 55:105–112.

Kaufman, M. P., and K. J. Rybicki. 1987. Discharge properties of group III and IV muscle afferents: their responses to mechanical and metabolic stimuli. *Circ.Res.* 61:160–165.

Kaufman, M. P., K. J. Rybicki, T. G. Waldrop, and G. A. Ordway. 1984a. Effect of ischemia on responses of group III and IV afferents to contraction. *J.Appl.Physiol.* 57:644–650.

Kaufman, M. P., K. J. Rybicki, T. G. Waldrop, G. A. Ordway, and J. H. Mitchell. 1984b. Effects of static and rhythmic twitch contractions on the discharge of group III and IV muscle afferents. *Cardiovasc.Res.* 18:663–668.

Kim, J. K., S. G. Hayes, A. E. Kindig, and M. P. Kaufman. 2007. Thin-fiber mechanoreceptors reflexly increase renal sympathetic nerve activity during static contraction. *Am.J.Physiol. Heart Circ.Physiol.* 292(2):H866–H873.

Kindig, A. E., S. G. Hayes, R. L. Hanna, and M. P. Kaufman. 2006. P2 antagonist PPADS attenuates responses of thin fiber afferents to static contraction and tendon stretch. *Am.J.Physiol. Heart Circ.Physiol.* 290(3):H1214–H1219.

Kindig, A. E., S. G. Hayes, and M. P. Kaufman. 2007. Blockade of purinergic 2 receptors attenuates the mechanoreceptor component of the exercise pressor reflex. *Am.J.Physiol. Heart Circ.Physiol.*

———. 2007. Purinergic 2 receptor blockade prevents the responses of group IV afferents to post-contraction circulatory occlusion. *J.Physiol.* 578(Pt.1):301–308.

Kindig, A. E., T. B. Heller, and M. P. Kaufman. 2005. VR-1 receptor blockade attenuates the pressor response to capsaicin but has no effect on the pressor response to contraction in cats. *Am.J.Physiol. Heart Circ.Physiol.* 288(4):H1867–H1873.

Kniffki, K-D., S. Mense, and R. F. Schmidt. 1978. Responses of group IV afferent units from skeletal muscle to stretch, contraction and chemical stimuli. *Exp.Brain Res.* 31:511–522.

Krishtal, O. A., S. M. Marchenko, and V. I. Pidoplichko. 1983. Receptor for ATP in the membrane of mammalian sensory neurones. *Neurosci.Lett.* 35:41–45.

Kruger, L., J. D. Silverman, P. W. Mantyh, C. Sternini, and N. C. Brecha. 1989. Peripheral patterns of calcitonin-gene-related peptide general somatic sensory innervation: cutaneous and deep terminations. *J.Comp.Neurol.* 280(2):291–302.

Kumazawa, T. N., and K. Mizumura. 1977. Thin-fibre receptors responding to mechanical, chemical and thermal stimulation in the skeletal muscle of the dog. *J.Physiol.* 273:179–194.

Lewis, C., S. Neidhart, C. Holy, R. A. North, G. Buell, and A. Surprenant. 1995. Coexpression of P2X$_2$ and P2X$_3$ receptor subunits can account for ATP-gated currents in sensory neurons. *Nature (Lond.)* 377:432–435.

Li, J., G. A. Hand, J. T. Potts, L. B. Wilson, and J. H. Mitchell. 1997. c-Fos expression in the medulla induced by static muscle contraction in cats. *Am.J.Physiol.* 272:H48–H56.

Li, J., and J. H. Mitchell. 2000. c-Fos expression in the midbrain periaqueductal gray during static muscle contraction. *Am.J.Physiol. Heart Circ.Physiol.* 279(6):H2986–H2993.

Light, A. R., R. W. Hughen, J. Zhang, J. Rainier, Z. Liu, and J. Lee. 2008. Dorsal root ganglion neurons innervating skeletal muscle respond to physiological combinations of protons, ATP, and lactate mediated by ASIC, P2X and TRPV1. *J.Neurophysiol.* 100:1184–1201.

MacLean, D. A., K. F. La Noue, K. S. Gray, and L. I. Sinoway. 1998. Effects of hindlimb contraction on pressor and muscle interstitital metabolite responses in the cat. *J.Appl. Physiol.* 85:1583–1592.

Mark, A. L., R. G. Victor, C. Nerhed, and B. G. Wallin. 1985. Microneurographic studies of the mechanisms of sympathetic nerve responses to static exercise in humans. *Circ.Res.* 57:461–469.

Matsukawa, K., P. T. Wall, L. B. Wilson, and J. H. Mitchell. 1990. Reflex responses of renal nerve activity during isometric muscle contraction in cats. *Am.J.Physiol.* 259:H1380–H1388.

———. 1994. Reflex stimulation of cardiac sympathetic nerve activity during static muscle contraction in cats. *Am.J.Physiol.* 267:H821–H827.

McCloskey, D. I., P. B. C. Matthews, and J. H. Mitchell. 1972. Absence of appreciable cardiovascular and respiratory responses to muscle vibration. *J.Appl.Physiol.* 33:623–626.

McCloskey, D. I., and J. H. Mitchell. 1972. Reflex cardiovascular and respiratory responses originating in exercising muscle. *J.Physiol.* 224:173–186.

McCord, J. L., S. G. Hayes, and M. P. Kaufman. 2008. Acid sensing ion and epithelial sodium channels do not contribute to the mechanoreceptor component of the exercise pressor reflex. *Am.J.Physiol. Heart Circ.Physiol.* 295:1017–1024.

Mense, S. 1977. Nervous outflow from skeletal muscle following chemical noxious stimulation. *J.Physiol.* 267:75–88.

————. 1981. Sensitization of group IV muscle receptors to bradykinin by 5-hydroxytryptamine and prostaglandin E-2. *Brain Res.* 225:95–105.

Mense, S., and A. D. III. Craig. 1988. Spinal and supraspinal terminations of primary afferent fibers from the gastrocnemius-soleus muscle in the cat. *Neuroscience* 26:1023–1035.

Mense, S., and H. Meyer. 1985. Different types of slowly conducting afferent units in cat skeletal muscle and tendon. *J.Physiol.* 363:403–417.

————. 1988. Bradykinin-induced modulation of the response behaviour of different types of feline group III and IV muscle receptors. *J.Physiol.* 398:49–63.

Mense, S., and R. F. Schmidt. 1974. Activation of group IV afferent units from muscle by algesic agents. *Brain Res.* 72:305–310.

Mense, S., and M. Stahnke. 1983. Responses in muscle afferent fibers of slow conduction velocity to contractions and ischemia in the cat. *J.Physiol.* 342:383–397.

Middlekauff, H. R., E. U. Nitzsche, C. K. Hoh, M. A. Hamilton, G. C. Fonarow, A. Hage, and J. D. Moriguchi. 2000. Exaggerated renal vasoconstriction during exercise in heart failure patients. *Circulation* 101(7):784–789.

Middlekauff, H. R., E. U. Niztsche, C. K. Hoh, M. A. Hamilton, G. C. Fonarow, A. Hage, and J. D. Moriguchi. 2001. Exaggerated muscle mechanoreflex control of reflex renal vasoconstriction in heart failure. *J.Appl.Physiol.* 90:1714–1719.

Middlekauff, H. R., E. U. Niztsche, A. H. Nguyen, C. K. Hoh, and G. G. Gibbs. 1997. Modulation of renal cortical blood flow during static exercise in humans. *Circ.Res.* 80:62–68.

Mitchell, J. H., M. P. Kaufman, and G. A. Iwamoto. 1983. The exercise pressor reflex: Its cardiovascular effects, afferent mechanisms, and central pathways. *Ann.Rev.Physiol.* 45:229–242.

Mizumura, K., and T. Kumazawa. 1976. Reflex respiratory response induced by chemical stimulation of muscle afferents. *Brain Res.* 109:402–406.

Molliver, D. C., D. C. Immke, L. Fierro, M. Pare, F. L. Rice, and E. W. McCleskey. 2005. ASIC3, an acid-sensing ion channel, is expressed in metaboreceptive sensory neurons. *Mol.Pain* 1:35.

Momen, A., U. A. Leuenberger, C. A. Ray, S. Cha, B. Handly, and L. I. Sinoway. 2003. Renal vascular responses to static handgrip: the role of the muscle mechanoreflex. *Am.J.Physiol. Heart Circ.Physiol.* 285:1247–1253.

Naves, L. A., and E. W. McCleskey. 2005. An acid-sensing ion channel that detects ischemic pain. *Braz.J.Med.Biol.Res.* 38(11):1561–1569.

Paintal, A. S. 1960. Functional analysis of group III afferent fibres of mammalian muscles. *J.Physiol.* 152:250–270.

Pan, H.-L., C. L. Stebbins, and J. C. Longhurst. 1993. Bradykinin contributes to the exercise pressor reflex: mechanism of action. *J.Appl.Physiol.* 75:2061–2068.

Person, R. J. 1989. Somatic and vagal afferent convergence on solitary tract neurons in cat: electrophysiological characteristics. *Neuroscience* 30:283–295.

Pickar, J. G., J. M. Hill, and M. P. Kaufman. 1994. Dynamic exercise stimulates group III muscle afferents. *J.Neurophysiol.* 71:753–760.

Ponikowski, P. P., T. P. Chua, D. P. Francis, A. Capucci, A. J. Coats, and M. F. Piepoli. 2001. Muscle ergoreceptor overactivity reflects deterioration in clinical status and cardio-respiratory reflex control in chronic heart failure. *Circulation* 104(19):2324–2330.

Price, M. P., S. L. McIlwrath, J. Xie, C. Cheng, J. Qiao, D. E. Tarr, K. A. Sluka et al. 2001. The DRASIC cation channel contributes to the detection of cutaneous touch and acid stimuli in mice. *Neuron.* 32(6):1071–1083.

Pryor, S. L., S. F. Lewis, R. G. Haller, L. A. Bertocci, and R. G. Victor. 1990. Impairment of sympathetic activation during static exercise in patients with muscle phosphorylase deficiency (McArdle's Disease). *J.Clin.Invest.* 85:1444–1449.

Reinöhl, J., U. Hoheisel, T. Unger, and S. Mense. 2003. Adenosine triphosphate as a stimulant for nociceptive and non-nociceptive muscle group IV receptors in the rat. *Neurosci.Lett.* 338:25–28.

Remme, W. J., Berg R. van den, M. Mantel, P. H. Cox, Hoogenhuyze Van, X. H. Krauss, C. J. Storm, and D. A. Kruyssen. 1986. Temporal relation of changes in regional coronary flow and myocardial lactate and nucleoside metabolism during pacing-induced ischemia. *Am.J.Cardiol.* 58(13):1188–1194.

Rotto, D. M., and M. P. Kaufman. 1988. Effects of metabolic products of muscular contraction on the discharge of group III and IV afferents. *J.Appl.Physiol.* 64:2306–2313.

Rotto, D. M., K. D. Massey, K. P. Burton, and M. P. Kaufman. 1989. Static contraction increases arachidonic acid levels in gastrocnemius muscles of cats. *J.Appl.Physiol.* 66:2721–2724.

Rotto, D. M., H. D. Schultz, J. C. Longhurst, and M. P. Kaufman. 1990. Sensitization of group III muscle afferents to static contraction by products of arachidonic acid metabolism. *J.Appl.Physiol.* 68:861–867.

Rotto, D. M., C. L. Stebbins, and M. P. Kaufman. 1989. Reflex cardiovascular and ventilatory responses to increasing H^+ activity in cat hindlimb muscle. *J.Appl.Physiol.* 67:256–263.

Rybicki, K. J., T. G. Waldrop, and M. P. Kaufman. 1985. Increasing gracilis interstitial potassium concentrations stimulates group III and IV afferents. *J.Appl.Physiol.* 58:936–941.

Saito, M. 1995. Differences in muscle sympathetic nerve response to isometric exercise in different muscle groups. *Eur.J.Appl.Physiol.* 70:26–35.

Saito, M., M. Naito, and T. Mano. 1990. Different responses in skin and muscle sympathetic nerve activity to static muscle contraction. *J.Appl.Physiol.* 69(6):2085–2090.

Sato, A., Y. Sato, and R. F. Schmidt. 1981. Heart rate changes reflecting modifications of efferent cardiac sympathetic outflow by cutaneous and muscle afferent volleys. *J.Auton.Nerv. Syst.* 4:231–241.

Sawynok, J., and A. Reid. 1997. Peripheral adenosine 5'-triphosphate enhances nociception in the formalin test via activation of a purinergic P(2X) receptor. *Eur.J.Pharmacol.* 330:115–121.

Sinoway, L., S. Prophet, I. Gorman, T. J. Mosher, J. Shenberger, M. Dolecki, R. Briggs, and R. Zelis. 1989. Muscle acidosis during static exercise is associated with calf vasoconstriction. *J.Appl.Physiol.* 66:429–436.

Sinoway, L. I., J. M. Hill, J. G. Pickar, and M. P. Kaufman. 1993. Effects of contraction and lactic acid on the discharge of group III muscle afferents in cats. *J.Neurophysiol.* 69:1053–1059.

Sinoway, L. I., K. J. Wroblewski, S. A. Prophet, S. M. Ettinger, K. S. Gray, S. K. Whisler, G. Miller, and R. L. Moore. 1992. Glycogen depletion-induced lactate reductions attenuate reflex responses in exercising humans. *Am.J.Physiol.* 263:H1499–H1505.

Smith, E. S., H. Cadiou, and P. A. McNaughton. 2007. Arachidonic acid potentiates acid-sensing ion channels in rat sensory neurons by a direct action. *Neuroscience* 145(2):686–698.

Smith, S. A., J. H. Mitchell, R. H. Naseem, and M. G. Garry. 2005a. Mechanoreflex mediates the exaggerated exercise pressor reflex in heart failure. *Circulation* 112(15):2293–2300.

Smith, S. A., M. A. Williams, J. H. Mitchell, P. P. Mammen, and M. G. Garry. 2005b. The capsaicin-sensitive afferent neuron in skeletal muscle is abnormal in heart failure. *Circulation* 111(16):2056–2065.

Stebbins, C. L., O. A. Carretero, T. Mindroiu, and J. C. Longhurst. 1990. Bradykinin release from contracting skeletal muscle of the cat. *J.Appl.Physiol.* 69:1225–1230.

Stebbins, C. L., Y. Maruoka, and J. C. Longhurst. 1988. Prostaglandins contribute to cardiovascular reflexes evoked by static muscular contraction. *Circ.Res.* 59:645–654.

Sterns, D. A., S. M. Ettinger, K. S. Gray, S. K. Whisler, T. J. Mosher, M. B. Smith, and L. I. Sinoway. 1991. Skeletal muscle metaboreceptor exercise responses are attenuated in heart failure. *Circulation* 84(5):2034–2039.

Sutherland, S. P., C. J. Benson, J. P. Adelman, and E. W. McCleskey. 2001. Acid-sensing ion channel 3 matches the acid-gated current in cardiac ischemia-sensing neurons. *Proc. Natl.Acad.Sci.U.S.A.* 98(2):711–716.

Symons, J. D., S. J. Theodossy, J. C. Longhurst, and C. L. Stebbins. 1991. Intramuscular accumulation of prostaglandins during static contraction of the cat triceps surae. *J.Appl. Physiol.* 71:1837–1842.

Thimm, F., and K. Baum. 1987. Response of chemosensitive nerve fibers of group III and IV to metabolic changes in rat muscles. *Pflügers Arch.* 410:143–152.

Tibes, U. 1977. Reflex inputs to the cardiovascular and respiratory centers from dynamically working canine muscles: some evidence for involvement of group III or IV nerve fibers. *Circ.Res.* 41:332–341.

Tsuchimochi, H., S. G. Hayes, J. L. McCord, and M. P. Kaufman. 2009. Both central command and exercise pressor reflex activate cardiac sympathetic nerve activity in decerebrate cats. *Am.J.Physiol. Heart Circ.Physiol.*

Ugawa, S., T. Ueda, Y. Ishida, M. Nishigaki, Y. Shibata, and S. Shimada. 2002. Amiloride-blockable acid-sensing ion channels are leading acid sensors expressed in human nociceptors. *J.Clin.Invest.* 110:1185–1190.

Victor, R. G., S. L. Pryor, N. H. Secher, and J. H. Mitchell. 1989a. Effects of partial neuromuscular blockade on sympathetic nerve responses to static exercise in humans. *Circ. Res.* 65(2):468–476.

Victor, R. G., D. M. Rotto, S. L. Pryor, and M. P. Kaufman. 1989b. Stimulation of renal sympathetic activity by static contraction: evidence for mechanoreceptor-induced reflexes from skeletal muscle. *Circ.Res.* 64:592–599.

Voilley, N., Weille J. de, J. Mamet, and M. Lazdunski. 2001. Nonsteroid anti-inflammatory drugs inhibit both the activity and the inflammation-induced expression of acid-sensing ion channels in nociceptors. *J.Neurosci.* 21(20):8026–8033.

von During, M., and K. H. Andres. 1990. Topography and ultrastructure of group III and IV nerve terminals of cat's gastrocnemius-soleus muscle. In *The Primary Afferent Neuron: A Survey of Recent Morpho-functional Aspects*, eds. Zenker, W., and W. L. Neuhuber, 35–41. (New York: Plenum).

Waldmann, R., F. Bassilana, Weille J. de, G. Champigny, C. Heurteaux, and M. Lazdunski. 1997. Molecular cloning of a non-inactivating proton-gated Na+ channel specific for sensory neurons. *J.Biol.Chem.* 272(34):20975–20978.

Waldmann, R., G. Champigny, E. Lingueglia, Weille De, Jr., C. Heurteaux, and M. Lazdunski. 1999. H(+)-gated cation channels. *Ann.N.Y.Acad.Sci.* 868:67–76.

Waldrop, T. G., K. J. Rybicki, and M. P. Kaufman. 1984. Chemical activation of group I and II muscle afferents has no cardiorespiratory effects. *J.Appl.Physiol.* 56:1223–1228.

Willis, W. D., Jr., X. Zhang, C. N. Honda, and G. J. Giesler, Jr. 2001. Projections from the marginal zone and deep dorsal horn to the ventrobasal nuclei of the primate thalamus. *Pain* 92(1–2):267–276.

Yagi, J., H. N. Wenk, L. A. Naves, and E. W. McCleskey. 2006. Sustained currents through ASIC3 ion channels at the modest pH changes that occur during myocardial ischemia. *Circ.Res.* 99(5):501–509.

13 Central Pain as a Thalamocortical Dysrhythmia

A Thalamic Efference Disconnection?

Kerry D. Walton and Rodolfo R. Llinás

CONTENTS

13.1 INTRODUCTION

The intrinsic electrical properties of neurons are presently considered as a salient parameter in brain function (Llinas 1988; Getting 1989; Connors and Gutnick 1990; Turrigiano et al. 1994; Margolis and Detwiler 2007). This is in contrast to the classical purely reflexological view, where neurons are considered to be passive agents that are activated or inhibited synaptically. This intrinsic functional view has been addressed in recent years in relation to thalamic neuron function and to its recurrent interaction with the cortex. Such a view is based upon single-cell neuronal electrophysiology (c.f. Llinas 1988; Steriade and Llinas 1988), and thalamocortical anatomy (Jones 2007). The neurological consequences of such a perspective (Llinas et al. 1999) have been corroborated by magnetoencephalography (MEG), single-cell intraoperative recordings, and encouraging surgical outcomes (Jeanmonod et al. 1996, 2001b, 2003). Moreover, as in the case of tinnitus, where a sound stimulus can suppress the centrally generated sensation (Coles and Hallam 1987), central pain can be modulated by peripheral stimulus (Somers and Somers 1999; Inui et al. 2006). From a functional imaging perspective, electroencephalogram (EEG) (Jeanmonod et al. 1993) and MEG data have shown the presence of a distinct increase in low-frequency activation in central pain (Schulman et al. 2005). In all patients with central, but not peripheral, pain, there was a second site of low-frequency oscillations that was localized to the mesial/orbito frontal and anterior cingulated cortices, as well as the temporal (insular) cortex. Central pain patients with these MEG characteristics did not respond to spinal cord stimulation. By contrast, patients without the frontal low-frequency component responded well to such stimulation.

In addition to clinical studies, direct experimental evidence for the functional organization of the thalamo-cortico-thalamo loop has been obtained in *in vitro* studies of rodent thalamocortical slices (Llinas et al. 2002) that have been extended to *in vivo* animal studies concerning neuropathic pain (Gerke et al. 2003; Kim et al. 2003) and thalamic deafferentation (Wang and Thompson 2008). These latter results have established a direct relationship between abnormal thalamic rhythmicity and the occurrence of central pain.

The relevance of an "essential thalamic structure" (i.e., a central generator) for neuropathic pain was first suggested by Head (Head and Holmes 1911). The findings summarized here extend this original proposal by addressing brain activity obtained from MEG, EEG, and preoperative unit recordings from patients with chronic neuropathic pain. We also briefly touch upon the contribution of animal studies to understanding the cellular and molecular components of neuropathic pain generation in the context of increased T-type calcium channel activty.

13.2 DEAFFERENTATION PAIN SYNDROMES

Spontaneous brain activity was recorded from three patients with deafferentation pain syndrome (phantom arm, braichial plexus avulsion, laterial thalamic lesion) using magnetoencephalography (MEG). Recordings were made while participants were alert with their eyes closed (Schulman et al. 2005).

13.2.1 Power Spectral Findings

The power spectra in these three patients showed a distinct increase in high theta-range (7–9 Hz) power. The spectra were similar to those of other thalamocortical dysrhythmia (TCD) disorders (Llinas et al. 2001). Calculation of the mean spectral energy (MSE) in two bands (7–9 Hz and 9–11 Hz) allows comparison of these results with those of other patient and control groups (Figure 13.1).

13.2.2 Localization of Theta Activity

Independent component (IC) analysis revealed somatotopically meaningful theta-range activity in two of these patients. In the patient with a right thalamus vascular lesion (who suffered from thalamic pain syndrome and tinnitus), theta activity was localized to the right sensorimotor cortex and superior temporal gyrus. In the patient with a left brachial plexus avulsion, theta-range activation was localized to the right somatosensory cortex as shown in the dorsal view of the brain in Figure 13.2. Theta activity was also present in the left temporal and bilateral mesial orbitofrontal cortices (Figure 13.2). This is significant considering that this person suffered from anxiety and depression as well as chronic pain. The limbic source distribution is consistent with reported structural and functional aberrations that have been identified in these disorders independently (Saxena et al. 1998; Mayberg 2000), and in the context of chronic pain in particular (Grachev et al. 2002). The presence of limbic sources in the same IC as aberrant somatosensory activation underscores the point

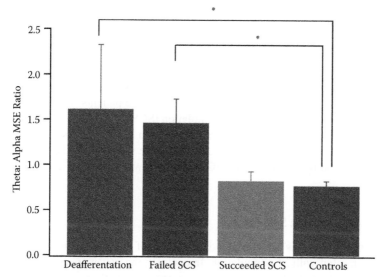

FIGURE 13.1 Comparisons of mean spectra energy (power ratio 7–9 Hz: 9–11 Hz) in three groups of patients and control group. The patients with successful SCS were not significantly different from the control group. In contrast, those in which the SCS did not bring relief and those with deafferentation pain had significantly more low-frequency activity than the control group. ($*P > 0.05$, **) (Modified from Schulman J.J., Zonenshayn M., Ramirez R.R., Ribary U., and Llinas R. 2005. Thalamocortical dysrhythmia syndrome: MEG imaging of neuro-pathic pain. *Thalamus and Related Systems* 3: 33–39.)

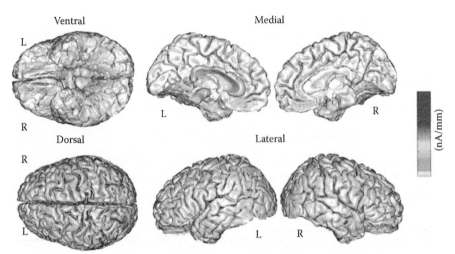

FIGURE 13.2   (see color insert following page 166) Example of localization of theta activity in a patient with brachial plexus avulsion pain. Activity is localized to the contralateral soma-tosensory cortex and bilaterally to the mesial orbitofrontal cortices. (Modified from Schulman J.J., Zonenshayn M., Ramirez R.R., Ribary U., and Llinas R. 2005. Thalamocortical dysrhyth-mia syndrome: MEG imaging of neuropathic pain. *Thalamus and Related Systems* 3: 33–39.)

that the affective component of the pain experience is physiologically tightly coupled with the sensory complaints. (Localization was not possible for the phantom pain subject due to metal artifacts.)

13.3   PERIPHERAL PAIN SYNDROMES

In this group patients experience chronic pain following injury to an extremity or to their back. MEG recordings were made in eight of these patients who had electrodes implanted in the spinal cord (SC) for stimulation (SCS) to alleviate the pain. Of this group, five reported a reduction in pain of more than 50% while the other three did not report improvement. The MEG recordings were made while the patients' eyes were closed (Schulman et al. 2005). (SC stimulation was turned off during the MEG recordings.)

13.3.1   MEG POWER SPECTRA

Theta range activity was seen in the power spectra of the three patients who did not receive pain relief with SC stimulation. In contrast, the power spectra from the five patients who obtained relief from SCS were comparable to those of healthy controls. This is illustrated when the MSE ratios are compared (Figure 13.1). These findings suggest that in patients in whom SCS is successful, the pathology is likely to be either spinal or peripheral, and the effectiveness of SCS is derived from the locally induced activation of inhibitory interneurons as described in the Gate-Control theory (Melzack and Wall 1965). In contrast, in patients in whom SCS fails, these findings suggest that the pathology is thalamocortical, and the distant induction of dorsal column depolarization provided by SCS is insufficient to effectively modify thalamocortical physiology.

13.3.2 LOCALIZATION OF THETA ACTIVITY

Independent component localization revealed somatotopically meaningful theta-range activity in two of the patients in the SCS failure group with back pain. There was bilateral theta activation in areas near the classical homoncular sensory representation of the trunk (Penfield 1958). Comparable independent components were not present in patients with successful SCS or in healthy, pain-free controls.

13.4 COMPLEX REGIONAL PAIN SYNDROME

MEG recordings were made in 11 patients with complex regional pain syndrome (CRPS) type I. This disorder is a chronic progressive disease characterized by severe pain, swelling, and changes in the skin in the region of pain. The absence of a nerve lesion distinguishes Type I from Type II CRPS.

13.4.1 POWER SPECTRA

As in the other MEG recordings of spontaneous activity in patients with neuropathic pain, the power spectra of these patients were characterized by the presence of activity in the theta range. Activity in the delta range was also marked in seven people in this group.

13.4.2 LOCALIZATION OF THETA AND DELTA ACTIVITY

Independent component analysis revealed that every patient had components with activation in the theta frequency range localized over the somatosensory cortex. These localizations were somatotopically meaningful with respect to their pain localization. In addition, every patient had component in the delta (4–8 Hz) frequency range that was localized bilaterally to mesial orbitofrontal cortex and temporal pole (Dubois et al. 2008).

13.5 HUMAN ELECTRICAL RECORDINGS

13.5.1 EEG AND FIELD POTENTIAL RECORDINGS

In agreement with the MEG findings summarized above, the power spectra of spontaneous cortical EEGs recorded from patients with chronic neuropathic pain were characterized by excess activity in the theta and beta frequency ranges compared to healthy controls (Stern et al. 2006; Boord et al. 2008). Activity was localized to several pain-associated areas including insula, anterior cingulate, prefrontal, and somatosensory cortices (Stern et al. 2006).

To examine the functional relationship between the EEG recordings and thalamus, field potential recordings were made from a region of the central lateral thalamic nucleus (Sarnthein and Jeanmonod 2008). Analysis of EEG and field potential power spectra revealed high temporal coherence in the theta band (6–9 Hz) in recordings from patients with neuropathic pain (Sarnthein and Jeanmonod 2008). Coherence between the activity of the tens of thousands of neurons seen by an EEG electrode and the estimated 5–10 thalamic neurons seen by the local field electrode is remarkable indeed. This finding supports the hypothesis that TCD is due to abnormal low-frequency activity in the thalamo-cortical loop, rather than in the thalamus or in the cortex alone.

13.5.2 PERIOPERATIVE UNIT RECORDINGS

Low-frequency bursting, consistent with the MEG and EEG findings, has been recorded from single neurons in the thalamus of patients with neurogenic pain during preoperative recordings (Modesti and Waszak 1975; Lenz et al. 1989; Rinaldi et al. 1991; Jeanmonod et al. 1993, 1996).

Although these bursts were elicited by spinal column stimulation in the earliest recording (Modesti and Waszak 1975), they have since been found to occur spontaneously. Single-cell recordings have been made from ventral posterior nucleus in thalamic regions related to the deafferented body area with neurogenic pain following spinal cord injury (Lenz et al. 1989), from the intralaminar nucleus in patients with deafferentation pain (Rinaldi et al. 1991), and from medial thalamus in a large group of patients with peripheral and central neurogenic pain related to injury or to cancer (Jeanmonod et al. 1993, 1996). Both spontaneous sporadic activity and spike bursts have been recorded. The rhythmically bursting units all discharged at 3–5 Hz while randomly bursting units and those characterized by sporadic spontaneous activity tended to be in this frequency range as well (Jeanmonod et al. 1993, 1996, 2001a). The firing pattern within each burst was also consistent. The first spike within the burst has the largest amplitude, and there is a positive correlation between the length of the first interspike interval and the number of spikes within the burst with a mean spike frequency of 206 Hz (Jeanmonod et al. 2001b).

13.6 ANIMAL STUDIES

An understanding of the cellular and molecular basis for TCD and its role in neuropathic pain has been pioneered by studies in animals. That thalamic neurons switch from tonic firing to bursting was first reported in the 1980s (Llinas and Jahnsen 1982; Carbone and Lux 1984; Jahnsen and Llinas 1984b) in animal studies. This bursting was elicited when the cells were hyperpolarized and were called low-threshold spikes (LTS). This bursting is supported by the activation of low-voltage activated (T-type, $Ca_v3.1$) calcium channels. That these LTS bursts may be the origin of neuropathic pain was hypothesized in 1992 in a rodent study (Roberts et al. 1992). Spontaneous oscillatory burst firing was recorded from thalamic neurons in rodents with allodynia following a spinal cord lesion (Gerke et al. 2003). The abnormal burst responses were absent in control animals. In addition to the thalamic dysrhythmia, rats with spinal cord lesions also demonstrated exaggerated vocal responses to normally innocuous mechanical skin stimulation. That this thalamic dysrhythmia may be due to deafferentation is supported by the finding of a delayed, marked increase in cortical theta rhythm and behavioral aberrations following experimentally induced lesions of the rostral pole of the thalamic reticular nucleus of rats (Marini et al. 2002). Finally, mice that lack the T-type calcium channel show a reduced behavioral response to pain and increased threshold for paw withdrawal to mechanical stimulation (Na et al. 2008).

13.7 THE THALAMOCORTICAL CIRCUIT

13.7.1 THALAMOCORTICAL GENERATION OF THETA ACTIVATION

In states of thalamocortical dysrhythmia, an ongoing theta-range thalamic activity serves as the trigger for cortical dysfunction. In the case of neurogenic pain, this self-sustaining generation of low-frequency oscillations results in a long-term pathological equilibrium in the cortical pain matrix. The generation of low-frequency activity by the thalamocortical circuit was first proposed in 1999 (Llinas et al. 1999) and has been developed since then based on findings in animal and intraoperative human recordings as summarized above. Changes in this circuit in neuropathic pain may be summarized in terms of three interconnected loops: the thalamo-thalamic, the specific (sensory) thalamo-cortico-specific thalamic, and the non-specific thalamo-cortico-thalamic.

Deafferentation of the specific and non-specific thalamic nuclei leads to hyperpolarization of these cells (Steriade et al. 1987). When they are hyperpolarized, thalamic neurons change from high-threshold tonic firing to low-threshold, theta-range oscillatory bursts (Llinas and Jahnsen 1982; Jahnsen and Llinas 1984a, b; Steriade et al. 1990) supported by T-type calcium currents. This shift to periodic bursting activity leads to a decrease in the excitatory input to the reticular thalamus (RT) and their subsequent hyperpolarization and low-frequency bursting (Steriade et al. 1993). Feedback of RT to the thalamic nuclei further supports the low-frequency bursts.

Specific thalamo-cortico-specific thalamic loop: Specific thalamic neurons send low-frequency input to the apical dendrites of layer IV and V pyramidal neurons. This reduced thalamic input leads to reduced firing rates at cortical level. Layer VI neurons feed back to the specific thalamus and RT. Layer V pyramidal neurons feed back to both specific and non-specific thalamus. In addition to pyramidal cells, specific thalamus also innervates cortical inhibitory interneurons leading to disinhibition in adjacent columns (see "edge effect," below).

Non-specific thalamo-cortico-thalamic loop: Bursting of non-specific thalamus innervates the apical dendrites of layer V pyramidal cells that feed back to both specific and non-specific thalamus, reinforcing the low-frequency thalamo-cortical-thalamo circuit.

13.7.2 THALAMOCORTICAL GENERATION OF GAMMA RANGE ACTIVITY, THE "EDGE EFFECT"

It is hypothesized that under normal conditions, a given region of cortex activated in the gamma range inhibits other gamma activity from occurring in its periphery. On the cortical level, during normal high-frequency cortical activity, Layer III–IV cortical interneurons release GABA onto neighboring cells, in a process termed *lateral inhibition* (Beierlein et al. 2000).

We have hypothesized that in thalamocortical dysrhythmia, where projections from thalamus entrain a core of low-frequency cortical activity, lateral inhibition is abolished due to the lower rate of firing. This phenomenon was first described in the retina by Hartline (Hartline 1967), who found that when a given region of *limulus* retina is activated, a reduction of lateral inhibition creates a physiological border between activated and silent zones.

Evidence from voltage-sensitive dye experiments of deafferented cortex supports the idea that disconnection-induced hyperpolarization of thalamocortical modules leads to the deinactivation of T-type Ca^{++} channels and low-threshold spike activity, accompanied by adjacent increases in high-frequency firing (Leznik et al. 2002).

The low-frequency activity due to asymmetric lateral inhibition forces adjacent cortical areas into high-frequency (gamma) oscillations. This "edge effect" (Llinas et al. 2005) has also been shown in tinnitus patients (Weisz et al. 2005; De Ridder et al. 2007), in central pain patients (Sarnthein and Jeanmonod 2008), and in an animal model (Ghazal et al. 2007). This abnormal gamma band in turn has been proposed as generating the positive symptoms of pain and allodynia (Schulman et al. 2005).

This hypothesis also finds support in a recent study of migraineurs (Coppola et al. 2007). They found that gamma band oscillations (GBO) evoked by visual stimuli differed in patients and controls in two respects: (1) There was an increased GBO response to the first two stimulus bands, but (2) there was a deficit in habituation to later stimuli in the patient group. They hypothesized that this was consistent with thalamic disconnection combined with a decreased cortical lateral inhibition. That is, an edge effect.

13.8 COMMENTS ON TREATMENT

13.8.1 THALAMIC LESIONS

Thalamic lesions have proven to be an effective treatment in patients with chronic pain who were resistant to therapy. That the rhythmic EEG and MEG activity and underlying thalamic bursting recorded perioperatively is an essential element in the generation and perception of neuropathic pain is supported by such results. Indeed, the bursting unit activity summarized above is co-localized with the most efficient therapeutic lesions. Targeted centrolateral nucleus lesions lead to 50%–100% pain relief in a majority of patients (Jeanmonod et al. 2001a, 2001b) with no somatosensory deficits in over 90% of the cases (Jeanmonod et al. 2001a). These lesions are thought to disrupt the synchronous, low-frequency activity in the thalamo-cortico-thalamo loop of which the posterior centrolateral nucleus is a part. These lesions have proven to be more beneficial to patients with intermittent pain or allodynia than to those with continuous pain.

13.8.2 STIMULATION

The use of deep brain stimulation (DBS) as a treatment in several disorders was recently reviewed (Kringelbach et al. 2007b). DBS has proven to be a successful treatment in some types of pain (Kumar et al. 1997; Nandi et al. 2003; Bittar et al. 2005a, 2005b; Rasche et al. 2006; Yamamoto et al. 2006; Kringelbach et al. 2007a) and other instances of TCD, particularly in Parkinson's disease (Benabid et al. 2003). The effectiveness of this therapy is consistent with the mechanisms postulated above: High-frequency stimulation could raise the resting potential of RT or intralaminar neurons above the level of T-type Ca^{++} channel deinactivation, and thalamic cells switch from bursting to tonic firing. The effectiveness of motor cortex stimulation in treating some cases of neuropathic pain (Carroll et al. 2000; Katayama et al. 2001a, 2001b) may be understood as resulting from top-down thalamic depolarization. It should be noted that the qualified success of this approach speaks to the challenge of modifying the thalamocortical loop using a depolarizer effectively situated outside the system. Recent advances using transcranial magnetic stimulation may similarly derive their effectiveness from a temporary rise in thalamocortical resting potential.

Consistent with this view and with the correlation between low-frequency activity and subjective pain experience is the fact that clinical improvement has been correlated with a reduction of low-frequency cortical activity. For example, reduction of neuropathic pain sensation following transcranial magnetic stimulation is concomitant with a reduction of low-frequency activity; pre-existing TCD electrical abnormalities have been attenuated after spinal cord stimulation leading to pain relief (Schulman et al. 2005); and the alleviation of phantom limb pain with DBS is correlated with a reduction of theta- and beta-band activity (Ray et al. 2009). These results are similar to those found in tinnitus cases (Richter et al. 2006; De Ridder et al. 2007). Moreover, such sound masking resulting in the disappearance of the tinnitus is accompanied by the marked reduction, or total abolition, of the low-frequency signature at the auditory cortex (Tyler 2000; Llinas et al. 2005).

13.8.3 UNSUCCESSFUL TREATMENTS

In addition to these successes, some of the therapeutic failures for chronic pain also lend support to the theory that deafferentation pain can be related to a centrally generated hyperpolarization. For example, the finding that sensory blockade is known to increase pain in some patients (Bonica 1991), coupled with the low success rate for surgical tractotomy (Tasker 1996), is consistent with a paradigm in which the additional sensory blockade deafferents the thalamus even further and adds to the disfacilitation that causes low-threshold spike bursts. Conversely, some patients with plexus avulsions report pain reduction with input to the affected limb. This presumably activates the thalamus, temporarily counteracting the deafferentation effects. These findings are supported by a model in which increased input to the thalamus restores a more depolarized CNS mode.

13.9 PAIN, ITS LOCALIZATION, AND SENSORY BINDING

A fundamental issue to be considered here is that of sensory binding and its functional correlates. This chapter presents a set of electrophysiological findings that indicate that pain, in this case central pain, is correlated with two distinct electrical components: one that defines the localization on the body of the pain experience and, as such, is *sui generis* for each patient; and one that is similar and omnipresent in all patients. While the former signature is indeed a description of location, the latter relates to the emotional sensation of pain, as it is common to all patients and is reduced by procedures that obliterate the emotional component of the sensation, that is, the hurt. It is very clear that the unpleasant sensation/emotion of being hurt happens without actual localization, as for instance with moral pain or as it happens with emotions such as fear. Having fear in one's hand would be of no survival value, but localizing pain is invaluable for survival. As such then a careful analysis of issues as temporal coherence between low frequencies in localized pain versus noncoherent activity between simultaneous but unrelated stimuli may go a long way in clarifying the electrophysiological mechanism for temporal binding.

REFERENCES

Beierlein M., Gibson J.R., and Connors B.W. 2000. A network of electrically coupled interneurons drives synchronized inhibition in neocortex. *Nature Neuroscience* 3: 904–910.

Benabid A.L., Vercucil L., Benazzouz A., Koudsie A., Chabardes S., Minotti L., Kahane P., Gentil M., Lenartz D., Andressen C., Krack P. and Pollak P. 2003. Deep brain stimulation: what does it offer? *Advances in Neurology* 91: 293–302.

Bittar R.G., Kar-Purkayastha I., Owen S.L., Bear R.E., Green A., Wang S., and Aziz T.Z. 2005a. Deep brain stimulation for pain relief: a meta-analysis. *Journal of Clinical Neuroscience* 12: 515–519.

Bittar R.G., Otero S., Carter H., and Aziz T.Z. 2005b. Deep brain stimulation for phantom limb pain. *Journal of Clinical Neuroscience* 12: 399–404.

Bonica J. 1991. Pain management: past and current status. In *Anesthesiology and Pain Management*, ed. Stanley, T., and Fine, P. Kluwer Academic, Boston.

Boord P., Siddall P.J., Tran Y., Herbert D., Middleton J., and Craig A. 2008. Electroencephalographic slowing and reduced reactivity in neuropathic pain following spinal cord injury. *Spinal Cord* 46: 118–123.

Carbone E., and Lux H.D. 1984. A low voltage-activated calcium conductance in embryonic chick sensory neurons. *Biophysical Journal* 46: 413–418.

Carroll D., Joint C., Maartens N., Shlugman D., Stein J., and Aziz T.Z. 2000. Motor cortex stimulation for chronic neuropathic pain: a preliminary study of 10 cases. *Pain* 84: 431–437.

Coles R.R., and Hallam R.S. 1987. Tinnitus and its management. *British Medical Bulletin* 43: 983–998.

Connors B.W., and Gutnick M.J. 1990. Intrinsic firing patterns of diverse neocortical neurons. *Trends in Neurosciences* 13: 99–104.

Coppola G., Ambrosini A., Di Clemente L., Magis D., Fumal A., Gerard P., Pierelli F., and Schoenen J. 2007. Interictal abnormalities of gamma band activity in visual evoked responses in migraine: an indication of thalamocortical dysrhythmia? *Cephalalgia* 27: 1360–1367.

De Ridder D., van der Loo E., Van der Kelen K., Menovsky T., van de Heyning P., and Moller A. 2007. Theta, alpha and beta burst transcranial magnetic stimulation: brain modulation in tinnitus. *International Journal of Medical Sciences* 4: 237–241.

Dubois M., Rojas-Soto D., Garcia J., Levacic D., Walton K., and Llinas R. 2008. *Abnormal brain activity in patients with complex regional pain syndrome (CRPS) type I.* In Abstract Viewer/Itinerary Planner, Program No 3463. Soc. for Neuroscience, Washington, D.C.

Gerke M.B., Duggan A.W., Xu L., and Siddall P.J. 2003. Thalamic neuronal activity in rats with mechanical allodynia following contusive spinal cord injury. *Neuroscience* 117: 715–722.

Getting P.A. 1989. Emerging principles governing the operation of neural networks. *Annual Review of Neuroscience* 12: 185–204.

Ghazal T., Moran K., Walton K., Llinas R., and Dubois M. 2007. *Patients with CRPS demonstrate thalamocortical dysrhythmia: a magneto-encephalographic (MEG) study.* In 23rd Annual Meeting, American Academy of Pain Medicine. New Orleans.

Grachev I.D., Fredrickson B.E., and Apkarian A.V. 2002. Brain chemistry reflects dual states of pain and anxiety in chronic low back pain. *J Neural Transm* 109: 1309–1334.

Hartline H.K. 1967. Visual receptors and retinal interaction. In *Nobel Lectures*. Elsevier, Amsterdam.

Head H., and Holmes G. 1911. Sensory disturbances from cerebral lesions. *Brain* 34: 102–254.

Inui K., Tsuji T., and Kakigi R. 2006. Temporal analysis of cortical mechanisms for pain relief by tactile stimuli in humans. *Cerebral Cortex* 16: 355–365.

Jahnsen H., and Llinas R. 1984a. Electrophysiological properties of guinea-pig thalamic neurones: an in vitro study. *J Physiol* 349: 205–226.

Jahnsen H., and Llinas R. 1984b. Voltage-dependent burst-to-tonic switching of thalamic cell activity: an in vitro study. *Archives Italiennes de Biologie* 122: 73–82.

Jeanmonod D., Magnin M., and Morel A. 1993. Thalamus and neurogenic pain: physiological, anatomical and clinical data.[erratum appears in *Neuroreport* 1993 Aug;4(8):1066]. *Neuroreport* 4: 475–478.

Jeanmonod D., Magnin M., and Morel A. 1996. Low-threshold calcium spike bursts in the human thalamus. Common physiopathology for sensory, motor and limbic positive symptoms. *Brain* 119: 363–375.

Jeanmonod D., Magnin M., Morel A., and Siegmund M. 2001a. Surgical control of the human thalamocortical dysrhythmia: I. Central lateral thalamotomy in neurogenic pain. *Thal Rel Sys* 1: 71–79.

Jeanmonod D., Magnin M., Morel A., Siegmund M., Cancro R., Lanz M., Llinas R., Ribary U., Kronberg E., Schulman J. & Zonenshayn M. 2001b. Thalamocortical dysrhythmia II. Clinical and surgical aspects. *Thal Rel Sys* 1: 245–254.

Jeanmonod D., Schulman J., Ramirez R., Cancro R., Lanz M., Morel A., Magnin M., Siegemund M., Kronberg E., Ribary U. and Llinas R. 2003. Neuropsychiatric thalamocortical dysrhythmia: surgical implications. *Neurosurgery Clinics of North America* 14: 251–265.

Jones E. 2007. *The Thalamus.* Cambridge University Press, Cambridge, UK.

Katayama Y., Yamamoto T., Kobayashi K., Kasai M., Oshima H., and Fukaya C. 2001a. Motor cortex stimulation for phantom limb pain: comprehensive therapy with spinal cord and thalamic stimulation. *Stereotactic and Functional Neurosurgery* 77: 159–162.

Katayama Y., Yamamoto T., Kobayashi K., Kasai M., Oshima H., and Fukaya C. 2001b. Motor cortex stimulation for post-stroke pain: comparison of spinal cord and thalamic stimulation. *Stereotactic and Functional Neurosurgery* 77: 183–186.

Kim D., Park D., Choi S., Lee S., Sun M., Kim C., and Shin H.S. 2003. Thalamic control of visceral nociception mediated by T-type Ca2+ channels. *Science* 302: 117–119.

Kringelbach M.L., Jenkinson N., Green A.L., Owen S.L., Hansen P.C., Cornelissen P.L., Holliday I.E., Stein J., and Aziz T.Z. 2007a. Deep brain stimulation for chronic pain investigated with magnetoencephalography. *Neuroreport* 18: 223–228.

Kringelbach M.L., Jenkinson N., Owen S.L., and Aziz T.Z. 2007b. Translational principles of deep brain stimulation. *Nature Reviews Neuroscience* 8: 623–635.

Kumar K., Toth C., and Nath R.K. 1997. Deep brain stimulation for intractable pain: a 15-year experience. *Neurosurgery* 40: 736–746; discussion 746–737.

Lenz F.A., Kwan H.C., Dostrovsky J.O., and Tasker R.R. 1989. Characteristics of the bursting pattern of action potentials that occurs in the thalamus of patients with central pain. *Brain Research* 496: 357–360.

Leznik E., Urbano F., and Llinas R. 2002. *Neurotransmitter modulation of high and low frequency inputs in somatosensory cortex: An in vitro optical imaging study.* In Society for Neuroscience. Abstract Viewer/Itinerary Planner, Program No. 651.12, Orlando.

Llinas R.R. 1988. The intrinsic electrophysiological properties of mammalian neurons: insights into central nervous system function. *Science* 242: 1654–1664.

Llinas R., and Jahnsen H. 1982. Electrophysiology of mammalian thalamic neurones in vitro. *Nature* 297: 406–408.

Llinas R.R., Leznik E., and Urbano F.J. 2002. Temporal binding via cortical coincidence detection of specific and nonspecific thalamocortical inputs: a voltage-dependent dye-imaging study in mouse brain slices. *Proceedings of the National Academy of Sciences of the United States of America* 99: 449–454.

Llinas R., Ribary U., Jeanmonod D., Cancro R., Kronberg E., Schulman J., Zonenshayn M., Magnin M., Morel A., and Siegmund M. 2001. Thalamocortical dysrhythmia I. Functional and imaging aspects. *Thal Rel Sys* 1: 237–244.

Llinas R.R., Ribary U., Jeanmonod D., Kronberg E., and Mitra P.P. 1999. Thalamocortical dysrhythmia: a neurological and neuropsychiatric syndrome characterized by magneto-encephalography. *Proceedings of the National Academy of Sciences of the United States of America* 96: 15222–15227.

Llinas R., Urbano F., Leznik E., Ramizeriz R., and Van Marle H. 2005. Rhythmic and dysrhythmic thalamocortical dynamics: GABA systems and the edge effect. *Trends in Neuroscience* 28: 325–333.

Margolis D.J., and Detwiler P.B. 2007. Different mechanisms generate maintained activity in ON and OFF retinal ganglion cells. *Journal of Neuroscience* 27: 5994–6005.

Marini G., Ceccarelli P., and Mancia M. 2002. Thalamocortical dysrhythmia and the thalamic reticular nucleus in behaving rats. *Clinical Neurophysiology* 113: 1152–1164.

Mayberg H.S. 2000. Depression. In *Human Brain Mapping: The Disorders*, ed. Mazziotta J., Toga A., and Frackowiak R.S.J., pp. 485–507. Academic Press, San Diego.

Melzack R., and Wall P.D. 1965. Pain mechanisms: a new theory. *Science* 150: 971–979.

Modesti L.M., and Waszak M. 1975. Firing pattern of cells in human thalamus during dorsal column stimulation. *Applied Neurophysiology* 38: 251–258.

Na S., Choi S., Kim J., Park J., and Shin H.-S. 2008. Attenuated neuropathic pain in CaV3.1 null mice. *Molecules and Cells* 25: 242–246.

Nandi D., Aziz T., Carter H., and Stein J. 2003. Thalamic field potentials in chronic central pain treated by periventricular gray stimulation—a series of eight cases. *Pain* 101: 97–107.

Penfield W. 1958. The excitable cortex in conscious man. In *The Sherrington lectures, 5*, pp. 17. C.C. Thomas, Springfield, Ill.

Rasche D., Rinaldi P.C., Young R.F., and Tronnier V.M. 2006. Deep brain stimulation for the treatment of various chronic pain syndromes. *Neurosurgical Focus* 21: E8.

Ray N., Jenkinson N., Kringelbach M., Hansen P., Pereira E., Brittain J., Holland P., Holliday I., Owen S., Stein J., and Aziz T. 2009. Abnormal thalamocortical dynamics may be altered by deep brain stimulation: using magnetoencephalography to study phantom limb pain. *J Clin Neurosci* 16: 32–36.

Richter G.T., Mennemeier M., Bartel T., Chelette K.C., Kimbrell T., Triggs W., and Dornhoffer J.L. 2006. Repetitive transcranial magnetic stimulation for tinnitus: a case study. *Laryngoscope* 116: 1867–1872.

Rinaldi P.C., Young R.F., Albe-Fessard D., and Chodakiewitz J. 1991. Spontaneous neuronal hyperactivity in the medial and intralaminar thalamic nuclei of patients with deafferentation pain. *Journal of Neurosurgery* 74: 415–421.

Roberts W.A., Eaton S.A., and Salt T.E. 1992. Widely distributed GABA-mediated afferent inhibition processes within the ventrobasal thalamus of rat and their possible relevance to pathological pain states and somatotopic plasticity. *Experimental Brain Research* 89: 363–372.

Sarnthein J., and Jeanmonod D. 2008. High thalamocortical theta coherence in patients with neurogenic pain. *Neuroimage* 39: 1910–1917.

Saxena S., Brody A.L., Schwartz J.M., and Baxter L.R. 1998. Neuroimaging and frontal-subcortical circuitry in obsessive-compulsive disorder. *Br J Psychiatry Suppl*: 26–37.

Schulman J.J., Zonenshayn M., Ramirez R.R., Ribary U., and Llinas R. 2005. Thalamocortical dysrhythmia syndrome: MEG imaging of neuropathic pain. *Thalamus and Related Systems* 3: 33–39.

Somers D.L., and Somers M.F. 1999. Treatment of neuropathic pain in a patient with diabetic neuropathy using transcutaneous electrical nerve stimulation applied to the skin of the lumbar region. *Physical Therapy* 79: 767–775.

Steriade M., Domich L., Oakson G., and Deschenes M. 1987. The deafferented reticular thalamic nucleus generates spindle rhythmicity. *Journal of Neurophysiology* 57: 260–273.

Steriade M., and Llinas R.R. 1988. The functional states of the thalamus and the associated neuronal interplay. *Physiological Reviews* 68: 649–742.

Steriade M., McCormick D.A., and Sejnowski T.J. 1993. Thalamocortical oscillations in the sleeping and aroused brain. *Science* 262: 679–685.

Steriade M., Pare D., Datta S., Oakson G., and Curro D.R. 1990. Different cellular types in mesopontine cholinergic nuclei related to ponto-geniculo-occipital waves. *J Neurosci* 10: 2560–2579.

Stern J., Jeanmonod D., and Sarnthein J. 2006. Persistent EEG overactivation in the cortical pain matrix of neurogenic pain patients. *Neuroimage* 31: 721–731.

Tasker R. 1996. Surgical approaches to chronic pain. In *Pain Management: Theory and Practice*, ed. Portenoy R., pp. 290–311. F.A. Davis, Philadelphia.

Turrigiano G., Abbott L.F., and Marder E. 1994. Activity-dependent changes in the intrinsic properties of cultured neurons. *Science* 264: 974–977.

Tyler R. 2000. *Tinnitus Handbook*. Singular Publishing Group, San Diego.

Wang G., and Thompson S.M. 2008. Maladaptive homeostatic plasticity in a rodent model of central pain syndrome: thalamic hyperexcitability after spinothalamic tract lesions. *Journal of Neuroscience* 28: 11959–11969.

Weisz N., Wienbruch C., Dohrmann K., and Elbert T. 2005. Neuromagnetic indicators of auditory cortical reorganization of tinnitus. *Brain* 128: 2722–2731.

Yamamoto T., Katayama Y., Obuchi T., Kano T., Kobayashi K., Oshima H., and Fukaya C. 2006. Thalamic sensory relay nucleus stimulation for the treatment of peripheral deafferentation pain. *Stereotactic and Functional Neurosurgery* 84: 180–183.

14 What Can Neuroimaging Tell Us about Central Pain?

D.S. Veldhuijzen, F.A. Lenz, S.C. LaGraize, and J.D. Greenspan

CONTENTS

14.1 INTRODUCTION

Central pain (CP) is defined by the International Association for the Study of Pain as "pain initiated or caused by a primary lesion or dysfunction of the CNS [central nervous system]" (Merskey and Bogduk 1994). A more recently proposed definition is "pain arising as a direct consequence of a lesion or disease affecting the central somatosensory system" (Treede et al. 2008). This newer refinement recognizes that an essential feature associated with CP is disturbance of part of the central somatosensory system. This may be the only universal feature that applies to CP conditions, which exhibit varied qualities across patients.

Similar to CP are two conditions that share almost all of the same features of CP, but the pathological sensation is described as something other than pain. Central dysesthesia involves the expression of unpleasant sensations (both spontaneous and evoked) that are described in terms other than pain. One may question whether this is simply a matter of semantics. However, experimental pain testing of such a patient did not reveal any abnormal responses to noxious thermal stimuli applied to unaffected body regions (Kim et al. 2007). Central neurogenic pruritis can also arise from CNS lesions, in which an itch sensation substitutes for the pathological pain sensation. Due to the scarcity of reports on these latter two phenomena, this chapter will focus on central pain. The interested reader on the topic of central pain is referred to two recent books, Canavero and Bonicalzi (2007) and Henry, Panju, and Yashpal (2007).

Prior to the 1990s, our knowledge of CP was derived from examinations of lesion type and location that produced CP and interpretation of the sensory abnormalities of CP patients. With the advent of functional neuroimaging, we have new tools to investigate CP. This chapter will focus on studies that have used functional neuroimaging to address mechanistic questions regarding central pain.

14.2 THE NATURE OF SENSORY ABNORMALITIES IN CENTRAL PAIN

CP patients inevitably demonstrate somatosensory abnormalities. When adequately tested, all CP patients show reduced sensitivity to some forms of somatic stimulation. A universal finding is a hyposensitivity to thermal and/or noxious stimuli. This does not mean that every CP patient shows sensory deficits for all types of thermal and noxious stimuli, but every CP patient shows hyposensitivity to at least some submodalities of thermal or noxious stimuli. Studies involving quantitative sensory testing (QST) of CP patients have reported cold hypoesthesia in 85%–91% of cases, warm hypoesthesia in 85%–100% of cases, cold pain hypoalgesia in approximately 45% of cases, and heat pain hypoalgesia in 7%–91% of cases (Andersen et al. 1995; Boivie, Leijon, and Johansson 1989; Greenspan et al. 2004; Leijon, Boivie, and Johansson 1989; Vestergaard et al. 1995). (This large range of values for heat pain hypoalgesia may reflect different criteria for identifying hypoalgesia rather than true differences in study samples. For instance, the largest percentage is based on reported abnormal thresholds for noxious temperature, for which the basis of abnormality is not explicitly described, and seems to include both heat and cold pain. The lowest rate is derived from a study using an explicit comparison to a normative data range.) Additionally, CP patients may or may not show deficits in tactile sensitivity, which has been reported to be between 23% and 52% of CP patients in the aforementioned studies. Thus, the fact that some aspect of thermal and/or pain perception is altered in every CP patient tested leads to the conclusion that the inciting CNS injury must encompass some component of the spino(trigemino)-thalamic-cortical system relaying information derived from A-delta and C-fiber afferents. Yet, the specific negative sensory signs show considerable variability across the CP patient population, suggesting that many variations exist in the pathophysiological processes responsible for CP.

Another important observation is that some CNS-injured patients may experience thermal hypoesthesia or hypoalgesia as a result of a CNS lesion without developing CP. This has been demonstrated for spinal cord (Ducreux et al. 2006; Finnerup et al. 2007), thalamic (Greenspan et al. 1997), and cortical lesion cases (Greenspan, Lee, and Lenz 1999). Even among CP patients, the degree of sensory deficit can range from none to complete. It is reasonable to ask whether the extent of sensory loss is related to the expression of CP. Based on the sample of 30 CP patients (mostly central post-stroke pain (CPSP) cases) evaluated with QST at our research center at the University of Maryland, we found no relationship between the extent of thermosensory loss (based on cool or warm thresholds) and the level of ongoing pain (as reflected in scores from the McGill Short Form; see Figure 14.1). Thus, these data revealed no parametric relationship between the extent of thermosensory deficit—presumably reflecting the extent of disruption of the spinothalamocortical system—and the magnitude of pain the patient suffers.

However, another study suggested that such a relationship does exist for a subset of spinal cord injury (SCI) patients (Ducreux, Attal, Parker, and Bouhassira 2006). In syringomyelia patients with spontaneous pain but no allodynia, the thermal sensory loss was significantly more asymmetrical, and there was a direct relationship between the extent of thermosensory deficits and the intensity of burning pain. In contrast, patients with allodynia had lesser thermal deficits, in terms of both magnitude and extent. In addition, the sensory deficits were different between patients with cold or tactile allodynia. Specifically, thermal deficits were less severe in patients with cold allodynia compared with those with tactile allodynia. These results suggest that the extent of spinothalamic tract (STT) injury may influence aspects of pain experienced by some types of CP patients.

14.3 IDENTIFYING COLD ALLODYNIA

Even though cold allodynia is considered a common feature of CP, it is not always present in CP patients. Several studies of consecutively recruited CP patients reported cold allodynia in 23%–56% of cases (Andersen, Vestergaard, Ingeman-Nielsen, and Jensen 1995; Attal et al. 2000; Boivie, Leijon, and Johansson 1989; Greenspan, Ohara, Sarlani, and Lenz 2004; Leijon, Boivie, and Johansson 1989; Vestergaard et al. 1995). Furthermore, there is considerable variability in the expression of cold allodynia, suggesting that there may be different neural mechanisms responsible for cold allodynia in different patients.

Cold allodynia is determined in the clinical setting by touching the patient with a cool/cold object (typically, room temperature metal). The concomitant tactile stimulus may be relevant to the allodynic sensation in this situation, especially if the person has a tactile allodynia. In an experimental testing situation, cold allodynia is determined most frequently from cold pain thresholds. These thresholds are assessed using a contact thermode held in place while the temperature is gradually decreased until the patient indicates a painful perception. Thus, in threshold determination, the cold stimulus is not abrupt, and the abrupt tactile contact is not contemporaneous with the cooling stimulus. These procedural differences may lead to different results. If there is no effective difference, a patient reporting pain with a cold object contact would be expected to have a cold pain threshold in the range of the contact probe, or higher. However, this expected result is not often observed.

Boivie et al. (Boivie, Leijon, and Johansson 1989) reported that while 5 of 22 CP patients exhibited cold allodynia in the clinical testing environment, none of them showed evidence of cold allodynia based on QST threshold testing. Similarly, Vestergaard et al. (1995) reported that 5 of 12 CP patients exhibited cold allodynia in the clinical testing environment, but none of them showed evidence of cold allodynia

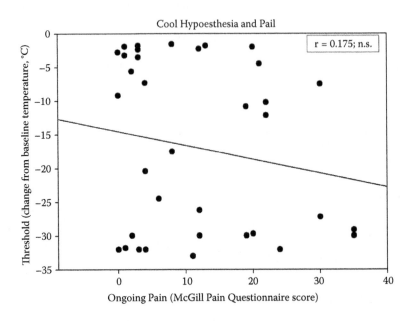

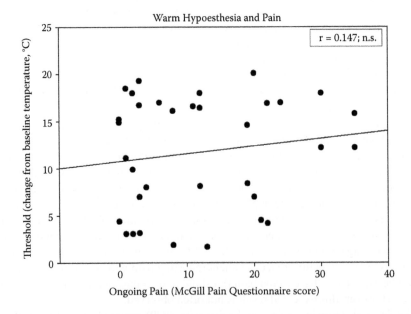

FIGURE 14.1 The relationship between thermal hypoesthesia and ongoing pain in a series of central pain patients. The correlations based on linear regression are not statistically significant.

based on QST threshold testing. Additionally, two other patients in this series had elevated (i.e., hypersensitive) cold pain thresholds, suggestive of cold allodynia based on QST, yet failed to show evidence of cold allodynia with clinical testing. A very fundamental question is whether only one or either method should be considered a valid indicator of cold allodynia.

We have also observed instances in our survey of CP patients when cold allo-dynia—defined by pain/dysesthesia derived from abrupt contact with a cold (20°C) object—did not necessarily predict a more sensitive cold pain threshold. However, reviewing our series of 23 CP patients tested in this manner, we did observe a significantly higher (more sensitive) cold pain threshold for CP patients with cold allodynia versus those without (Whitney rank sum test, p=0.048), despite a clear overlap in threshold values between the two groups (Figure 14.2). These observations serve to illustrate that the method by which cold allodynia is assessed can make a difference in the determination of the presence or absence of cold allodynia

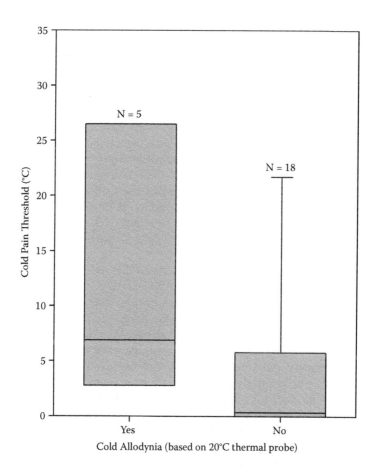

FIGURE 14.2   Box plots show the distributions of cold pain thresholds for CP patients with and without cold allodynia.

on a case-by-case basis. Although at a group level, cold pain threshold has some relationship to clinically determined cold allodynia.

14.4 CENTRAL PAIN MECHANISMS: ONGOING PAIN

Although a lesion in the spino(trigemino)-thalamo-cortical pathway is necessary for CP to develop, no relation between size or location of a lesion and presence of CP or pain intensity has been found (Andersen et al. 1995; Boivie, Leijon, and Johansson 1989; Bowsher, Leijon, and Thoumas 1998; Leijon, Boivie, and Johansson 1989; Vestergaard et al. 1995). Additionally, studies on SCI patients show that the STT was equally affected in pain-free patients and patients with CP (Ducreux et al. 2006; Finnerup et al. 2003). These data demonstrate that lesions involving the spino-thalamo-cortical pathway, while necessary, are not sufficient to explain the development of CP. Moreover, CP does not appear to depend upon a critical extent"of SCI injury.

Hyperexcitability of neurons after deafferentation has been proposed as a mechanism for CP. Supporting this idea, low-threshold spike (LTS) bursts are increased in neurons in the region of Ventral caudal (Vc) thalamus in awake patients with CP versus patients with movement disorders (Jeanmonod, Magnin, and Morel 1996; Lenz et al. 1989, 1994). Similar results were found in animal models of SCI, including monkeys (Weng et al. 2000) and rodents (Hains et al. 2003). In contrast, another report found no differences in the frequency of thalamic bursts rate, or LTS bursts in particular, between chronic pain patients versus movement disorder patients (Radhakrishnan et al. 1999). However, in this last study most of the neuronal recordings were made outside of Vc in various types of pain patients, mostly peripheral neuropathy patients. Thus, it is not clear that these burst patterns are particularly related to CP. Regardless of these electrophysiological findings, electrical stimulation in the area of Vc evoked pain more commonly in CP patients with allodynia versus those without allodynia (Davis et al. 1996; Lenz et al. 1998). These observations suggest that functional reorganization involving Vc is likely to contribute to the ongoing pain and allodynia of CP.

A magnetic resonance spectroscopy study found that metabolite concentrations of the neuronal marker N-acetyl aspartate (NAA) and the glial marker myo-inositol (Ins) in the thalamus differed between patients with or without CP after SCI (Pattany et al. 2002). Mean NAA concentrations and the NAA/Ins ratio were significantly lower for pain patients, compared with pain-free patients, and mean Ins concentrations were higher for pain patients versus pain-free patients, which approached significance. Furthermore, NAA concentrations were negatively correlated with pain intensity, and Ins was positively correlated with pain intensity in the pain group. No significant differences were found between the right and left thalamus, but it is unclear if pain was localized unilaterally in these cases. These results reflect dysfunction or loss of neurons in the thalamus to a greater extent in central pain SCI patients versus pain-free SCI patients.

Functional imaging studies have reported primarily (but not exclusively) thalamic hypoactivity for CP patients, in the absence of provocative stimulation. Two positron emission tomography (PET) case studies reported decreased cerebral blood flow (CBF) in the ipsilesional thalamus in CP patients during rest (Cahana et al. 2004; Peyron et al. 1995). The specific thalamic nuclei could not be identified due to the limited spatial resolution of these PET studies. This thalamic hypoactivity was reversed by motor cortex stimulation (Peyron et al. 1995), or by repeated cycles of daily intravenous lidocaine infusion (Cahana et al. 2004). Several studies have noted a similar decrease in thalamic perfusion during chronic neuropathic pain not limited to CP, such that the hemithalamus contralateral to the affected body region exhibited substantially lower CBF than the ipsilateral hemithalamus (Di Piero et al. 1991; Hsieh et al. 1995; Iadarola et al. 1995). However, this story is more complicated. One single-photon emission CT (SPECT) study of an SCI CP case reported distinct differences in signal depending upon the patient's pain level, noting increases in signal for the thalamus bilaterally during the experience of high-intensity pain versus low-intensity pain (Ness et al. 1998). Recently, PET results from CP patients showed that unilateral thalamic hypoperfusion may be in medial as well as lateral thalamus (Casey 2007).

Several SPECT and PET studies of CPSP patients have demonstrated thalamic hyperactivity in response to stimulation of the allodynic side compared to the nonallodynic side, or to patients without allodynia, or both (Cesaro et al. 1991; Ducreux et al. 2006; Peyron et al. 1998, 2000). These data are compatible with the hypothesis that the resting state hypoactive thalamus is a result of a loss of γ-aminobutyric acid (GABA)-inhibitory neurons that are normally tonically active to some degree and thereby control the excitability of thalamic neurons (Casey 2007). It is important to note that cerebral metabolism, as measured by PET or SPECT, reflects both inhibitory and excitatory synaptic activity. Thus, reduced CBF in the thalamus of CP patients may reflect the loss of inhibitory synaptic activity, perhaps related to actual neuron loss, as suggested in the previously described magnetic resonance spectroscopy study (Pattany et al. 2002). This could be the case even with the observation of greater spontaneous spiking of thalamic neurons, given that such activity is not metabolically demanding, per se (Viswanathan and Freeman 2007). It also follows that inhibitory neurons would be selectively lost because the thalamus is hyperactive in response to painful stimulation.

Another possibility is that the thalamus undergoes adaptive changes over time. A suggestive SPECT study in patients with complex regional pain syndrome found thalamic hyperperfusion for patients with symptoms for only 3 to 7 months, but thalamic hypoperfusion for patients with long-term symptoms (24–36 months) (Fukumoto et al. 1999). It remains to be determined if the same process occurs with CP.

Even with the thalamus having a key role in CP, it does not function in isolation. It is certainly the case that cortical functional reorganization plays some role in CP. A few PET and SPECT studies have measured regional blood flow or glucose metabolism asymmetries in CP patients. While thalamic hypometabolism or reduced blood flow has routinely been reported, parietal cortical effects range from no discernable change (LaTerre, De Volder, and Goffinet 1988; Peyron et al. 1995), to hypometabolism (Canavero et al. 1995; Lee, Choi, and Chung 1989; Peyron et al. 1995), to hypermetabolism (Hirato et al. 1993; Ness et al. 1998). It should be noted that instances of no apparent parietal cortical changes have been associated with extra-thalamic lesions, while lesions that included the thalamus were more often associated with parietal hypometabolism, and one report of parietal hypermetabolism. Thus, direct thalamic involvement appears to have measureable effects upon parietal cortical function (as one would most likely predict), yet CP can occur without apparent parietal cortical abnormalities assessed with PET or SPECT.

Two studies used [11C] diprenorphine PET to assess whether there were significant differences in regional opiate receptor binding between CP patients and healthy controls (Jones et al. 2004; Willoch et al. 2004). In both studies, significantly reduced binding was found for the CP patients in thalamus, inferior parietal cortex (including secondary somatosensory cortex—S2), insular and lateral prefrontal cortices, and along the midline in anterior and posterior cingulate cortices. It should be noted that these common results were found despite a wide range of lesion locations among the CP patients. Thus, it appears that CP is associated with reduced opioid-binding capacity at several thalamic and cortical sites.

14.5 CEREBRAL MECHANISMS OF COLD ALLODYNIA IN CENTRAL PAIN

A prominent theory of cold allodynia is that it results from a disinhibition of the medial nociceptive processing system as a consequence of damage or dysfunction of the lateral thermo-nociceptive processing system (Craig 1998; Craig 2007). Although a version of this disinhibition hypothesis was proposed a century ago (Head and Holmes 1911), more recent versions propose that specific brain structures—most notably the medial thalamus and ACC—are essentially involved in CP, and cold allodynia in particular. A few detailed neuroimaging studies provide evidence that changes in the organization of the brain occur in patients with cold allodynia. A PET study reported on stimulation with a cool, rubbing stimulus (ice in a plastic container) in the area affected by allodynia in patients with CP due to lateral medullary (Wallenberg) syndrome (Peyron et al. 1998). This stimulus produced activation of several contralateral structures that were not significantly activated with the same stimuli applied to the opposite body site, including the lateral region of the contralateral thalamus, S1/M1, S2/inferior parietal lobule (bilaterally), and the frontal inferior gyrus. Another PET study of a patient with severe central pain and cold allodynia secondary to a subnuclear thalamic infarct exhibited a similar pattern of activation with exposure to a cold (20°C) water bath (Kim et al. 2007).

In a recent fMRI study of syringomyelia patients, cold allodynia was associated with activation in contralateral mid/posterior insula, ACC, bilateral anterior insula, S2/inferior parietal cortex, ipsilateral frontal gyrus, and contralateral supplementary motor cortex (Ducreux et al. 2006). All three aforementioned studies found that brain activation patterns in response to cool allodynic stimulation were out of proportion to the applied stimulus and were more closely comparable to those elicited by painful stimulation in healthy controls. The Peyron et al. (1998) and Kim et al. (2007) studies failed to find ACC activation with the allodynic stimulation, despite the fact that ACC activation is commonly found in the response to acute painful stimuli in normal controls (Apkarian et al. 2005). Also, a combined PET and fMRI study of an unusual patient with a bifocal lesion involving the parietal cortex and

ACC showed that this patient still developed cold allodynia, while no hyperactivity was found in the spared portion of the ACC (Peyron et al. 2000). In contrast, another CP case study found that the ACC, associative areas in the parietal cortex, and the putamen were specifically activated when a sensation of pain was evoked with cold stimulation (Seghier et al. 2005). One recent fMRI study evaluated experimentally induced cold allodynia in healthy volunteers with the topical application of menthol, which sensitized the skin to otherwise innocuous cold stimuli (Seifert and Maihofner 2007). The "allodynic" cold cerebral responses were compared with normal cold pain responses, using stimuli that produced equal perceived intensity in both conditions. The allodynic pain produced greater activation in dorsolateral prefrontal cortex (DLPFC), bilateral anterior insula, and the brainstem. This limited set of results precludes any generalizations regarding the CNS mechanisms specifically related to cold allodynia, including the role of the ACC. The varied results do suggest that multiple CNS mechanisms may lead to cold allodynia, rather than a single pathophysiological mechanism.

14.6 CEREBRAL MECHANISMS OF TACTILE ALLODYNIA IN CENTRAL PAIN

Tactile hypoesthesia and tactile allodynia, as measured by application of von Frey monofilaments and moving brushes, are common clinical features of CP. Lesions of the dorsal columns are associated with deficits in tactile sensation, whereas no deficits in tactile sensations are found with lesions of the STT when sparing the dorsal columns (Nathan, Smith, and Cook 1986). Therefore, reduced tactile thresholds are likely due to decreased transmission of stimuli through the dorsal column-medial lemniscal-thalamocortical pathway. There is no evidence to suggest that disruption of the dorsal column-medial lemniscal system contributes to CP in general or tactile allodynia in particular. One study of SCI patients found that patients with and without pain had similar reductions of mechanical detection thresholds at body sites below injury level (Finnerup et al. 2007). A study of CPSP patients reported that tactile allodynia was more often associated with normal tactile thresholds than it was with tactile hypoesthesia (Greenspan et al. 2004). These results suggest that tactile allodynia is more likely to result from abnormalities in cortical processing of the information conveyed by a functional dorsal column-medial lemniscal system, rather than a dysfunctional dorsal column-medial lemniscal system per se. Related to this, activation of afferents known to project through the dorsal column-medial lemniscal system was found to elicit unpleasant dysesthesias in stroke patients with post-stroke dysesthesias, a variant of CPSP (Triggs and Beric 1994).

Few imaging studies have examined cortical activation associated with tactile allodynia. One fMRI study specifically evaluated cerebral responses to tactile allodynia derived from capsaicin treatment in healthy individuals (Baron et al. 1999). This study reported that only S1 and S2 activation were associated with nonpainful mechanical stimulation using von Frey filaments. When stimulating the area of secondary mechanical hyperalgesia, significant activation was found in the prefrontal cortex, and in the middle and inferior frontal gyri, without a difference in activation for S1 and S2. No significant ACC activation was observed for either stimulus condition.

A few studies have examined the cerebral responses to tactile allodynic stimuli in patients with peripheral neuropathic pain. A single-case fMRI study described a patient with constant burning pain and tactile allodynia following peroneal nerve damage (Hofbauer, Olausson, and Bushnell 2006). Despite reduced tactile sensitivity, touch evoked deep pain. When the tactile allodynic response on the affected foot was compared to brushing on the nonaffected foot, more activated brain regions were found associated with tactile allodynia. With tactile allodynic stimulation, significantly greater activation was found in S2, ipsilateral anterior insula, and ACC. No activation was found in S1 or ipsilateral posterior insula after stimulation of the affected foot, despite finding that these regions were activated following brushing of the nonaffected foot. In another fMRI study, punctate mechanical stimulation of the affected side of CRPS patients evoked hyperalgesia and greater than normal activity in many nociceptive-processing brain regions, including contralateral S1, bilateral S2, bilateral insula, superior frontal cortex, middle frontal cortex, inferior frontal cortex, inferior parietal lobule, parietal association cortex, and anterior and posterior ACC (Maihofner et al. 2005). Differential activation upon dynamically evoked allodynia was also found in a PET study in a sample of patients with upper- and lower-limb traumatic peripheral nerve injury, many of whom reported ongoing pain and all of whom showed brush-evoked allodynia (Witting et al. 2006). Brush-evoked allodynia in the affected limb yielded significant activation in different regions compared to the nonaffected limb, including contralateral orbitofrontal cortex, ipsilateral anterior insular cortex, and cerebellum. In contrast, brushing of normal skin induced activation in contralateral S1 and posterior parietal cortex that was not found during allodynic stimulation.

The only imaging study of tactile allodynia in CP patients demonstrated that tactile allodynia produced a pattern of brain activation distinct from that of cold allodynia, in patients with syringomyelia (Ducreux et al. 2006). Tactile stimuli consisted of repetitive stroking with a soft brush applied in the allodynic area of patients with and without CP and normal controls. The pattern of activation associated with tactile allodynia was compared to nonpainful brushing in normal volunteers and syringomyelia patients without pain. In all groups, activation was observed in the contralateral S1, contralateral S2, and inferior and superior parietal areas. Activation specific to allodynia was elicited in the contralateral thalamus, bilateral middle frontal gyrus, caudate nucleus, and supplementary motor areas. Tactile allodynia was not associated with activation in the insula or anterior cingulate, despite the fact that these brain regions were activated with cold allodynic stimulation in this same patient group.

14.7 CONCLUSIONS

Functional neuroimaging studies have suggested critical involvement of various thalamic and cortical structures in CP. The specific roles that the different brain regions play in CP and the specific pathophysiological processes involved remain to be determined from future research. It is inappropriate to draw any more than tentative conclusions based on data available to date. A frequent observation from PET and SPECT studies is a thalamic hypometabolism in the painful resting state. At the same time, allodynic stimulation can evoke a stronger thalamic signal than normal. Both observations can be explained by a partially denervated thalamus that also has major disruption of GABA-mediated inhibition. Although functional imaging studies cannot provide sufficiently detailed information to differentiate specific thalamic nuclei, there is some evidence that the lateral and posterior portions of thalamus are more altered in imaging studies.

The cortical regions associated with CP are varied across studies. Even considering different aspects of CP separately (ongoing pain, cold allodynia, and tactile allodynia), it is not evident which cortical areas are physiologically altered in CP and which have an essential role in its development or maintenance. Many studies have indicated that parietal cortex (S1 in particular) is functionally altered in CP; however, some studies fail to find evidence of such abnormalities. Other brain regions that are sometimes found altered in CP include frontal cortical regions (including M1), cingulate cortex, insular cortex, and basal ganglia. It is quite likely that factors such as lesion location, lesion extent, time elapsed since injury, and yet-to-be-determined physiological factors determine CP expression and the precise physiological consequences. More definitive understanding of CP mechanisms will await future work with larger samples, in which functional neuroimaging is combined with other detailed evaluations of clinical phenomena, perceptual alterations, and other measures of brain structure and function.

ACKNOWLEDGMENT

Portions of this work were supported by the National Institute of Neurological Disorders, National Institutes of Health (NS-39337, NS-38493).

REFERENCES

Andersen, G., K. Vestergaard, M. Ingeman-Nielsen, and T. S. Jensen. 1995. Incidence of central post-stroke pain. *Pain* 61: 187–193.

Apkarian, A. V., M. C. Bushnell, R. D. Treede, and J. K. Zubieta. 2005. Human brain mechanisms of pain perception and regulation in health and disease. *Eur. J. Pain* 9 (4): 463–484.

Attal, N., V. Gaudé, L. Brasseur, M. Dupuy, F. Guirimand, F. Parker, and D. Bouhassira. 2000. Intravenous lidocaine in central pain—A double-blind, placebo-controlled, psychophysical study. *Neurology* 54 (3): 564–574.

Baron, R., Y. Baron, E. Disbrow, and T. P. L. Roberts. 1999. Brain processing of capsaicin-induced secondary hyperalgesia—A functional MRI study. *Neurology* 53 (3): 548–557.

Boivie, J., G. Leijon, and I. Johansson. 1989. Central post-stroke pain—A study of the mechanisms through analyses of the sensory abnormalities. *Pain* 37: 173–185.

Bowsher, D., G. Leijon, and K. A. Thoumas. 1998. Central poststroke pain—Correlation of MRI with clinical pain characteristics and sensory abnormalities. *Neurology* 51 (5): 1352–1358.

Cahana, A., A. Carota, M. L. Montadon, and J. M. Annoni. 2004. The long-term effect of repeated intravenous lidocaine on central pain and possible correlation in positron emission tomography measurements. *Anesthesia Analgesia* 98 (6): 1581–1584.

Canavero, S. and V. Bonicalzi. 2007. *Central Pain Syndrome*. Cambridge: Cambridge University Press.

Canavero, S., V. Bonicalzi, C. A. Pagni, G. Castellano, R. Merante, S. Gentile, G. B. Bradac, et al. 1995. Propofol analgesia in central pain: Preliminary clinical observations. *Journal of Neurology* 242 (9): 561–567.

Casey, K. L. 2007. Pathophysiology of central poststroke pain: The contribution of functional imaging and a hypothesis. In *Central Neuropathic Pain: Focus on Poststroke Pain*, eds. J. L. Henry, A. Panju, and K. Yashpal, 115–131. Seattle: IASP Press.

Cesaro, P., M. W. Mann, J. L. Moretti, G. Defer, B. Roualdès, J. P. Nguyen, and J. D. Degos. 1991. Central pain and thalamic hyperactivity: A single photon emission computerized tomographic study. *Pain* 47: 329–336.

Craig, A. D. 1998. A new version of the thalamic disinhibition hypothesis of central pain. *Pain Forum* 7: 1–14.

———. 2007. Mechanisms of thalamic pain. In *Central Neuropathic Pain: Focus on Poststroke Pain*, eds. J. L. Henry, A. Panju, and K. Yashpal, 81–99. Seattle: IASP Press.

Davis, K. D., Z. H. T. Kiss, R. R. Tasker, and J. O. Dostrovsky. 1996. Thalamic stimulation-evoked sensations in chronic pain patients and in nonpain (movement disorder) patients. *Journal of Neurophysiology* 75: 1026–1037.

Di Piero, V., A. K. P. Jones, F. Iannotti, M. Powell, D. Perani, G. L. Lenzi, and R. S. J. Frackowiak. 1991. Chronic pain: A PET study of the central effects of percutaneous high cervical cordotomy. *Pain* 46: 9–12.

Ducreux, D., N. Attal, F. Parker, and D. Bouhassira. 2006. Mechanisms of central neuropathic pain: a combined psychophysical and fMRI study in syringomyelia. *Brain* 129 (4): 963–976.

Finnerup, N. B., I. L. Johannesen, A. Fuglsang-Frederiksen, F. W. Bach, and T. S. Jensen. 2003. Sensory function in spinal cord injury patients with and without central pain. *Brain* 126 (Pt 1): 57–70.

Finnerup, N. B., L. Sorensen, F. Biering-Sorensen, I. L. Johannesen, and T. S. Jensen. 2007. Segmental hypersensitivity and spinothalamic function in spinal cord injury pain. *Experimental Neurology* 207 (1): 139–149.

Fukumoto, M., T. Ushida, V. S. Zinchuk, H. Yamamoto, and S. Yoshida. 1999. Contralateral thalamic perfusion in patients with reflex sympathetic dystrophy syndrome. *Lancet* 354 (9192): 1790–1791.

Greenspan, J. D., S. E. Joy, S. L. B. McGillis, C. M. Checkosky, and S. J. Bolanowski. 1997. A longitudinal study of somesthetic perceptual disorders in an individual with a unilateral thalamic lesion. *Pain* 72 (1–2): 13–25.

Greenspan, J. D., R. R. Lee, and F. A. Lenz. 1999. Pain sensitivity alterations as a function of lesion location in the parasylvian cortex. *Pain* 81 (3): 273–282.

Greenspan, J. D., S. Ohara, E. Sarlani, and F. A. Lenz. 2004. Allodynia in patients with poststroke central pain (CPSP) studied by statistical quantitative sensory testing within individuals. *Pain* 109 (3): 357–366.

Hains, B. C., J. P. Klein, C. Y. Saab, M. J. Craner, J. A. Black, and S. G. Waxman. 2003. Upregulation of sodium channel Nav1.3 and functional involvement in neuronal hyperexcitability associated with central neuropathic pain after spinal cord injury. *Journal of Neuroscience* 23 (26): 8881–8892.

Head, H., and G. Holmes. 1911. Sensory disturbances from cerebral lesions. *Brain* 34: 102–254.

Henry, J. L., A. Panju, and K. Yashpal. 2007. *Central Neuropathic Pain: Focus on Poststroke Pain*. Seattle: IASP Press.

Hirato, M., S. Horikoshi, Y. Kawashima, K. Satake, T. Shibasaki, and C. Ohye. 1993. The possible role of the cerebral cortex adjacent to the central sulcus for the genesis of central (thalamic) pain—a metabolic study. *Acta Neurochir.* 58 (Suppl.): 141–144.

Hofbauer, R. K., H. W. Olausson, and M. C. Bushnell. 2006. Thermal and tactile sensory deficits and allodynia in a nerve-injured patient: a multimodal psychophysical and functional magnetic resonance imaging study. *Clin J Pain* 22 (1): 104–108.

Hsieh, J. C., M. Belfrage, S. Stone-Elander, P. Hansson, and M. Ingvar. 1995. Central representation of chronic ongoing neuropathic pain studied positron emission tomography. *Pain* 63: 225–236.

Iadarola, M. J., M. B. Max, K. F. Berman, M. G. Byas-Smith, R. C. Coghill, R. H. Gracely, and G. J. Bennett. 1995. Unilateral decrease in thalamic activity observed with positron emission tomography in patients with chronic neuropathic pain. *Pain* 63: 55–64.

Jeanmonod, D., M. Magnin, and A. Morel. 1996. Low-threshold calcium spike bursts in the human thalamus—Common physiopathology for sensory, motor and limbic positive symptoms. *Brain* 119 (2): 363–375.

Jones, A. K. P., H. Watabe, V. J. Cunningham, and T. Jones. 2004. Cerebral decreases in opioid receptor binding in patients with central neuropathic pain measured by [11C]diprenorphine binding and PET. *European Journal of Pain* 8 (5): 479–485.

Kim, J. H., J. D. Greenspan, R. C. Coghill, S. Ohara, and F. A. Lenz. 2007. Lesions limited to the human thalamic principal somatosensory nucleus (ventral caudal) are associated with loss of cold sensations and central pain. *J. Neurosci.* 27 (18): 4995–5004, PM:17475808.

LaTerre, E. C., A. G. De Volder, and A. M. Goffinet. 1988. Brain glucose metabolism in thalamic syndrome. *Journal of Neurology, Neurosurgery and Psychiatry* 51: 427–428.

Lee, M. S., I. S. Choi, and T. S. Chung. 1989. Thalamic syndrome and cortical hypoperfusion on technitium-99m HM–PAO brain SPECT. *Yonsei. Med. J.* 30: 151–157.

Leijon, G., J. Boivie, and I. Johansson. 1989. Central post-stroke pain—neurological symptoms and pain characteristics. *Pain* 36: 13–25.

Lenz, F. A., R. H. Gracely, F. H. Baker, R. T. Richardson, and P. M. Dougherty. 1998. Reorganization of sensory modalities evoked by microstimulation in region of the thalamic principal sensory nucleus in patients with pain due to nervous system injury. *Journal of Comparative Neurology* 399 (1): 125–138.

Lenz, F. A., H. C. Kwan, J. O. Dostrovsky, and R. R. Tasker. 1989. Characteristics of the bursting pattern of action potentials that occurs in the thalamus of patients with central pain. *Brain Research* 496: 357–360.

Lenz, F. A., H. C. Kwan, R. Martin, R. Tasker, R. T. Richardson, and J. O. Dostrovsky. 1994. Characteristics of somatotopic organization and spontaneous neuronal activity in the region of the thalamic principal sensory nucleus in patients with spinal cord transection. *Journal of Neurophysiology* 72: 1570–1587.

Maihofner, C., C. Forster, F. Birklein, B. Neundorfer, and H. O. Handwerker. 2005. Brain processing during mechanical hyperalgesia in complex regional pain syndrome: a functional MRI study. *Pain* 114 (1–2): 93–103.

Merskey, H. and N. Bogduk. 1994. *Classification of Chronic Pain.* Seattle: IASP Press.

Nathan, P. W., M. C. Smith, and A. W. Cook. 1986. Sensory effects in man of lesions of the posterior columns and of some other afferent pathways. *Brain* 109: 1003–1041.

Ness, T. J., E. C. San Pedro, J. S. Richards, L. Kezar, H. G. Liu, and J. M. Mountz. 1998. A case of spinal cord injury-related pain with baseline rCBF brain SPECT imaging and beneficial response to gabapentin. *Pain* 78 (2): 139–143.

Pattany, P. M., R. P. Yezierski, E. G. Widerstrom-Noga, B. C. Bowen, A. Martinez-Arizala, B. R. Garcia, and R. M. Quencer. 2002. Proton magnetic resonance spectroscopy of the thalamus in patients with chronic neuropathic pain after spinal cord injury. *AJNR Am J Neuroradiol.* 23 (6): 901–905.

Peyron, R., L. Garcia-Larrea, M. P. Deiber, L. Cinotti, P. Convers, M. Sindou, F. Mauguière, and B. Laurent. 1995. Electrical stimulation of precentral cortical area in the treatment of central pain: Electrophysiological and PET study. *Pain* 62: 275–286.

Peyron, R., L. García-Larrea, M. C. Grégoire, P. Convers, F. Lavenne, L. Veyre, J. C. Froment, F. Mauguière, D. Michel, and B. Laurent. 1998. Allodynia after lateral-medullary (Wallenberg) infarct—A PET study. *Brain* 121 (2): 345–356.

Peyron, R., L. García-Larrea, M. C. Grégoire, P. Convers, A. Richard, F. Lavenne, F. G. Barral, F. Mauguière, D. Michel, and B. Laurent. 2000. Parietal and cingulate processes in central pain. A combined positron emission tomography (PET) and functional magnetic resonance imaging (fMRI) study of an unusual case. *Pain* 84 (1): 77–87.

Radhakrishnan, V., J. Tsoukatos, K. D. Davis, R. R. Tasker, A. M. Lozano, and J. O. Dostrovsky. 1999. A comparison of the burst activity of lateral thalamic neurons in chronic pain and non-pain patients. *Pain* 80 (3): 567–575.

Seghier, M. L., F. Lazeyras, P. Vuilleumier, A. Schnider, and A. Carota. 2005. Functional magnetic resonance imaging and diffusion tensor imaging in a case of central poststroke pain. *The Journal of Pain* 6 (3): 208–212.

Seifert, F., and C. Maihofner. 2007. Representation of cold allodynia in the human brain—A functional MRI study. *NeuroImage* 35 (3): 1168–1180.

Treede, R. D., T. S. Jensen, J. N. Campbell, G. Cruccu, J. O. Dostrovsky, J. W. Griffin, P. Hansson, R. Hughes, T. Nurmikko, and J. Serra. 2008. Neuropathic pain: redefinition and a grading system for clinical and research purposes. *Neurology* 70 (18): 1630–1635.

Triggs, W. J., and A. Beric. 1994. Dysaesthesiae induced by physiological and electrical activation of posterior column afferents after stroke. *Journal of Neurology, Neurosurgery and Psychiatry* 57: 1077–1080.

Vestergaard, K., J. Nielsen, G. Andersen, M. Ingeman-Nielsen, L. Arendt-Nielsen, and T. S. Jensen. 1995. Sensory abnormalities in consecutive, unselected patients with central post-stroke pain. *Pain* 61: 177–186.

Viswanathan, A., and R. D. Freeman. 2007. Neurometabolic coupling in cerebral cortex reflects synaptic more than spiking activity. *Nat Neurosci* 10 (10): 1308–1312.

Weng, H. R., J. I. Lee, F. A. Lenz, A. Schwartz, C. Vierck, L. Rowland, and P. M. Dougherty. 2000. Functional plasticity in primate somatosensory thalamus following chronic lesion of the ventral lateral spinal cord. *Neuroscience* 101 (2): 393–401.

Willoch, F., F. Schindler, H. J. Wester, M. Empl, A. Straube, M. Schwaiger, B. Conrad, and T. R. Tolle. 2004. Central poststroke pain and reduced opioid receptor binding within pain processing circuitries: a [11C]diprenorphine PET study. *Pain* 108 (3): 213–220.

Witting, N., R. C. Kupers, P. Svensson, and T. S. Jensen. 2006. A PET activation study of brush-evoked allodynia in patients with nerve injury pain. *Pain* 120 (1–2): 145–154.

15 Human Brain Imaging Studies of Chronic Pain
Translational Opportunities

A. Vania Apkarian

CONTENTS

The advent of non-invasive human brain imaging technologies provided the opportunity for direct examination of the human brain. This occurred about 15 years ago with the related expectation that we were at the threshold of a revolution in our understanding of chronic pain. This expectation remains largely unfulfilled, although much has been published in the topic. Here we concentrate mainly on our own work in the topic, arguing in general that the subject of brain mechanisms of chronic pain remains in its infancy mainly because of a heavy emphasis in the field on studying nociception rather than chronicity of pain.

The definition of chronic pain, namely, pain sustained beyond the healing process [53], says nothing regarding underlying mechanisms and perhaps suggests that mechanisms similar to acute pain are maintained for a longer period. Early brain imaging attempts to characterize chronic pain adopted the same methods used for studying nociception, where acute painful stimuli of various dimensions were applied to clinical chronic pain conditions seeking to observe differences in brain activity. The assumption was that the interaction between chronic and acute pain could be identified at the level of brain activity. Often, such studies were conducted without even performing simple psychophysical tests to determine whether the procedure gives rise to differences in pain perception [3]. More recent studies along these lines of thought have actually demonstrated that fibromyalgia and chronic back pain (CBP) patients report pressure pain at lower stimulus amplitudes, and demonstrate that related brain activity, even after correcting for this difference in sensitivity, showed increased responses in brain regions involved in acute pain perception [29;31;32]. The weakness of this approach is its lack of specificity. The stimulus location seems to have no relationship to the sites where chronic pain is being felt, for example, relative to the tender points in fibromyalgia patients. Also the change in pressure pain seems to be similar in both CBP and fibromyalgia even though the first group does not report generalized body pain while the latter does, that is, the procedure does not distinguish between chronic pain conditions that have distinct clinical reports. As animal studies of peripheral and spinal cord mechanisms of chronic pain emphasize peripheral and central sensitization, other studies have focused on brain activity for brief sensitization paradigms. Most notably, capsaicin-induced sensitization and brain activity for painful stimuli applied within this field have been proposed as models for events that may reflect chronic pain conditions [13;43;44;50;51;76;82]. Such studies equate chronic pain to brain processes that underlie sensitization that only lasts for a few minutes to a few hours, and assume that reversible mechanisms for transient sensitization adequately model chronic pain conditions that commonly can result in a state of suffering that could last a lifetime.

Rather than building surrogate models of chronic pain in humans, we have taken the approach of studying these clinical conditions directly. Over the last 10 years we have attempted to generate methodologies with which properties of various chronic pain conditions can be examined. To test the validity of these approaches we have then tested the reverse translational approach of applying the information garnered in humans as to its applicability to rodent models of chronic pain, specifically with the notion of developing new drug therapies for chronic pain.

15.1 PSYCHOLOGICAL, COGNITIVE, AND SENSORY ABNORMALITIES

Depression is ranked as one of the strongest predictors for low back pain. This association is observed by multiple studies, with odds ratios increasing with intensity of back pain and severity of depression [55;60]. In order to investigate the predictive power of baseline depression on the transition from acute to chronic pain (3 months post-acute back pain), a recent prospective study evaluated the direct and indirect effects of multiple parameters on chronic pain severity and disability [81]. The model only accounted for 26% of the variance in chronic pain. Acute pain intensity did not directly predict pain three months later, and baseline pain beliefs failed to predict chronic pain magnitude. Despite these relatively weak relationships to chronic pain, the authors argued that their findings support the growing literature contending that progression to chronic pain is more dependent on psychosocial and occupational factors than on medical characteristics of the spinal condition. In general, a long series of studies now describe psychosocial and psychological factors in predicting functional and social disability, where the interrelationship between ratings of catastrophizing, pain-related fear of (re-) injury, depression, disability, and pain severity are studied and modeled in combination with demographics in various chronic pain conditions. Although these factors may be associated with pain in certain individuals, attempts to create models of CBP based upon them have been unproductive [38;47;74;79]; for further details see [2]. It is now being recognized that psychosocial factors constitute "non-negligible risks" for the development of low back pain [17] and cannot account for how or why a patient transitions into the chronic pain state.

We have examined cognitive and sensory properties of chronic pain patients, with the simple notion that living with chronic pain may impart a cost in such processing. A long list of cognitive abnormalities has been described in chronic pain patients. The most noteworthy are attentional and memory deficits [18;68]. However, little effort has been placed in differentiating such deficits based on chronic pain type. We reported that, in contrast to matched healthy controls, CBP and complex regional pain syndrome (CRPS) patients are significantly impaired on an emotional decision-making task [5]. Moreover, the performance of CBP patients was highly correlated with their verbal report of pain at the time of performing the task. In contrast to CBP, CRPS patients' performance was not modified when their pain was manipulated using a sympathetic block. The latter implies that the brain mechanisms underlying the two types of chronic pain, or the impact of each condition on the brain, may be distinct and thus also distinctly modulate emotional states. It should be noted that the CRPS patients were tested on a long battery of other cognitive tasks as well, and their performance on these was not different from healthy control subjects. Two important conclusions are drawn from these observations. Firstly, cognitive disruption in chronic pain is specific to the type of test administered, implying impact on unique brain circuitry. Secondly, there are important differences between chronic pain conditions, even on the same cognitive task, suggesting that each condition may underlie unique brain activity/reorganization, which complicates our understanding of these conditions as it demands that we study each and every chronic pain regarding its cognitive and brain signature. More importantly, the implication that distinct chronic pain conditions have unique brain signatures open the possibility that each one of them may be understood in its own right, enabling the development of novel, specific therapies for each. This theme will be repeated several times below, emphasizing that brain-derived parameters repeatedly indicate the specific properties of distinct chronic pain conditions.

It should be emphasized that the impact of chronic pain does not only result in deficits. When CBP patients were contrasted to healthy subjects as to gustatory abilities, we were able to show decreased threshold to gustatory detection and increased sensitivity to supra-threshold tastants for all modalities examined [69]. Thus, these CBP patients are generally more sensitive in taste perception. It is possible that this ability is a predisposing factor. In fact, all the brain and cognitive changes we describe here are cross-sectional studies, and thus in all cases the distinction between predisposition and a consequence to chronic pain remains unsettled. More likely, increased taste sensitivity is a result of brain representation/reorganization as a consequence of living with CBP. We ascribe the deficit in emotional decision-making in CBP and CRPS as a consequence of the representation/reorganization of brain activity between the lateral and medial prefrontal cortex.

In fact, our observation of activity in medial prefrontal cortex in CRPS [7] was the initial impetus for testing chronic pain patients with the "gambling test." We also think that the increased taste sensitivity is a direct result of changes in activity/connectivity of the insular cortex in CBP. Our earlier observation that insular cortex activity increases in CBP patients in proportion to the number of years they are in chronic pain [10], and given that parts of the insular cortex are considered primary gustatory cortex, led us to the hypothesis that gestation may be different in CBP. Therefore, the cognitive and sensory changes we have observed in chronic pain patients are derived from observations regarding cortical representation and reorganization, and thus these domains are complementary to each other, reinforcing the idea that the brain abnormalities do result in very specific changes in information processing.

15.2 BRAIN METABOLITES

We published the first study showing that brain chemistry is abnormal in CBP patients as compared to matched healthy controls, using magnetic resonance spectroscopy (MRS) [33]. Our study revealed decreased N-acetyl-aspartate in the lateral prefrontal cortex, as well as correlations between brain regional chemistry and clinical parameters of pain duration, intensity, and McGill Pain Questionnaire dimensions. We found that the relative concentrations of chemicals in the cingulate cortex and thalamus reflected pain duration (in opposite directions). Moreover, chemical concentrations were found to positively correlate with sensory, affective, and intensity ratings of CBP. A small number of similar studies have been published since, in a number of chronic pain conditions [23;59;70], yet the topic remains in its infancy. It is likely that the method would provide clinically important information regarding various chronic pain conditions, especially as it is becoming an important biomarker in neurodegenerative conditions. For example, metabolic changes are observed in presymptomatic mutation carriers years before onset of Alzheimer's disease [30], suggesting that metabolic markers may also be useful in predicting predisposition to chronic pain. A recent study in fibromyalgia patients indicates that MRS may also be useful in assessing levels of glutamate in the brain, and further that this signal seems to be modulated with clinical parameters as well as with acute painful stimulation [40]. Even though assessment of metabolic signals seems promising, it does suffer from important weaknesses, the main difficulty being the lack of standardized methods for localizing brain regions studied. Thus, reproducing results even in the same subject remains problematic. Moreover, the MRS signal is contaminated with the properties of the tissue examined since the concentration of all metabolites is influenced by the proportion of CSF to white to gray matter within any region examined, and corrections for such contaminations remain inadequate. Current MRS acquisition methods are time-consuming and only enable collecting a small number of single voxels; multi-voxel MRS in turn suffers from more severe tissue cross-contamination artifacts. Given that chronic pain patients find it uncomfortable to remain immobile, head position cannot be assumed to be fixed within and especially across patients and controls. There is no question that MRS signals are distorted by head motion, yet there are currently no acceptable means of correcting for this artifact. Thus, in general, technical difficulties complicate the implementation and interpretation of results obtained by this approach.

An alternative approach is the use of positron emission tomography (PET) to examine binding changes for various ligands in chronic pain. With this approach a recent study identified mu-opiate binding decreases in fibromyalgia, with the decrease being related to the pain characteristics in a number of brain regions [39]. Dopamine (D) release in the basal ganglia is also disrupted in fibromyalgia patients [78] and in those with burning mouth syndrome and atypical facial pain [34;35]. Additionally, D2 binding in the basal ganglia has been proposed as a marker for diagnosis and treatment of chronic pain [36]. A recent review article discusses the impact as well as limitation of PET opiate receptor binding studies in general [41], emphasizing its impact on acute and chronic pain conditions. The authors conclude that a major obstacle in the field is the limited number of tracers available and their binding specificity for subclasses of opiate receptors.

15.3 SPONTANEOUS PAIN

The primary complaint of chronic pain patients is spontaneous pain. The large majority of such patients seek health care to relieve pain that is ongoing, that fluctuates unpredictably, and results in decreased quality of life and increased anxiety and depression. When a physician asks the patient to rate his/her pain, the physician is specifically attempting to capture the intensity of spontaneous pain. Similarly, in the large majority of clinical drug trials, the most commonly used primary outcome is a visual analogue scale of spontaneous pain intensity. Thus, therapies for pain relief by and large have also targeted the diminution or silencing of spontaneous pain. This is an important issue given that there are currently no convincing methods for studying spontaneous pain in animal models. Moreover, perhaps by borrowing from animal studies, human brain imaging studies (outside of our lab) have only studied stimulus-evoked brain activity for acute and chronic pain.

We recently revealed that the spontaneous pain of chronic pain patients fluctuates in the scale of seconds to minutes, that these fluctuations are distinct for various chronic pain conditions, and that normal healthy subjects are unable to mimic them [22]. Participants were instructed to continuously rate their subjective assessment of the intensity of pain. The primary observation of this study is the notion that spontaneous pain fluctuates enough that simply monitoring it for 10 minutes is sufficient to characterize and use its properties to distinguish between types of chronic pain. We observed that the fluctuations of spontaneous pain do not possess stable mean or variance, implying that these time series can be characterized better by a non-linear, fractal analysis. To this end, we applied time and frequency domain techniques to characterize variability of pain ratings with a single parameter: fractal dimension, D. We demonstrated that D is distinct between types of chronic pain, and from ratings of thermal stimulation and of imagined pain; and that there is a correspondence between D for pain ratings and D for brain activity in CBP patients using fMRI. This study remains the only one where spontaneous pain fluctuations at such time scales have been characterized. In this study we showed that post-herpetic neuralgia (PHN) patients' fluctuations of spontaneous pain had temporal properties distinct from that of CBP patients. If we make the simple assumption that the temporal fluctuations of spontaneous pain are a reflection of the interaction between peripheral nociceptive activity and CNS processes, then these results suggest that these interactions are distinct between PHN and CBP, implying that central processes involved should be unique for each.

15.4 BRAIN ACTIVITY DURING CHRONIC PAIN

Once the temporal properties of spontaneous pain were characterized, it provided the background upon which we could begin examining related brain activity using fMRI. We have now used this approach to study CBP [10], PHN [25;28], pelvic pain (PP), and osteoarthritis (OA) [12]. The studies in CBP, PHN, and PP examined brain activity for spontaneous pain, while in OA spontaneous pain showed very little temporal modulation, forcing us to instead examine brain activity for pressure applied to the painful knee joint. In PHN, we studied brain activity for spontaneous pain as well as for touch-evoked pain (tactile allodynia) in the same group of patients. In CBP, PHN, and OA we also tested the modulation of resultant brain activity with the use of therapy, testing the specificity of brain activity in relation to pain relief with therapy.

Using non-invasive brain imaging (fMRI) in combination with online ratings of fluctuations of spontaneous pain, we identified the brain activity idiosyncratic to CBP [10]. The data were analyzed using two different vectors: (1) when ratings of spontaneous pain were high in contrast to low, and (2) when ratings of spontaneous pain were rapidly increasing in contrast to all other times. The brain activity obtained, after subtracting a visual rating task that corrects for the cognitive, evaluative, and motor confounds, differed greatly for the two conditions. During epochs when pain was high, activation of the medial prefrontal cortex (mPFC) was most robust, with less activity seen in the amygdala and the ventral striatum. However, for periods when pain was rapidly increasing, the insula, anterior cingulate cortex (ACC), multiple cortical parietal regions, and the cerebellum became activated. In the same study, using the same procedures (continuous ratings of perceived pain and subtraction of a visual control), we identified brain activity in back pain patients and healthy controls for an acute thermal stimulus applied to the back. The results showed no difference between patients and healthy controls for brain regions activated during acute thermal pain stimulation of the back. This activity pattern closely matched brain activity observed in earlier studies regarding acute pain in healthy subjects [3] and was similar also to the activity we observed for spontaneous pain for the contrast of rapidly increasing pain. We also studied two separate groups of CBP patients using two MRI magnets and in both groups identified the mPFC as the primary region activated for high pain. Moreover, in both groups mPFC activity was strongly correlated with pain intensity. Moreover, the insula activity, during the increasing phase of pain, predicted the duration of pain in years. In contrast, anxiety or depression levels were not related to brain activations identified in relation to spontaneous pain. These results imply that spontaneous CBP engages the limbic emotional-mentalizing regions of the brain into a state of continued negative emotions (suffering) regarding the self, punctuated by occasional nociceptive inputs that perpetuate the state. The sustained prefrontal activity is most likely related to the maladaptive psychological and behavioral cost associated with chronic pain.

Essentially the same approach was used to study brain activity for spontaneous pain in PHN patients [25]. Overall brain activity for spontaneous pain of PHN involved affective and sensory-discriminative areas (thalamus, primary and secondary somatosensory, insula and anterior cingulate cortices), as well as areas involved in emotion, hedonics, reward, and punishment (ventral striatum, amygdala, orbital frontal cortex, and ventral tegmental area). Thus, in PHN more extensive brain regions are involved in spontaneous pain than in CBP, yet similar limbic and prefrontal regions are also observed for both conditions. PHN is the prototypical neuropathic chronic pain condition as it clearly involves peripheral nerve injury. In contrast, CBP is far more complex a condition and can involve multiple sources of nociceptive inputs, including muscle, fascia, joints, and nerve injury. In fact, the majority of CBP cases are idiopathic—that is, we have no clue concerning the source of the peripheral nociceptive signal. A simple, legitimate question is, why should CBP involve a limited prefrontal/limbic brain activity while PHN encompasses a wider brain circuitry including many areas seen for acute pain?

We now have preliminary results for brain activity for spontaneous pain in a small group of PP patients. The results again indicate an activity pattern quite different from both CBP and PHN. We have yet to understand the underlying processes and parameters that are dictating these diverse brain activations for seemingly similar pain conditions. The PP condition can be thought of as a visceral chronic pain, whereas CBP and PHN are certainly dominantly somatic in origin. Yet, brain imaging studies for acute visceral and somatic pain show very small differences [72]. Why are the chronic conditions resulting in such diverse activations? It needs to be emphasized that it is unlikely that these patterns are a result of random sampling of diverse conditions where, depending on the number of patients included, one would observe distinct patterns of brain activity. On the contrary, the activity patterns we observe for each of these conditions, be it CBP, PHN, or PP, seem homogeneous in that the brain activity patterns for each condition are robustly reproducible. For example, we have examined CBP brain activity in three separate studies and using 1.5 T magnet and 3.0 T scanners, and in each case the activity patterns are essentially identical. Moreover, if we subdivide any given group data and compare subgroupings, the activations show very similar results to the whole data set. We have recently reported on brain activity for knee OA [11]. The study contrasted spontaneous pain for CBP to knee pressure-induced brain activity for knee OA. OA painful mechanical knee stimulation was associated with bilateral activity in the thalamus, secondary somatosensory, insular, and cingulate cortices, and unilateral activity in the putamen and amygdala. There was no brain activity overlap between knee OA and CBP spontaneous pain, which again was mainly associated with mPFC activity. In knee OA we were unable to study spontaneous pain because these patients reported minimal spontaneous pain and even when present this pain was for the most part constant. The source of the brain activity differences between these two groups remains unclear. It may be partially due to the type of pain studied, spontaneous versus knee pressure, and partly due to underlying mechanisms (inflammatory vs. other sources) as well as other sources. Still, the knee OA activity best resembles activity observed for acute pain, and yet when OA and CBP patients are examined using questionnaire-based outcomes, they cannot be differentiated from each other [20].

The issue of brain activity and its dependence on pain modality was directly tested in PHN patients, where we studied spontaneous pain and touch-evoked pain (dynamic tactile allodynia) in the same group of patients [25;27]. Essentially the same brain regions were activated for spontaneous and touch-evoked pain. However, within each of the brain regions distinct subportions were associated with each modality, with minimal overlap between the conditions. One worries about cross-contamination of activity from one modality into the other. Yet, we had taken multiple steps to correct for such confounds. Thus, at least in PHN we can state that subtle brain activity differences across the same brain regions can give rise to either perception of spontaneous or touch-evoked pain.

Another important issue that needs highlighting is that across the diverse chronic pain conditions we have characterized (CBP, PHN, OA, and PP), there is a large spectrum of brain activity patterns, with some resembling acute pain more than others; and in all cases limbic and prefrontal cortical activity is observed repeatedly, and especially for the more neuropathic conditions. These observations contradict the standard notion that chronic pain is mainly a consequence of peripheral and spinal cord sensitization, in which case the expected brain activity would simply be an enhancement of activity of brain regions involved in acute pain. Novel activations in limbic and prefrontal cortical regions instead imply that the sensory/emotional construct of the pain is distinct between various chronic pain conditions and that the interaction between pain and emotional and hedonic circuitry must be considered as part of the definition of these pain conditions. Our confidence regarding the importance of activated brain regions in chronic pain stems mainly from correlational analyses, where various clinical parameters were tested as to their relationship to observed brain activity. For example, in CBP we could relate the visual analog scale of back pain intensity on the day of scan to mPFC activity and relate the duration of chronic pain in number of years living with the condition with insular activity. Similarly, limbic brain activity in PHN patients was tightly correlated to questionnaire-based outcomes regarding the neuropathic properties of their pain. Thus, fundamental clinical properties of these chronic pain conditions are strongly tied to the specific brain activity underlying at least spontaneous pain.

15.5   BRAIN ACTIVITY FOR THERAPIES FOR CHRONIC PAIN

As chronic pain remains mostly intractable, its modulation by therapeutic approaches is complicated and hard to tackle in relation to brain activity. However, we have now demonstrated that therapeutic manipulations are a powerful method with which we can at least improve our confidence that observed brain activity is relevant to the conditions being studied, and perhaps even advance new knowledge of the efficacy and route of action for some of these manipulations. Pharmacology-based fMRI has been commented on in the past [15;77]. Here we will concentrate on using this tool in manipulating chronic pain. To simplify analysis and minimize contamination of pain-related brain activity with direct drug effects, we have opted to study drugs that for the most part have minimal central effects. Thus, we have used 5% lidocaine patch as our main tool for reducing pain locally. As the patch is applied to the skin, blood levels of lidocaine are small, and thus the effects must be considered primarily local. We have also tested a cox2-inhibitor, which should also be primarily acting on the local inflamed tissue. However, in this case we cannot rule out a spinal cord effect as well.

In a psoriatic arthritis patient whose pain was adequately managed by a cox2-inhibitor, we examined the effects of this drug on pain relief in relation to brain activity [9]. The effort was to first demonstrate that single-subject studies of chronic pain are feasible and useful, especially when coupled with drug manipulation. To generate enough data to perform statistical testing the subject was scanned four times, after he had stopped his medication for 24 hours. Then he ingested a single dose of the drug and was scanned again six more times at different time delays from ingestion. Brain activity was determined for joint pressure pain ratings. The brain activity for palpating the painful joints included bilateral insula, thalamus, and secondary somatosensory cortex, as well as contralateral primary somatosensory cortex and mid-anterior cingulate. This activation pattern is very similar to that seen for acute pain in healthy subjects, suggesting that psoriatic pain is akin to acutely activating nociceptors. It is also very similar to the activity in knee OA, although the latter involves more limbic activations as well. Regarding the effects of the cox2-inhibitor, we observed a decrease in joint palpation pain, and this tightly correlated with decreased activity in the insula and secondary somatosensory cortex. This study demonstrates the feasibility of studying effects of a single dose of an analgesic on brain activity for a clinical pain condition in an individual subject. The methodology provides an objective approach that may be used for drug development and testing effects of drugs in individual cases.

We have now reported changes in brain activity with topical lidocaine patch use in CBP, OA, and PHN patients [12;25;27]. The main outcome of these studies is the observation that modulation of chronic pain with this manipulation results in decreased brain activity in specific regions, within the set of regions identified as being active for the pain condition studied, providing an additional line of evidence that the identified brain areas are in fact involved in the perception associated with each condition. It should be emphasized that for each of the conditions, a unique set of brain regions were modulated with lidocaine therapy. We think these results point to the critical nodes of the brain circuitry involved in the pain perception for each condition, suggesting that the cognitive/emotional/sensory properties of each condition are unique. From the viewpoint of developing new therapies for these conditions, these differential responses provide clues concerning molecular pathways and neurotransmitters that may be manipulated for each case. The main weakness of these studies is the fact that they were all open-labeled. Thus, we cannot rule out placebo effects. As such, these studies do not really give us new clues as to the efficacy of the treatment. Instead, they provide a tool with which the properties of the brain activations can be explored pharmacologically.

15.6   BRAIN MORPHOLOGICAL CHANGES WITH CHRONIC PAIN

In 2004 we published the first brain morphometric study showing anatomical evidence for brain atrophy in CBP patients [6]. This result has now been replicated in CBP and other types of chronic pain conditions [48;65;66]. Notably we were able to show that these morphological changes are correlated with the clinical parameters of the condition. Neocortical gray matter volume, after correcting for intracranial volume, age, and sex, was significantly less in CBP patients than in matched controls. Moreover, this parameter showed dependence on pain duration, with similar slopes for patients with and without neuropathic (radicular) back pain, but only significantly for the neuropathic back pain group. When the same data were analyzed to directly compare regional gray matter differences between CBP patients and controls, two brain areas showed the most robust difference: dorsolateral lateral prefrontal cortex (DLPFC) and right thalamus. When we studied the DLPFC further in relationship to clinical parameters, gray matter density was found to be dependent on the presence and type (neuropathic or non-neuropathic) of CBP. Thus, regional gray matter changes are related to pain characteristics, and this pattern is different for neuropathic compared with non-neuropathic types. This dissociation is consistent with extensive clinical data showing that neuropathic pain conditions are more debilitating and have a stronger negative affect [19], and we suggested that this difference is directly attributable to the larger decrease in gray matter density in the DLPFC of neuropathic CBP patients.

We recently reported on gray matter morphological changes in CRPS [26] studied in the same lab using similar imaging and data analysis techniques as we had done in CBP. The results indicate that brain atrophy for the two clinical conditions affect non-overlapping brain regions, and yet in both the changes are correlated with duration and/or intensity of the pain. In the CRPS patients, diffusion tensor imaging (DTI) analyses were used to examine the relationship between gray matter decreased density and white matter connectivity. The study generally indicates that brain regions where gray matter is reduced are also accompanied with a general decrease in white matter connectivity, although in some cases this was also accompanied with target-specific increased connectivity as well. This is the first study linking gray matter changes to white matter properties, and the results are consistent with the general idea of loss of neurons but also suggest that it is a result of competitive reorganization of connectivity across brain regions. More importantly, this study demonstrates the power of combining various brain anatomical imaging techniques to begin to unravel the processes underlying brain reorganization with chronic pain.

15.7   PUTTING ALL THE HUMAN OBSERVATIONS TOGETHER

Above we reviewed cognitive abnormalities, brain activity patterns, and brain morphometry and connectivity changes observed in various chronic pain conditions. To what extent are these observations complementary and inter-related or contradict each other? The brain metabolic study suggested cell loss in the lateral PFC in CBP, and in fact we could observe morphologically that the main brain region showing atrophy in CBP is lateral PFC. Moreover, brain activity in CBP was mainly localized to medial PFC. Given that lateral and medial PFC inhibit each other and functionally compete with each other, the activation in medial PFC may be a consequence of the atrophy in lateral PFC. Both CBP and CRPS show deficits in emotional decision making but with distinct properties. The brain morphological results are consistent with this, as both groups show brain regional atrophy but involving distinct regions. Importantly, in CBP the lateral PFC shows atrophy, but in CRPS the medial PFC shows atrophy. These brain regions compete with each other—the former is associated to cognitive and memory-related tasks, and the latter is linked to emotional and self-related tasks. Perhaps the strongest evidence for consistency across these measures is the extent to which each of them is repeatedly observed to correlate with clinical parameters associated with distinct conditions. Moreover, for the same pharmacological treatment (topical lidocaine), pain relief in distinct chronic pain conditions involves distinct brain regional activity decreases.

15.8   ANIMAL MODELS AND CORRELATES FOR HUMAN CHRONIC PAIN

Animal models advanced over the last 20 years have revolutionized our understanding of chronic pain mechanisms. However, this work has for the most part concentrated on delineating abnormalities in afferent sensory inputs, spinal cord reorganization as a result of neuropathic or inflammatory injury, and changes in descending modulatory circuitry. All of this implicitly assumes that the role of the cortex in such conditions is a passive reflection of events occurring in the spinal cord. The above-reviewed human brain imaging studies, however, indicate an active role of the cortex in the processing of pain. In light of these new findings, and consistent with them, there is a growing literature of animal studies focusing on the full role of the central nervous system in chronic pain.

Recent animal studies show that cortical manipulations can modulate pain behavior [8;37;45;46;67]. Results emphasize the role of the insula, anterior cingulate, mPFC, and amygdala in pain, which are limbic structures with strong interconnectivity. Particularly relevant is a study by Johansen and Fields [46] demonstrating that anterior cingulate activity is necessary and sufficient for noxious stimuli to produce an aversive memory, via a glutamate-mediated neuronal activation. Anatomically, the anterior cingulate and mPFC are in close proximity to one another and tightly interconnected. It is possible that the two structures are involved in different phases of acquisition and extinction of pain-related memories. Researchers have also demonstrated that the NR2B component of the n-methyl-D-aspartic acid (NMDA) receptor undergoes transient upregulation within the anterior cingulate in rats following an inflammatory injury, and administration of NR2B receptor-selective antagonists inhibit behavioral responses to peripheral inflammation [80]. Two recent studies have further implicated anterior cingulate in contextual fear memory acquisition [52;83], as well as the amygdala and hippocampus, and show NMDA in anterior cingulate is critical for fear acquisition. A similar study examined plasticity of amygdala central nucleus neurons following induction of arthritis and showed that pain-related synaptic plasticity is accompanied by protein kinase A (PKA)-mediated, enhanced NMDA-receptor function and increased phosphorylation of NMDA-receptor 1 (NR1) subunits [14]. These results provide solid evidence that NMDA receptors undergo long-term plastic changes in the brain after injury and contribute to persistent pain by changing neuronal activity. Consistent and complementary to these results, we have evidence that rats with neuropathic injury show increased expression of cytokines in the prefrontal cortex and thalamus/striatum [4].

A nagging issue regarding existing animal models of chronic pain remains their correspondences to the human conditions. There are no reliable tests for establishing such correspondences. For example, chronic constriction injury (CCI) or spared nerve injury (SNI) models are equated to human CRPS or CBP by different investigators using arbitrary criteria. One hopes that future animal functional brain imaging studies may provide direct links between the models and human pain conditions. Recent studies have started exploring this issue, using other approaches. An animal model was recently developed to test gambling-like behavior in rodents [58]. The paradigm was developed to enable performing rodent studies for emotional decision making, in analogy to the human study where CBP and CRPS patients show impairments [5]. As this task requires intact medial PFC in humans, the rodent study examined changes in decision making in healthy animals and in animals with orbital frontal lesions, and demonstrated that the lesion enhances risky behavior in similarity to the human results. Moreover, the authors have preliminary results showing that rodents with inflammatory or neuropathic injuries also exhibit enhanced risky behavior, thus establishing a behavioral correspondence between rodents with persistent pain and human chronic pain patients on emotional decision making. Increased anxiety is an important comorbidity in patients with chronic pain. The first study on its prevalence in rodents with neuropathic injury examined its presence and modulation by morphine and gabapentin [62]. Importantly the authors show that some neuropathic models induce anxiety (CCI) but not others (partial sciatic nerve ligation), the implication being that human chronic pain conditions should also be differentiable along anxiety dimension. Other investigators have started differentiating the emotional component of neuropathic pain from its sensory properties [49]. Bushnell and colleagues (IASP abstract) recently demonstrated a direct correspondence of brain atrophy in rodents following neuropathic injury, where they tracked brain morphological changes in SNI animals and observed that months after the injury the animals exhibit prefrontal cortex atrophy as well as increased anxiety. We have previously shown that lateral thalamic electrophysiological properties of groups of neighboring neurons show multiple signs of reorganization minutes after a peripheral nerve injury and the process is sustained for the duration of monitoring [16]. More recently a study in Martina's lab demonstrated morphological and electrophysiological reorganization of medial PFC in SNI rats a week following the peripheral injury [54]. Although the body of literature is small, important advances are being made in relating the fundamental observations in human chronic pain to animal models. These approaches promise to lay a new foundation for developing objective methods with which human and rodent brain properties can be equated for specific chronic pain conditions.

15.9   NOVEL PHARMACOTHERAPY BASED ON THE ROLE OF THE CORTEX IN CHRONIC PAIN

The human studies regarding spontaneous pain suggest medial PFC and amygdala to be importantly involved in chronic pain, and also imply that lateral PFC atrophy may underlie the enhanced activity in medial PFC. Moreover, the rodent studies suggest that NMDA transmission in the cingulate cortex may play a critical role in neuropathic painlike behavior. These observations provide the notion that D-cycloserine (DCS) may be an antineuropathic drug acting through these cortical sites. We tested this by examining the effects of treatment of neuropathic injured rats with DCS [57]. DCS given systemically or centrally enhances cognitive processes [71], improves attention and memory [42;73], and facilitates fear extinction [61;75] through *de novo* memory trace formation involving NMDA plasticity [21;56;64]. We tested the effects of DCS on chronic neuropathic pain behavior, hypothesizing that it should enhance the extinction of pain-related memories and, thus, exhibit antinociceptive properties for neuropathic pain. The main finding was that repeated treatment with DCS, a partial agonist at the strychnine-insensitive glycine-recognition site on the NMDA receptor complex [24], reduces tactile sensitivity and protective paw posturing in rat models of neuropathic pain. This antinociception was dose-dependent and increased in efficacy for up to three weeks. Upon cessation of treatment, DCS effects on pain behavior persisted for a duration proportional to the length of treatment. When antinociception was assessed by measuring changes in mechanical sensitivity, the effect of the treatment was relatively small; however, a much larger effect of DCS treatment was revealed when antinociception was assessed by an operant stimulus avoidance task. Moreover, selective infusions of DCS into mPFC and amygdala (but not the spinal cord, thalamus, insula, or occipital cortices) were antinociceptive and dependent on NMDA receptor availability. We presume that DCS-induced reinforcement of NMDA-receptor mediated transmission within mPFC works to disengage spontaneous pain from its associations previously formed in learning and memory. To our knowledge this is the first drug therapy study directly derived from human brain imaging results, and thus its success reinforces the tenant that the human results can be used to study molecular pathways in rodents for efficacy in chronic pain. The effects of DCS remain to be studied in humans, and until we observe an effect in humans these results remain mostly of theoretical interest.

15.10   TOWARD A NEW THEORY OF CHRONIC PAIN

In light of the above briefly outlined results, chronic pain can no longer be viewed as a pure perceptual state, that is, persistence of pain. Instead it needs to incorporate emotional suffering and related behavioral/cognitive/hedonic modifications that in turn also modify decision making and behavior as a consequence of reorganization of brain circuitry. Overall, the brain activity patterns and the changes in morphology and connectivity show a general picture that more closely resembles the addicted brain and provides no evidence for increased sensory processing of pain. Reorganization and representation primarily involve limbic and prefrontal brain circuitry and minimally impinge on sensory properties of pain. We presume and hypothesize that these changes are a reflection of both the suffering and coping strategies that impinge on learning and memory and on hedonics of everyday experience. The cumulative cost of these behaviors and their specific interaction with peripheral and spinal cord reorganization as a consequence of the specific nociceptive barrage associated with the condition provide a unique cortical imprint for each chronic pain condition, which continues to reorganize due to the dynamics of the interaction of the presence of pain with everyday experiences.

We have proposed a new theory regarding mechanisms of transition to chronic pain based on the interaction between pain and memory, and based on the accumulating evidence of circuits involved or reorganized with distinct chronic pain conditions [1;2]: we argue that the state of the brain's emotional and motivational circuitry determines the suffering of chronic pain. Emotional suffering with chronic pain is manifested by increased anxiety, depression, and dramatically reduced quality of life, as well as cognitive and behavioral impairments. Clinicians treating chronic pain patients commonly observe that this suffering is maintained even when the intensity of chronic pain is reduced by therapies. It is also a common observation that patients complain far more, and in disproportion to the pain intensity, about the emotional load associated with chronic pain. Moreover, in many cases the emotional suffering is maintained even though the peripheral signs of the injury, and thus a source of nociceptive activity, have long disappeared. Furthermore, there is ample evidence that drugs that are highly effective in treating acute pain, like aspirin and opiates, show little or no effect on treating chronic pain. Non-invasive human brain imaging studies have provided the opportunity to directly peer into the brain of chronic pain patients. These studies show no evidence of increased nociceptive representation but rather point to enhanced activity in the emotional and motivational cortical-limbic circuitry. Therefore, we theorize that identifying and manipulating processes underlying the emotional suffering (cortical-limbic circuitry) should be more successful in treating chronic pain than the standard approaches that have been tested for decades and that have concentrated on the source of nociceptive signals in the skin and spinal cord.

Since the work of Pavlov it has been known that pain is a potent stimulus for creating memories. It induces single-event learning, and associated memories can last for the rest of one's life. These properties have been used in the field of learning and memory for more than 100 years. Surprisingly they have had little impact on pain research. The human brain imaging studies also point to reorganization of cortical-limbic circuitry that seems specific for distinct chronic pain conditions. Hence the suffering of chronic pain is likely the consequence of plastic changes in cortical-limbic processing leading to new learning and memories that are mediated through emotional and motivational associations with the persistent pain. Regarding the involvement of cortico-limbic circuits in behavior we adopt the strong position, best formulated by E. Rolls [63], stating: "operation of the brain to evaluate rewards and punishers is the fundamental solution of the brain to interfacing sensory systems to action selection and execution systems." Thus, we conclude that the reorganization of this circuitry in chronic pain affects emotions, decisions, and behavior.

Given our theoretical position we suggest that translational studies should be targeted to these novel mechanisms:

1. New drug development for chronic pain conditions should be based on the processes/pathways/molecules *derived directly from human data* (top-down translational approach based on brain imaging).
2. The tight relationship between pain and learning suggests that circuits involved in such processes are candidate *targets* that can be studied as to their role in chronic pain, and in an effort to explore new therapies.
3. Even though there are growing human brain imaging and animal model results showing either disruption or involvement of brain learning circuitry in specific chronic pain conditions, all of such studies have explored the brain after chronic pain was established. Therefore, there is a dire need for studies exploring mechanisms and circuits for the *transition* to chronic pain.

REFERENCES

1. Apkarian AV. Pain perception in relation to emotional learning. *Curr Opin Neurobiol* 2008;18:464–468.
2. Apkarian AV, Baliki MN, Geha PY. Towards a theory of chronic pain. *Prog Neurobiol* 2008;87:81–97.
3. Apkarian AV, Bushnell MC, Treede RD, Zubieta JK. Human brain mechanisms of pain perception and regulation in health and disease. *Eur J Pain* 2005;9:463–484.
4. Apkarian AV, Lavarello S, Randolf A, Berra HH, Chialvo DR, Besedovsky HO, Del Rey A. Expression of IL-1 beta in supraspinal brain regions in rats with neuropathic pain. *Neurosci Lett* 2006;407:176–181.
5. Apkarian AV, Sosa Y, Krauss BR, Thomas PS, Fredrickson BE, Levy RE, Harden R, Chialvo DR. Chronic pain patients are impaired on an emotional decision-making task. *Pain* 2004;108:129–136.
6. Apkarian AV, Sosa Y, Sonty S, Levy RE, Harden R, Parrish T, Gitelman D. Chronic back pain is associated with decreased prefrontal and thalamic gray matter density. *J Neurosci* 2004;24:10410–10415.
7. Apkarian AV, Thomas PS, Krauss BR, Szeverenyi NM. Prefrontal cortical hyperactivity in patients with sympathetically mediated chronic pain. *Neurosci Lett* 2001;311:193–197.
8. Baliki M, Al Amin HA, Atweh SF, Jaber M, Hawwa N, Jabbur SJ, Apkarian AV, Saade NE. Attenuation of neuropathic manifestations by local block of the activities of the ventrolateral orbito-frontal area in the rat. *Neuroscience* 2003;120:1093–1104.
9. Baliki M, Katz J, Chialvo DR, Apkarian AV. Single subject pharmacological-MRI (phMRI) study: Modulation of brain activity of psoriatic arthritis pain by cyclooxygenase-2 inhibitor. *Mol Pain* 2005.
10. Baliki MN, Chialvo DR, Geha PY, Levy RM, Harden RN, Parrish TB, Apkarian AV. Chronic pain and the emotional brain: specific brain activity associated with spontaneous fluctuations of intensity of chronic back pain. *J Neurosci* 2006;26:12165–12173.
11. Baliki MN, Geha PY, Jabakhanji R, Harden N, Schnitzer TJ, Apkarian AV. A preliminary fMRI study of analgesic treatment in chronic back pain and knee osteoarthritis. *Mol Pain* 2008;4:47.

12. Baliki MN, Geha PY, Jabakhanji R, Harden N, Schnitzer TJ, Apkarian AV. A preliminary fMRI study of analgesic treatment in chronic back pain and knee osteoarthritis. *Mol Pain* 2008;4:47.

13. Baron R, Baron Y, Disbrow E, Roberts TP. Brain processing of capsaicin-induced secondary hyperalgesia: a functional MRI study. *Neurology* 1999;53:548–557.

14. Bird GC, Lash LL, Han JS, Zou X, Willis WD, Neugebauer V. Protein kinase A-dependent enhanced NMDA receptor function in pain-related synaptic plasticity in rat amygdala neurones. *J Physiol* 2005;564:907–921.

15. Borsook D, Becerra L, Hargreaves R. A role for fMRI in optimizing CNS drug development. *Nat Rev Drug Discov* 2006;5:411–424.

16. Bruggemann J, Galhardo V, Apkarian AV. Immediate reorganization of the rat somatosensory thalamus after partial ligation of sciatic nerve. *J Pain* 2001;2:220–228.

17. Clays E, De Bacquer D, Leynen F, Kornitzer M, Kittel F, De Backer G. The impact of psychosocial factors on low back pain: longitudinal results from the Belstress study. *Spine* 2007;32:262–268.

18. Dick BD, Rashiq S. Disruption of attention and working memory traces in individuals with chronic pain. *Anesth Analg* 2007;104:1223–1229, tables.

19. Dworkin RH. An overview of neuropathic pain: syndromes, symptoms, signs, and several mechanisms. *Clin J Pain* 2002;18:343–349.

20. Dworkin RH, Jensen MP, Gammaitoni AR, Olaleye DO, Galer BS. Symptom profiles differ in patients with neuropathic versus non-neuropathic pain. *J Pain* 2007;8:118–126.

21. Falls WA, Miserendino MJ, Davis M. Extinction of fear-potentiated startle: blockade by infusion of an NMDA antagonist into the amygdala. *J Neurosci* 1992;12:854–863.

22. Foss JM, Apkarian AV, Chialvo DR. Dynamics of pain: fractal dimension of temporal variability of spontaneous pain differentiates between pain states. *J Neurophysiol* 2006;95:730–736.

23. Fukui S, Matsuno M, Inubushi T, Nosaka S. N-Acetylaspartate concentrations in the thalami of neuropathic pain patients and healthy comparison subjects measured with (1) H-MRS. *Magn Reson Imaging* 2006;24:75–79.

24. Furukawa H, Gouaux E. Mechanisms of activation, inhibition and specificity: crystal structures of the NMDA receptor NR1 ligand-binding core. *EMBO J* 2003;22:2873–2885.

25. Geha PY, Baliki MN, Chialvo DR, Harden RN, Paice JA, Apkarian AV. Brain activity for spontaneous pain of postherpetic neuralgia and its modulation by lidocaine patch therapy. *Pain* 2007;128:88–100.

26. Geha PY, Baliki MN, Harden RN, Bauer WR, Parrish TB, Apkarian AV. The brain in chronic CRPS pain: abnormal gray-white matter interactions in emotional and autonomic regions. *Neuron* 2008;60:570–581.

27. Geha PY, Baliki MN, Wang X, Harden RN, Paice JA, Apkarian AV. Brain dynamics for perception of tactile allodynia (touch-induced pain) in postherpetic neuralgia. *Pain* 2008;138:641–656.

28. Geha PY, Baliki MN, Wang X, Harden RN, Paice JA, Apkarian AV. Brain dynamics for perception of tactile allodynia (touch-induced pain) in postherpetic neuralgia. *Pain* 2008.

29. Giesecke T, Gracely RH, Grant MA, Nachemson A, Petzke F, Williams DA, Clauw DJ. Evidence of augmented central pain processing in idiopathic chronic low back pain. *Arthritis Rheum* 2004;50:613–623.

30. Godbolt AK, Waldman AD, MacManus DG, Schott JM, Frost C, Cipolotti L, Fox NC, Rossor MN. MRS shows abnormalities before symptoms in familial Alzheimer disease. *Neurology* 2006;66:718–722.

31. Gracely RH, Grant MA, Giesecke T. Evoked pain measures in fibromyalgia. *Best Pract Res Clin Rheumatol* 2003;17:593–609.

32. Gracely RH, Petzke F, Wolf JM, Clauw DJ. Functional magnetic resonance imaging evidence of augmented pain processing in fibromyalgia. *Arthritis Rheum* 2002;46:1333–1343.

33. Grachev ID, Fredrickson BE, Apkarian AV. Abnormal brain chemistry in chronic back pain: an in vivo proton magnetic resonance spectroscopy study. *Pain* 2000;89:7–18.

34. Hagelberg N, Forssell H, Aalto S, Rinne JO, Scheinin H, Taiminen T, Nagren K, Eskola O, Jaaskelainen SK. Altered dopamine D2 receptor binding in atypical facial pain. *Pain* 2003;106:43–48.

35. Hagelberg N, Forssell H, Rinne JO, Scheinin H, Taiminen T, Aalto S, Luutonen S, Nagren K, Jaaskelainen S. Striatal dopamine D1 and D2 receptors in burning mouth syndrome. *Pain* 2003;101:149–154.

36. Hagelberg N, Jaaskelainen SK, Martikainen IK, Mansikka H, Forssell H, Scheinin H, Hietala J, Pertovaara A. Striatal dopamine D2 receptors in modulation of pain in humans: a review. *Eur J Pharmacol* 2004;500:187–192.

37. Han JS, Neugebauer V. mGluR1 and mGluR5 antagonists in the amygdala inhibit different components of audible and ultrasonic vocalizations in a model of arthritic pain. *Pain* 2005;113:211–222.

38. Harris IA, Young JM, Rae H, Jalaludin BB, Solomon MJ. Factors associated with back pain after physical injury: a survey of consecutive major trauma patients. *Spine* 2007;32:1561–1565.

39. Harris RE, Clauw DJ, Scott DJ, McLean SA, Gracely RH, Zubieta JK. Decreased central mu-opioid receptor availability in fibromyalgia. *J Neurosci* 2007;27:10000–10006.

40. Harris RE, Sundgren PC, Pang Y, Hsu M, Petrou M, Kim SH, McLean SA, Gracely RH, Clauw DJ. Dynamic levels of glutamate within the insula are associated with improvements in multiple pain domains in fibromyalgia. *Arthritis Rheum* 2008;58:903–907.

41. Henriksen G, Willoch F. Imaging of opioid receptors in the central nervous system. *Brain* 2008;131:1171–1196.

42. Hughes RN. Responsiveness to brightness change in male and female rats following treatment with the partial agonist of the N-methyl-D-aspartate receptor, D-cycloserine. *Behav Brain Res* 2004;152:199–207.

43. Iadarola MJ, Berman KF, Zeffiro TA, Byas-Smith MG, Gracely RH, Max MB, Bennett GJ. Neural activation during acute capsaicin-evoked pain and allodynia assessed with PET. *Brain* 1998;121(Pt 5):931–947.

44. Iannetti GD, Zambreanu L, Wise RG, Buchanan TJ, Huggins JP, Smart TS, Vennart W, Tracey I. Pharmacological modulation of pain-related brain activity during normal and central sensitization states in humans. *Proc Natl Acad Sci USA* 2005;102:18195–18200.

45. Jasmin L, Rabkin SD, Granato A, Boudah A, Ohara PT. Analgesia and hyperalgesia from GABA-mediated modulation of the cerebral cortex. *Nature* 2003;424:316–320.

46. Johansen JP, Fields HL. Glutamatergic activation of anterior cingulate cortex produces an aversive teaching signal. *Nat Neurosci* 2004;7:398–403.

47. Keeley P, Creed F, Tomenson B, Todd C, Borglin G, Dickens C. Psychosocial predictors of health-related quality of life and health service utilisation in people with chronic low back pain. *Pain* 2007.

48. Kuchinad A, Schweinhardt P, Seminowicz DA, Wood PB, Chizh BA, Bushnell MC. Accelerated brain gray matter loss in fibromyalgia patients: premature aging of the brain? *J Neurosci* 2007;27:4004–4007.

49. Lagraize SC, Borzan J, Peng YB, Fuchs PN. Selective regulation of pain affect following activation of the opioid anterior cingulate cortex system. *Exp Neurol* 2006;197:22–30.

50. Maihofner C, Handwerker HO. Differential coding of hyperalgesia in the human brain: a functional MRI study. *Neuroimage* 2005;28:996–1006.

51. Maihofner C, Schmelz M, Forster C, Neundorfer B, Handwerker HO. Neural activation during experimental allodynia: a functional magnetic resonance imaging study. *Eur J Neurosci* 2004;19:3211–3218.
52. Malin EL, McGaugh JL. Differential involvement of the hippocampus, anterior cingulate cortex, and basolateral amygdala in memory for context and footshock. *Proc Natl Acad Sci USA* 2006;103:1959–1963.
53. Merskey H, Bogduk N. *Classification of chronic pain*. Seattle: IASP Press, 1994.
54. Metz AE, Yau HJ, Centeno MV, Apkarian AV, Martina M. Morphological and functional reorganization of rat medial prefrontal cortex in neuropathic pain. *Proc Natl Acad Sci USA* 2009;106:2423–2428.
55. Meyer T, Cooper J, Raspe H. Disabling low back pain and depressive symptoms in the community-dwelling elderly: a prospective study. *Spine* 2007;32:2380–2386.
56. Milad MR, Quirk GJ. Neurons in medial prefrontal cortex signal memory for fear extinction. *Nature* 2002;420:70–74.
57. Millecamps M, Centeno MV, Berra HH, Rudick CN, Lavarello S, Tkatch T, Apkarian AV. D-cycloserine reduces neuropathic pain behavior through limbic NMDA-mediated circuitry. *Pain* 2006.
58. Pais-Vieira M, Lima D, Galhardo V. Orbitofrontal cortex lesions disrupt risk assessment in a novel serial decision-making task for rats. *Neuroscience* 2007;145:225–231.
59. Pattany PM, Yezierski RP, Widerstrom-Noga EG, Bowen BC, Martinez-Arizala A, Garcia BR, Quencer RM. Proton magnetic resonance spectroscopy of the thalamus in patients with chronic neuropathic pain after spinal cord injury. *AJNR Am J Neuroradiol* 2002;23:901–905.
60. Reid MC, Williams CS, Gill TM. The relationship between psychological factors and disabling musculoskeletal pain in community-dwelling older persons. *J Am Geriatr Soc* 2003;51:1092–1098.
61. Richardson MP, Strange BA, Dolan RJ. Encoding of emotional memories depends on amygdala and hippocampus and their interactions. *Nat Neurosci* 2004;7:278–285.
62. Roeska K, Doods H, Arndt K, Treede RD, Ceci A. Anxiety-like behaviour in rats with mononeuropathy is reduced by the analgesic drugs morphine and gabapentin. *Pain* 2008;139:349–357.
63. Rolls ET. *Emotion explained*. Oxford: Oxford University Press, 2005.
64. Santini E, Ge H, Ren K, Pena dO, Quirk GJ. Consolidation of fear extinction requires protein synthesis in the medial prefrontal cortex. *J Neurosci* 2004;24:5704–5710.
65. Schmidt-Wilcke T, Leinisch E, Ganssbauer S, Draganski B, Bogdahn U, Altmeppen J, May A. Affective components and intensity of pain correlate with structural differences in gray matter in chronic back pain patients. *Pain* 2006;125:89–97.
66. Schmidt-Wilcke T, Leinisch E, Straube A, Kampfe N, Draganski B, Diener HC, Bogdahn U, May A. Gray matter decrease in patients with chronic tension type headache. *Neurology* 2005;65:1483–1486.
67. Senapati AK, Lagraize SC, Huntington PJ, Wilson HD, Fuchs PN, Peng YB. Electrical stimulation of the anterior cingulate cortex reduces responses of rat dorsal horn neurons to mechanical stimuli. *J Neurophysiol* 2005;94:845–851.
68. Sjogren P, Christrup LL, Petersen MA, Hojsted J. Neuropsychological assessment of chronic non-malignant pain patients treated in a multidisciplinary pain centre. *Eur J Pain* 2005;9:453–462.
69. Small DM, Apkarian AV. Increased taste intensity perception exhibited by patients with chronic back pain. *Pain* 2006;120:124–130.
70. Sorensen L, Siddall PJ, Trenell MI, Yue DK. Differences in metabolites in pain-processing brain regions in patients with diabetes and painful neuropathy. *Diabetes Care* 2008;31:980–981.

71. Stromme JT, Myhrer T. Impaired visual memory in rats reared in isolation is reversed by D-cycloserine in the adult rat. *Eur J Pharmacol* 2002;437:73–77.

72. Verne GN, Robinson ME, Price DD. Representations of pain in the brain. *Curr Rheumatol Rep* 2004;6:261–265.

73. Vertes RP. Interactions among the medial prefrontal cortex, hippocampus and midline thalamus in emotional and cognitive processing in the rat. *Neuroscience* 2006;142:1–20.

74. Vlaeyen JW, Linton SJ. Fear-avoidance and its consequences in chronic musculoskeletal pain: a state of the art. *Pain* 2000;85:317–332.

75. Walker DL, Ressler KJ, Lu KT, Davis M. Facilitation of conditioned fear extinction by systemic administration or intra-amygdala infusions of D-cycloserine as assessed with fear-potentiated startle in rats. *J Neurosci* 2002;22:2343–2351.

76. Wiech K, Seymour B, Kalisch R, Stephan KE, Koltzenburg M, Driver J, Dolan RJ. Modulation of pain processing in hyperalgesia by cognitive demand. *Neuroimage* 2005;27:59–69.

77. Wise RG, Tracey I. The role of fMRI in drug discovery. *J Magn Reson Imaging* 2006;23:862–876.

78. Wood PB, Patterson JC, Sunderland JJ, Tainter KH, Glabus MF, Lilien DL. Reduced presynaptic dopamine activity in fibromyalgia syndrome demonstrated with positron emission tomography: a pilot study. *J Pain* 2007;8:51–58.

79. Woods MP, Asmundson GJ. Evaluating the efficacy of graded in vivo exposure for the treatment of fear in patients with chronic back pain: a randomized controlled clinical trial. *Pain* 2007.

80. Wu LJ, Toyoda H, Zhao MG, Lee YS, Tang J, Ko SW, Jia YH, et al. Upregulation of forebrain NMDA NR2B receptors contributes to behavioral sensitization after inflammation. *J Neurosci* 2005;25:11107–11116.

81. Young CC, Greengerg MA, Nicassio PM, Harpin RE, Hubbard D. Transition from acute to chronic pain and disability: a model including cognitive, affective, and trauma factors. *Pain* 2007.

82. Zambreanu L, Wise RG, Brooks JC, Iannetti GD, Tracey I. A role for the brainstem in central sensitisation in humans. Evidence from functional magnetic resonance imaging. *Pain* 2005;114:397–407.

83. Zhao MG, Toyoda H, Lee YS, Wu LJ, Ko SW, Zhang XH, Jia Y, et al. Roles of NMDA NR2B subtype receptor in prefrontal long-term potentiation and contextual fear memory. *Neuron* 2005;47:859–872.

16 Consideration of Pharmacokinetic Pharmacodynamic Relationships in the Discovery of New Pain Drugs

Garth T. Whiteside and Jeffrey D. Kennedy

CONTENTS

16.1 INTRODUCTION

In this chapter we seek to describe the stages of drug discovery, focusing on the utility of animal pain models and of pharmacokinetic/pharmacodynamic relationships (PK/PD) at each stage. In simple terms, the study of PK and PD in drug discovery is often paired and described in reciprocal terms, where PK is the analysis of how the body affects a drug, while PD is the analysis of how a drug affects the body. PK is defined by how a compound is absorbed, distributed, metabolized, and excreted; and PD is the measure of a compound's ability to interact with its intended target leading to a biologic effect. In this chapter, we briefly describe the stages of drug discovery and the process of defining structure–activity relationship (SAR) both *in vitro*, through optimization of several key characteristics collectively referred to as pharmaceutical profiling, and *in vivo*, through the combined use of PK assessment and animal pain models to assess compound efficacy, noting the types and end-points employed, and why it is important to pair these efficacy models with models of side effects. We follow this with data investigating penetration of compounds into the brain versus that in spinal cord and also correlate efficacy in rodent pain models with clinical efficacy. We hope to convey the importance of PK and of PK/PD relationships in the process of developing pain drugs [aspects of the PK/PD relationships described in this chapter have been published elsewhere (Whiteside et al. 2008)].

16.2   STAGES OF DRUG DISCOVERY

The drug discovery process progresses via a number of distinct stages. Different companies assign different nomenclature to each stage and also differ in the require-ments for transitioning drug candidates between the various stages. Nonetheless, the process can be described in a generic way and will help inform the utility of animal pain models during these stages.

16.2.1   TARGET VALIDATION

Target validation describes the process by which an endogenous molecule, process, or pathway is linked to a disease state or animal model of disease (O'Connell and Roblin 2006). While identification provides potential targets, validation demonstrates that they are an integral part of a disease model and ideally of the disease itself. For drug discovery purposes, one must also determine if the molecule, process, or pathway can be manipulated in order to halt or reverse the disease or alleviate the symptoms. Targets are most commonly proteins such as G-protein coupled receptors, ion chan-nels, enzymes, or transporters. There are various strategies for validating targets. Target validation efforts can utilize samples and tissues of fluids from humans or animals, or alternatively, whole animals can be used. Validation frequently follows a genetic, proteomic, or pharmacological route. The genetic approach involves knock-out technology, where a gene is deleted allowing the consequences to be examined in a whole animal, or knockdown approaches, where the mRNA is interfered with so that no or reduced levels of protein result. The proteomic approach manipulates the protein target directly using neutralizing antibodies or more novel alternative approaches such as laser inactivation of target proteins (Smith 2003). Finally, a phar-macological approach generally uses a small molecule that interacts with the target. The small molecule may be a drug used to validate a target in a new disease, or it may be a tool molecule that has inappropriate characteristics to become a drug but may be useful to validate a target in an animal model. None of the above techniques is flawless; for example, knockout animals are subject to developmental compensation; proteomic approaches are limited by the available technology and ability of reagents such as antibodies to penetrate target tissues; and small molecule pharmacology is often confounded by lack of selectivity of the available small molecules used as tool reagents. For this reason, often multiple or combined approaches are used in order to gain confidence that a target is worth further investigation toward lengthy and expensive clinical trials.

16.2.2   HIT-TO-LEAD AND LEAD OPTIMIZATION

Validated targets are progressed through the drug discovery process. Most commonly, an *in vitro* assay is built employing the target of interest, and then a library of molecules is screened in a high-throughput fashion to determine "hits" that manipulate the target in the desired way. These hits are often intrinsically flawed or not "druglike," but through an iterative chemistry process, additional molecules are synthesized that improve on the inadequacies. Animal models are often employed at this stage to ensure that improvements translate into an acceptable pharmacokinetic and efficacy profile *in vivo*.

16.2.3   SCREENING

Following lead optimization efforts, multiple potential drug candidates are proposed. These compounds are often assessed for pharmacokinetic and efficacy profiles in a screening mode. This mode is designed to generate a small amount of data on a large number of compounds, for example, testing a single dose or time point in a single model. The purpose of this activity is to find the molecule with the best *in vivo* profile and, therefore, one that is most likely to succeed.

16.2.4   DEVELOPMENT CANDIDATE PROFILING

The screening phase allows selection of a small number of drug candidates that are then more completely profiled in multiple models and with additional pharmacokinetic data and sometimes safety/toxicology data. At this stage, extensive dose response curves (DRCs) are generated for both efficacy and plasma exposure, and frequently in the neuroscience/pain setting, brain exposure. The DRC allows estimation of quantitative measures of efficacy and potency such as the MED (minimally effective dose) or the ED^{50} (defined in this case as the dose that causes 50% of a maximal effect in 100% of the population). Various formulations also may be employed here to see how they affect exposure/efficacy (such as immediate release vs. sustained release formulations). Finally, the effect of chronic dosing of a compound is investigated to ensure that tachyphylaxis (tolerance) or accumulation (compound buildup) does not occur with repeated administration. In addition to considering efficacy, side effect tests are performed to ensure that the conclusion of efficacy is not confounded by the presence of sedation or other non-specific effects. Side-effect profiling can also allow a pre-clinical therapeutic index (TI) to be calculated, defined as the lowest dose that is efficacious compared with the lowest dose that causes an unwanted side effect. Ideally, the TI should be as large as possible, although certain disease areas, such as oncology, will tolerate TI values less than 10. In all of the above endeavors, consideration of pharmacokinetics is paramount. These parameters are discussed in more detail later. Initially, efficacy is measured against dose, but with the addition of pharmacokinetic data, one can consider efficacy against drug concentration or exposure, ideally in the tissue of interest.

16.3   ROLE OF SMALL ANIMAL MODELS IN PAIN DRUG DISCOVERY

The use of animals as living test tubes always has been one of the primary vehicles of learning and insight into the biological sciences. Despite claims to the contrary, there has yet to be developed a combination of *in vitro* molecular, cellular assays or even computer-based modeling approaches and experimental paradigms that can faithfully predict human biology. Though limitations and imperfections exist, animals serve a noble function as the gateway through which researchers have gained otherwise unattainable knowledge of human physiology, pathology, and disease progression. From the laborious practice of backcrossing used to create the many genetically distinct strains that laid the foundation for current research, the huge leap in our knowledge that resulted from harnessing molecular biology tools in the late 20th century has tremendously increased the power of animal models, particularly rodents, to help reveal insights into the vast complexity of many interacting biological systems within an intact organism. Researchers now have the ability to introduce, delete, and modify activity of virtually any gene of interest to study its function *in vivo*. In the realm of pharmaceutical research, this practice enlightens not only target validation efforts, as already discussed, but also facilitates a more complete understanding of drug characteristics and allows investigations of larger scope where entire pathways can be isolated and interrogated for identification of new potential drug targets (Werner 2008).

As in other disease areas, the use of animals in pain research has been invaluable in gaining an understanding of the types of nerves, receptors, ion channels, mediators, and biochemical pathways involved in the initiation, transmission, transduction, perception, and regulation of pain. From the earliest, basic measures of nociceptive pain as modeled in rats by pinch or thermal sensitivity assays such as the hot plate and tail flick, over the years a large number of animal pain models have been developed to assess pain due to inflammation, viral infection, nerve injury, and cancer, including metastatic bone pain, and arthropathies. Many of these models afford face, construct, and predictive validity based on similarity to human chronic pain conditions that manifest similar pathobiology and allow the assessment of pain in the context of multiple, clinically used endpoints, including heat, cold, pressure, and light touch. In addition, though more work is needed, new and improved disease-relevant models are being developed, such as models of diabetic neuropathic pain, chemotherapy- or retroviral drug-induced pain that hope to further bridge the gap between animals and humans. In pharmaceutical research, a suite of these models is typically used in order to assess activity of putative drug candidates. In addition to gaining mechanistic understanding of heretofore unprecedented targets, the advantage of a strategy that evaluates drugs in a variety of models and endpoints lies in gaining information that will help guide identification of both the most relevant and broadest clinical indications where a drug might show efficacy.

In general terms, the methodologies employed for assessing pain in animals can be broken down into endpoints and models. Endpoints are the tests conducted to ascertain the extent of pain and most commonly describe either spontaneous pain-related behaviors or thresholds to a ramping stimulus. Pain-related behaviors, such as biting, licking, guarding, and flinching are absent, or minimal, in normal animals and are only elicited upon establishment of a model. An evoked stimulus-response measurement consists of an application of a stimulus of increasing intensity, which is commonly thermal or mechanical in nature, followed by measurement of a threshold or latency at which the animal displays nocifensive behavior. When any of these stimulus-response measurements is applied to normal animals, it constitutes a measure of normal nociception and can be used to assess the effect of frank analgesics (defined as those that inhibit non-pathological, nociceptive pain) such as opioids and local anesthetics. Hargreave's apparatus, von Frey fibers, hot plate, tail-flick, tail-dip, and Randall-Selitto apparatus are all tools for applying a ramping stimulus to evoke a response (Campbell and Meyer 2006; Honore 2006; Sullivan et al. 2007; Valenzano et al. 2005).

Models describe manipulations of animals that are performed in order to generate a pain state, which is commonly manifest as behavioral hypersensitivity such as hyperalgesia, defined as a heightened sensitivity to a painful stimulus; allodynia, a painful response to a normally non-painful stimulus; or both and/or spontaneous pain behavior. Commonly used models can be broken down into three main groups; the first involves local injection of a pain-causing substance, such as capsaicin, bradykinin, or dilute acid. The second involves injection of substances, either locally or systemically, that cause an inflammatory response and pain subsequent to the inflammation. Examples of such substances include carrageenan, zymosan, and Freund's complete adjuvant. The final group involves injury to the nervous system by direct mechanical, metabolic, or chemical means. Examples of each include spinal nerve ligation (mechanical), streptozotocin treatment (metabolic), and taxol treatment (chemical). Each of these models is generally paired with one or a number of endpoints. In this way, the extent of the hypersensitivity can be measured, and reversal of pain back to "normal" levels by pharmacological intervention can be assessed. Effective treatments are known as anti-hyperalgesics or anti-allodynics, depending on the stimulus modality they reverse; however, it is important to note that the frank analgesics mentioned above will also reverse pain in these hypersensitivity assays.

Beyond primary considerations of achieving efficacy across a number of animal pain models (ideally, both inflammatory and neuropathic) at doses that will allow scaling to a feasible human dosage, of major importance to pharmaceutical research efforts, is developing drug candidates that demonstrate a wide separation between efficacy and side effects. This is important on at least two levels. First, from a clinical perspective, one of the chief complaints and unmet needs resulting from currently approved chronic pain therapies is that many are associated with dose-limiting side effects. These can manifest in many forms, but CNS-related effects such as dizziness, sedation, somnolence, nausea, and "foggy thinking" are commonly cited and are most debilitating. In addition, it is important to understand to what extent measures of efficacy might be influenced by side effects that could impair the animal's ability to respond, leading to the potential for scoring a drug as a false-positive. Lack of attention to this potential confound has likely contributed to the poor translation of new candidate drugs into successful clinical therapies. Among models commonly employed to assess CNS-related side effects are the accelerating rotarod (Valenzano et al. 2005), beam-walking (Goldstein and Davis 1990), and open-field movement (Visser et al. 2006) where animals are evaluated for activities requiring motor coordination, cognition, and motivation. Additional models are used to evaluate the potential for specific mechanism-based compound liabilities such as catalepsy for cannabinoids (Fox et al. 2001). Other models of drug-mediated side effects are also commonly used and include GI transit time (Bohn and Raehal 2006; Galligan and Burks 1983) and hypothermia (Baker and Meert 2002; Handler et al. 1992), applicable to the study of opiate and cannabinoid-like drugs, as well as changes in cardiovascular parameters such as heart rate and blood pressure. Finally, a number of models are used when a mechanism is known or suspected to be linked to abuse potential, including conditioned place preference, drug discrimination, and self-administration (Gardner 2005).

While the drug discovery paradigm is built to maximize the chances for sucessssful clinical translation once a compound is advanced for *in vivo* pain models evaluation, the number of uncontrollable variables increase, which can impact observed efficacy. Analyzing reasons why compounds fail to achieve *in vivo* efficacy is often difficult given the large number of PK parameters that come into play in assessing the ultimate, physiologically meaningful interaction of drug with its target, an analysis compounded by the knowledge that no two mechanisms are likely to demand of a small molecule the exact same combination of PK characteristics to yield efficacy. But before a compound ever advances to *in vivo* testing, a number of parameters are assessed, collectively referred to as *in vitro* pharmaceutical profiling, that help predict the potential of a drug candidate to reach its intended *in vivo* target. Though not all-inclusive, chief among these are determinations of molecular weight, solubility, lipophilicity, ligand efficiency (LE, a calculated ratio relating molecular weight to potency), plasma protein binding, interactions with cytochrome P_{450} (CYP) isoenzymes (both induction and inhibition), ability to migrate across a cell barrier (CaCo, MDCK cell lines, and/or artificial lipid membranes are often used to address potential for absorption across the gut and blood–brain barrier), and stability (measured either as half-life of a compound mixed with human or rodent serum and in solutions of varying pH simulating stomach or GI tract environments, or when incubated with liver microsomes or hepatocytes as an early metabolism read-out). Microsomal stability analyses are often conducted across up to four species (rat, dog, non-human primate, and human) and combined with a metabolite identification to understand the potential primary routes of metabolism and major metabolites generated; depending on proportion and stability, these may be synthesized separately and evaluated for potency, efficacy, and potential toxicities. A detailed description of these *in vitro* assays and ranges of acceptable performance values is beyond the scope of this chapter (Di and Kerns 2005); however, behavior of potential drug candidates in each assay in addition to other *in vitro* assays demonstrating on-target activity and selectivity versus potential side effect or toxicity-inducing receptors/pathways together build the compound SAR profile (defined as minor changes in the structure of a compound that lead to a predictable change in biological activity). This is a time- and labor-intensive effort that often spans several years from identification of the first compound serving as a template for building improved characteristics.

Once a small molecule compound meets *in vitro* criteria for advancement into animal pain models for efficacy evaluation, a major hurdle remains in the extent to which the compound is able to be absorbed across (most often) the GI tract into the systemic circulation and then distributed in the tissues such that it reaches the target organ of interest and achieves sufficient concentrations over a period of time long enough to impact the targeted mechanism. In some cases, compounds must be able to cross several barriers, including the very non-permissive blood–brain barrier as discussed below. The extent to which a compound satisfies these demands denotes its PK/PD profile (Csajka and Verotta 2006), which will be discussed in greater detail below. As SAR is developed, compounds are often tested by non-clinically viable or commercially less favored (e.g., subcutaneous, intraperitoneal, intrathecal, intravenous) routes in order to establish proof-of-concept and PD effects, especially when the target mechanism is not well validated and/or compound PK characteristics are poor such as in the early stages of drug discovery.

16.4 WHICH PK PARAMETERS TO CONSIDER

A full PK profile is usually established before compounds undergo extensive pain model testing; the importance of individual parameters varies depending on the type of mechanism being targeted and its location, whether peripheral, central (brain and spinal cord), or both. In addition to favorable bioavailability following oral dosing, other key determinants are the maximum plasma concentration (Cmax), time at which the maximum concentration is attained (Tmax), exposure measured as area under the curve (AUC), and half-life ($t_{1/2}$), all assessed from plasma samples. Other measures include the volume of distribution (Vss, a measure of tissue penetration), the rate of compound clearance (Clp), and brain exposure. The latter is important when maximum efficacy at certain pain pathway targets requires central nervous system (CNS) exposure and is often expressed as a ratio with plasma exposure; B/P ratios of 0.5 or greater are desirable. For a given mechanism, efficacy in one model (e.g., acute inflammatory pain such as the carrageenan model) might correlate more closely with one parameter than another (e.g., maximum concentration, or C_{max}, vs. total exposure or AUC_{0-24}). The opposite might be true when interrogating the same mechanism in the context of neuropathic pain models; these variables are usually manifested as different MED values. Additional factors in the determination of PK/PD relationships include the timing of dosing relative to model induction, the endpoint chosen (i.e., touch, pressure, thermal), and the frequency of dosing. Typically, efficacy in pain models is reported following a single drug dose; however, as many pain conditions are chronic in nature, repeated dosing of potential drug candidates should become more standard practice, although care must be taken to ensure that improved efficacy following repeated dosing is due to sustained or enhanced PD effects and not altered PK parameters, such as accumulation, leading to increased drug $t_{1/2}$ or exposure.

While exceptions exist, in general and based partially on pragmatic considerations, drugs with PK/PD relationships suitable for further advancement into late-stage pre-clinical toxicology models and ultimately clinical trials are those showing significant efficacy in pain models compared with a positive control or standard-of-care drug at oral doses below 10 mg/kg, oral bioavailability of 20% or more, systemic exposures at least 10-fold greater than the *in vitro* ED_{50} or IC_{50} value, and PD effects that are maintained for multiple hours. In addition, besides improved efficacy, in order to differentiate new drugs in development from older pain drugs currently in use, greater separation of efficacy from dose-limiting side effects is desired; while many of these drugs have exposure ratios (ER, the efficacious exposure compared to exposure producing toxic effects) of near 1, some of the mechanisms being targeted by current drug discovery efforts offer the promise of achieving ERs of 50- to 100-fold or greater.

16.5 BRAIN AND SPINAL CORD PENETRATION

Compounds often need to penetrate the CNS in order to be effective analgesics. Frequently, the site of action is exclusively the spinal cord or both the spinal cord and higher centers in the brain. Both spinal cord and brain are protected by a selective barrier, the blood–brain barrier (BBB), which is composed of capillary endothelial cells connected by tight junctions acting in concert with resident brain cells such as astrocytes (Calabria and Shusta 2006; de Boer and Gaillard 2006). It serves as a physical barrier and active filter that restricts and regulates penetration of molecules into and out of the brain in order to maintain brain homeostasis. Likewise, the spinal cord is equipped with the blood–spinal cord barrier (BSCB) that serves the same function and maintains the cord fluid microenvironment within a narrow limit (Sharma 2005). Lipid-soluble substances, including gases and alcohol, pass through these barriers easily, and water can enter via diffusion; however, in general, under normal conditions, large molecules are excluded (Deli et al. 2005). Small molecules such as electrolytes, glucose, and amino acids enter the brain through energy-dependent selective transport mechanisms (Strbian et al. 2008). Conversely, efflux transporters, such as P-glycoprotein, actively pump small molecules out of CNS tissue (Bostrom et al. 2005).

During the drug discovery process, the extent to which a compound penetrates the brain is assessed. This commonly occurs during the target validation phase, for tool compounds, and during the screening and profiling phases for potential development candidates. The concentration of compound is determined in homogenates of whole brain, and this is assumed to correlate with penetration into the spinal cord. We sought to confirm this assumption using three clinically relevant, pharmaceutically active small molecules with low, intermediate, and high penetration of the CNS: atenolol (a peripheralized beta-blocker), morphine (a centrally active analgesic), and oxycodone (a highly permeable, centrally active analgesic). Following dosing, plasma was collected, the animals were trans-cardially perfused, brains and spinal cord were removed, and the concentration of each compound was determined

(Table 16.1). The ratio of concentration in the brain to concentration in the CNS tissue was calculated and is presented in Table 16.1B. As expected, substantial levels of all compounds were observed in plasma 30 minutes post-dosing. Also in line with expectations, atenolol had the lowest concentration in the CNS, oxycodone had the highest concentration, and morphine was intermediate (Table 16.1). When comparing brain to spinal cord, the ratios were very similar. There was a small increase in spinal cord/plasma ratio for both morphine and oxycodone and a small decrease for atenolol as compared with brain/plasma ratio (Table 16.2). Overall, we can say that the level of penetration into the spinal cord is comparable to level of penetration of the brain. In pragmatic terms this means that brain levels can be used as a surrogate for spinal cord levels.

We were also interested in the effect of perfusion on the concentration of compounds in the CNS. Traditionally, animals are transcardially perfused to remove residual blood from the brain prior to PK analysis. As blood contained within the vasculature in the brain will contain compound, it can be considered as a contaminant that results in an overestimate of the compound concentration in the CNS. For this reason, perfusion is undertaken, which removes the contaminating blood; it is, however, possible that via an osmotic effect compound could be drawn out of the CNS tissue, leading to an underestimation of CNS levels. The true concentration therefore would be between these two extremes. In order to determine the extent of any separation between perfused and non-perfused CNS tissue, we utilized the same three compounds as described above and compared penetration into the spinal cord and brain between perfused and non-perfused animals. Perfusion resulted in a reduced concentration of compound in the brain and spinal cord in all cases except for morphine penetration into the brain, which increased slightly (Table 16.1). Considering tissue/plasma ratios, these also show a decreased ratio in all cases except for oxycodone in the brain, which increased slightly (Table 16.2). Overall, the changes were quite small in magnitude and would not lead to an inappropriate

TABLE 16.1
Concentration of Atenolol, Morphine, and Oxycodone in the Plasma, Spinal Cord, and Brain

	Atenolol	Morphine	Oxycodone
Plasma conc (ng/ml)	3330 ± 321	2927 ± 272	1022 ± 65
Spinal cord (ng/g)	115 ± 34	298 ± 55	1829 ± 92
Brain (ng/g)	52 ± 7	357 ± 75	2583 ± 256

(B) Plasma/Tissue Ratio for Concentration of Atenolol, Morphine, and Oxycodone

	Atenolol	Morphine	Oxycodone
Spinal cord: plasma ratio	0.035	0.102	1.790
Brain: plasma ratio	0.016	0.122	2.527

Note: All compounds were administered i.p. (atenolol 10 mg/kg, morphine 10 mg/kg, and oxycodone 5 mg/kg) and samples were taken 30 minutes post dosing.

TABLE 16.2

Concentration of Atenolol, Morphine, and Oxycodone in the Plasma, Spinal Cord, and Brain of Perfused versus Non-Perfused Animals

	Atenolol	Morphine	Oxycodone
Plasma conc (ng/ml)	8207 ± 1482	1157 ± 384	664 ± 47
Spinal cord (ng/g)	212 ± 98	443 ± 134	2332 ± 466
Brain (ng/g)	105 ± 28	214 ± 46	1481 ± 363
(Perfused) plasma conc (ng/ml)	7010 ± 984	1217 ± 450	582 ± 111
Perfused spinal cord (ng/g)	129 ± 68	224 ± 68	1295 ± 397
Perfused brain (ng/g)	62 ± 12	222 ± 66	1374 ± 332

(B) Plasma/Tissue Ratio for Concentration of Atenolol, Morphine, and Oxycodone of Perfused versus Non-Perfused Animals

	Atenolol	Morphine	Oxycodone
Spinal cord: plasma ratio	0.030	0.366	3.512
Brain: plasma ratio	0.015	0.228	2.230
Perfused Spinal cord: plasma ratio	0.016	0.195	2.225
Perfused brain: plasma ratio	0.008	0.186	2.360

Note: All compounds were administered i.p. (atenolol 10 mg/kg, morphine 10 mg/kg, and oxycodone 5 mg/kg) and samples were taken 30 minutes post dosing.

classification of a compound as being peripheral, semipermeable, or highly brain-permeable. Therefore, we can conclude that use of either perfused or non-perfused CNS tissue is appropriate for determination of brain penetration.

16.6 PK/PD MODELING

The goal of PK/PD modeling is to correlate the concentration of a compound at a particular place in the body to the effect that that drug is causing. Ideally, the site at which drug is measured is also the site of action. As described above for analgesics, this is frequently the brain or spinal cord. While this is possible in animals, it is obviously not possible for humans, where inferences have to be made based on drug levels in the plasma or the cerebrospinal fluid, which may more closely correlate with levels in CNS tissue. Understanding and modeling PK/PD in one species can then allow predictions to be made for other species. Rodents are most commonly used in pre-clinical drug discovery; however, more frequently, PK/PD data also are being generated from larger species as described in a separate chapter. Comparing PK/PD relationships across multiple species allows one to estimate the PK/PD relationship in humans prior to clinical trials; however, ultimately one must complete the circle and compare how the PK/PD relationship in humans compares with that in the pre-clinical test species in order to better inform future decisions.

For clinical compounds directed toward the treatment of pain, we have compared plasma exposures at clinical maintenance doses to plasma exposures in rats at the minimally efficacious exposure (Whiteside et al. 2008). The majority of our animal efficacy, side effect, and pharmacokinetic data cited were generated in-house at Wyeth (and supplemented from the literature), while the human efficacious doses and exposures were found via literature search. The acute, inflammatory, and nerve injury models selected for analysis were chosen based upon their common use in the industrial setting for drug discovery (Iyengar et al. 2004; Jarvis et al. 2002; Sullivan et al. 2007; Valenzano et al. 2005). In addition, AUC was used as the measure of drug exposure and in order to make appropriate comparisons between species single-dose pharmacokinetic data for both rat and human data. Furthermore, identical AUC measures (e.g., AUC [0–infinity] or AUC [0–12 h]) in animals and humans were used for each compound analyzed to ensure that the comparisons were based on similar data sets. Clinical treatment of moderate to severe acute pain, such as that caused by a surgical incision, continues to be dominated by opioids (Leykin et al. 2007). As such, we have summarized preclinical and clinical data for morphine and oxycodone as prototypic opioid agonists. Efficacy data for morphine in the hot plate assay were generated in-house using methods described in detail elsewhere (Whiteside et al. 2005). Efficacy data for oxycodone were identified from literature reports that utilize equivalent methodology (Lemberg et al. 2006) to those used in-house. Stated MEDs, shown in Tables 16.3–16.5, are doses that do not produce statistically significant motor deficits in our in-house rotarod assay of ataxia, using methodology as previously described (Valenzano et al. 2005). All in-house pharmacokinetic studies were conducted according to previously described methods (Sullivan et al. 2007). In acute pain, the efficacious exposure in rats for morphine is 3 times greater than that observed in humans (Table 16.3). In contrast, the efficacious exposure for oxycodone is almost 40-fold greater than that observed in humans. Considering Cmax, the ratios are reversed, with a 51-fold higher exposure in rats as compared with humans for morphine, while the efficacious concentration for oxycodone in rats is 0.8 times that in humans. It is worth noting that the human exposures were determined from immediate release and sustained release formulations for morphine and oxycodone, respectively, which may make the human/rat correlation less accurate than using similar formulations for both. This most likely explains why comparing exposures for morphine and plasma concentration for oxycodone yield very close ratios (2.9 and 0.8, respectively), whereas the reverse yields ratios that do not approximate. Beyond this caveat, the observed difference may be due to species differences in metabolism, brain and tissue penetration, plasma protein binding, or other factors altering availability of compound at the target tissue.

It is noteworthy that rat exposures are often described at the MED, for a single administration, while clinical exposures are described at the maintenance dose based on repeated administration. The MED is the lowest dose that elicits a statistically significant effect; it is, therefore, by definition, an effect of limited magnitude. This is likely to be in contrast to a maintenance dose in patients, which is expected to produce an effect large enough such that the patient realizes a substantial benefit. This discrepancy may result in the rat efficacious exposure under-estimating the exposure necessary to maintain efficacy in humans. In addition, efficacy in animals is based upon single acute dosing, while that in man is typically based upon chronic administration; thus, chronic dosing in pre-clinical studies may improve the predictivity of

TABLE 16.3

Comparison of the Pharmacokinetic/Pharmacodynamic Relationship of Acute Pain Drugs in Rats and Humans

Exposure

Compound	Human Daily Dose (mg)	Human mg/kg Maintenance Dose	Rat MED Hot Plate (mg/kg)	Rat Exposure (AUC; ng*h/mL) @ MED	Human Exposure (AUC; ng*h/mL)	Exposure Ratio (rat/human)	Source
Morphine	60	0.9	3	799	279	2.9	rat: Wyeth in-house; human: Anonymous 2007
Oxycodone	160	2.3	0.6	71100	1856	38	rat: Wyeth in-house, Huang et al. 2005 human: Anonymous 2007

Concentration

Compound	Human Daily Dose (mg)	Human mg/kg Maintenance Dose	Rat MED hot plate (mg/kg)	Rat Cmax (ng/mL) @ MED	Human Cmax (ng/mL)	Concentration Ratio (rat/human)	Source
Morphine	60	0.9	3	976	19	51	rat: Wyeth in-house human: Anonymous 2007
Oxycodone	160	2.3	0.6	123	156	0.8	rat: Wyeth in-house human: Anonymous 2007

Abbreviations: mg, milligram; kg, kilogram; MED, minimum efficacious dose; AUC, area under the curve; ng, nanogram; h, hour; mL, milliliter.

Note: All rat data were generated following subcutaneous administration. All human data were generated following oral administration and based on 70 kg body weight. Rat efficacy data for morphine were generated in-house at Wyeth, and oxycodone were literature derived (Lemberg et al. 2006). AUC for all studies is AUC (0–infinity). Extrapolations of pharmacokinetic data assume linearity. Morphine rat exposure extrapolated from data after 10 mg/kg dose and Cmax extrapolated from 1 mg/kg dose. Oxycodone rat exposure extrapolated from data after 5 mg/kg dose and Cmax extrapolated from 2 mg/kg dose.

Source: Whiteside, G. T., Adedoyin, A., Leventhal, L., 2008. Predictive validity of animal pain models? A comparison of the pharmacokinetic–pharmacodynamic relationship for pain drugs in rats and humans. *Neuropharmacology 54, 767–775.* With permission.

TABLE 16.4
Comparison of the Pharmacokinetic/Pharmacodynamic Relationship of Inflammatory Pain Drugs in Rats and Humans

Exposure

Compound	Human Daily Dose (mg)	Human mg/kg Maintenance Dose	Rat MED FCA (mg/kg)	Rat Exposure (AUC; ng*h/mL) @ MED	Human Exposure (AUC; ng*h/mL)	Exposure Ratio (rat/human)	Source
Celecoxib	200	3	10	9200	6564	1.4	rat: Guirguis et al. 2001 human: Paulson et al. 2001
Indomethacin	50	1	3	35407	8710	4	rat: Kim and Ku 2000 human: Khosravan et al. 2006

Concentration

Compound	Human Daily Dose (mg)	Human mg/kg Maintenance Dose	Rat MED FCA (mg/kg)	Rat Cmax (ng/mL) @ MED	Human Cmax (ng/mL)	Concentration Ratio (rat/human)	Source
Celecoxib	200	3	10	1880	806	2.3	rat: Guirguis et al. 2001 human: Paulson et al. 2001
Indomethacin	50	1	3	3853	2760	1.4	rat: Kim and Ku 2000 human: Khosravan et al. 2006

Abbreviations: FCA, Freund's complete adjuvant; mg, milligram; kg, kilogram; MED, minimum efficacious dose; AUC, area under the curve; ng, nanogram; h, hour; mL, milliliter.

Note: All compounds were administered orally, and human data are based on a 70 kg body weight. Rat efficacy data were generated in-house at Wyeth. AUC data for celecoxib is AUC (0–infinity) and indomethacin is AUC (0–12 h). Extrapolations of pharmacokinetic data assume linearity. Celecoxib rat exposure extrapolated from data after 5 mg/kg dose. Indomethacin rat exposure extrapolated from data after 22.5 mg/kg dose.

Source: Whiteside, G. T., Adedoyin, A., Leventhal, L., 2008. Predictive validity of animal pain models? A comparison of the pharmacokinetic–pharmacodynamic relationship for pain drugs in rats and humans. *Neuropharmacology* 54, 767–775. With permission.

TABLE 16.5
Comparison of the Pharmacokinetic/Pharmacodynamic Relationship of Neuropathic Pain Drugs in Rats and Humans

Compound	Human Daily Dose Range (mg) [Maintenance]	Human mg/kg Maintenance Dose	Rat MED SNL (mg/kg)	Rat Exposure (AUC; ng*h/mL) @ MED	Human Exposure (AUC; ng*h/mL)	Exposure Ratio (rat/human)	Source
				Exposure			
Duloxetine	40–120 [60]	0.9	30	8673	584	15	rat: Wyeth in-house human: Chan et al. 2007
Gabapentin	300–3600 [1800]	26.0	100	146000	125370	1.2	rat: Radulovic et al. 1995 human: Gidal et al. 1998
Lamotrigine	100–500 [200]	2.9	10	208200	69754	3	rat: Castel-Branco et al. 2004 human: Theis et al. 2005
Carbamazepine	100–1200 [1200]	17.0	100	55780	14120	4	rat: Chen et al. 2002 human: Theis et al. 2005
Milnacipran	50–150 [50]	0.7	30	6732	939	7	rat: Wyeth in-house human: Puozzo et al. 2006
Amitriptyline	10–150 [150]	2.1	>100	>2526	3540	not able to determine	rat: Wyeth in-house human: Park et al. 2003

Compound	Human Daily Dose Range (mg) [Maintenance]	Human mg/kg Maintenance Dose	Rat MED SNL (mg/kg)	Rat Cmax (ng/mL) @ MED	Human Cmax (ng/mL)	Exposure Ratio (rat/human)	Source
				Concentration			
Duloxetine	40–120 [60]	0.9	30	1439	39	37.2	rat: Wyeth in-house human: Chan et al. 2007
Gabapentin	300–3600 [1800]	26	100	32800	11940	2.7	rat: Radulovic et al. 1995 human: Gidal et al. 1998

Lamotrigine	100–500 [200]	2.9	10	5420	4479	1.2	rat: Castel-Branco et al. 2004 human: Theis et al. 2005
Carbamazepine	100–1200 [1200]	17	100	8550	3320	2.6	rat: Chen et al. 2002 human: Theis et al. 2005
Milnacipran	50–150 [50]	0.7	30	1678	144	11.7	rat: Wyeth in-house human: Puozzo et al. 2006
Amitriptyline	10–150 [150]	2.1	>100	>232	109	not able to determine	rat: Wyeth in-house human: Park et al. 2003

Abbreviations: SNL, spinal nerve ligation; mg, milligram; kg, kilogram; MED, minimum efficacious exposure; AUC, area under the curve; ng, nanogram; h, hour; mL, milliliter.

Note: All compounds were administered orally except for lamotrigine, which was administered intraperitoneally. Human data are based on a 70 kg body weight. Rat efficacy data were generated in-house at Wyeth. AUC for all compounds in AUC (0–infinity) except for human lamotrigine, carbamazepine, and amitriptyline data that is AUC (0–12 h), AUC (0–24 h), and AUC (0–96 h), respectively. Extrapolations of pharmacokinetic data assume linearity. Gabapentin rat exposure extrapolated from data after 50 mg/kg dose and human exposure from data after 600 mg dose. Lamotrigine rat exposure extrapolated from data after 20 mg/kg dose. Amitriptyline human exposure extrapolated from data after 50 mg dose.

Source: Whiteside, G. T., Adedoyin, A., Leventhal, L., 2008. Predictive validity of animal pain models? A comparison of the pharmacokinetic–pharmacodynamic relationship for pain drugs in rats and humans. *Neuropharmacology* 54, 767–775. With permission.

the models and exposure. The data in Table 16.3 focus on the relationship between rat and human efficacious plasma concentrations and drug exposures. We can conclude from the table that overall, efficacious drug exposure in the rat approximates to efficacious exposure in humans. Although the routes of administration differ, it is assumed that efficacious exposure and plasma concentration are independent of route of administration; however, comparisons based on dose cannot be made, since the routes of administration between rat and human are not consistent (subcutaneous vs. oral, respectively). While the hot plate assay is commonly used as a measure of acute pain, it is actually more an assay for normal nociceptive pain and, as such, may be only predictive for a subgroup of treatments such as opioids and local anesthetics. Therefore, caution is warranted in using this model to predict clinical efficacy in conditions such as post-operative pain (also referred to as acute pain).

Pain relief for patients with inflammatory diseases, such as rheumatoid arthritis, is largely based upon the use of nonsteroidal anti-inflammatory drugs (NSAIDs). Included in this group are the troubled COX-2 inhibitors; although celecoxib is still marketed for the treatment of pain (Leykin et al. 2007), patient use has radically declined (Bresalier et al. 2005); rofecoxib was voluntarily withdrawn in 2004 (Bresalier et al. 2005), and the FDA rejected etoricoxib in 2007 (Fitzgerald 2007). As such, we have summarized pre-clinical and clinical data for celecoxib. In addition, we show data for indomethacin as a prototypic NSAID. Efficacy data for both compounds in the Freund's Complete Adjuvant (FCA) model of chronic inflammatory pain with the Randall-Selitto endpoint were generated in-house using methods that are described in detail elsewhere (Valenzano et al. 2005). All in-house pharmacokinetic studies were conducted according to previously described methods (Sullivan et al. 2007). In inflammatory pain, the efficacious dose, plasma concentration, and exposure are less than 5-fold higher in humans as compared with rats for both celecoxib and indomethacin (Table 16.4). The data in Table 16.4 focus on the relationship between rat and human efficacious plasma concentrations and drug exposures. We can conclude from the table that, firstly, the rat model of chronic inflammatory pain predicts efficacious exposure in humans and, secondly, efficacious dose, plasma concentration, and exposure in the rat approximates to the efficacious exposure in humans.

There have been relatively few drugs approved for use in the treatment of neuropathic pain compared with other diseases. These include the anticonvulsant gabapentin (for post-herpetic neuralgia), the anticonvulsant pregabalin (for post-herpetic neuralgia, diabetic neuropathy, and fibromyalgia), the anticonvulsant carbamazepine (for trigeminal neuralgia), the local anesthetic lidocaine (topically for post-herpetic neuralgia), and the antidepressant duloxetine (for diabetic neuropathy). Other drugs, including opioids (although controversy exists as to their effectiveness), additional anticonvulsants such as lamotrigine, additional antidepressants such as amitriptyline and milnacipran, and the calcium channel blocker ziconitide (given intrathecally), also are used off-label. We focused on a subset of these approved and off-label treatments. Efficacy data for all compounds in the spinal nerve ligation (SNL) model of neuropathic pain with Randall-Selitto endpoint were generated in-house using methods that are described in detail elsewhere (Leventhal et al. 2007; Valenzano et al. 2005). We focused our analysis on the SNL model of neuropathic pain as this is commonly used for pre-clinical drug screening of compounds for neuropathic pain indications (Iyengar et al. 2004; Jarvis et al. 2002; Sullivan et al. 2007; Valenzano et al. 2005). All in-house pharmacokinetic studies were conducted according to previously described methods (Sullivan et al. 2007).

In neuropathic pain, the minimal efficacious exposure in rats for all compounds is between 1- and 15-fold greater than that observed in humans (Table 16.5). Interestingly, the efficacious exposure for gabapentin is almost identical in rat and humans. The efficacious exposure for amitriptyline could not be determined, because no efficacy was observed in the SNL model (highest dose tested was 100 mg/kg, p.o.). Ratios based on plasma concentration confirmed the ratios based on exposure; the concentration ratio for all compounds, except for duloxetine, was from 1–12. Duloxetine resulted in the most extreme values in both cases with exposure and concentration ratios of 15 and 37, respectively. In contrast to exposure, the efficacious doses for all compounds, except gabapentin and lamotrigine, were more than 15-fold greater in rat than human; however, the difference overall between rat and human for exposure was within 15-fold. All compounds investigated required higher plasma concentrations and exposures in the rat to achieve efficacy as compared with humans. In addition, the efficacious dose for all but two of the compounds was considerably higher in rat compared with human. The data in Table 16.5 focus on the relationship between rat and human efficacious plasma concentrations and drug exposures. We can conclude from the table that, firstly, the SNL rat model of neuropathic pain predicts efficacious exposure in humans and, secondly, efficacious plasma concentration and exposure and in the rat approximates to efficacious plasma concentration and exposure in humans.

16.7 CONCLUDING REMARKS

In this chapter we have sought to describe the stages of drug discovery and the utility of animal pain models at each stage. We then identify PK parameters that are important during discovery of new pain therapies and present data investigating CNS penetration. We finish with an investigation of the correlative relationships between rat pain models and patients for a number of pain states using efficacious exposure data. Our goal was to convey the importance of PK and of PK/PD relationships in the process of developing pain drugs. It is the failure of compounds with otherwise spectacular *in vitro* potency, selectivity, and other pharmaceutical characteristics to succeed at achieving an adequate PK/PD profile that vexes medicinal chemists and biologists alike and impedes the development of new therapies. This is perhaps no more evident than in the field of pain drug development, where the rate of translation of drugs from the pre-clinical to the clinical arena has been poor, despite the large number of mechanisms that have been interrogated over the past decade.

Despite these difficulties, basic and clinical research scientists alike remain encouraged by ever-increasing insights obtained from genetic studies identifying new targets and patient subtypes that might allow more focused treatment options, cell and molecular biology studies that reveal in greater detail the mechanistic complexity and interplay involved in the generation and perception of pain, and pre-clinical efforts to build additional pain models with even greater predictive validity that should further reduce the gap between bench and bedside. Current treatment options are restricted to opioids, antidepressants, anticonvulsants, NSAIDs, and other analgesics, all of which suffer either from limited efficacy, efficacy in the presence of CNS-related or systemic side effects and toxicities or the development of tolerance and dependence that negatively impact quality of life, or development of tolerance and dependence that limit long-term use in chronic pain. As standards of care (SOC), these drugs leave much room for improvement, hence the seemingly low hurdle for the uptake of new drugs currently in clinical evaluation to only show better than 50% pain reduction in greater than 50% of patients treated. Despite this, few new drugs are being approved, which may speak, in part, to the redundancy and overlap of pain mechanisms and their regulation across the many clinical presentations of this disease class and a perhaps naive assumption that any one target might dominate such a critically important and finely-tuned system.

We believe a new approach to pain drug development is warranted, one embracing more flexibility and willingness to investigate clinical strategies focused on combining mechanisms, be they new ones with older SOC drugs or building portfolios of drugs, each perhaps with modest efficacy that when prescribed together could lead to more significant relief in the absence of unwanted side effects. This will require a paradigm shift on many levels, including on the part of basic scientists who continue to take a "magic bullet" approach to mechanistic studies of pain and clinical scientists who are bound (and often defeated) by the need to clearly separate efficacy of individul candidate drugs from that of comparator SOC drugs and placebo effects. Such a strategy will require greater engagement and dialogue with regulatory agencies to help overcome inherent complexities in trial design and the approval process for poly-pharmacy approaches. Finally, and importantly, this approach will be facilitated by providing better education to clinicians, health care providers, and patients so that more strategic prescribing and utilization of drugs can be fostered.

REFERENCES

Anonymous, 2007. *Physician's Desk Reference.* Thomson Scientific & Healthcare, Florence, KY.

Baker, A. K., Meert, T. F., 2002. Functional effects of systemically administered agonists and antagonists of mu, delta, and kappa opioid receptor subtypes on body temperature in mice. *Journal of Pharmacology and Experimental Therapeutics* 302, 1253–1264.

Bohn, L. M., Raehal, K. M., 2006. Opioid receptor signaling: Relevance for gastrointestinal therapy. *Current Opinion in Pharmacology* 6, 559–563.

Bostrom, E., Simonsson, U. S. H., Hammarlund-Udenaes, M., 2005. Oxycodone pharmacokinetics and pharmacodynamics in the rat in the presence of the P-glycoprotein inhibitor PSC833. *Journal of Pharmaceutical Sciences* 94, 1060–1066.

Bresalier, R. S., Friedewald, V. E., Rakel, R. E., Roberts, W. C., Williams, G. W., 2005. The editor's roundtable: Cyclooxygenase-2 inhibitors and cardiovascular risk. *American Journal of Cardiology* 96, 1589–1604.

Calabria, A. R., Shusta, E. V., 2006. Blood–brain barrier genomics and proteomics: Elucidating phenotype, identifying disease targets and enabling brain drug delivery. *Drug Discovery Today* 11, 792–799.

Campbell, J. N., Meyer, R. A., 2006. Mechanisms of neuropathic pain. *Neuron* 52, 77–92.

Castel-Branco, M. M., Falcao, A. C., Figueiredo, I. V., Macedo, T. R. A., Caramona, M. M., 2004. Lamotrigine kidney distribution in male rats following a single intraperitoneal dose. *Fundamental & Clinical Pharmacology* 18, 51–55.

Chan, C., Yeo, K. P., Pan, A. X., Lim, M., Knadler, M. P., Small, D. S., 2007. Duloxetine pharmacokinetics are similar in Japanese and Caucasian subjects. *British Journal of Clinical Pharmacology* 63, 310–314.

Chen, L. C., Chen, Y. F., Chou, M. H., Lin, M. F., Yang, L. L., Yen, K. Y., 2002. Pharmacokinetic interactions between carbamazepine and the traditional Chinese medicine Paeoniae Radix. *Biological & Pharmaceutical Bulletin* 25, 532–535.

Csajka, C., Verotta, D., 2006. Pharmacokinetic–pharmacodynamic modelling: History and perspectives. *Journal of Pharmacokinetics and Pharmacodynamics* 33, 227–279.

de Boer, A. G., Gaillard, P. J., 2006. Blood–brain barrier dysfunction and recovery. *Journal of Neural Transmission* 113, 455–462.

Deli, M. A., Abraham, C. S., Kataoka, Y., Niwa, M., 2005. Permeability studies on in vitro blood–brain barrier models: Physiology, pathology, and pharmacology. *Cellular and Molecular Neurobiology* 25, 59–127.

Di, L., Kerns, E. H., 2005. Application of pharmaceutical profiling assays for optimization of drug-like properties. *Current Opinion in Drug Discovery & Development* 8, 495–504.

Fitzgerald, G. A., 2007. COX-2 in play at the AHA and the FDA. *Trends in Pharmacological Sciences* 28, 303–307.

Fox, A., Kesingland, A., Gentry, C., McNair, K., Patel, S., Urban, L., James, I., 2001. The role of central and peripheral cannabinoid(1) receptors in the antihyperalgesic activity of cannabinoids in a model of neuropathic pain. *Pain* 92, 91–100.

Galligan, J. J., Burks, T. F., 1983. Centrally mediated inhibition of small intestinal transit and motility by morphine in the rat. *Journal of Pharmacology and Experimental Therapeutics* 226, 356–361.

Gardner, E. L., 2005. Endocannabinoid signaling system and brain reward: Emphasis on dopamine. *Pharmacology Biochemistry and Behavior* 81, 263–284.

Gidal, B. E., Maly, M. M., Kowalski, J. W., Rutecki, P. A., Pitterle, M. E., Cook, D. E., 1998. Gabapentin absorption: Effect of mixing with foods of varying macronutrient composition. *Annals of Pharmacotherapy* 32, 405–409.

Goldstein, L. B., Davis, J. N., 1990. Beam-walking in rats—studies towards developing an animal-model of functional recovery after brain injury. *Journal of Neuroscience Methods* 31, 101–107.

Guirguis, M. S., Sattari, S., Jamali, F., 2001. Phaumacokinetics of celecoxib in the presence and absence of interferon-induced acute inflammation in the rat: Application of a novel HPLC assay. *Journal of Pharmacy and Pharmaceutical Sciences* 4, 1–6.

Handler, C. M., Geller, E. B., Adler, M. W., 1992. Effect of mu-selective, kappa-selective, and delta-selective opioid agonists on thermoregulation in the rat. *Pharmacology Biochemistry and Behavior* 43, 1209–1216.

Honore, P., 2006. Behavioral assessment of neuropathic pain in preclinical models. *Drug Development Research* 67, 302–307.

Huang, L., Edwards, S. R., Smith, M. T., 2005. Comparison of the pharmacokinetics of oxycodone and noroxycodone in male dark agouti and Sprague-Dawley rats: Influence of streptozotocin-induced diabetes. *Pharmaceutical Research* 22, 1489–1498.

Iyengar, S., Webster, A. A., Hemrick-Luecke, S. K., Xu, J. Y., Simmons, R. M. A., 2004. Efficacy of duloxetine, a potent and balanced serotonin-norepinephrine reuptake inhibitor in persistent pain models in rats. *Journal of Pharmacology and Experimental Therapeutics* 311, 576–584.

Jarvis, M. F., Burgard, E. C., McGaraughty, S., Honore, P., Lynch, K., Brennan, T. J., Subieta, A., et al., 2002. A-317491, a novel potent and selective nonnucleotide antagonist of P2X(3) and P2X(2/3) receptors, reduces chronic inflammatory and neuropathic pain in the rat. *Proceedings of the National Academy of Sciences of the United States of America* 99, 17179–17184.

Khosravan, R., Wu, J. T., Joseph-Ridge, N., Vernillet, L., 2006. Pharmacokinetic interactions of concomitant administration of febuxostat and NSAIDs. *Journal of Clinical Pharmacology* 46, 855–866.

Kim, J. Y., Ku, Y. S., 2000. Enhanced absorption of indomethacin after oral or rectal administration of a self-emulsifying system containing indomethacin to rats. *International Journal of Pharmaceutics* 194, 81–89.

Lemberg, K. K., Kontinen, V. K., Siiskonen, A. O., Viljakka, K. M., Yli-Kauhaluoma, J. T., Korpi, E. R., Kalso, E. A., 2006. Antinociception by spinal and systemic oxycodone: Why does the route make a difference? In vitro and in vivo studies in rats. *Anesthesiology* 105, 801–812.

Leventhal, L., Smith, V., Hornby, G., Andree, T. H., Brandt, M. R., Rogers, K. E., 2007. Differential and synergistic effects of selective norepinephrine and serotonin reuptake inhibitors in rodent models of pain. *Journal of Pharmacology and Experimental Therapeutics* 320, 1178–1185.

Leykin, Y., Pellis, T., Ambrosio, C., 2007. Highlights in postoperative pain treatment. *Expert Review of Neurotherapeutics* 7, 533–545.

O'Connell, D., Roblin, D., 2006. Translational research in the pharmaceutical industry: from bench to bedside. *Drug Discovery Today* 11, 833–838.

Park, E. S., Lee, D. S., Kwon, S. Y., Chi, S. C., 2003. A new formulation of controlled release amitriptyline pellets and its in vivo/in vitro assessments. *Archives of Pharmacal Research* 26, 569–574.

Paulson, S. K., Vaughn, M. B., Jessen, S. M., Lawal, Y., Gresk, C. J., Yan, B., Maziasz, T. J., Cook, C. S., Karim, A., 2001. Pharmacokinetics of celecoxib after oral administration in dogs and humans: Effect of food and site of absorption. *Journal of Pharmacology and Experimental Therapeutics* 297, 638–645.

Puozzo, C., Hermann, P., Chassard, D., 2006. Lack of pharmacokinetic interaction when switching from fluoxetine to milnacipran. *International Clinical Psychopharmacology* 21, 153–158.

Radulovic, L. L., Turck, D., Vonhodenberg, A., Vollmer, K. O., McNally, W. P., Dehart, P. D., Hanson, B. J., Bockbrader, H. N., Chang, T., 1995. Disposition of gabapentin (neurontin) in mice, rats, dogs, and monkeys. *Drug Metabolism and Disposition* 23, 441–448.

Sharma, H. S., 2005. Pathophysiology of blood–spinal cord barrier in traumatic injury and repair. *Current Pharmaceutical Design* 11, 1353–1389.

Smith, C., 2003. Drug target validation: Hitting the target. *Nature* 422, 341–347.

Strbian, D., Durukan, A., Pitkonen, M., Marinkovic, I., Tatlisumak, E., Pedrono, E., Abo-Ramadan, U., Tatlisumak, T., 2008. The blood–brain barrier is continuously open for several weeks following transient focal cerebral ischemia. *Neuroscience* 153, 175–181.

Sullivan, N., Leventhal, L., Cummons, T. A., Smith, V. A., Sun, S. C., Harrison, J., Lu, P., et al., 2007. Pharmacological characterization of the muscarinic agonist, WAY-132983, in in vitro and in vivo models of chronic pain. *Journal of Pharmacology and Experimental Therapeutics* in press.

Theis, J. G., Sidhu, J., Palmer, J., Job, S., Bullman, J., Ascher, J., 2005. Lack of pharmacokinetic interaction between oxcarbazepine and lamotrigine. *Neuropsychopharmacology* 30, 2269–2274.

Valenzano, K. J., Tafesse, L., Lee, G., Harrison, J. E., Boulet, J. M., Gottshall, S. L., Mark, L., et al., 2005. Pharmacological and pharmacokinetic characterization of the cannabinoid receptor 2 agonist, GW405833, utilizing rodent models of acute and chronic pain, anxiety, ataxia and catalepsy. *Neuropharmacology* 48, 658–672.

Visser, L., van den Bos, R., Kuurman, W. W., Kas, M. J. H., Spruijt, B. M., 2006. Novel approach to the behavioural characterization of inbred mice: Automated home cage observations. *Genes Brain and Behavior* 5, 458–466.

Werner, T., 2008. Bioinformatics applications for pathway analysis of microarray data. *Current Opinion in Biotechnology* 19, 50–54.

Whiteside, G. T., Adedoyin, A., Leventhal, L., 2008. Predictive validity of animal pain models? A comparison of the pharmacokinetic–pharmacodynamic relationship for pain drugs in rats and humans. *Neuropharmacology* 54, 767–775.

17 Large Animal Models for Pain Therapeutic Development

Darrell A. Henze and Mark O. Urban

CONTENTS

The humane use of preclinical animal models plays a critical role both in understanding the basic biology of pain as well as in the development of therapeutic treatments to alleviate pain. Clinically relevant pain is the result of complex processes involving peripheral transduction and transmission as well as central modulation and processing that leads to the final conscious sensation of pain. Much has been learned about the mechanisms underlying the transduction and transmission of the pain signal within the nervous system through the use of cellular, biochemical, and molecular techniques (Millan 1999; Scholz and Woolf 2007; Zeilhofer 2005). However, understanding the actual experience of pain will always require an intact organism that can integrate the full range of external stimuli and internal cognitive and emotional states that drive and modulate pain.

Rodent models of pain have historically played a dominant role in the study of pain mechanisms (Negus et al. 2006; Walker et al. 1999). There are many good reasons for this, including the practicalities of low cost, simplified ethical concerns, and the scientific value of having a large database of prior research to provide context for new findings. A large historical database is particularly important in the field of drug discovery and development since the sensitivity and predictive validity of animal models can only be established through extensive testing in many contexts. For these reasons, the rat and mouse models will continue to be the workhorses driving

371

basic research as well as drug discovery. Unfortunately, there are many ways that the biology of rodents may fail to accurately predict the biology and pharmacology of clinical pain conditions in humans (Blackburn-Munro 2004; Le Bars et al. 2001). Given the high cost of developing new therapeutics (Adams and Brantner 2006), there is a growing need to validate biological and pharmacological findings in non-rodent species that, while perhaps less tractable than rodents, address known or ill-defined differences between mice and men. It is hoped that through the humane study and evaluation of pain states in higher order preclinical species, we can better predict whether biological mechanisms and specific compounds have relevance for clinical pain. Ultimately, well-validated pain models in non-rodent species could enhance the speed, reduce the costs, and increase the probability of the successful development of new analgesic therapeutics offering enhanced efficacy and reduced adverse effects.

This chapter will review the gaps in current pain research using rodent models that may potentially be addressed using preclinical animal models of pain in higher species. Additionally, we will review the pain models and assessments that have been developed to date in "higher" and larger species and highlight areas where there is a need for the development of new models and methods of pain assessment.

17.1 THE IMPORTANCE OF PREDICTIVE ANIMAL MODELS FOR DRUG DISCOVERY AND DEVELOPMENT

While it is always exciting to see a compound with a novel mechanism of action show great efficacy in a rodent model of inflammatory or neuropathic pain, the obvious ultimate goal is to identify novel compounds that are found to be safe and effective in humans. A major challenge faced in the development of new analgesics is establishing confidence in the predictive validity of preclinical models of pain. There are many examples where compounds that have demonstrated efficacy and tolerability in rodent models have failed to show sufficient efficacy or safety in the clinic (e.g., neurokinin-1 [NK-1] receptor antagonists; Hill 2000). Given the uncertainty about the predictive validity of the rodent models to determine efficacy and safety of compounds associated with novel mechanisms, it is logical to suggest that novel potential analgesic compounds should be tested in several clinical pain populations to evaluate safety and efficacy. Counterbalancing this logic, there is an ethical and fiscal responsibility to test new compounds in the clinic only under circumstances where there are preclinical data supporting safety and efficacy. Given that the success rate of bringing a new drug to market is less than 10% (Kola and Landis 2004), there clearly is a need to identify and validate preclinical model systems that will increase the probability of success. There has been a recent increased focus on assessing the predictive validity of preclinical model, and what must be done to understand and improve upon the current state (Negus et al. 2006; Whiteside et al. 2008). It is important to note that although existing animal models appear to do very well at predicting true positives, there is less confidence about their ability to predict false-positive and false-negative outcomes (Rice et al. 2008).

17.2 THE POTENTIAL VALUE OF LARGE ANIMAL MODELS

The preclinical animal models that are used to test novel compounds and mechanisms need to be able to predict accurately both the clinical efficacy and adverse effects of any new therapeutic. In most cases, safety studies are completed in both rodents as well as in some other larger species such as the dog or non-human primate. Since there are many rodent models of pain, it is relatively easy to calculate a therapeutic index in the rat or mouse. Unfortunately, there are very few validated large animal pain models in dogs or primates that would allow a similar therapeutic index calculation to be made in these species that may be more relevant to the human case.

One obvious reason why large animal models have a real potential to add value to the drug development process is that they are phylogenetically closer to humans than rodents. This is important for several reasons. The first is that at the molecular level many pharmaceutical targets have species specific variations in sequences or expression patterns that result in differences in affinity or potency for the target and functional importance of the target. In certain circumstances, it may not be possible to identify compounds that have identical or even similar affinity for the human versus rodent variants (e.g., calcitonin gene-related peptide (CGRP) receptor antagonists; Salvatore et al. 2008), in which case, higher order species such as the non-human primate will usually share a greater sequence homology with the human form. The second important reason is that at a systems level, the pharmacological response and network connectivity of the central nervous system varies. For example, pregabalin demonstrates robust efficacy in rodent models of neuropathic pain within 1 to 2 hours but it often takes days to weeks for robust efficacy in humans (Arezzo et al. 2008; Field et al. 1999). A third reason phylogenetic differences may matter is that organisms likely have experienced very different evolutionary pressures as they developed behavioral expressions of pain. These differences may confound reasonable comparisons of the amount of pain any given organism is experiencing. Rice et al. have pointed out that animals such as rats that often live in a hostile environment are usually viewed as potential prey and therefore may process and express pain in a very different way than predators, such as canines or primates, who live typically in a less hostile environment (Rice et al. 2008). Finally, it is well known that different species can metabolize compounds in different ways and different rates, with dogs and monkeys often being better of predictors of human metabolism (Lin 1999; Ward et al. 2005). Therefore, better estimates of efficacy and safety often can be obtained in large animals since sustained compound exposure can be an important contributor to the efficacy profile for a chronic pain drug.

A second general reason that large animal models benefit drug development is that they are, in fact, large and thus more suitable for certain types of pain measurements. For example, the increasing use of functional imaging technologies to reveal patterns of brain activation in pain states is much easier to accomplish in a large species such as a dog or primate. In addition, certain forms of chronic pain, such as osteoarthritis, appear to occur more naturally in large animals compared to lower species. This is particularly useful both for the ability to test potential therapeutics on naturally occurring diseases, as well as for the ability to test alternate delivery methods such as topical delivery on a joint of similar size and geometry to the human joint. However, note that the relatively large size of dogs and primates has a number of disadvantages, mostly due to the need for increased resources in housing and maintenance, as well as the increased amounts of test compound required to achieve desired exposures.

17.3   MODELS OF ACUTE NOCICEPTION

While pathological pain, particularly chronic pain, is detrimental for an organism, acute pain is an important factor in protecting organisms from harm. Therefore it is important that there are animal models that can provide insight into the impact of analgesic compounds on acute nociception. In rodents, acute nociception is usually assessed using either a noxious heat stimulus (e.g., hot plate, tail withdrawal, Hargreaves test) or mechanical stimulus (Randall-Selitto test) applied to either a hindpaw or tail. While there are no reported quantitative assays to assess acute mechanical nociception in large animals, there are two assays that have been reported to assess acute thermal nociception: a canine thermal escape assay and a primate tail withdrawal assay. These assays may be used to evaluate whether novel compounds will result in blunting of the protective responses to noxious heat, with the primate model being of particular value due to the similarity of physiology compared to humans.

The canine thermal escape assay was recently described by Yaksh and colleagues (Wegner et al. 2008; Table 17.1). In this assay, a dog is trained to rest in a sling with its hindpaws resting on a clear plate. A high-powered lamp is then turned on to apply a heat stimulus to the anterior third of the metatarsal paw pad, and the latency from when the light is turned on is measured. Normal dogs withdraw their paws at about 9.3 seconds, which corresponds to a plate temperature of ~50°C. A variety of opioids were tested in this assay, and each showed a significant effect on the escape latency. A dose of 1 mg/kg (IV) morphine produced a latency very close to the cutoff-latency of 20 seconds. The sedating alpha2 adrenergic agonist dexmedetomidine (0.1 mg/kg) also produced a maximal possible effect. However, the phenothiazine sedative acepromazine (0.1 mg/kg) did not produce any increase in withdrawal latency, suggesting that sedative effects alone do not account for the increases in latency observed with opiate analgesics.

Similar in concept, the primate tail withdrawal assay has been used to assess sensitivity to acute application of noxious thermal stimuli (Dykstra and Woods 1986; Table 17.1). In this assay, a monkey is trained to sit in a primate chair. The distal tip of the tail is shaved and immersed in a water bath at 50°C or 55°C, and the time required for the subject to withdraw the tail is measured. A normal monkey will withdraw its tail from 55°C water in about 2–3 seconds, but a prior dose of 10 mg/kg morphine will increase the latency to longer than the cutoff time of 20 seconds. In contrast, monkeys that appeared sedated to the point of sleep with 30 mg/kg pentobarbital still withdrew their tails within 2–4 seconds, again suggesting that the increases in latency are not due to general sedative effects of the morphine (Dykstra and Woods 1986). One utility of this model for therapeutic development is to determine whether analgesic compounds of interest possess potentially undesirable effects on acute nociceptive thresholds and heat sensitivity. For example, the analgesic gabapentin has no effect on acute heat nociceptive thresholds in this assay (Figure 17.1A), although it

TABLE 17.1

Large Animal Models of Acute Nociception

Species	Pain Stimulus	Assessment Method	Pharmacology Tested	Active?	Reference
Dog	Noxious heat (~ 50° C) to paw	Time to paw withdrawal	Morphine (1 mg/kg)	Y	Wegner et al. 2008
			Buprenorphine (0.03 mg/kg)	Y	Wegner et al. 2008
			Butorphanol (0.4 mg/kg)	Y	Wegner et al. 2008
			Fentanyl (0.01 mg/kg)	Y	Wegner et al. 2008
			Hydromorphone (0.2 mg/kg)	Y	Wegner et al. 2008
			Dexmedetomidine (0.1 mg/kg)	Y	Wegner et al. 2008
			Acepromazine (0.1 mg/kg)	N	Wegner et al. 2008
Non-human primate	Tail immersion in hot water (> ~ 50° C)	Time to tail withdrawal	Morphine (0.1–10 mg/kg)	Y	Dykstra and Woods 1986
			Pentobarbitol (30 mg/kg)	N	
			Phencyclidine (1 mg/kg)	Y	

is effective in inhibiting warm allodynia following sensitization of the tail skin with capsaicin (see below; Figure 17.1B). These findings are consistent with the known analgesic properties of gabapentin in terms of its selective effects on persistent pain associated with injury-induced sensitization (Urban et al. 2005).

17.4 MODELS OF ACUTE INFLAMMATORY PAIN

Clinically, acute inflammatory pain occurs in response to tissue injury due to surgery, burn, or the inflammatory response to an infection. Similar to acute nociception, acute inflammatory pain also provides a protective role, encouraging protective behaviors following a tissue injury. However, post-surgical pain and burn pain are two clinically important inflammatory pain states often targeted for therapeutic intervention. Some of the most common rodent models of acute inflammatory pain are the carrageenan model (Hargreaves et al. 1988), ultraviolet-B radiation burn (Bishop et al. 2007), formalin model (Dubuisson and Dennis 1977), and hindpaw incision model (Brennan et al. 1996). Both canine and primate models have been described that are at least in part analogous to these rodent models.

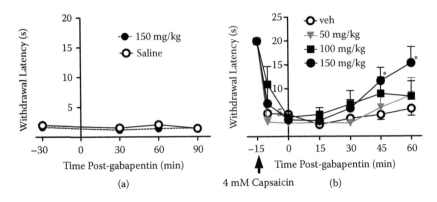

FIGURE 17.1 (A) Effect of gabapentin (150 mg/kg, s.c.) in the rhesus monkey acute noxious heat tail withdrawal assay. Application of a noxious heat stimulus (52°C water) to the tail of rhesus monkeys results in a tail withdrawal response within 1–3 seconds. A high dose of gabapentin (150 mg/kg) has no effect on acute noxious heat sensitivity. (B) Effect of gabapentin (50–150 mg/kg, s.c.) in the rhesus monkey capsaicin pain model. Capsaicin (4mM) application to the tail of rhesus monkeys produces warm allodynia observed as a decrease in tail withdrawal latency from a warm water stimulus (42°C). The highest dose of gabapentin (150 mg/kg) effectively reverses capsaicin-induced allodynia. These studies were performed at the AAALAC accredited Merck Research Laboratories (MRL) site in West Point, Pennsylvania, were approved by the MRL Institutional Animal Care and Use Committee, and were in accordance with The Guide for the Care and Use of Laboratory Animals.

A commonly used canine model of inflammatory synovitis is achieved by intra-articular injections of uric acid crystals leading to inflammation and pain (Table 17.2). The unilateral injection of uric acid crystals into the hind limb stifle (knee) results in an acute synovitis and quantifiable lameness on the injected side (McCarty, Jr. et al. 1966). The lameness is typically quantified through a gait analysis using a force plate apparatus. The observed reduced weight loading of the affected limb is interpreted as reflecting a painful state of the joint since there is no structural change in the injected joint. The peak change in gait occurs approximately 4 hours post treatment and resolves within 24 hours (Rumph et al. 1993). A number of studies have demonstrated that non-steroidal anti-inflammatory drugs (NSAIDs; carprofen, etodolac, ketoprofen, and meloxicam), cyclo-oxygenase 2 (Cox2) inhibitors (deracoxib, firocoxib, ML-1,785,713), opioids (Butorphanol), and ketamine all show efficacy in this model (Borer et al. 2003; Cross et al. 1997; Drag et al. 2007; Hamilton et al. 2005; Hazewinkel et al. 2003; McCann et al. 2004; Millis et al. 2002). Additionally, in agreement with clinical studies demonstrating lack of efficacy of NK-1 antagonists (Boyce and Hill 2000), there is a report that an NK-1 antagonist failed to show efficacy in this model (Punke et al. 2007). Taken together, these data may support use of this canine model as predictive of the human clinical response for therapeutics targeting inflammatory pain.

In primates, a variation of the tail immersion assay described above has been used to evaluate inflammation-induced thermal allodynia (Table 17.2). The standard tail immersion procedure described above is modified by injecting the tail of the monkey

TABLE 17.2

Large Animal Acute Inflammatory Pain Models

Species	Pain Stimulus/ Model	Assessment Method	Pharmacology Tested	Active?	Reference
Dog	Urate crystal injection into knee	Force-plate gait analysis	Carprofen (2.2–4.4 mg/kg)	Y	Drag et al. 2007, Millis et al. 2002
		Numerical rating scales	Etodolac (17 mg/kg)	Y	Borer et al. 2003
			Ketoprofen (0.25–0.75 mg/kg)	Y	Hazewinkel et al. 2003
			Meloxicam (0.1–0.5 mg/kg)	Y	Cross et al. 1997, Drag et al. 2007, Borer et al. 2003
			Deracoxib (0.3–10 mg/kg)	Y	Drag et al. 2007; Millis et al. 2002
			Firocoxib (5 mg/kg)	Y	Drag et al. 2007
			Cox-2 inhib ML-1,785,713	Y	McCann et al. 2004
			Butorphanol (0.2 mk/kg)	Y	Borer et al. 2003
			Ketamine (2 mg/kg epidural)	Y	Hamilton et al. 2005
			NK-1 antagonist (3 mg/kg)	N	Punke et al. 2007
Non-human primate	Carrageenan injected into tail, then tail immersed in warm water	Latency to tail withdrawal	Ketorolac (0.3–3 mg/kg)	Y	Ko and Lee 2002
			Fentanyl (0.001–0.018 mg/kg)	Y	Ko and Lee 2002
			U50488 (0.01–0.18 mg/kg)	Y	Ko and Lee 2002
			BW373U86 (0.1–1.8 mg/kg)	Y	Ko and Lee 2002
			Naproxen	Y	Hawkinson et al. 2007
			BK1 antagonist (ELN441958)	Y	Hawkinson et al. 2007
Pig	Ultraviolet-B irradiation	Dermal flare	Lidocaine	Y	Rukweid et al. 2008

with carrageenan, and the water temperature is reduced to a normally non-noxious temperature of 42°C. Under these conditions, a normal monkey will not withdraw its tail, but a monkey receiving a carrageenan injection will rapidly withdraw its tail due to the increased thermal sensitivity associated with the inflammatory response (Ko and Lee 2002). This model is sensitive to NSAIDs, opioids, and the Bradykinin receptor 1 (BK1) antagonist ELN441958 (Hawkinson et al. 2007; Ko and Lee 2002); and although the BK1 antagonist has not been clinically tested, the NSAID and opioid efficacy observed in this model suggests that it does have a positive predictive power for clinical inflammatory pain.

A recent report has begun to characterize a novel porcine ultraviolet-B (UV-B) irradiation model (Rukwied et al. 2008; Table 17.2). The UV-B irradiation model has already been developed in both rat (Bishop et al. 2007; Davies et al. 2005) and human (Harrison et al. 2004; Sycha et al. 2003). Because the pig has skin that is physiologically very similar to human skin, it may provide an excellent translational model system to study inflammatory pain associated with the skin. The initial description of the model examined physiological changes in erythema (flare or dermovasodilation) that are driven by activation of nociceptive C-fibers (Rukwied et al. 2008). This readout, while not evaluating a pain response per se, does allow for the pharmacodynamic assessment of compounds that are intended to modulate C-fiber nociceptor activation. Pharmacological validation of this idea was supported by showing that 1% lidocaine injected intracutaneously blocked the heat-induced flare response in UV-B irradiated skin (Rukwied et al. 2008).

17.5   MODELS OF CHRONIC INFLAMMATORY PAIN

Clinically, chronic inflammatory pain, particularly pain due to osteoarthritis (OA), accounts for the largest single population of patients seeking analgesic therapies. Although NSAIDS and Cox2 inhibitors are included in the current standard of care, patients continue to seek more effective and safer treatments. There are a number of rodent models of chronic inflammatory pain that are commonly used to support the development of new therapeutics. Common rodent experimentally induced arthritis models include Complete Freund's Adjuvant (CFA)-induced arthritis of the paw (Stein et al. 1988) and monoiodoacetate-induced arthritis of the knee (Fernihough et al. 2004). In these models, the pain responses are typically measured by assessing mechanical and/or thermal hypersensitivity of a hindpaw. While there are a number of large animal models of OA and rheumatoid arthritis, there is a surprising lack of data describing and validating the measurement of pain and efficacy of analgesics in these models.

The most commonly described large animal model of experimental induced osteoarthritis is the canine anterior (or cranial) cruciate ligament (ACL) transection of the hind limb stifle (knee) joint (Table 17.3). There are a number of different surgical methods used to rupture the ligament (Lopez et al. 2003; Marshall and Chan 1996; McDevitt et al. 1977; Pond and Nuki 1973) with various levels of secondary surgical trauma; but in all cases, once the ligament is transected, the joint immediately becomes unstable. Over weeks to months (Budsberg 2001; Dedrick et al. 1993; O'Connor et al. 1993), the joint shows signs of progressive OA including osteophyte

TABLE 17.3

Large Animal Chronic Inflammatory Pain Models

Species	Pain Stimulus/ Model	Assessment Method	Pharmacology Tested	Active?	Reference
Dog	Anterior cruciate ligament transection	Force-plate gait analysis	Firocoxib (5.8 mg/kg)	N (force plate) Y (subjective rating)	Steiner et al. in press
Dog	Natural osteoarthritis	Canine brief pain inventory scale	Carprofen	Y	Brown et al. 2008

growth, cartilage and meniscal damage, and ultimately changes in subchondral bone, thus sharing most features with human OA. One of the typical behavioral features of this model is that the dog shows a variable lameness of the affected limb. The degree of lameness is typically assessed using a graded or continuous rating scale used by an expert observer, or a more quantitative gait analysis is performed using a force-plate apparatus. It is often the case that lameness is interpreted and referred to as a behavioral correlate of pain in the joint, and the quantitative approach of the gait analysis would appear to make this a very good model to study OA pain. Despite these attractive features, there is very little activity in the literature that attempts to validate pain assessment methods to study analgesics in this model.

A very recent paper using ACL transection reported testing the selective Cox2 inhibitor firocoxib in this model out to 18 weeks post lesion. Surprisingly, the study found no effects of treatment to enhance use and recovery of the affected limb assessed using a force-plate analysis. It was concluded that the deficits observed in the force-plate analysis were primarily due to the reduced structural stability of the lesioned joint but were not due to pain per se. Interestingly, animals in the study that were treated with firocoxib required statistically less rescue medication (butorphanol) in the days immediately after the ligament rupture. The need for rescue medication was determined by treatment group-blinded veterinary staff making subjective assessments of lameness on a five-point scale (Steiner et al. 2009). Taken together, these data suggest that the use of the force-plate analysis to read out pain is most likely confounded by structural effects in this model at time points out to 18 weeks post lesion. It is possible that force-plate analysis may be a more accurate measure of pain at longer time points post lesion (e.g., 1–2 years), since the relative contributions of structural effects versus direct pain effects may be different as the joint damage progresses over time (Budsberg 2001; Dedrick et al. 1993; O'Connor et al. 1993). The rescue medication data of Steiner et al. also suggest that there is a pain phenotype that can be observed and that the development and validation of specific rating scales may provide a way to assess pain in this model. Note that the lack of validity of the force-plate analysis in the early period following the ACL transection model contrasts with its apparent validity when used in the canine urate crystal induced synovitis model described above. This further underscores the hypothesis that models that include structural changes in the joint may confound pain assessment via force-plate analysis.

In addition to experimentally induced arthritis, both dogs and primates naturally develop osteoarthritis of the knee and hip joints as they age. Large-breed companion dogs are particularly susceptible to hip dysplasia and subsequent arthritic sequelae (Smith et al. 2001). The main challenge in attempting to take advantage of these populations of animals to study pain and the effects of specific analgesic agents is to have objective measures of the ongoing pain experienced by these animals.

Pain due to osteoarthritis in large-breed companion dogs is a common ailment seen in veterinary clinics. As a result, numerous randomized, placebo-controlled studies have been carried out to determine the potential benefit of various NSAIDs and Cox2 inhibitors to treat these animals (e.g., Budsberg et al. 1999; Peterson and Keefe 2004; Ryan et al. 2006; Vasseur et al. 1995). These veterinary studies use a variety of different subjective rating scales reported by the investigators as well as the dog owners. In some cases, a force-plate analysis is also done. In each case, the therapeutic was found to provide improvement over the placebo group using whichever measure was utilized, suggesting that this is a robust method to assess analgesic efficacy in natural model of OA in a large animal species. There are several challenges in leveraging these observations to enable studies in companion dogs for making critical go/no-go decisions about novel therapeutics ultimately intended for use in humans. The first challenge is that there does not appear to be any generally accepted and validated assessment instrument for the pain experienced in these animals. It is not clear how sensitive the various subjective scales and force-plate analyses are to therapeutic treatment, and obviously, the most sensitive assessment measures are desired. For example, natural osteoarthritis has an impact on joint structure. Since changes in force-plate analysis are probably a convolution of both structural deficits and painful sensation (see above), force-plate analysis may not be the most sensitive measure to assess pain alone. The second challenge is understanding whether this model and methods of pain assessment can generalize across different analgesic classes outside of the NSAID and Cox2 inhibitors. Ideally, work should be carried out to determine which assessment instruments are the most specific and sensitive to a variety of analgesic treatments. This type of information is critical to enable conclusive go/no-go decisions for a new therapeutic approach following testing in canines early in the development process.

Toward that end, at least two groups have begun developing and validating standardized pain assessment instruments for canines suffering from pain due to either osteoarthritis or bone cancer (Brown et al. 2007; Brown et al. 2008; Wiseman-Orr et al. 2004, 2006). These groups have followed sound and previously described methods for developing new tools for observers to assess the subjective states of others. Wiseman-Orr and colleagues have focused on the owners' evaluation of the behavioral expression of affective states and developed an extensive questionnaire of 109 descriptors that are rated in a scale of 1 to 7 (the Glasgow University Veterinary School questionnaire, GUVQuest). In contrast, Brown and colleagues focused on transforming a known human clinical instrument, the brief pain inventory (BPI), into a form compatible for use in dogs, the canine brief pain inventory (CBPI). The final form is an 11-question instrument that is analogous to the BPI scale often used in human clinical trials. A strong advantage of both the GUVQuest and the CBPI is that they report the dog owner's assessment of the dog's pain over a period of time living with the animal. The simple fact that the pain assessment is not based on a single time period associated with a visit to the vet clinic increases the likelihood that this instrument is a more sensitive measure of functional outcome in real life. Currently, there are no published studies using the GUVQuest to assess its sensitivity to analgesic treatment. In contrast, the CBPI has been demonstrated to be able to detect the analgesic effect of carprofen after a 2-week course of treatment (Brown et al. 2008; Table 17.3). Hopefully further work will continue to describe the utility of these instruments, taking care to assess the sensitivity and predictive validity of the instruments to predict human clinical efficacy.

The existence of natural OA in primates has been documented in free-ranging rhesus monkeys after about 12 years of age (Chateauvert et al. 1989; Chateauvert et al. 1990; Kessler et al. 1986). Unfortunately, there have been no published results of any attempts to objectively quantify the pain that is expected to be associated with natural OA in primates. Therefore, although this animal model has very good face validity as a preclinical model of natural OA, any consideration of use for assessing the effectiveness of analgesics will require the development of validated pain assessment techniques. Perhaps some of the rating and measurement tools being used and developed in the clinic or in canines could be leveraged to jump start development of assessment tools for use in natural disease in primate species.

17.6 MODELS OF CHRONIC NEUROPATHIC PAIN

Patients suffering from painful diabetic polyneuropathy (DPN) represent the largest clinical population of neuropathic pain patients in need of effective analgesics. The development of new therapeutics for DPN is primarily supported using rodent models of neuropathic pain including the spinal nerve ligation (SNL; Kim and Chung 1992) and chronic constriction injury models (CCI; Bennett and Xie 1988), although other rodent models of neuropathic pain are used as well (for review see Beggs and Salter 2006). The attractiveness of these models involving peripheral nerve ligation is that the surgical procedure is relatively straightforward and there is a good correlation between preclinical and clinical pharmacology for known analgesics. Although there is a rodent model of experimentally induced diabetes and associated diabetic neuropathy induced by streptozotocin treatment, there is a high degree of morbidity associated with this model, and there are questions regarding whether this is a true model of diabetic neuropathic pain (Bramwell et al. 2007; Fox et al. 1999). Additional rodent models of diabetic neuropathic pain include use of rats or mice that have a genetic predisposition for diabetes, including Zucker diabetic fatty rats and db/db leptin receptor-deficient mice (Obrosova et al. 2007; Sugimoto et al. 2008; Vareniuk et al. 2007). However, these models have not yet been fully validated and characterized in terms of their pharmacology and ability to predict analgesic effects.

There has been very limited exploration in developing neuropathic pain models in large animals. While there are no published reports describing canine experimental models of neuropathic pain, there have been a limited set of studies that have described a version of the rodent SNL model applied to non-human primates. Chung and colleagues ligated the L7 spinal nerve in rhesus monkeys, and ~14 days post lesion described the alteration in activity of spinothalamic tract (STT) neurons (Palecek et al. 1992) and behavior (Carlton et al. 1994). They found that following the L7 ligation, STT neurons showed increased activity in response to mechanical and thermal (both hot and cold) stimuli that were nominally non-noxious in the normal state (Palecek et al. 1992). In addition, the animals exhibited behavioral responses consistent with mechanical and thermal (both hot and cold) allodynia or hyperalgesia (Carlton et al. 1994). Interestingly, one additional monkey underwent a procedure analogous to the rodent CCI model and failed to develop symptoms of neuropathic pain. Upon post-mortem analysis, it was concluded that the larger size and thicker protective sheath of the primate spinal nerve prevented the nerve pathology typically observed in the rodent version of the CCI model (Palecek et al. 1992) Two additional studies using the L7 ligation model explored whether intraspinal administration of potentially analgesic compounds modulated the changes in STT neuron activity in the L7 ligated state. It was found that intraspinal administration of either the N-methyl-D-aspartate (NMDA)-antagonist dextrorphan (Carlton et al. 1997) or the kainate GluR5 receptor antagonist LY382884 (Palecek et al. 2004) reduced the activity of STT cells in both the normal and ligated state. Other investigators have used a primate L6 ligation model to induce a neuropathic state (Ali et al. 1999). In this study, a skin nerve preparation was prepared 14–21 days post ligation and used to study the excitability of the peripheral nociceptors. It was found that there was an increase in the spontaneous activity of C-fibers recorded from ligated animals. The C-fibers were also more sensitive to independent application of alpha-1 or alpha-2 adrenergic agonists. However, any behavioral effects associated with the neuropathic state were not reported in that study.

Unfortunately, the primate nerve ligation models described above, while interesting, are very labor intensive and raise concerns about the humane and ethical use of primate species in laboratory research. It is certain that it would require a significant effort to fully understand and validate these models as predictive preclinical models of neuropathic pain. An unevaluated or tested alternative to the surgical models described above would be to determine if a naturally expressed form of neuropathic pain is expressed in diabetic non-human primates. Spontaneously type-II diabetic rhesus monkeys show significant signs of distal peripheral neuropathy (Pare et al. 2007) that would be expected to be associated with neuropathic pain if the proper assessment methods were developed and employed. There is a significant opportunity to explore and develop useful pain assessment tools in a primate with naturally occurring diabetic neuropathy.

17.7 ALGOGENIC/PHARMACODYNAMIC MODELS

The various pain models described above, in part, are attractive because they all have some degree of face validity with at least one clinically relevant pain patient population. However, this can be a two-edged sword in that the preclinical models also may share many of the complexities of the human clinical conditions that may confound understanding of the actual target-based activity of a specific analgesic compound in development. For the successful development of novel analgesics, it is important to understand if a compound is actually engaging the therapeutic target, regardless of demonstrated analgesic efficacy in a specific pain model. Fortunately, there are a number of large animal model systems that have been described that can be used to explore specific pain therapeutic targets and how they contribute to physiological and pathological pain.

One common approach is to build upon the primate tail immersion acute nociceptive or inflammatory pain model described above to create an algogen-sensitized tail immersion-withdrawal assay (Table 17.4). Similar to the inflammation-driven hypersensitivity induced by carrageenan injection into the tail, the injection or topical application of a variety of algogenic substances can lead to increased sensitivity to what is otherwise a non-painful stimulus. This approach has been shown to work for the selective transient receptor potential vanilloid 1 (TRPV1) agonist capsaicin

TABLE 17.4
Large Animal Algogenic/Pharmacodynamic Pain Models

Species	Pain Stimulus/ Model	Assessment Method	Pharmacology Tested	Active?	Reference
Primate	Injected capsaicin	Tail withdrawal latency	Fentanyl (0.003–0.1 mg)	Y	Ko et al. 1998
			DAMGO (0.001–0.03 mg)	Y	Ko et al. 1998
Primate	Topical capsaicin	Tail withdrawal latency	Kappa opioid agonist—U69,593 (0.01–0.1 mg/kg)	Y	Butelman et al. 2003
			Ketamine (0.32–1.8 mg/kg)	Y	Butelman et al. 2003
			MK-801 (0.032–0.056 mg/kg)	Y	Butelman et al. 2003
			Gabapentin (50–150 mg/kg)	Y	Choong et al. 2007; see Figure 17.1
Primate	Injected prostaglandin-E2	Tail withdrawal latency	Morphine (0.32–3.2 mg/kg)	Y	Negus et al. 1993
			BW373U86 (0.18–0.56 mg/kg)	N	Butelman et al. 1995
Primate	Injected bradykinin	Tail withdrawal latency	BW373U86 (0.18–0.56 mg/kg s.c.)	Y	Butelman et al. 1995
Primate	Topical capsaicin	Vasodilation	CGRP receptor antagonist BIBN4096BS (0.03 mg/kg)	Y	Hershey et al. 2005

(Butelman et al. 2003; Ko et al. 1998), the prostaglandin receptor agonist prostaglandin-E2 (Negus et al. 1993), and the non-selective bradykinin receptor agonist bradykinin (Butelman et al. 1995). In each case, when the agonist is either injected into or topically applied to the skin of the tail, the monkey will rapidly withdraw its tail from what would otherwise be a non-noxious thermal stimulus of 38°C–46°C. The hypersensitivity induced by these algogens can be inhibited by a variety of analgesic agents including various opioids and NMDA receptor antagonists (Butelman et al. 1995, 2003; Ko et al. 1998; Negus et al. 1993). It is interesting to note that the delta-opioid receptor agonist BW373U86 was effective in reversing bradykinin but not prostaglandin-E2 induced-hypersensitivity, suggesting that this approach may potentially be used to distinguish subtle differences in pain processing.

Capsaicin-induced allodynia and hyperalgesia is a commonly used model of experimental pain that has been used in rodents, human subjects, and non-human primates to evaluate pharmacodynamic antinociceptive effects of compounds of interest (Dirks et al. 2002; Joshi et al. 2006; Ko et al. 1998). Interestingly, a recent study by Joshi et al. (2006) suggested that this model may be used as a surrogate model of neuropathic pain, given the similar pharmacology observed in rodent models of capsaicin-induced and neuropathy-induced pain. Consistent with this notion, the anticonvulsant gabapentin is also efficacious in the rhesus monkey capsaicin pain model, suggesting this experimental pain model in non-human primates may also possibly be used as a surrogate model of neuropathic pain (Figure 17.1B). However, more studies will need to be performed to gain a better understanding of the pharmacology associated with this model and how closely the pharmacology tracks to mechanisms associated with neuropathic pain.

In addition to measuring stimulus-evoked reflex withdrawal responses, another way to assess the pharmacodynamic effects of compounds of interest is to take advantage of the axon reflex-induced flare or vasodilation response. It is commonly observed that direct activation of C-fibers with agents such as capsaicin leads to an increase in dermal vasodilation in the region surrounding the application of capsaicin (Willis 1999). Thus, compounds that target a mechanism that is involved in the detection and propagation of the capsaicin stimulus (e.g., TRPV1) or the expression of the vasodilation response (e.g., CGRP) may alter the capsaicin-induced vasodilation response (e.g., see Hershey et al. 2005; Table 17.4).

17.8   SUMMARY

The development and use of rodent pain models has undoubtedly contributed to our understanding of the pathophysiology and pharmacology associated with acute and chronic pain states. However, despite significant research efforts using these models, few novel analgesics have been developed to date based on novel mechanisms identified in these models. Although many potential reasons for this can be identified, it seems clear that a disproportionate amount of research efforts have focused on rodent pain models as opposed to models in higher species, despite the advantages of using large animal models in terms of human translation. The relative paucity of large-animal pain model data to date illustrates the need to better characterize the existing models in terms of methods of pain assessment and pharmacology. In other cases, there are minimal or no large animal pain models that correspond to certain prevalent pain conditions such as in the case of painful diabetic polyneuropathy. The further characterization and development of large animal pain models that more closely represent the natural underlying disease in humans (e.g., osteoarthritis, diabetic neuropathy) will likely offer greater insight into novel mechanisms responsible for chronic pain and will increase the probability of success to develop novel therapeutics to treat pain.

REFERENCES

Adams, C. P., and V. V. Brantner. 2006. Estimating the cost of new drug development: is it really 802 million dollars? *Health Aff. (Millwood.)* 25, no. 2:420–428.

Ali, Z., M. Ringkamp, T. V. Hartke, H. F. Chien, N. A. Flavahan, J. N. Campbell, and R. A. Meyer. 1999. Uninjured C-fiber nociceptors develop spontaneous activity and alpha-adrenergic sensitivity following L6 spinal nerve ligation in monkey. *J. Neurophysiol.* 81, no. 2:455–466.

Arezzo, J. C., J. Rosenstock, L. Lamoreaux, and L. Pauer. 2008. Efficacy and safety of pregabalin 600 mg/d for treating painful diabetic peripheral neuropathy: a double-blind placebo-controlled trial. *BMC. Neurol.* 8:33.

Beggs, S., and M. W. Salter. 2006. Neuropathic pain: symptoms, models, and mechanisms. *Drug Dev. Res.* 67, no. 4:289–301.

Bennett, G. J., and Y. K. Xie. 1988. A peripheral mononeuropathy in rat that produces disorders of pain sensation like those seen in man. *Pain* 33, no. 1:87–107.

Bishop, T., D. W. Hewson, P. K. Yip, M. S. Fahey, D. Dawbarn, A. R. Young, and S. B. McMahon. 2007. Characterisation of ultraviolet-B-induced inflammation as a model of hyperalgesia in the rat. *Pain* 131, no. 1–2:70–82.

Blackburn-Munro, G. 2004. Pain-like behaviours in animals—how human are they? *Trends Pharmacol. Sci.* 25, no. 6:299–305.

Borer, L. R., J. E. Peel, W. Seewald, P. Schawalder, and D. E. Spreng. 2003. Effect of carprofen, etodolac, meloxicam, or butorphanol in dogs with induced acute synovitis. *Am. J. Vet. Res.* 64, no. 11:1429–1437.

Boyce, S., and Hill R.G. 2000. Discrepant results from preclinical and clinical studies on the potential of substance P–receptor antagonist compounds as analgesics. In *Proc 9th World Congress on Pain*, edited by Devor, M. ISAP Press.

Bramwell, S., J. Jink, L. Corradini, H. Rees, S. England, I. Machin, and G. Burgess. Streptozotocin-induced mechanical hyperalgesia in the rat is not a model of diabetic neuropathy. 2007 Neuroscience Meeting Planner Online. [Program No. 181.8]. 2007. Ref Type: Abstract.

Brennan, T. J., E. P. Vandermeulen, and G. F. Gebhart. 1996. Characterization of a rat model of incisional pain. *Pain* 64, no. 3:493–501.

Brown, D. C., R. C. Boston, J. C. Coyne, and J. T. Farrar. 2007. Development and psychometric testing of an instrument designed to measure chronic pain in dogs with osteoarthritis. *Am. J. Vet. Res.* 68, no. 6:631–637.

Brown, D. C., R. C. Boston, J. C. Coyne, and J. T. Farrar. 2008. Ability of the canine brief pain inventory to detect response to treatment in dogs with osteoarthritis. *JAVMA—Journal of the American Veterinary Medical Association* 233, no. 8:1278–1283.

Budsberg, S. C. 2001. Long-term temporal evaluation of ground reaction forces during development of experimentally induced osteoarthritis in dogs. *Am. J. Vet. Res.* 62, no. 8:1207–1211.

Budsberg, S. C., S. A. Johnston, P. D. Schwarz, C. E. DeCamp, and R. Claxton. 1999. Efficacy of etodolac for the treatment of osteoarthritis of the hip joints in dogs. *J. Am. Vet. Med. Assoc.* 214, no. 2:206–210.

Butelman, E. R., J. W. Ball, T. J. Harris, and M. J. Kreek. 2003. Topical capsaicin-induced allodynia in unanesthetized primates: pharmacological modulation. *J. Pharmacol. Exp. Ther.* 306, no. 3:1106–1114.

Butelman, E. R., S. S. Negus, M. B. Gatch, K. J. Chang, and J. H. Woods. 1995. BW373U86, a delta-opioid receptor agonist, reverses bradykinin-induced thermal allodynia in rhesus monkeys. *Eur. J. Pharmacol.* 277, no. 2–3:285–287.

Carlton, S. M., H. A. Lekan, S. H. Kim, and J. M. Chung. 1994. Behavioral manifestations of an experimental model for peripheral neuropathy produced by spinal nerve ligation in the primate. *Pain* 56, no. 2:155–166.

Carlton, S. M., H. Rees, K. Gondesen, and W. D. Willis. 1997. Dextrorphan attenuates responses of spinothalamic tract cells in normal and nerve-injured monkeys. *Neurosci. Lett.* 229, no. 3:169–172.

Chateauvert, J., K. P. Pritzker, M. J. Kessler, and M. D. Grynpas. 1989. Spontaneous osteoarthritis in rhesus macaques. I. Chemical and biochemical studies. *J. Rheumatol.* 16, no. 8:1098–1104.

Chateauvert, J. M., M. D. Grynpas, M. J. Kessler, and K. P. Pritzker. 1990. Spontaneous osteoarthritis in rhesus macaques. II. Characterization of disease and morphometric studies. *J. Rheumatol.* 17, no. 1:73–83.

Cross, A. R., S. C. Budsberg, and T. J. Keefe. 1997. Kinetic gait analysis assessment of meloxicam efficacy in a sodium urate-induced synovitis model in dogs. *Am. J. Vet. Res.* 58, no. 6:626–631.

Davies, S. L., C. Siau, and G. J. Bennett. 2005. Characterization of a model of cutaneous inflammatory pain produced by an ultraviolet irradiation-evoked sterile injury in the rat. *J. Neurosci. Methods* 148, no. 2:161–166.

Dedrick, D. K., S. A. Goldstein, K. D. Brandt, B. L. O'Connor, R. W. Goulet, and M. Albrecht. 1993. A longitudinal study of subchondral plate and trabecular bone in cruciate-deficient dogs with osteoarthritis followed up for 54 months. *Arthritis Rheum.* 36, no. 10:1460–1467.

Dirks, J., K. L. Petersen, M. C. Rowbotham, and J. B. Dahl. 2002. Gabapentin suppresses cutaneous hyperalgesia following heat-capsaicin sensitization. *Anesthesiology* 97, no. 1:102–107.

Drag, M., B. N. Kunkle, D. Romano, and P. D. Hanson. 2007. Firocoxib efficacy preventing urate-induced synovitis, pain, and inflammation in dogs. *Vet. Ther.* 8, no. 1:41–50.

Dubuisson, D., and S. G. Dennis. 1977. The formalin test: a quantitative study of the analgesic effects of morphine, meperidine, and brain stem stimulation in rats and cats. *Pain* 4, no. 2:161–174.

Dykstra, L. A., and J. H. Woods. 1986. A tail withdrawal procedure for assessing analgesic activity in rhesus monkeys. *J. Pharmacol. Methods* 15, no. 3:263–269.

Fernihough, J., C. Gentry, A. Malcangio, A. Fox, J. Rediske, T. Pellas, B. Kidd, S. Bevan, and J. Winter. 2004. Pain related behaviour in two models of osteoarthritis in the rat knee. *Pain* 112, no. 1–2:83–93.

Field, M. J., S. McCleary, J. Hughes, and L. Singh. 1999. Gabapentin and pregabalin, but not morphine and amitriptyline, block both static and dynamic components of mechanical allodynia induced by streptozocin in the rat. *Pain* 80, no. 1–2:391–398.

Fox, A., C. Eastwood, C. Gentry, D. Manning, and L. Urban. 1999. Critical evaluation of the streptozotocin model of painful diabetic neuropathy in the rat. *Pain* 81, no. 3:307–316.

Hamilton, S. M., S. A. Johnston, and R. V. Broadstone. 2005. Evaluation of analgesia provided by the administration of epidural ketamine in dogs with a chemically induced synovitis. *Vet. Anaesth. Analg.* 32, no. 1:30–39.

Hargreaves, K., R. Dubner, F. Brown, C. Flores, and J. Joris. 1988. A new and sensitive method for measuring thermal nociception in cutaneous hyperalgesia. *Pain* 32, no. 1:77–88.

Harrison, G. I., A. R. Young, and S. B. McMahon. 2004. Ultraviolet radiation-induced inflammation as a model for cutaneous hyperalgesia. *J. Invest. Dermatol.* 122, no. 1:183–189.

Hawkinson, J. E., B. G. Szoke, A. W. Garofalo, D. S. Hom, H. Zhang, M. Dreyer, J. Y. Fukuda, et al. 2007. Pharmacological, pharmacokinetic, and primate analgesic efficacy profile of the novel bradykinin B1 Receptor antagonist ELN441958. *J. Pharmacol. Exp. Ther.* 322, no. 2:619–630.

Hazewinkel, H. A., W. E. van den Brom, L. F. Theijse, M. Pollmeier, and P. D. Hanson. 2003. Reduced dosage of ketoprofen for the short-term and long-term treatment of joint pain in dogs. *Vet. Rec.* 152, no. 1:11–14.

Hershey, J. C., H. A. Corcoran, E. P. Baskin, C. A. Salvatore, S. Mosser, T. M. Williams, K. S. Koblan, R. J. Hargreaves, and S. A. Kane. 2005. Investigation of the species selectivity of a nonpeptide CGRP receptor antagonist using a novel pharmacodynamic assay. *Regul. Pept.* 127, no. 1–3:71–77.

Hill, R. 2000. NK1 (substance P) receptor antagonists—why are they not analgesic in humans? *Trends Pharmacol. Sci.* 21, no. 7:244–246.

Joshi, S. K., G. Hernandez, J. P. Mikusa, C. Z. Zhu, C. Zhong, A. Salyers, C. T. Wismer, P. Chandran, M. W. Decker, and P. Honore. 2006. Comparison of antinociceptive actions of standard analgesics in attenuating capsaicin and nerve-injury-induced mechanical hypersensitivity. *Neuroscience* 143, no. 2:587–596.

Kessler, M. J., J. E. Turnquist, K. P. Pritzker, and W. T. London. 1986. Reduction of passive extension and radiographic evidence of degenerative knee joint diseases in cage-raised and free-ranging aged rhesus monkeys (Macaca mulatta). *J. Med. Primatol.* 15, no. 1:1–9.

Kim, S. H., and J. M. Chung. 1992. An experimental model for peripheral neuropathy produced by segmental spinal nerve ligation in the rat. *Pain* 50, no. 3:355–363.

Ko, M.-C., E. R. Butelman, and J. H. Woods. 1998. The role of peripheral mu opioid receptors in the modulation of capsaicin-induced thermal nociception in rhesus monkeys. *J. Pharmacol. Exp. Ther.* 286, no. 1:150–156.

Ko, M.-C. and H. Lee. 2002. An experimental model of inflammatory pain in monkeys: comparison of antinociceptive effects of opioids and NSAIDs against carageenan-induced thermal hyperalgesia. In *10th World Congress on Pain* (Seattle: ISAP Press).

Kola, I., and J. Landis. 2004. Can the pharmaceutical industry reduce attrition rates? *Nat. Rev. Drug Discov.* 3, no. 8:711–715.

Le Bars, D., M. Gozariu, and S. W. Cadden. 2001. Animal models of nociception. *Pharmacol. Rev.* 53, no. 4:597–652.

Lin, J. H. 1999. Role of pharmacokinetics in the discovery and development of indinavir. *Adv. Drug Deliv. Rev.* 39, no. 1–3:33–49.

Lopez, M. J., D. Kunz, R. Vanderby, Jr., D. Heisey, J. Bogdanske, and M. D. Markel. 2003. A comparison of joint stability between anterior cruciate intact and deficient knees: a new canine model of anterior cruciate ligament disruption. *J. Orthop. Res.* 21, no. 2:224–230.

Marshall, K. W., and A. D. Chan. 1996. Arthroscopic anterior cruciate ligament transection induces canine osteoarthritis. *J. Rheumatol.* 23, no. 2:338–343.

McCann, M. E., D. R. Andersen, D. Zhang, C. Brideau, W. C. Black, P. D. Hanson, and G. J. Hickey. 2004. In vitro effects and in vivo efficacy of a novel cyclooxygenase-2 inhibitor in dogs with experimentally induced synovitis. *Am. J. Vet. Res.* 65, no. 4:503–512.

McCarty, D. J., Jr., P. Phelps, and J. Pyenson. 1966. Crystal-induced inflammation in canine joints. I. An experimental model with quantification of the host response. *J. Exp. Med.* 124, no. 1:99–114.

McDevitt, C., E. Gilbertson, and H. Muir. 1977. An experimental model of osteoarthritis; early morphological and biochemical changes. *J. Bone Joint Surg. Br.* 59, no. 1:24–35.

Millan, M. J. 1999. The induction of pain: an integrative review. *Prog. Neurobiol.* 57, no. 1:1–164.

Millis, D. L., J. P. Weigel, T. Moyers, and F. C. Buonomo. 2002. Effect of deracoxib, a new COX-2 inhibitor, on the prevention of lameness induced by chemical synovitis in dogs. *Vet. Ther.* 3, no. 4:453–464.

Negus, S. S., E. R. Butelman, Y. Al, and J. H. Woods. 1993. Prostaglandin E2-induced thermal hyperalgesia and its reversal by morphine in the warm-water tail-withdrawal procedure in rhesus monkeys. *J. Pharmacol. Exp. Ther.* 266, no. 3:1355–1363.

Negus, S. S., T. W. Vanderah, M. R. Brandt, E. J. Bilsky, L. Becerra, and D. Borsook. 2006. Preclinical assessment of candidate analgesic drugs: recent advances and future challenges. *J. Pharmacol. Exp. Ther.* 319, no. 2:507–514.

O'Connor, B. L., D. M. Visco, K. D. Brandt, M. Albrecht, and A. B. O'Connor. 1993. Sensory nerves only temporarily protect the unstable canine knee joint from osteoarthritis. Evidence that sensory nerves reprogram the central nervous system after cruciate ligament transection. *Arthritis Rheum.* 36, no. 8:1154–1163.

Obrosova, I. G., O. Ilnytska, V. V. Lyzogubov, I. A. Pavlov, N. Mashtalir, J. L. Nadler, and V. R. Drel. 2007. High-fat diet induced neuropathy of pre-diabetes and obesity: effects of "healthy" diet and aldose reductase inhibition. *Diabetes* 56, no. 10:2598–2608.

Palecek, J., P. M. Dougherty, S. H. Kim, V. Paleckova, H. Lekan, J. M. Chung, S. M. Carlton, and W. D. Willis. 1992. Responses of spinothalamic tract neurons to mechanical and thermal stimuli in an experimental model of peripheral neuropathy in primates. *J. Neurophysiol.* 68, no. 6:1951–1966.

Palecek, J., V. Neugebauer, S. M. Carlton, S. Iyengar, and W. D. Willis. 2004. The effect of a kainate GluR5 receptor antagonist on responses of spinothalamic tract neurons in a model of peripheral neuropathy in primates. *Pain* 111, no. 1–2:151–161.

Pare, M., P. J. Albrecht, C. J. Noto, N. L. Bodkin, G. L. Pittenger, D. J. Schreyer, X. T. Tigno, B. C. Hansen, and F. L. Rice. 2007. Differential hypertrophy and atrophy among all types of cutaneous innervation in the glabrous skin of the monkey hand during aging and naturally occurring type 2 diabetes. *J. Comp. Neurol.* 501, no. 4:543–567.

Peterson, K. D., and T. J. Keefe. 2004. Effects of meloxicam on severity of lameness and other clinical signs of osteoarthritis in dogs. *J. Am. Vet. Med. Assoc.* 225, no. 7:1056–1060.

Pond, M. J., and G. Nuki. 1973. Experimentally-induced osteoarthritis in the Dog. *Ann. Rheum. Dis.* 32:387–388.

Punke, J. P., A. L. Speas, L. R. Reynolds, R. F. Claxton, and S. C. Budsberg. 2007. Kinetic gait and subjective analysis of the effects of a tachykinin receptor antagonist in dogs with sodium urate-induced synovitis. *Am. J. Vet. Res.* 68, no. 7:704–708.

Rice, A. S., D. Cimino-Brown, J. C. Eisenach, V. K. Kontinen, M. L. Lacroix-Fralish, I. Machin, J. S. Mogil, and T. Stohr. 2008. Animal models and the prediction of efficacy in clinical trials of analgesic drugs: a critical appraisal and call for uniform reporting standards. *Pain* 139, no. 2:243–247.

Rukwied, R., M. Dusch, M. Schley, E. Forsch, and M. Schmelz. 2008. Nociceptor sensitization to mechanical and thermal stimuli in pig skin in vivo. *Eur. J. Pain* 12, no. 2:242–250.

Rumph, P. F., S. A. Kincaid, D. K. Baird, J. R. Kammermann, D. M. Visco, and L. F. Goetze. 1993. Vertical ground reaction force distribution during experimentally induced acute synovitis in dogs. *Am. J. Vet. Res.* 54, no. 3:365–369.

Ryan, W. G., K. Moldave, and D. Carithers. 2006. Clinical effectiveness and safety of a new NSAID, firocoxib: a 1,000 dog study. *Vet. Ther.* 7, no. 2:119–126.

Salvatore, C. A., J. C. Hershey, H. A. Corcoran, J. F. Fay, V. K. Johnston, E. L. Moore, S. D. Mosser, et al. 2008. Pharmacological characterization of MK-0974 [N-[(3R,6S)-6-(2,3-difluorophenyl)-2-oxo-1-(2,2,2-trifluoroethyl)azepan-3-yl]-4-(2-oxo-2,3-dihydro-1H-imidazo[4,5-b]pyridin-1-yl)piperidine-1-carbox amide], a potent and orally active calcitonin gene-related peptide receptor antagonist for the treatment of migraine. *J. Pharmacol. Exp. Ther.* 324, no. 2:416–421.

Scholz, J., and C. J. Woolf. 2007. The neuropathic pain triad: neurons, immune cells and glia. *Nat. Neurosci.* 10, no. 11:1361–1368.

Smith, G. K., P. D. Mayhew, A. S. Kapatkin, P. J. McKelvie, F. S. Shofer, and T. P. Gregor. 2001. Evaluation of risk factors for degenerative joint disease associated with hip dysplasia in German Shepherd dogs, Golden Retrievers, Labrador Retrievers, and Rottweilers. *J. Am. Vet. Med. Assoc.* 219, no. 12:1719–1724.

Stein, C., M. J. Millan, and A. Herz. 1988. Unilateral inflammation of the hindpaw in rats as a model of prolonged noxious stimulation: alterations in behavior and nociceptive thresholds. *Pharmacol. Biochem. Behav.* 31, no. 2:455–51.

Steiner, S., Y. Lu, J. P. Menetski, M. Milovancev, B. Nemke, and M. D. Markel. 2009. Effects of firocoxib therapy on progression of osteoarthris in a canine model. *Curr. Orthop. Practice.* 20, no. 3:304–312.

Sugimoto, K., I. B. Rashid, K. Kojima, M. Shoji, J. Tanabe, N. Tamasawa, T. Suda, and M. Yasujima. 2008. Time course of pain sensation in rat models of insulin resistance, type 2 diabetes, and exogenous hyperinsulinaemia. *Diabetes Metab. Res. Rev.* 24, no. 8:642–650.

Sycha, T., B. Gustorff, S. Lehr, A. Tanew, H. G. Eichler, and L. Schmetterer. 2003. A simple pain model for the evaluation of analgesic effects of NSAIDs in healthy subjects. *Br. J. Clin. Pharmacol.* 56, no. 2:165–172.

Urban, M. O., K. Ren, K. T. Park, B. Campbell, N. Anker, B. Stearns, J. Aiyar, M. Belley, C. Cohen, and L. Bristow. 2005. Comparison of the antinociceptive profiles of gabapentin and 3-methylgabapentin in rat models of acute and persistent pain: implications for mechanism of action. *J. Pharmacol. Exp. Ther.* 313, no. 3:1209–1216.

Vareniuk, I., I. A. Pavlov, V. R. Drel, V. V. Lyzogubov, O. Ilnytska, S. R. Bell, J. Tibrewala, J. T. Groves, and I. G. Obrosova. 2007. Nitrosative stress and peripheral diabetic neuropathy in leptin-deficient (ob/ob) mice. *Exp. Neurol.* 205, no. 2:425–436.

Vasseur, P. B., A. L. Johnson, S. C. Budsberg, J. D. Lincoln, J. P. Toombs, J. G. Whitehair, and E. L. Lentz. 1995. Randomized, controlled trial of the efficacy of carprofen, a nonsteroidal anti-inflammatory drug, in the treatment of osteoarthritis in dogs. *J. Am. Vet. Med. Assoc.* 206, no. 6:807–811.

Walker, K., A. J. Fox, and L. A. Urban. 1999. Animal models for pain research. *Mol. Med. Today* 5, no. 7:319–321.

Ward, K. W., R. Nagilla, and L. J. Jolivette. 2005. Comparative evaluation of oral systemic exposure of 56 xenobiotics in rat, dog, monkey and human. *Xenobiotica* 35, no. 2:191–210.

Wegner, K., K. A. Horais, N. A. Tozier, M. L. Rathbun, Y. Shtaerman, and T. L. Yaksh. 2008. Development of a canine nociceptive thermal escape model. *J. Neurosci. Methods* 168, no. 1:88–97.

Whiteside, G. T., A. Adedoyin, and L. Leventhal. 2008. Predictive validity of animal pain models? A comparison of the pharmacokinetic-pharmacodynamic relationship for pain drugs in rats and humans. *Neuropharmacology* 54, no. 5:767–775.

Willis, W. D. 1999. Dorsal root potentials and dorsal root reflexes: a double-edged sword. *Exp. Brain Res.* 124, no. 4:395–421.

Wiseman-Orr, M. L., A. M. Nolan, J. Reid, and E. M. Scott. 2004. Development of a questionnaire to measure the effects of chronic pain on health-related quality of life in dogs. *Am. J. Vet. Res.* 65, no. 8:1077–1084.

Wiseman-Orr, M. L., E. M. Scott, J. Reid, and A. M. Nolan. 2006. Validation of a structured questionnaire as an instrument to measure chronic pain in dogs on the basis of effects on health-related quality of life. *Am. J. Vet. Res.* 67, no. 11:1826–1836.

Zeilhofer, H. U. 2005. Synaptic modulation in pain pathways. *Rev. Physiol. Biochem. Pharmacol.* 154:73–100.

18 Drug Discovery and Development for Pain

Sandra R. Chaplan, William A. Eckert III, and Nicholas I. Carruthers

CONTENTS

18.1 INTRODUCTION

Two decades ago, systemic drugs indicated for pain belonged roughly to three mechanistic classes: the opioids, the nonselective NSAIDs, and the anticonvulsants, the latter class represented by a single member, carbamazepine. As of this writing, there are approximately 10 classes of drugs approved for use in the management of pain in the United States and Canada. Recent additions to the pharmacopoeia for pain exemplify two significantly differing pathways for bringing new pain therapies to market.

On the one hand, there are drugs that have been in clinical use for some time for other indications, which have been shown to have analgesic efficacy and subsequently have obtained additional label indications for the treatment of pain. Examples in this category include the alpha-2 adrenoreceptor agonist clonidine (launched as an antihypertensive in 1966; gained approval for epidural use in the treatment of cancer pain 1996), and the anticonvulsant gabapentin (launched as an anticonvulsant 1994; gained approval for neuropathic pain in 2002). A number of additional marketed drugs have repeatedly demonstrated therapeutic efficacy in pain states in controlled studies but lack specific label approval for this indication. For example, random-ized, controlled studies have demonstrated the utility of older tricyclic antidepres-sants including amitriptyline, nortriptyline, and desipramine in chronic pain states; due to the preponderance of supportive evidence these drugs are considered useful pain therapeutics.

Ongoing advances in the understanding of pain mechanisms continue to reveal opportunities for the creative use of drugs of known pharmacology. In addition, many older, marketed drugs reveal a surprising richness of pharmacology when studied using new methods. Due to their potential for benefit, and also due in no small part to the lengthy and complex process required for the development and regulatory approval of novel drugs, there are many obvious advantages to the study of approved medications for the treatment of pain. The ability to conduct clinical trials with a substance that is already approved for use in humans provides for immediate test-ing of a hypothesis. Furthermore, since approved drugs have by definition already amassed the requisite preclinical and clinical safety data, positive results from such trials lead to a much shorter path to new regulatory approval, and thus the consider-able costs of years of study involved in novel compound discovery and development are avoided.

There are, however, limitations to this approach. Existing drugs may not have the ideal pharmacological or pharmacokinetic properties required. In addition, while the cost outlay of repurposing approved drugs is far less than that of inventing new ones, the return on investment may be poor. Market protection for drugs with older patents may be limited, and from a purely business standpoint, cost analyses of investing in new research for such older drugs may be unfavorable. In addition, as safety regulation becomes more stringent, pharmaceutical companies may be reluctant to assume safety liabilities of drugs that were originally approved under less rigorous requirements.

On the other hand, recent advances in research have resulted in the development of entirely novel classes of drugs, specifically developed as analgesics and brought forward with pain as an initial indication. These include COX2-selective inhibitors (celecoxib launched 1999 for pain and inflammation), the N-type calcium channel inhibitor ziconotide (launched 2005), the triptan class of serotonin receptor subtype 1B/D agonists (sumatriptan launched early 1990s), and newer cannabinoids such as Sativex, launched in Canada 2005 for the treatment of pain and spasticity due to multiple sclerosis.

Because of the lengthy and complex evaluation required to bring a compound to market, the failure rate of new compounds is extremely high, as are the costs of devel-opment. The current estimate of the expenditure required to bring a new compound to market hovers near US $1 billion.[1,2] Hundreds of thousands of compounds may be screened in order to identify one or two compounds that are considered suitable clinical candidates. Industry estimates are that these clinical candidates have about a 90% chance of failure during the most costly stage of evaluation, clinical trials. In general, across therapeutic areas, reasons for failure include disappointing pharma-cokinetic properties in humans,[3] unacceptable clinical safety profiles,[4] and lack of clinical efficacy.[5] Compounds may also fail to reach the marketplace due to other advances in the therapeutic field (for example, the development of selective sero-tonin reuptake inhibitors supplanting the tricyclic antidepressants) or commercial reasons, such as the inability of a company to support an orphan drug program. Nevertheless, drug discovery for pain is an exciting, fast-paced, and rewarding field. Both the investigation of the analgesic effects of marketed drugs and the pioneering of novel drug classes are active areas of research and development: a survey of industry pipeline databases reveals more than 25 different mechanistic classes of compounds in various stages of evaluation.

18.2 OVERVIEW OF THE PROCESS OF DRUG DISCOVERY

Drug discovery and development are heavily multidisciplinary undertakings and can be broadly divided into preclinical and clinical phases.

- **Preclinical research** requires a minimum of approximately 3 to 4 years and consists of target identification, lead identification, and lead optimization, including preclinical safety and toxicology investigation.
- **Clinical evaluation** of proposed new drugs is closely regulated by governmental authorities. Evaluation of safety and efficacy prior to approval for launch of a new drug is divided into three phases as described below.
  - Investigational new drug (IND) (or Clinical Trial Authorization (CTA) filed
    - Phase 1 or first in human (FIH), safety assessment first dose in human, healthy volunteers *not* patients; assessment of a biomarker, if applicable
    - Phase 2: limited efficacy, patients
    - Phase 3: efficacy, safety, patients
  - NDA filed, drug approved
  - Drug approval and launch
  - Phase 4, postmarketing studies (optional)

From first human dosing to New Drug Application (NDA) filing is a process that takes 5 to 8 years on average.

The discovery of drugs takes place in the preclinical phase and can be further divided as follows.

18.3 TARGET SELECTION/TARGET IDENTIFICATION

The role of folk medicine in drug discovery and the central place that natural products have played cannot be understated. The concept of a molecular receptor or target within the body that is specifically modulated by an exogenously dosed drug is quite recent. The origins of all drug development stretch back before recorded human history to the trial-and-error selection of plant products with beneficial properties. We can only imagine how long it must have taken our ancestors to discover that extracts of poppy, or willow bark, could be useful; however, we can imagine that they would have devoted considerable effort to finding analgesic treatments. The utility of *Papaver somniferum* poppy derivatives and *Salix alba* willow bark derivatives is recorded in some of the earliest Egyptian and Greek texts.[6,7] The identification of their molecular targets represents a series of classic case studies in pharmacology, ultimately leading to the cloning of the mu opioid receptor [8–11] and the cyclooxygenase COX enzymes[12,13,8] in the 1970s.

Therapeutic approach: In the early days of drug discovery, preclinical efficacy models were introduced as a means of compound selection. These models were developed using drugs with known efficacy in humans, studying their quantifiable effects in laboratory animals, or on isolated organs in tissue baths, and applying this knowledge in the selection of new compounds. The effects of modification of a chemical structure could be thus observed in their totality, but without an appreciation of their complex effects on specificity for a given molecular target, metabolic stability, or tissue distribution.

The therapeutic approach remains extremely useful where preclinical models with high predictive value exist,[14] and particularly where some understanding of the structure–activity relationship (SAR) required for efficacy can be derived using *in vitro* tools. Many drugs developed in this way are still in widespread use. For example, the semi-synthetic and the synthetic mu opioids were developed using this approach, years before the cloning of the mu receptor, including fentanyl, synthesized in 1960 by the late Dr. Paul Janssen and still one of the most potent and selective opioids clinically available.

The development of gabapentin, and the subsequent development of pregabalin, is another more recent illustration of this approach. Gabapentin was first developed as a GABA analog, and selected on the basis of its activity in *in vivo* epilepsy models, and was brought to market in 1993 as an anticonvulsant with add-on efficacy in partial/complex epileptiform disorders.[15,16] Observations in clinical practice led to identification of its utility in refractory pain states,[17–20] then to research in preclinical models supporting this indication[21,22] and ultimately to controlled clinical trials leading to regulatory approval in 2002 for the pain of post-herpetic neuralgia.[23,24] With the experience gained from developing gabapentin, additional 3-substituted GABA analogs were profiled in the laboratory. Pregabalin was identified as an analog with notable potency in the mouse maximal electroshock-induced seizure assay, and subsequently shown to be active in neuropathic pain assays.[16] Thus, the development of these two important neuropathic pain drugs followed a therapeutic pathway, rather than a molecular target pathway. Importantly, in 1996, approximately 5 years after the identification of pregabalin, a CNS binding site for gabapentin was identified, consisting of the $\alpha 2\delta_1$ calcium channel accessory protein.[25] The mechanism whereby this interaction results in analgesia and the functional significance of this binding are still not entirely clear.[26,27]

The overall business model of pharmaceutical companies has in most cases evolved toward molecular target-directed drug discovery. Currently, most but not all drug development projects are centered on a target whose molecular identity has been deliberately selected *a priori*. This method of identifying candidate compounds fits best with the capabilities of modern pharmaceutical research facilities (see section Lead Identification and Optimization below). The ability to screen large compound collections for structures that selectively interact with the target, using recombinantly expressed target protein in binding assays and immortalized cell lines, is one of the many advantages to this approach. Knowledge of the molecular target also enables work in transgenic animals to further assess the primary and secondary pharmacology of compounds in development. Selection of a molecular target may be based on its validation in preclinical or clinical pain states, or may be on theoretical grounds. Expansion of the knowledge of the molecular pharmacology of known target families has provided numerous promising target candidates; for review see e.g., Dray (2008).[28] Pharmacological investigations of the pathways providing components of nociceptive processing (in other words, hypothesis-driven explorations of the known molecular pharmacology of pain) have arguably been the most directed and productive of the methods of selecting novel targets. The development of high-throughput technologies for systematically scanning the genome/proteome has given the ability to launch major unbiased exploratory efforts using comparisons of gene expression profiles between normal and chronic pain states.[29] In addition, these technologies have enabled the identification of numerous potential therapeutic targets in the so-called orphan receptors, ion channels, and enzymes and are the foundation of subsequent efforts to identify their respective ligands/substrates and physiological functions.

18.4   TARGET VALIDATION

A fully validated target has undergone proof of concept in man, meaning that pharmacological manipulation of the target has been demonstrated to achieve the desired endpoint and to be physiologically tolerable. Such clinical validation comes in one of three ways: through folk use of a natural product, through clinical observations of the analgesic benefits of a drug used primarily for another indication (e.g., gabapentin, amitriptyline), or through a successful clinical trial of a drug tareting a novel pain mechanism (examples below).

While many research laboratories are willing and eager to develop drugs for as-yet clinically unvalidated targets, target validation is extremely useful at the preclinical level to support compound development efforts. Evidence from the phenotype of knockout or transgenic animals in support of the validity of a target is desirable when available. Pharmacological demonstration that the target is relevant to pain pathways is critical. The ability to modulate the target *in vivo* and record an effect on pain in a preclinical species provides a means to correlate the pharmacokinetic/pharmacodynamic relationship of compounds in development. A preclinical *in vivo* model may also provide for development of a clinical biomarker or imaging strategy, that is, a means of acutely measuring a dose-response to the pharmacological effects of the compound that can be used as a practical tool in the clinic to gauge whether an exposure in the presumed efficacious range has been attained. Notwithstanding some debate about the ability of preclinical models to predict success in clinical trials, preclinical models of pain have very high "face validity," or commonsense resemblance to the clinical phenomenon in man, a decided advantage in the search for novel therapies.[31–32]

One example of a target that was preclinically validated in pain models and has since been successfully validated in clinical populations is the N-type voltage-dependent calcium channel (N-VDCC). The N-VDCC blocker ziconotide (Prialt®) is one of the first examples of prospective development of a compound with a novel molecular analgesic mechanism. Compound binding to the spinal cord dorsal horn led investigators to hypothesize that the target in question was involved in pain pathways. Initial proof of concept in animal models showed the lack of utility in acute pain but showed efficacy in tonic and neuropathic pain models. Clinical trials have been positive in pain states of multiple etiologies in man, and ziconotide is approved for intrathecal delivery for management of severe chronic pain refractory to other treatments.[35, 36]

The 5HT1B/D receptor agonists exemplify novel target validation using an exclusively mechanistic set of preclinical models. The development of the triptans is a remarkable example of hypothesis-based thinking that has proven to be successful in the clinic. Behavioral animal models of migraine pain were not described at that time, and are as yet not widely validated. Sumatriptan was originally developed using preclinical models of inhibition of evoked cerebral vasodilatation based on the presumed contribution of this phenomenon to migraine pain. Later preclinical experiments showed that triptans prevented other clinically relevant migraine-associated phenomena, including calcitionin-gene related peptide and substance P release produced by electrical stimulation.[37]

Despite the fundamental similarities in mammalian physiology and pharmacology, pharmacogenomic and other differences between species commonly studied in the laboratory and used in drug development are widely appreciated, and some targets that are valid in rodents have been shown to fall short of expected efficacy in humans. In particular, experience with the development of neurokinin-1 (NK-1) receptor antagonists has illuminated gaps in our understanding of pain at a theoretical level. Although compounds have been effective in multiple preclinical pain models, modest efficacy at best has been seen in clinical trials. The basis for this difference remains the subject of debate. A lack of selectivity for the neurokinin receptors may have impeded the evaluation of the first antagonist widely used in preclinical work. Important species differences are suggested by the observations that the role of substance P differs between humans and rodents: Rodent neurons secrete substance P as a result of noxious stimulation, which contributes to neurogenic effects including vascular permeability, edema, and inflammation. Human neurons appear not to release substance P in response to the varieties of stimuli used to simulate neurogenic inflammation. Although humans do display neurogenic inflammation, the composition of the neurosecretion is different from that of the rat.

18.5 LEAD IDENTIFICATION

18.5.1 Assay Development and Compound Screening

Contemporary compound screening relies on high-throughput methodologies. The goal is to gain a rapid understanding of the chemical SAR of compound interaction with the unique target, and thus enable rational medicinal chemistry strategies for optimizing target interaction.

Challenges in assay development are common, and may include a lack of positive control compounds to use as tools to develop an assay at a novel target, or an incomplete understanding of the native configuration of a receptor or ion channel complex and thus of its *in vivo* pharmacology. High-throughput assays need to be robust enough to be adapted to automated formats, need to provide a signal that can be readily quantified by a machine, and must be able to be performed in a time frame compatible with the workflow of a department.

Automated machinery may handle multiples of 96-well formats; 384-well format is becoming increasingly common, and higher multiples are not unheard of. The higher the density of wells on a compound screening plate, the smaller the volume of each well, so accuracy/precision of measurements becomes quite challenging; this does minimize reagent consumption though. Large compound libraries can be screened in a matter of days or weeks using this approach.

Challenges inherent in compound libraries include how to obtain sufficient variety and diversity of high-quality compounds to provide for adequate chances to identify structures that interact with the target, quality control of the long-term stability and purity of the library elements, accurate dispensing into the tiny wells of high-content screening plates, and correct tracking of the identities of a large number of compounds through the assay and data analysis stages.

18.5.2 LEAD IDENTIFICATION AND OPTIMIZATION

Completion of a successful high-throughput screening campaign is expected to produce a number of "hits," or compounds with significant affinity for the target and with sufficient medicinal chemistry tractability to allow for exploration of their SAR and improvement of any suboptimal properties. Problems may include lack of potency; lack of selectivity over other targets of pharmacological significance; and lack of druglike properties such as acceptable solubility, chemical stability, or ability to be readily absorbed from the digestive tract and distributed into appropriate bodily compartments. Compounds may exhibit rapid metabolism, or other sources of poor pharmacokinetics, interference with CYP450 metabolic enzyme function, problems related to elimination, or unacceptable toxicity/safety profiles. Some of these problems can be identified using specific *in vitro* tests, but some must still be screened for using more broadly aimed physiological assays, including observation for signs of toxicity after repeated administration to animals.

Since the move away from *in vivo* screening in the last 20 years, the medicinal chemist is now quite often faced with highly potent molecules (*in vitro*) that are not druglike. In the old days, *in vivo* models only identified compounds with druglike pharmacokinetic properties; compounds with poor properties just didn't work.

It is the object of a lead optimization campaign to identify and improve upon undesirable properties of potent compounds, by delineating an SAR and altering the properties of the molecule to make it more acceptably druglike. If the crystal structure of the target is available, or if techniques such as structure–activity determination by nuclear magnetic resonance (NMR) or site-directed mutagenesis experiments to build models (for review see Powers)[49] can be applied, these additional rational principles can help improve the interaction with the target.

Particular challenges may be associated with different families of targets. For example, the absence of natural ligands, close structural homology, and large numbers of related genes has made identification of potent and selective voltage-gated ion channel modulating compounds historically difficult. Some targets have problematic CNS side effects that have been difficult to eliminate through compound optimization, whether for improved selectivity or for CNS exclusion. Examples of the latter include NMDA receptor antagonists[51,52] and selective kappa opioid receptor agonists[53]; in both cases, clinical acceptability of drugs in these classes has been largely limited by their unacceptable CNS effects.

18.6 CANDIDATE SELECTION

Compounds with acceptable profiles become clinical candidates. In order to support further evaluation, a clinical candidate compound must be produced in large quantities, or "scaled up." Compound synthesis at the bench research level is typically linear and designed to allow specific modifications to explore SAR, but for large-scale synthesis, these algorithms must often be substantially revised. In addition, the process must be conducted in accordance with exacting Good Manufacturing Practice (GMP) standards[55,56] in order that the compound with which toxicological studies are conducted is as free of contaminants as possible and most closely resembles the substance that will be used in human subjects. Predominant considerations for scale-up synthesis are aimed toward the fewest synthetic steps, the highest product yields, convergent reactions (where multiple components are synthesized in parallel and joined in a modular fashion), and minimization of costs. Clinical candidates must undergo rigorous toxicological studies in at least two species prior to dosing in man. These studies call for prolonged administration of high doses of (impurity-free) compound. Acceptable outcome of toxicological evaluation allows for the filing of an application to regulatory authorities for an IND/CTA.

18.6.1 PHASE I: FIRST IN MAN

Phase I Trials: Initial studies to determine the metabolism and pharmacologic actions of drugs in humans, the side effects associated with increasing doses, and to gain early evidence of effectiveness; may include healthy participants and/or patients….In Phase I trials, researchers test an experimental drug or treatment in a small group of people (20–80) for the first time to evaluate its safety, determine a safe dosage range, and identify side effects. (http://clinicaltrials.gov/ct2/info/understand)

Phase I studies are usually divided into single ascending dose (SAD) followed by multiple ascending dose (MAD) studies to determine the safe and tolerable limits of administration.

Increasingly, attention is paid to the effort to incorporate the study of biomarkers, or clinical/laboratory findings that signal whether a pharmacologically efficacious dose has been attained, into early clinical trials. No physiological biomarker exists for pain, but pharmacological biomarker strategies are often possible based on the mechanism of the compound in question (*ex vivo* demonstration of inhibition of COX2 in non-platelet blood elements, for example).

Demonstrations of receptor occupancy are highly useful biomarkers in healthy volunteer studies. Numerous positron emission tomography (PET) studies have been conducted using radiolabeled NK1 ligands in human volunteers. It must be pointed out that PET ligands for novel targets, at the same time that they add tremendous value to drug development programs, call for the invention of novel (radiolabeled) compounds. Radiolabel incorporation into compounds of interest is likely to alter critical properties, from target affinity to pharmacokinetics. The identification of a good PET ligand for a novel target represents a parallel but separate drug discovery project in its own right, and, such projects are expensive and time-consuming. Additional brain imaging techniques are being pioneered as well; functional magnetic resonance imaging (fMRI) to examine regional changes in cerebral blood flow that are indicative of localized alterations in neuronal activity patterns can be used to observe patterns associated with analgesia as well as aversive subjective effects. In some cases, with adequate safety data, it is possible to incorporate proof of pharmacological activity into Phase I studies with the use of various pain threshold tests in healthy compound- versus placebo-treated volunteers.

18.6.2 PHASE II

Phase II Trials: Controlled clinical studies conducted to evaluate the effectiveness of the drug for a particular indication or indications in patients with the disease or condition under study and to determine the common short-term side effects and risks.... In Phase II trials, the experimental study drug or treatment is given to a larger group of people (100–300) to see if it is effective and to further evaluate its safety." (http://clinicaltrials.gov/ct2/info/understand)

There are particular challenges associated with efficacy trials in pain drug development. One challenge is the selection of an appropriate patient population. Ethical study design considerations related to the use of placebo versus active treatment as a control are significant. Compared with other therapeutic areas where determination of efficacy requires prolonged use of a drug in order to observe its effect on disease manifestations, pain trials can be comparatively short and the primary outcome measure (patients' reports of pain relief) can be straightforward. However, measurement of the intensity of pain in the clinic is hampered by a pervasive anxiety that subjective pain reports may be subject to influence by environmental and experiential influences other than the test drug, and a desire for more objective organic measures of pain has long existed. The mechanics of analgesic clinical trial design are such that dropout rates can be high enough to render trials uninterpretable. Regulatory guidelines for the design of efficacy trials in analgesic development are still in development, meaning that to some degree, clinical trials are conducted at risk.

18.6.3 PHASE III

Phase III Trials: Expanded controlled and uncontrolled trials after preliminary evidence suggesting effectiveness of the drug has been obtained, and are intended to gather additional information to evaluate the overall benefit–risk relationship of the drug and provide and adequate basis for physician labeling....In Phase III trials, the experimental study drug or treatment is given to large groups of people (1,000–3,000) to confirm its effectiveness, monitor side effects, compare it to commonly used

treatments, and collect information that will allow the experimental drug or treatment
to be used safely. (http://clinicaltrials.gov/ct2/info/understand)

Again, trial design and outcome measurement can be challenging. The significant
placebo effect expected in pain trials is difficult to minimize and may obscure the
identification of all but very substantial drug effects. Statistical imputation methods
for handling patients who do not complete these trials remain a controversial topic.

18.7   NDA

For decades, the regulation and control of new drugs in the United States has been
based on the New Drug Application (NDA). Since 1938, every new drug has been the
subject of an approved NDA before U.S. commercialization. The NDA application is
the vehicle through which drug sponsors formally propose that the FDA approve a new
pharmaceutical for sale and marketing in the United States. The data gathered during
the animal studies and human clinical trials of an Investigational New Drug (IND)
become part of the NDA.

The goals of the NDA are to provide enough information to permit FDA reviewers
to reach the following key decisions:

- Whether the drug is safe and effective in its proposed use(s), and whether
  the benefits of the drug outweigh the risks.
- Whether the drug's proposed labeling (package insert) is appropriate, and
  what it should contain.
- Whether the methods used in manufacturing the drug and the controls used
  to maintain the drug's quality are adequate to preserve the drug's identity,
  strength, quality, and purity.
- The documentation required in an NDA is supposed to tell the drug's
  whole story, including what happened during the clinical tests, what the
  ingredients of the drug are, the results of the animal studies, how the drug
  behaves in the body, and how it is manufactured, processed and packaged.
  (http://www.fda.gov/Drugs/DevelopmentApprovalProcess/HowDrugsare
  DevelopedandApproved/ApprovalApplications/NewDrugApplicationNDA/
  default.htm)

While some experience has been gained from the development of neuropathic pain
drugs, specific guidelines to enable label indications for specific forms of pain, such
as visceral pain, are still to come.

18.8   CONCLUSION

Pain is a very active field of drug discovery. As of this writing, hundreds of compounds are listed in various stages of development for analgesic uses. The diversity
of targeted mechanisms is substantial. Based on statistical predictions, few new com-

pounds are likely to make it to the market; still, the wealth of active target-directed drug discovery efforts is impressive.

These experiences reflect really only a few decades of combined preclinical and clinical research investment, and basic mechanisms of nociception and hyperalgesia remain incompletely elucidated. The more the basic science in the field continues to provide scientific guidance for the selection of novel targets, the more new developments in pain therapeutics can be anticipated.

REFERENCES

1. DiMasi JA, Hansen RW, Grabowski HG. The price of innovation: new estimates of drug development costs. *J Health Econ.* Mar 2003;22(2):151–185.
2. Adams CP, Brantner VV. Estimating the cost of new drug development: is it really 802 million dollars? *Health Aff (Millwood).* Mar–Apr 2006;25(2):420–428.
3. Fredheim OM, Moksnes K, Borchgrevink PC, Kaasa S, Dale O. Clinical pharmacology of methadone for pain. *Acta Anaesthesiol Scand.* 2008 Aug;52(7):879–89. Review.
4. Hermann M, Ruschitzka FT. The hypertension peril: lessons from CETP inhibitors. *Curr Hypertens Rep* 2009;11(1):76–80.
5. Kristen N, Davis J. Pfizer discontinues development of two phase 3 compounds. Feb 24, 2009. BusinessWire-New York. http://www.businesswire.com/news/home/20090224005779/en.
6. Brownstein MJ. A brief history of opiates, opioid peptides, and opioid receptors. *Proc. Natl. Acad. Sci. USA.* 1993. 90:5391–5393.
7. Vane JR. The fight against rheumatism: from willow bark to Cox-1 sparing drugs. *J Physiol. Pharmacol.* 2000. Dec;51(4 Pt 1):573–586. Review.
8. Pert CB, Snyder SH. Opiate receptor: demonstration in nervous tissue. *Science.* Mar 9, 1973;179(77):1011–1014.
9. Chen Y, Mestek A, Liu J, Yu L. Molecular cloning of a rat kappa opioid receptor reveals sequence similarities to the mu and delta opioid receptors. *Biochem J.* Nov 1, 1993;295 (Pt 3):625–628.
10. Fukuda K, Kato S, Mori K, Nishi M, Takeshima H. Primary structures and expression from cDNAs of rat opioid receptor delta- and mu-subtypes. *FEBS Lett.* Aug 2, 1993;327(3):311–314.
11. Wang JB, Imai Y, Eppler CM, Gregor P, Spivak CE, Uhl GR. mu opiate receptor: cDNA cloning and expression. *Proc Natl Acad Sci USA.* Nov 1, 1993;90(21):10230–10234.
12. Ferreira SH, Moncada S, Vane JR. Further experiments to establish that the analgesic action of aspirin-like drugs depends on the inhibition of prostaglandin biosynthesis. *Br J Pharmacol.* Mar 1973;47(3):629P–630P.
13. Yokoyama C, Tanabe T. Cloning of human gene encoding prostaglandin endoperoxide synthase and primary structure of the enzyme. *Biochem Biophys Res Commun.* Dec 15 1989;165(2):888–894.
14. Willner P. The validity of animal models of depression. *Psychopharmacology (Berl).* 1984;83(1):1–16.
15. Taylor CP. Emerging perspectives on the mechanism of action of gabapentin. *Neurology.* Jun 1994;44(6 Suppl 5):S10–16; discussion S31–12.
16. Bryans JS, Wustrow DJ. 3-substituted GABA analogs with central nervous system activity: a review. *Med Res Rev.* Mar 1999;19(2):149–177.
17. Segal AZ, Rordorf G. Gabapentin as a novel treatment for postherpetic neuralgia. *Neurology.* Apr 1996;46(4):1175–1176.
18. Rosner H, Rubin L, Kestenbaum A. Gabapentin adjunctive therapy in neuropathic pain states. *Clin J Pain.* Mar 1996;12(1):56–58.

19. Mellick LB, Mellick GA. Successful treatment of reflex sympathetic dystrophy with gabapentin. *Am J Emerg Med.* Jan 1995;13(1):96.

20. Mellick GA, Mellicy LB, Mellick LB. Gabapentin in the management of reflex sympathetic dystrophy. *J Pain Symptom Manage.* May 1995;10(4):265–266.

21. Hwang JH, Yaksh TL. Effect of subarachnoid gabapentin on tactile-evoked allodynia in a surgically induced neuropathic pain model in the rat. *Reg Anesth.* May–Jun 1997;22(3):249–256.

22. Shimoyama N, Shimoyama M, Davis AM, Inturrisi CE, Elliott KJ. Spinal gabapentin is antinociceptive in the rat formalin test. *Neurosci Lett.* Jan 24, 1997;222(1):65–67.

23. Rowbotham M, Harden N, Stacey B, Bernstein P, Magnus-Miller L. Gabapentin for the treatment of postherpetic neuralgia: a randomized controlled trial. *JAMA.* Dec 2, 1998;280(21):1837–1842.

24. Backonja M, Beydoun A, Edwards KR, et al. Gabapentin for the symptomatic treatment of painful neuropathy in patients with diabetes mellitus: a randomized controlled trial. *JAMA.* Dec 2 1998;280(21):1831–1836.

25. Gee NS, Brown JP, Dissanayake VU, Offord J, Thurlow R, Woodruff GN. The novel anticonvulsant drug, gabapentin (Neurontin), binds to the alpha2delta subunit of a calcium channel. *J Biol Chem.* Mar 8, 1996;271(10):5768–5776.

26. Mortell KH, Anderson DJ, Lynch JJ, 3rd, et al. Structure-activity relationships of alpha-amino acid ligands for the alpha2delta subunit of voltage-gated calcium channels. *Bioorg Med Chem Lett.* Mar 1, 2006;16(5):1138–1141.

27. Maneuf YP, Luo ZD, Lee K. alpha2delta and the mechanism of action of gabapentin in the treatment of pain. *Semin Cell Dev Biol.* 2006 Oct;17(5):565–70. Epub 2006 Sep 24. Review.

28. Dray A. Neuropathic pain: emerging treatments. *Br J Anaesth* 2008;101(1):48–58.

29. Kamme F, Erlander MG. Global gene expression analysis of single cells. *Curr Opin Drug Discov Devel.* 2003 Mar;6(2):231–236. Review.

30. Kontinen VK, Meert TF. Predictive validity of neuropathic models in pharmacological studies with a behavioral outcome in rat: a systematic review in Proceedings of the 10th World Congress on Pain. Editors Dostrovsky JO, Carr DB, Koltzenburg M. 2003. *IASP Press.*

31. Rice AS, Cimino-Brown D, Eisenach JC, Kontinen VK, Lacroix-Fralish ML, Machin I; Preclinical Pain Consortium, Mogil JS, Stöhr T. Animal models and the prediction of efficacy in clinical trials of analgesic drugs: a critical appraisal and call for uniform reporting standards. *Pain.* 2008 Oct 15;139(2):243–7. Epub 2008 Sep 23. Review.

32. Whiteside GT, Adedoyin A, Leventhal L. Predictive validity of animal pain models? A comparison of the pharmacokinetic-pharmacodynamic relationship for pain drugs in rats and humans. *Neuropharmacology.* 2008 Apr;54(5):767–75. Epub Jan 12, 2008. Review.

33. Chaplan SR, Pogrel JW, Yaksh TL. Role of voltage-dependent calcium channel subtypes in experimental tactile allodynia. *J Pharmacol Exp Ther.* Jun 1994;269(3):1117–1123.

34. Malmberg AB, Yaksh TL. Voltage-sensitive calcium channels in spinal nociceptive processing: blockade of N- and P-type channels inhibits formalin-induced nociception. *J Neurosci.* Aug 1994;14(8):4882–4890.

35. Staats PS, Yearwood T, Charapata SG, et al. Intrathecal ziconotide in the treatment of refractory pain in patients with cancer or AIDS: a randomized controlled trial. *JAMA.* Jan 7 2004;291(1):63–70.

36. Williams JA, Day M, Heavner JE. Ziconotide: an update and review. *Expert Opin Pharmacother.* 2008 Jun;9(9):1575–1583. Review.

37. Arvieu L, Mauborgne A, Bourgoin S, Oliver C, Feltz P, Hamon M, Cesselin F. Sumatriptan inhibits the release of CGRP and substance P from the rat spinalcord. *Neuroreport.* Aug 12, 1996;7(12):1973–1976.

38. Herbert MK, Holzer P. [Why are substance P(NK1)-receptor antagonists ineffective in pain treatment?]. *Anaesthesist.* Apr 2002;51(4):308–319.

39. Hill R. NK1 (substance P) receptor antagonists—why are they not analgesic in humans? *Trends Pharmacol Sci.* Jul 2000;21(7):244–246.

40. Laird JM, Hargreaves RJ, Hill RG. Effect of RP 67580, a non-peptide neurokinin1 receptor antagonist, on facilitation of a nociceptive spinal flexion reflex in the rat. *Br J Pharmacol.* Jul 1993;109(3):713–718.

41. Laird JM, Roza C, De Felipe C, Hunt SP, Cervero F. Role of central and peripheral tachykinin NK1 receptors in capsaicin-induced pain and hyperalgesia in mice. *Pain.* Feb 1, 2001;90(1–2):97–103.

42. Dionne RA, Max MB, Gordon SM, et al. The substance P receptor antagonist CP-99,994 reduces acute postoperative pain. *Clin Pharmacol Ther.* Nov 1998;64(5):562–568.

43. Nagahisa A, Asai R, Kanai Y, et al. Non-specific activity of (+/-)-CP-96,345 in models of pain and inflammation. *Br J Pharmacol.* Oct 1992;107(2):273–275.

44. Nagahisa A, Asai R, Kanai Y, et al. Non-specific activity of (+/-)CP-96,345 in models of pain and inflammation. *Regul Pept.* Jul 2, 1993;46(1–2):433–436.

45. Petersen LJ, Winge K, Brodin E, Skov PS. No release of histamine and substance P in capsaicin-induced neurogenic inflammation in intact human skin in vivo: a microdialysis study. *Clin Exp Allergy.* Aug 1997;27(8):957–965.

46. Schmelz M, Luz O, Averbeck B, Bickel A. Plasma extravasation and neuropeptide release in human skin as measured by intradermal microdialysis. *Neurosci Lett.* Jul 18, 1997;230(2):117–120.

47. Hutter MM, Wick EC, Day AL, et al. Transient receptor potential vanilloid (TRPV-1) promotes neurogenic inflammation in the pancreas via activation of the neurokinin-1 receptor (NK-1R). *Pancreas.* Apr 2005;30(3):260–265.

48. Gorse AD. Diversity in medicinal chemistry space. *Curr Top Med Chem.* 2006;6(1):3–18.

49. Powers R. Applications of NMR to structure-based drug design in structural genomics. *J Struct Funct Genomics.* 2002;2(2):113–123. Review.

50. England S. Voltage-gated sodium channels: the search for subtype-selective analgesics. *Expert Opin Investig Drugs.* Dec 2008;17(12):1849–1864.

51. Forst T, Smith T, Schütte K, Marcus P, Pfützner A; CNS 5161 Study Group. Dose escalating safety study of CNS 5161 HCl, a new neuronal glutamate receptor antagonist (NMDA) for the treatment of neuropathic pain. *Br J Clin Pharmacol.* 2007 Jul;64(1):75–82.

52. Muir KW, Grosset DG, Gamzu E, Lees KR. Pharmacological effects of the non-competitive NMDA antagonist CNS 1102 in normal volunteers. *Br J Clin Pharmacol.* 1994 Jul;38(1):33–38.

53. Pande AC, Pyke RE, Greiner M, Wideman GL, Benjamin R, Pierce MW. Analgesic efficacy of enadoline versus placebo or morphine in postsurgical pain. *Clin Neuropharmacol.* Oct 1996;19(5):451–456.

54. Pande AC, Pyke RE, Greiner M, Cooper SA, Benjamin R, Pierce MW. Analgesic efficacy of the kappa-receptor agonist, enadoline, in dental surgery pain. *Clin Neuropharmacol* 1996;19(1):92–97.

55. Ng R. Good manufacturing practice: Regulator requirements, in *Drugs: From discovery to approval.* 2004. John Wiley & Sons Inc. Hoboken, NJ. Chp 9: 211–246.

56. Ng R. Good manufacturing practice: Drug manufacturing, in *Drugs: From discovery to approval.* 2004. John Wiley & Sons Inc. Hoboken, NJ. Chp 10: 247–280.

57. Del Rosario RB, Mangner TJ, Gildersleeve DL, et al. Synthesis of a nonpeptide carbon-11 labeled substance P antagonist for PET studies. *Nucl Med Biol.* May 1993;20(4):545–547.

58. Livni E, Babich JW, Desai MC, et al. Synthesis of a 11C-labeled NK1 receptor ligand for PET studies. *Nucl Med Biol.* Jan 1995;22(1):31–36.

59. Bergstrom M, Fasth KJ, Kilpatrick G, et al. Brain uptake and receptor binding of two [11C]labelled selective high affinity NK1-antagonists, GR203040 and GR205171—PET studies in rhesus monkey. *Neuropharmacology.* Feb 14, 2000;39(4):664–670.

60. Hargreaves R. Imaging substance P receptors (NK1) in the living human brain using positron emission tomography. *J Clin Psychiatry.* 2002;63 Suppl 11:18–24.

61. Bender D, Olsen AK, Marthi MK, Smith DF, Cumming P. PET evaluation of the uptake of N-[11C]methyl CP-643,051, an NK1 receptor antagonist, in the living porcine brain. *Nucl Med Biol.* Aug 2004;31(6):699–704.

62. Solin O, Eskola O, Hamill TG, et al. Synthesis and characterization of a potent, selective, radiolabeled substance-P antagonist for NK1 receptor quantitation: ([18F]SPA-RQ). *Mol Imaging Biol.* Nov-Dec 2004;6(6):373–384.

63. Gao M, Mock BH, Hutchins GD, Zheng QH. Synthesis and initial PET imaging of new potential NK1 receptor radioligands 1-[2-(3,5-bis-trifluoromethyl-benzyloxy)-1-phe-nyl-ethyl]-4-[11C]methyl-pip erazine and {4-[2-(3,5-bis-trifluoromethyl-benzyloxy)-1-phenyl-ethyl]-piperazine-1-yl} -acetic acid [11C]methyl ester. *Nucl Med Biol.* Jul 2005;32(5):543–552.

64. Furmark T, Appel L, Michelgard A, et al. Cerebral blood flow changes after treatment of social phobia with the neurokinin-1 antagonist GR205171, citalopram, or placebo. *Biol Psychiatry.* Jul 15, 2005;58(2):132–142.

65. Hietala J, Nyman MJ, Eskola O, et al. Visualization and quantification of neurokinin-1 (NK1) receptors in the human brain. *Mol Imaging Biol.* Jul–Aug 2005;7(4):262–272.

66. Nyman MJ, Eskola O, Kajander J, et al. Gender and age affect NK1 receptors in the human brain—a positron emission tomography study with [18 F]SPA-RQ. *Int J Neuropsychopharmacol.* Mar 30, 2006:1–11.

67. Borsook D, Becerra L, Hargreaves R. A role for fMRI in optimizing CNS drug development. *Nat Rev Drug Discov.* Apr 7, 2006;5(5):411–425.

68. McQuay HJ, Edwards JE, Moore RA. Evaluating analgesia: the challenges. *Am J Ther.* May–Jun 2002;9(3):179–187.

60. Dworkin RH, Farrar JT. Research design issues in pain clinical trials. *Neurology.* Dec 29, 2005;65(12_suppl_4):S1–S2.

70. Barden J, Edwards JE, Mason L, McQuay HJ, Moore RA. Outcomes in acute pain trials: systematic review of what was reported? *Pain.* Jun 2004;109(3):351–356.

71. Turk DC, Dworkin RH, Allen RR, et al. Core outcome domains for chronic pain clinical trials: IMMPACT recommendations. *Pain.* Dec 2003;106(3):337–345.

72. Raskin P, Donofrio PD, Rosenthal NR, et al. Topiramate vs placebo in painful diabetic neuropathy: analgesic and metabolic effects. *Neurology.* Sep 14, 2004;63(5):865–873.

73. Harden RN, Gracely RH, Carter T, Warner G. The placebo effect in acute headache management: ketorolac, meperidine, and saline in the emergency department. *Headache.* Jun 1996;36(6):352–356.

74. Price DD, Milling LS, Kirsch I, Duff A, Montgomery GH, Nicholls SS. An analysis of factors that contribute to the magnitude of placebo analgesia in an experimental paradigm. *Pain.* Nov 1999;83(2):147–156.

75. De Pascalis V, Chiaradia C, Carotenuto E. The contribution of suggestibility and expectation to placebo analgesia phenomenon in an experimental setting. *Pain.* Apr 2002;96(3):393–402.

76. Vase L, Riley JL, 3rd, Price DD. A comparison of placebo effects in clinical analgesic trials versus studies of placebo analgesia. *Pain.* Oct 2002;99(3):443–452.

77. Charron J, Rainville P, Marchand S. Direct comparison of placebo effects on clinical and experimental pain. *Clin J Pain.* Feb 2006;22(2):204–211.

Index

Printed and bound by CPI Group (UK) Ltd, Croydon, CR0 4YY

21/10/2024

01777044-0015